The Cadherin Superfamily

Shintaro T. Suzuki • Shinji Hirano
Editors

The Cadherin Superfamily

Key Regulators of Animal Development and Physiology

Springer

Editors
Shintaro T. Suzuki
Department of Bioscience
Kwansei Gakuin University
Sanda
Hyogo, Japan

Shinji Hirano
Department of Cell Biology
Kansai Medical University
Hirakata
Osaka, Japan

ISBN 978-4-431-56031-9 ISBN 978-4-431-56033-3 (eBook)
DOI 10.1007/978-4-431-56033-3

Library of Congress Control Number: 2016939963

Printed on acid-free paper

This Springer imprint is published by Springer Nature
The registered company is Springer Japan KK

Foreword

Intercellular interactions are prime determinants of the form and function of tissues in all multicellular organisms. It is therefore no wonder that searching for the molecular bases of these interactions has preoccupied developmental and cell biologists for over a half-century. Classical embryological experiments during the 1950s and 1960s by pioneers such as Holtfreter, Moscona, Sperry, and Steinberg led to the idea of selective affinities among cells as a major driving force in morphogenesis. Their compelling hypotheses motivated initial attempts to isolate "adhesion molecules" by the next generation of biochemists, including Glaser, Gottlieb, Lilien, Marchase, Roseman, and Steinberg (Gottlieb and Glaser 1980). Sadly, these forays fared poorly: biochemical methods of the time were inadequate to deal with molecules embedded in membranes and present in small amounts.

Then, suddenly, everything changed. In 1977, Masatoshi Takeichi demonstrated that cells bear two adhesion systems, calcium-dependent and calcium-independent, and that calcium protects the calcium-dependent adhesion molecule from proteolysis (Takeichi 1977). His critical insights suggested a strategy for purifying the adhesion molecules, and within a few years, his group and several others had identified what we now know to be the founding member of the cadherin superfamily, E-cadherin or cdh1. Also in 1977, Edelman's group described the first immunoglobulin superfamily adhesion molecule, N-CAM (Thiery et al. 1977).

These landmark discoveries led to a decade of rapid progress. Takeichi soon identified two additional cadherins (N- and P-) to accompany the original E; Edelman and others identified an additional immunoglobulin superfamily member, Ng-CAM. As molecular biology came of age, their genes were cloned, enabling many additional members to be identified by homology. Other families followed later—for example Eph kinases and ephrins, semaphorins and plexins, Robos and Slits, and leucine-rich repeat proteins. Intercellular binding partners were also identified—for example, the catenins—leading to the realization that proteins initially viewed as merely adhesive are in fact sophisticated signal transducers.

Most important, once key molecules and reagents were in hand, it became possible to return to the organism, an enterprise that began in the 1980s and has picked up steam ever since. In this regard, the cadherins, to which this book is devoted, have arguably had the greatest impact on our understanding of morphogenesis and physiology. We now know that the mammalian genome encodes over 100 members of the cadherin superfamily, divided into around 10 subfamilies. Their structures and signaling mechanisms vary, but they are united by the presence of extracellular cadherin (EC) domains that mediate adhesion. Together, they have been implicated in processes ranging from gastrulation to tumor invasion to the formation and maintenance of tight and adherens junctions. In the development of the nervous system—my own area of interest—cadherins cast a particularly long shadow, being critical for, among other phenomena, neural crest migration, neurogenesis, neuronal arrangement, axon outgrowth, dendritic patterning, synapse formation, synaptic specificity, and synaptic plasticity. For those of us who believe that genetics provides the best barometer of functional importance, the striking phenotypes of many mouse mutants provides unassailable evidence of the crucial roles that cadherins play.

In this volume, Shintaro T. Suzuki and Shinji Hirano, themselves important contributors to our understanding of cadherins, have assembled an up-to-date summary of our current knowledge about this extraordinary and critically important family of cell surface proteins. The first two chapters give us a broad overview of the history, diversity, and evolution of the cadherin superfamily (Suzuki and Hirano; Hulpiau et al.). The next three chapters (Gumbiner; Shapiro; Tan et al.) discuss the structure of and signaling by the classical cadherins, so-named because they include the original members identified by Takeichi along with their closest relatives. They remain the best-understood subfamily. The chapter by Fujiwara et al. complements these with a thorough review of nectins, a set of adhesion molecules that interact and cooperate with classical cadherins. Five of the subsequent chapters summarize what is known about the other cadherin subfamilies (Chidgey and Garrod; Mah and Weiner; Jontes; Shi et al.; Imai-Okano and Hirano). Finally, four chapters focus on roles in particular tissues and diseases, looking broadly at the involvement of multiple subfamilies (Brayshaw and Price; El-Amraoui and Petit; Albrecht et al.; Hirano and Imai-Okano).

All in all, this book provides the best way, at the moment, for molecular, cell, and developmental biologists to appreciate the current state of knowledge about the cadherins. And the news is good. We are, of course, far from understanding how intercellular interactions shape the organism. Nonetheless, as we approach the 40thanniversary of the first description of a cadherin, those of us who stand in Takeichi's shadow can take some satisfaction in the insights that have been gained and the prospects for more to come.

Center for Brain Science
Harvard University
Cambridge, MA, USA

Joshua R. Sanes, PhD

References

Gottlieb DI, Glaser L (1980) Cellular recognition during neural development. Annu Rev Neurosci 3:303–318

Takeichi M (1977) Functional correlation between cell adhesive properties and some cell surface proteins. J Cell Biol. 75:464–474

Thiery JP, Brackenbury R, Rutishauser U, Edelman GM (1977) Adhesion among neural cells of the chick embryo. II. Purification and characterization of a cell adhesion molecule from neural retina. J Biol Chem 252:6841–6845

Preface

Cadherins were initially identified as Ca^{2+}-dependent cell–cell adhesion proteins that were involved in the formation and maintenance of tissue structure after extensive searching for adhesion molecules. Since then, a large number of studies of classical cadherins have been carried out and most of the fundamental questions of classical cadherins have been clarified. Now it is widely accepted that cadherins constitute a large superfamily and are involved in various biological processes not only in the formation and maintenance of tissue architectures but also in diversified signaling processes. However, the research has also revealed many contradictory results and raised new questions. It is clear that the entire field is not so simple as once thought and many unsettled questions remain. Hence, we think it is an appropriate time to publish a book about cadherins in which the current status of cadherin research is reviewed.

The principal aim of this book is to furnish an overview of the entire field of cadherin research and to provide the current basic concept of cadherins for a wide range of readers from beginners to researchers in the field. We invited leading researchers to cover various aspects of the cadherin superfamily including the history of cadherin research, basic properties of classical cadherins as well as non-classical cadherins, cadherin-associated proteins, and the roles of cadherins in health and diseases. In addition, this book presents some contradictory results and important unanswered questions, and the authors propose their working hypotheses or future directions, to inspire future studies, especially by new participants in the field. We hope this book will provide useful information and guidance for all readers.

Finally, we would like to acknowledge all the contributors to this book for their generous acceptance of authorship and their great efforts. In addition, we are grateful to Prof. Joshua Sanes for his kind agreement to write a Foreword to this book at an extremely busy time for him. We also thank Dr. Misato Kochi,

Ms. Yoshiko Shikano, and Mr. Kaoru Hashimoto of the Springer Japan team for their proposal of this book and great assistance in handling and editing of manuscripts.

Hyogo, Japan Shintaro T. Suzuki
Osaka, Japan Shinji Hirano

Contents

Part I Overview

1 **Introduction** 3
Shintaro T. Suzuki and Shinji Hirano

2 **Evolution of Cadherins and Associated Catenins** 13
Paco Hulpiau, Ismail Sahin Gul, and Frans van Roy

Part II Classical Cadherins

3 **Classical Cadherins** 41
Barry M. Gumbiner

4 **Structure and Function of Cadherin Extracellular Regions** 71
Lawrence Shapiro

5 **Regulation of Cadherin–Catenin Biology by Mechanical Force and Phosphorylation** 93
Jiongyi Tan, Blair W. Benham-Pyle, William I. Weis, and W. James Nelson

6 **Cooperative Roles of Nectins with Cadherins in Physiological and Pathological Processes** 115
Takeshi Fujiwara, Akira Mizoguchi, and Yoshimi Takai

Part III Atypical Cadherins

7 **Desmosomal Cadherins** 159
Martyn Chidgey and David Garrod

8 **Clustered Protocadherins** 195
Kar Men Mah and Joshua A. Weiner

9 The Nonclustered Protocadherins 223
James D. Jontes

10 Seven-Pass Transmembrane Cadherin CELSRs, and Fat4 and Dchs1 Cadherins: From Planar Cell Polarity to Three-Dimensional Organ Architecture 251
Dongbo Shi, Masaki Arata, Tadao Usui, Toshihiko Fujimori, and Tadashi Uemura

11 Various Atypical Cadherins: T-Cadherin, RET, Calsyntenin, and 7D-Cadherin 277
Keiko Imai-Okano and Shinji Hirano

Part IV Cadherins in Tissues and Diseases

12 Cadherins in Neural Development 315
Lewis L. Brayshaw and Stephen R. Price

13 Cadherins in the Auditory Sensory Organ 341
Aziz El-Amraoui and Christine Petit

14 Cadherins in Cancer 363
Lauren V. Albrecht, Kathleen J. Green, and Adi D. Dubash

15 Cadherin-Related Diseases 399
Shinji Hirano and Keiko Imai-Okano

Index 423

Part I
Overview

Chapter 1
Introduction

Shintaro T. Suzuki and Shinji Hirano

Abstract Complex organisms are composed of small units named cells, which are regularly assembled to form tissues, organs, and finally individual organisms. To elucidate the mechanism of the structure formation, Townes and Holtfreter proposed a famous model called selective adhesion hypothesis in 1955. However, the molecular mechanism had remained elusive for a long time. Eventually, Takeichi clarified the Ca^{2+}-dependent cell adhesion mechanism and his group identified the responsible protein, cadherin. Since then, many cadherins were identified from various tissues of different organisms. Now, it is well accepted that many cadherins play various roles in a variety of tissues and organs, although the details are still uncharacterized.

Keywords Actin • Cadherins • Ca^{2+}-dependent cell adhesion • Catenins • Cell adhesion proteins • Tissue formation • Zipper model

The invention of the microscope and the subsequent development of histological methods have established one fundamental concept in biology: complex organisms are composed of small units named cells, which are regularly assembled to form tissues, organs, and finally individual organisms. This concept promptly raised the important question of how such intricate structures are generated and various hypotheses have been postulated. Among them, Townes and Holtfreter (1955) proposed a famous model called the selective adhesion hypothesis based on the observation that the same type of cells aggregate together and different types of cells were segregated and formed a tissue-like structure when dissociated cells were allowed to associate (Fig. 1.1).

S.T. Suzuki (✉)
Department of Bioscience, School of Science and Technology, Kwansei Gakuin University, 2-1 Gakuen, Sanda-Shi, Hyogo-Ken 669-1337, Japan
e-mail: stsuzuki@kwansei.ac.jp

S. Hirano
Department of Cell Biology, Kansai Medical University, 2-5-1 Shinmachi, Hirakata, Osaka 573-1010, Japan
e-mail: s-hirano@hirakata.kmu.ac.jp

S.T. Suzuki, S. Hirano (eds.), *The Cadherin Superfamily*,
DOI 10.1007/978-4-431-56033-3_1

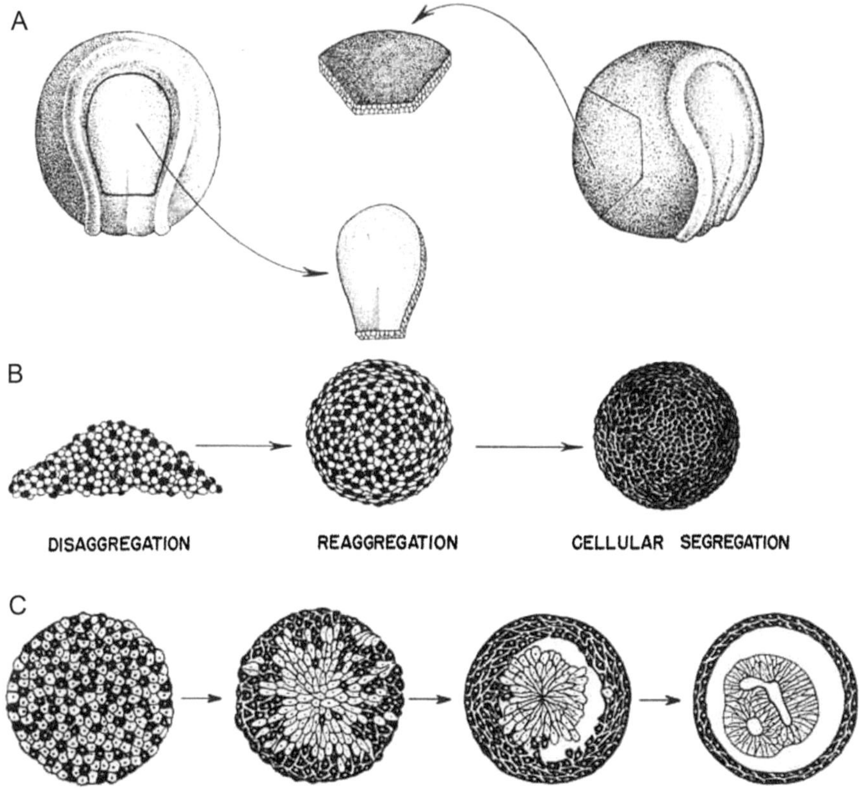

Fig. 1.1 Specific adhesion hypothesis for tissue formation by Townes and Holtfreter. (**a**) Townes and Holtfreter proposed the famous hypothesis for tissue formation based on the following experiments. Two segments were dissected from different germ layers of amphibian embryos (**b**) and dissociated. Then, the cells were mixed and incubated on a shaker. The resultant cells formed aggregates and two cell populations were gradually segregated. (**c**) Moreover, two cell populations formed two different tissue-like structures after a prolonged incubation. (This is copied from the paper by Townes and Holtfreter (1955) with permission from John Wiley and Sons)

Then, the next question was what molecule or protein was responsible for the cell adhesion activity. However, it had been a hard task and remained unsolved for a long time. Finally, one promising method using a function-inhibitory antibody became available: the cell adhesion-inhibitory antibody should bind to the cell adhesion molecule and could immunoprecipitate it, because the antibody inhibits the cell adhesion activity. Indeed, this was proved to be a very powerful method and has been successfully applied for the identification of various functional molecules such as different receptors. Edelman's group and others extensively searched for the cell adhesion proteins using the method (Brackenbury et al. 1977) and finally identified a cell surface protein named NCAM from nervous tissue (Rutishauser et al. 1978) that was later shown to belong to the immunoglobulin superfamily.

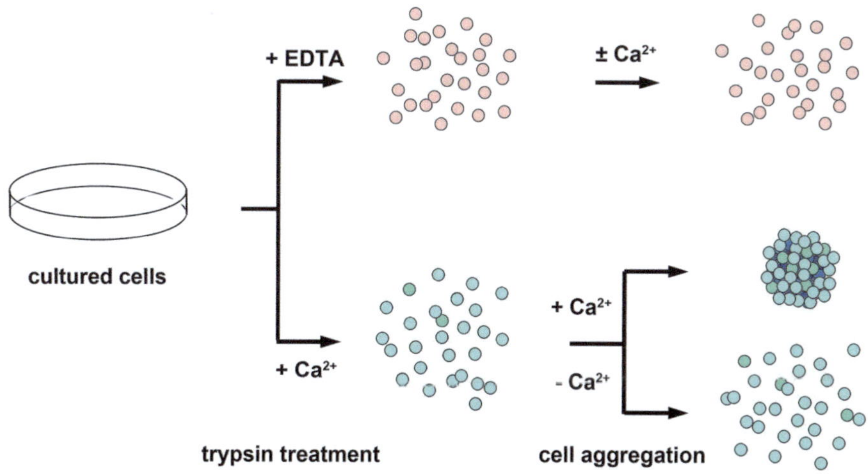

Fig. 1.2 Ca^{2+} protection of cell adhesion activity from trypsin digestion and Ca^{2+}-dependent cell–cell adhesion mechanism. Cell–cell aggregation activity is destroyed by trypsin digestion in the presence of EDTA, whereas the activity is protected from trypsin digestion in the presence of Ca^{2+}. By using this property, Takeichi's group identified the Ca^{2+}-dependent cell adhesion mechanism

On the other hand, Takeichi discovered an important property of cell adhesion activity in 1977: cell adhesion activity of cultured cells was easily destroyed by trypsin treatment, but Ca^{2+}-dependent cell adhesion activity was well protected from the digestion in the presence of Ca^{2+}. Based on these results, he classified cell adhesion mechanisms into two types, Ca^{2+}-dependent cell adhesion activity and Ca^{2+}-independent cell adhesion activity (Takeichi 1977) (Fig. 1.2). Then, his group positively identified a reasonable candidate for the responsive protein using this property (Urushihara and Takeichi 1980), which they later named cadherin. It should be emphasized that this property is so powerful that it has been used for the identification of cadherin and/or cadherin activity, because the property is very unique and is shared among most members of the cadherin family as described later.

After they identified the candidate, cloning technology had gradually become available for many investigators. Naturally, his group isolated the cDNA for the protein using the method and then transfected a fibroblast cell line L cell with the cDNA, because the cells lack Ca^{2+}-dependent cell adhesion activity. Fortunately, they could easily reconstitute the Ca^{2+}-dependent cell adhesion activity in L cells, because L cells could support the cell adhesion activity mediated by cadherin. Thus, they finally confirmed that the protein they identified was responsible for the Ca^{2+}-dependent cell adhesion activity (Nagafuchi et al. 1987).

Once cadherin was identified, different groups started to characterize the properties of cadherin and provided various evidence for the importance of Ca^{2+}-dependent cell adhesion by cadherin. As described above, cadherins mediate cell adhesion activity and Hyafil et al. (1981) reported that uvomorulin, another name for cadherin, was responsible for the compaction process of morula, in which process the cells became tightly bound together. On the other hand, Volk and

Geiger (1984) showed that cadherins were localized at adherens junctions which had been known as a major cell–cell junction structure with association with actin fibers from morphological studies. Because of these results, Ca^{2+}-dependent cell adhesion mediated by cadherins became gradually recognized as a major cell adhesion mechanism in epithelium.

Despite the apparent importance of cadherins, however, no genetic disease related to cadherins had been known for a while since the discovery of cadherin, putting cadherin research into an odd situation. However, a knockout experiment by Kemler's group (1994) gave a clear answer to this question that E-cadherin knockout ES cells could not develop embryos: The E-cadherin-ablated morula could not perform compaction and could not proceed further, indicating clearly a vital role of E-cadherin in early development of vertebrates (Larue et al. 1994). Furthermore, various diseases and conditional knockout studies related to cadherins have gradually been accumulated, showing the importance of cadherins. Now, nobody doubts the importance of cadherins in vertebrates.

A next question was the mechanism of the cadherin function. It was known from early investigations that immunoprecipitation of cadherin brought down several proteins. Ozawa et al. extensively examined the proteins and revealed that they bound to the specific site of the cadherin cytoplasmic domain and were essential for the strong cell adhesion activity (Ozawa et al. 1989; Ozawa et al. 1990). They named the proteins catenins and established the current popular model of cadherin function: cadherins are firmly associated with β-catenin that binds to actin filaments via α-catenin and the resultant complex exerts strong cell adhesion activity (Fig. 1.3a). Now, it is known that the cytoplasmic domain of cadherin directly or indirectly interacts with a variety of other proteins (Lecuit and Yap 2015), which may be the structural base of divergent functions of cadherin as discussed later.

Since the discovery of cadherin, a puzzling problem remained for a while that no one could show the formation of the extracellular domain aggregates or even the dimers, despite the notion that the domains should be responsible for the cell adhesion by cadherins. Gradually, however, various results indicated that the extracellular domains indeed had interaction activity. The problem was that the interactions were very weak and the strong cell adhesion activity still remained unexplained. Then, x-ray crystallography revealed the structural basis for the cadherin interaction (Shapiro et al. 1995): an extracellular domain of cadherin homophilically interacted with another extracellular domain at their N-terminal cadherin repeats, forming trans dimers or trans interaction. Simultaneously, the dimers interacted side by side, forming cis dimers or cis interaction. The two interactions formed threadlike arrangements of cadherin proteins (Fig. 1.3b). This is the famous zipper model of cadherin interaction. Intriguingly, this model does not directly require Ca^{2+} for adhesion activity, although the formation of active conformation does require Ca^{2+}.

It is generally thought that the basic mechanism of cadherin function has been elucidated. Interestingly, however, some studies revealed another aspect of functions of classical-type cadherins. Different cadherins are expressed in lower organisms as described later, but most of their functions have not been characterized well.

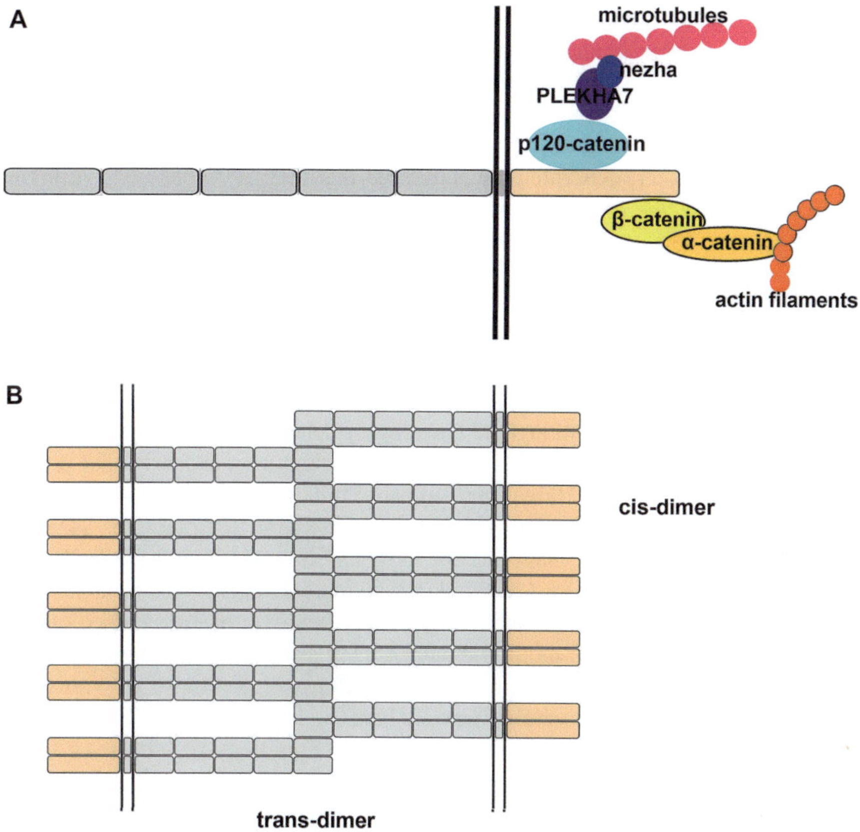

Fig. 1.3 Cell adhesion complex of classical cadherins. (**a**) Basic cell adhesion complex of classical cadherins is composed of cadherin and catenins (α-, β-, γ-, and p120-catenins). The complex also includes other proteins as well (not shown). (**b**) The cadherin complex forms *cis*-dimers on one cell, which then interact with the other *cis*-dimers on the adjacent cells (*trans*-dimers); thereby a gigantic cell adhesion complex was formed at the cell–cell contact sites (zipper model)

In *C. elegans*, however, an E-cadherin homologue was identified and was reported to be apparently dispensable for the formation of cell adhesion in the embryos (Costa et al. 1998). Furthermore, several groups reported that E-cadherin was not necessarily required for the formation of cell junctions in intestinal epithelia of vertebrate embryos, although E-cadherin was essential for the formation of the proper epithelium (Bondow et al. 2012; Schneider et al. 2010). Possibly, cadherins might have gradually obtained the vital function during the evolution of vertebrates and some members of the immunoglobulin superfamily might have substituted a part of the functions in these processes.

From the beginning of cadherin study, expression of multiple cadherins was postulated to explain the formation of complex tissue structure in higher organisms. Indeed, soon after the identification of a cadherin, Takeichi's group discovered two additional cadherins (Hatta and Takeichi 1986; Nose and Takeichi 1986), showing that they actually constitute a protein family. They named these cadherins E-cadherin, N-cadherin, and P-cadherin, respectively. On the other hand, Suzuki and his colleagues (1991) identified various cadherins by a degenerate PCR approach using the primers corresponding to the highly conserved cytoplasmic region and classified them into type I and type II cadherins based on the amino acid sequences. Later, genomic data confirmed that they are indeed two different groups. Cadherin-like proteins were also found in desmosomes (Koch et al. 1990; Collins et al. 1991). These desmosomal cadherins named desmogleins and desmocollins are usually classified as the members of an independent subfamily. However, they actually belong to a subgroup of type I cadherins, judging from the amino acid sequences and gene structures. Indeed, γ-catenin or plakoglobin, a homologue of β-catenin, binds to the regions of these proteins corresponding to the β-catenin binding sites of classical cadherins.

Cadherins were initially thought to have highly conserved cytoplasmic domains. However, Mahoney et al. (1991) identified in Drosophila a cadherin-like protein named Fat of which the extracellular domain contained multiple repeats of the cadherin-specific motif found in the extracellular domains of vertebrate cadherins, whereas the cytoplasmic domain did not show any homology with the cadherins. On the other hand, Sano et al. (1993) identified many cadherin-like proteins from different organisms by a degenerate PCR approach using the primers corresponding to the extracellular cadherin motif and named them protocadherins. The molecular structures of the cadherin-like proteins are different from those of vertebrate cadherins; especially invertebrate proteins are highly divergent. Furthermore, some of the cytoplasmic domains had a significant homology with those of vertebrate cadherins. It is clear that a large number of various types of proteins containing the cadherin-specific extracellular domain motif constitute a large cadherin superfamily (Fig. 1.4). Hence, cadherins should be defined by the cadherin-specific motif found in the extracellular domains and the proteins that contain the cytoplasmic domain homologous to those of vertebrate classical cadherins should be defined as classical-type cadherins.

Protocadherins constitute the largest subfamily among the cadherin family in vertebrates. Protocadherins were originally named for nonclassical cadherins, but now most of the investigators use the term for a specific group of cadherins in vertebrates that contain six or seven repeats of the cadherin motif. We follow this naming in this book. Protocadherins are subdivided into the clustered protocadherins and the nonclustered protocadherins that are further classified into δ-protocadherins and others. It should be mentioned that clustered protocadherins have unique gene structure wherein the gene contains tandem repeats of the extracellular domain, transmembrane domain, and a part of the cytoplasmic domain and one cytoplasmic domain, which was initially predicted by cloning experiments (Obata et al. 1998) and was later confirmed by genome project data (Wu and Maniatis 2000).

Because a vast number of cadherins have been identified, the role of these cadherins is an interesting research field, especially the role of nonclassical cadherins. Recent studies have clearly shown that these cadherins play a variety

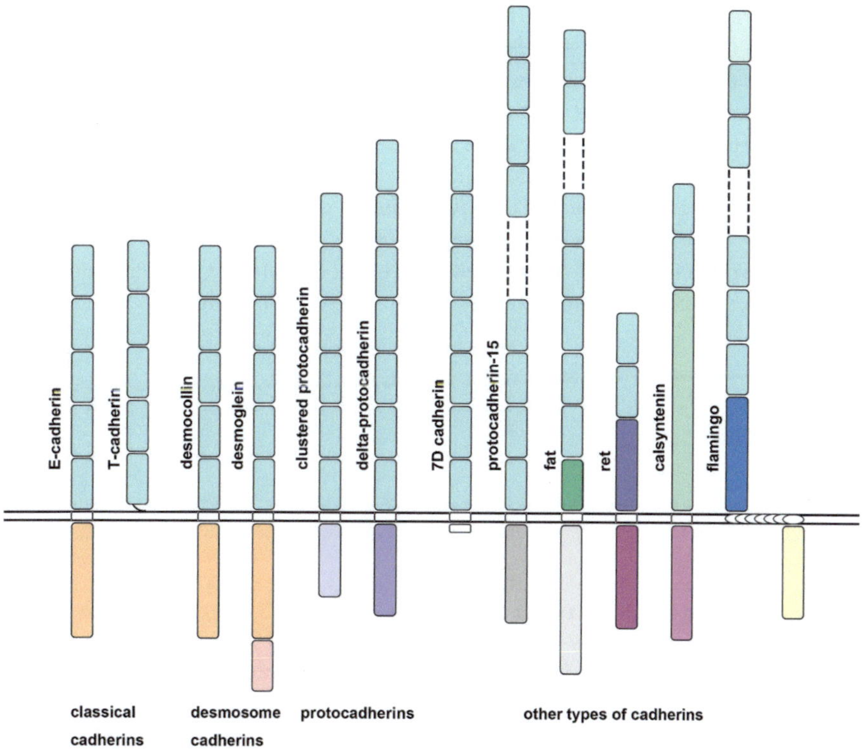

Fig. 1.4 Cadherin superfamily. Cadherins are defined as the proteins that contain the unique cadherin motif in their extracellular domains. The cytoplasmic domains are highly variable. Cadherins constitute a superfamily that includes classical cadherins, desmosomal cadherins, protocadherins, and various atypical cadherins. Desmosomal cadherins and T-cadherin are closely related to classical cadherins

of functions. Nonclassical cadherin Fat was initially isolated as a tumor suppressor in Drosophila (Mahoney et al. 1991). The mechanism of the function has remained an enigma for a long time. Recently, Fat was shown to be involved in the Hippo signaling pathway (Cho et al. 2006). Another nonclassical cadherin, Flamingo, is known to play a role in planner cell polarity formation (Usui et al. 1999). Cadherin 23 and protocadherin 15 are involved in the hearing process in the inner ear (Ahmed et al. 2001; Bolz et al. 2001; Di Palma et al. 2001).

Another interesting research field is the role of cadherins in the nervous system. The central nervous system in higher organisms is thought to function based on the specific network made by neurons called neural circuits. Intriguingly, the circuits are rather flexible and the properties continuously change responding to stimuli from the outside world. The molecular mechanism of the formation and modification of the circuits is a central issue and it still has many unknowns. From the beginning of the cadherin study, cadherins were postulated to be involved in the formation and function of the nervous system because of their specific interaction.

Indeed, various cadherins were isolated from the nervous system based on this assumption (Suzuki et al. 1991; Sano et al. 1993) and recent studies have revealed that various types of cadherins are involved in the processes such as development of nervous tissues, formation of the circuits, and formation and modification of synapses (Hirano and Takeichi 2012). Especially, protocadherins are attracting the interest of various investigators, inasmuch as they contain a large number of proteins with specific interaction activities and are mainly expressed in the nervous system (Chen and Maniatis 2013). They are now postulated to be involved in the formation of specific neural circuits and synapse modification with a vast number of specific interactions by protocadherins and are actively investigated.

References

Ahmed ZM, Riazuddin S, Bernstein SL, Ahmed Z, Khan S, Griffith AJ, Morell RJ, Friedman TB, Riazuddin S, Wilcox ER (2001) Mutations of the protocadherin gene PCDH15 cause Usher syndrome type 1F. Am J Hum Genet 69:25–34

Bolz H, von Brederlow B, Ramirez A, Bryda EC, Kutsche K, Nothwang HG, Seeliger M, Salcedo-Cabera MC, Vila MC, Molina OP et al (2001) Mutation of CDH23, encoding a new member of the cadherin gene family, causes Usher syndrome type 1D. Nat Genet 27:108–112

Bondow BJ, Faber ML, Wojta KJ, Walker EM, Battle MA (2012) E-cadherin is required for intestinal morphogenesis in the mouse. Dev Biol 371:1–12

Brackenbury R, Thiery JP, Rutishauser U, Edelman GM (1977) Adhesion among neural cells of the chick embryo. I. An immunological assay for molecules involved in cell-cell binding. J Biol Chem 252:6835–6840

Chen WV, Maniatis T (2013) Clustered protocadherins. Development 140:3297–3302

Cho E, Feng Y, Rauskolb C, Maitra S, Fehon R, Irvine KD (2006) Delineation of a fat tumor suppressor pathway. Nat Genet 38:1142–1150

Collins JE, Legan PK, Kenny TP, MacGarvie J, Holton JL, Garrod DR (1991) Cloning and sequence analysis of desmosomal glycoproteins 2 and 3 (desmocollins): cadherin-like desmosomal adhesion molecules with heterogeneous cytoplasmic domains. J Cell Biol 113:381–391

Costa M, Raich W, Agbunag C, Leung B, Hardin J, Priess JR (1998) A putative catenin-cadherin system mediates morphogenesis of the *Caenorhabditis elegans* embryo. J Cell Biol 141:297–308

Di Palma F, Holme RH, Bryda EC, Belyantseva IA, Pellegrino R, Kachar B, Steel KP, Noben-Trauth K (2001) Mutations in Cdh23, encoding a new type of cadherin, cause stereocilia disorganization in waltzer, the mouse model for Usher syndrome type 1D. Nat Genet 27:103–107

Hatta K, Takeichi M (1986) Expression of N-cadherin adhesion molecules associated with early morphogenetic events in chick development. Nature 320:447–449

Hirano S, Takeichi M (2012) Cadherins in brain morphogenesis and wiring. Physiol Rev 92:597–634

Hyafil F, Babinet C, Jacob F (1981) Cell-cell interactions in early embryogenesis: a molecular approach to the role of calcium. Cell 26:447–454

Koch PJ, Walsh MJ, Schmelz M, Goldschmidt MD, Zimbelmann R, Franke WW (1990) Identification of desmoglein, a constitutive desmosomal glycoprotein, as a member of the cadherin family of cell adhesion molecules. Eur J Cell Biol 53:1–12

Larue L, Ohsugi M, Hirchenhain J, Kemler R (1994) E-cadherin null mutant embryos fail to form a trophectoderm epithelium. Proc Natl Acad Sci U S A 91:8263–8267

Lecuit T, Yap AS (2015) E-cadherin junctions as active mechanical integrators in tissue dynamics. Nat Cell Biol 17:533–539

Mahoney PA, Weber U, Onofrechuk P, Biessmann H, Bryant PJ, Goodman CS (1991) The fat tumor suppressor gene in Drosophila encodes a novel member of the cadherin gene superfamily. Cell 67:853–868

Nagafuchi A, Shirayoshi Y, Okazaki K, Yasuda K, Takeichi M (1987) Transformation of cell adhesion properties by exogenously introduced E-cadherin cDNA. Nature 329:341–343

Nose A, Takeichi M (1986) A novel cadherin cell adhesion molecule: its expression patterns associated with implantation and organogenesis of mouse embryos. J Cell Biol 103:2649–2658

Obata S, Sago H, Mori N, Davidson M, St John T, Suzuki ST (1998) A common protocadherin tail: multiple protocadherins share the same sequence in their cytoplasmic domains and are expressed in different regions of brain. Cell Adhes Commun 6:323–333

Ozawa M, Baribault H, Kemler R (1989) The cytoplasmic domain of the cell adhesion molecule uvomorulin associates with three independent proteins structurally related in different species. EMBO J 8:1711–1717

Ozawa M, Ringwald M, Kemler R (1990) Uvomorulin-catenin complex formation is regulated by a specific domain in the cytoplasmic region of the cell adhesion molecule. Proc Natl Acad Sci U S A 87:4246–4250

Rutishauser U, Thiery JP, Brackenbury R, Edelman GM (1978) Adhesion among neural cells of the chick embryo. III. Relationship of the surface molecule CAM to cell adhesion and the development of histotypic patterns. J Cell Biol 79:371–381

Sano K, Tanihara H, Heimark RL, Obata S, Davidson M, St John T, Taketani S, Suzuki S (1993) Protocadherins: a large family of cadherin-related molecules in central nervous system. EMBO J 12:2249–2256

Schneider MR, Dahlhoff M, Horst D, Hirschi B, Trulzsch K, Muller-Hocker J, Vogelmann R, Allgauer M, Gerhard M, Steininger S, Wolf E, Kolligs F (2010) A key role for E-cadherin in intestinal homeostasis and Paneth cell maturation. PLoS One 5:e14325

Shapiro L, Fannon AM, Kwong PD, Thompson A, Lehmann MS, Grübel G, Legrand JF, Als-Nielsen J, Colman DR, Hendrickson WA (1995) Structural basis of cell-cell adhesion by cadherins. Nature 374:327–337

Suzuki S, Sano K, Tanihara H (1991) Diversity of the cadherin family: evidence for eight new cadherins in nervous tissue. Cell Regul 2:261–270

Takeichi M (1977) Functional correlation between cell adhesive properties and some cell surface proteins. J Cell Biol 75:464–474

Townes PL, Holtfreter J (1955) Directed movements and selective adhesion of embryonic amphibian cells. J Exp Zool 128:53–120

Urushihara H, Takeichi M (1980) Cell-cell adhesion molecule: identification of a glycoprotein relevant to the Ca^{2+}–independent aggregation of Chinese hamster fibroblasts. Cell 20:363–371

Usui T, Shima Y, Shimada Y, Hirano S, Burgess RW, Schwarz TL, Takeichi M, Uemura T (1999) Flamingo, a seven-pass transmembrane cadherin, regulates planar cell polarity under the control of Frizzled. Cell 98:585–595

Volk T, Geiger B (1984) A 135-kd membrane protein of intercellular adherens junctions. EMBO J 3:2249–2260

Wu Q, Maniatis T (2000) Large exons encoding multiple ectodomains are a characteristic feature of protocadherin genes. Proc Natl Acad Sci U S A 97:3124–3129

Chapter 2
Evolution of Cadherins and Associated Catenins

Paco Hulpiau, Ismail Sahin Gul, and Frans van Roy

Abstract During more than 600 Ma of multicellular animal evolution, the cadherin superfamily has become strikingly diverse, both structurally and functionally. Cadherins are typically transmembrane proteins with an ectodomain comprising so-called cadherin repeats. Cadherins are involved in cell–cell recognition, intercellular adhesion, and associated signaling, and are major players in morphogenesis and tissue behavior. Members of the three major cadherin families (cadherins, protocadherins, and cadherin-related proteins) differ in many aspects from each other. E-cadherin is the best-studied family member. Its cytoplasmic domain binds armadillo catenins, which form linkages to the cytoskeleton and trigger complex signaling pathways. Alpha-catenins play complementary roles. Even basal animals such as placozoans and cnidarians express several distinct cadherins and catenins, and their study may identify paradigms for ancient though crucial biological processes. The complex domain compositions of the different superfamily members and their respective functionalities appear to be key features of the emergence of multicellular animal life. Moreover, the origin of vertebrates coincided with a large increase in the number of cadherins and armadillo proteins, including modern molecules such as contemporary "classical" cadherins, clustered protocadherins and plakophilins. Although much needs to be learned about the biology of cadherins, the steadily increasing knowledge on cadherins is fascinating and points to key roles in many biological processes and in several important pathologies. This chapter focuses on the evolutionary relationships between different cadherin family members. The aim is to contribute to a deeper insight into their versatile roles in metazoans, and to foster further research on this remarkable superfamily.

Keywords Metazoans • Protein superfamily • Molecular evolution • Gene duplication • Cell-cell adhesion • Cadherin repeat • Protocadherins • Cadherin-related proteins • Armadillo catenins • α-catenin

---

P. Hulpiau • I.S. Gul • F. van Roy (✉)
Inflammation Research Center, VIB, Technologiepark 927, B-9052 Ghent (Zwijnaarde), Belgium

Department of Biomedical Molecular Biology, Ghent University, Ghent, Belgium
e-mail: frans.vanroy@ugent.be

S.T. Suzuki, S. Hirano (eds.), *The Cadherin Superfamily*,
DOI 10.1007/978-4-431-56033-3_2

2.1 Cadherins in Multicellular Animals

2.1.1 General Structure and Nomenclature of Cadherins Throughout Evolution

E-cadherin (CDH1) is the prototype of cadherins, which are typically transmembrane proteins characterized by consecutive extracellular cadherin (EC) repeats in their single extracellular domain. E-cadherin was initially named L-CAM, cell-CAM120/80, or uvomorulin by its discoverers (Bertolotti et al. 1980; Damsky et al. 1981; Peyrieras et al. 1983). Its current name was given by Masatoshi Takeichi and colleagues on the basis of its calcium-dependent cell–cell adhesion properties in epithelial cells (Yoshida-Noro et al. 1984; Nagafuchi et al. 1987). Since those early days, the number of identified cadherins has grown tremendously, and the human cadherin superfamily now has 114 discrete members. Moreover, even ancestral metazoans show a high diversity in cadherins, but nonmetazoan organisms such as plants and fungi have no genuine cadherins.

Here we use the term "cadherin superfamily" because it has become clear that in addition to the gene/protein family of so-called classical cadherins, there are also several more distantly related cadherin gene/protein families. The most important homologous proteins are the protocadherins and the cadherin-related members (Hulpiau and van Roy 2009, 2011). The cadherin nomenclature has not always been straightforward (for examples, see Table 2 in Hulpiau and van Roy 2009). For instance, the cadherin-related family members 1 and 2 (CDHR1 and CDHR2) were originally named, respectively, protocadherin 21 (PCDH21) and protocadherin 24 (PCDH24), yet they are not protocadherins but cadherin-related family members. For clarity, we refer to cadherins and cadherin-related molecules by their official nomenclature, but we also mention common (or preferred) names and aliases between brackets.

The classification of cadherins into different families and subfamilies of the cadherin superfamily is based on evolutionary, structural, and functional criteria, as explained below. By definition, the ectodomains of cadherins contain at least two consecutive EC repeats, generally comprising highly conserved Ca^{2+}-binding residues at the interface connecting them. Each EC repeat is about 110 amino acids long and contains an immunoglobulin-like fold consisting of seven β strands forming two β sheets. The number of ECs in mammalian cadherin ectodomains can vary from 2 up to 34 in an uninterrupted stretch. However, both classical cadherins and protocadherins have ectodomains with 5–7 ECs. More ancestral cadherins have additional domains in their extracellular part, of which laminin-G (LamG) and epidermal growth factor (EGF)-like domains occur most frequently. These additional domains are located between the EC repeats and the membrane-spanning region. The N-terminal end of the ectodomain is typically responsible for homophilic or heterophilic interactions. The intracellular regions of cadherins and cadherin-related molecules are less conserved, except for some binding motifs shared within specific cadherin families. For example, two catenin-binding motifs in classical cadherins bind the armadillo proteins β-catenin and p120ctn (see below).

Because these cytoplasmic domains are highly diverse, they can be used efficiently to subdivide the cadherin superfamily into different families and subfamilies. On the other hand, the most amino-terminal ECs (called EC1) as well as longer stretches of ECs have been used successfully to analyze the overall evolution of cadherins in the whole spectrum of metazoans. Here, we have used a stretch of multiple amino-terminal ECs (as indicated in the figure legends) to analyze evolutionary relationships and to build phylogenetic trees. First, we found it useful and phylogenetically correct to subdivide the cadherin superfamily into a cadherin major branch (CMB) and a cadherin-related major branch (CrMB). The latter comprises the large protocadherin (PCDH) family and many other cadherin-related proteins. In spite of their molecular and functional diversity, CMB members have conserved motifs for binding cytoplasmic interaction partners of the armadillo family, of which beta-catenin, plakoglobin, and p120 catenin are the best studied (see also below). However, there are exceptions, such as CDH16 and CDH17 (7D subfamily) with a very small cytoplasmic domain, CDH13 (or T-cadherin), which is membrane anchored by a GPI structure, and flamingo/CELSR members with a 7TM domain and a cytoplasmic domain lacking clear motifs.

The ongoing sequencing of many metazoan genomes makes it possible to study in detail the evolution of complex protein families such as the cadherin superfamily. In particular, genomes of basal organisms, such as the sea anemone *Nematostella vectensis* and the primitive placozoan *Trichoplax adhaerens*, can give us clues about the molecular tools acquired by animals nearly 600 Ma ago, about how a bilaterian body becomes organized, and about how extremely complex neural systems like those of higher metazoans evolved. In the genome of *T. adhaerens* only eight cadherins were identified (Hulpiau and van Roy 2011). The cnidarian *N. vectensis* has 16 putative cadherin and cadherin-related genes, a number similar to the 17 cadherins in *Drosophila melanogaster*, the 12 in *Caenorhabditis elegans*, and the 14 in the sea urchin *Strongylocentrotus purpuratus* (Hulpiau and van Roy 2011). Higher vertebrates contain many more cadherins, for example, 114 in human and 119 in mouse. A two-step expansion of cadherin superfamily members is illustrated in Fig. 2.1, where the number of different cadherins remains fairly constant from placozoans to nonvertebrate deuterostomes (comprising cephalochordates, urochordates, and echinoderms), but expands significantly from a single classical cadherin and a single ancestral protocadherin in the last common ancestor of vertebrates to more than 20 classical cadherins and more than 50 protocadherins in mammals. This evolution can be explained in part by two whole-genome duplications (WGD) during vertebrate evolution; this is known as Ohno's 2R hypothesis (Kasahara 2007). However, the appearance of completely new cadherin families, such as most classical cadherins and the clustered protocadherins, is difficult to explain solely by the 2R hypothesis.

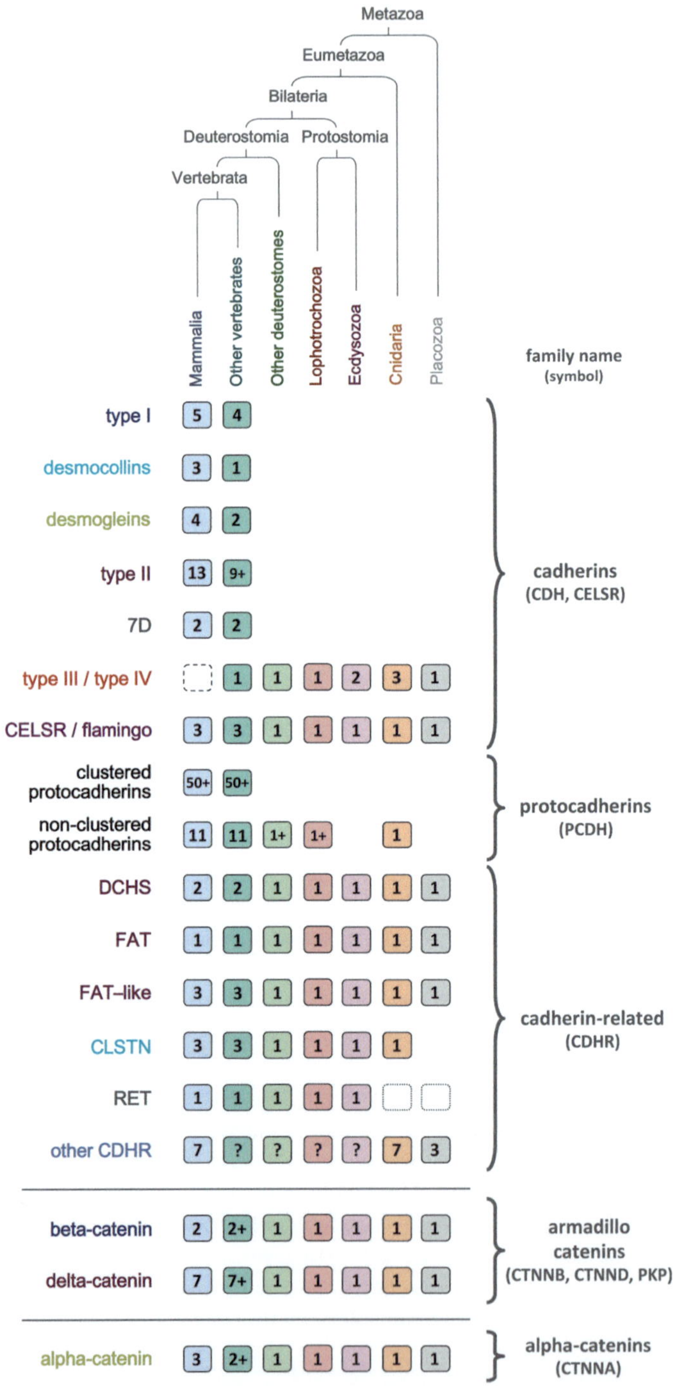

	Mammalia	Other vertebrates	Other deuterostomes	Lophotrochozoa	Ecdysozoa	Cnidaria	Placozoa
type I	5	4					
desmocollins	3	1					
desmogleins	4	2					
type II	13	9+					
7D	2	2					
type III / type IV		1	1	1	2	3	1
CELSR / flamingo	3	3	1	1	1	1	1
clustered protocadherins	50+	50+					
non-clustered protocadherins	11	11	1+	1+		1	
DCHS	2	2	1	1	1	1	1
FAT	1	1	1	1	1	1	1
FAT–like	3	3	1	1	1	1	1
CLSTN	3	3	1	1	1	1	
RET	1	1	1	1	1		
other CDHR	7	?	?	?	?	7	3
beta-catenin	2	2+	1	1	1	1	1
delta-catenin	7	7+	1	1	1	1	1
alpha-catenin	3	2+	1	1	1	1	1

Fig. 2.1 Number of cadherin superfamily members throughout evolution. Numbers were collected from the literature based on cadherin repertoires from species with genomic sequences available: *Homo sapiens* and *Mus musculus* for Mammalia; *Gallus gallus*, *Xenopus tropicalis*, and

2.1.2 The Cadherin Major Branch (CMB) and Its C-1 Sub-Branch

On the basis of our previous analysis of mainly EC1 sequences, we have proposed a subdivision of the CMB into two large sub-branches, C-1 and C-2 (Hulpiau and van Roy 2009). In view of the standard ectodomain length of 5 ECs (4 typical ECs and 1 terminal EC), we and others (Sotomayor et al. 2014) have meanwhile extended the reference sequence for phylogenetic analysis of CMB cadherins to blocks of 4 ECs. The sub-branch C-1 comprises the type-I or classical cadherins, the type-II or atypical cadherins, the desmosomal cadherins with desmocollins and desmogleins, and the 7D-cadherin family (Fig. 2.2). The majority of C-1 family members have five EC repeats. The 7D cadherins CDH16 (KSP-cadherin) and CDH17 (LI-cadherin) are notable exceptions as they have two additional EC repeats originating from duplication of the first two EC repeats. All these cadherins are confined to vertebrates (Fig. 2.1). The five type-I cadherins, CDH1 (E-cadherin, epithelial), CDH2 (N-cadherin, neuronal), CDH3 (P-cadherin, placental), CDH4 (R-cadherin, retinal), and CDH15 (M-cadherin, myotubule), share a highly conserved Trp at position 2, used for an adhesion mechanism by strand swapping in either cis interactions with the EC1 of a fellow cadherin in the same plasma membrane, or in trans interactions with the EC1 of a cadherin on the opposite cell surface (Harrison et al. 2011; see also chapter by L. Shapiro). The five type-I cadherins also share two highly conserved cytoplasmic domains used for association with armadillo proteins of the p120 and beta-catenin subfamilies (Fig. 2.2A).

As outlined in Fig. 2.2, the type-II cadherin subfamily consists of 13 members: CDH5 (VE-cadherin, vascular endothelium), CDH6 (K-cadherin, fetal kidney), CDH7, CDH8, CDH9 (T1-cadherin, testis), CDH10 (T2-cadherin, testis), CDH11 (OB-cadherin, osteoblast), CDH12, CDH18, CDH19, CDH20, CDH22, and CDH24. Like type-I cadherins, they are all present in vertebrates (Fig. 2.1), and they have five ECs in their ectodomains and two highly conserved armadillo–catenin binding domains in the cytoplasmic domains. Among the obvious differences is the presence of a second highly conserved amino acid in the EC1, Trp4, in addition to Trp2 (Fig. 2.2A). Indeed, the structure of homophilically interacting type-II cadherins differs significantly from that of type-I cadherins, and a so-called strand swap dimer is formed by virtue of these two conserved Trp residues (Patel et al. 2006).

The ectodomains of desmocollins and desmogleins are quite similar to those of type-I cadherins, including the conserved Trp2, but their cytoplasmic domains differ significantly (Fig. 2.2A). This makes sense, as the macromolecular

Fig. 2.1 (continued) Danio rerio for other vertebrates; *Ciona intestinalis*, *Branchiostoma floridae*, and *Strongylocentrotus purpuratus* for other deuterostomes; *Aplysia californica* for Lophotrochozoa; *Caenorhabditis elegans* and *Drosophila melanogaster* for Ecdysozoa; *Nematostella vectensis* for Cnidaria; *Trichoplax adhaerens* for Placozoa

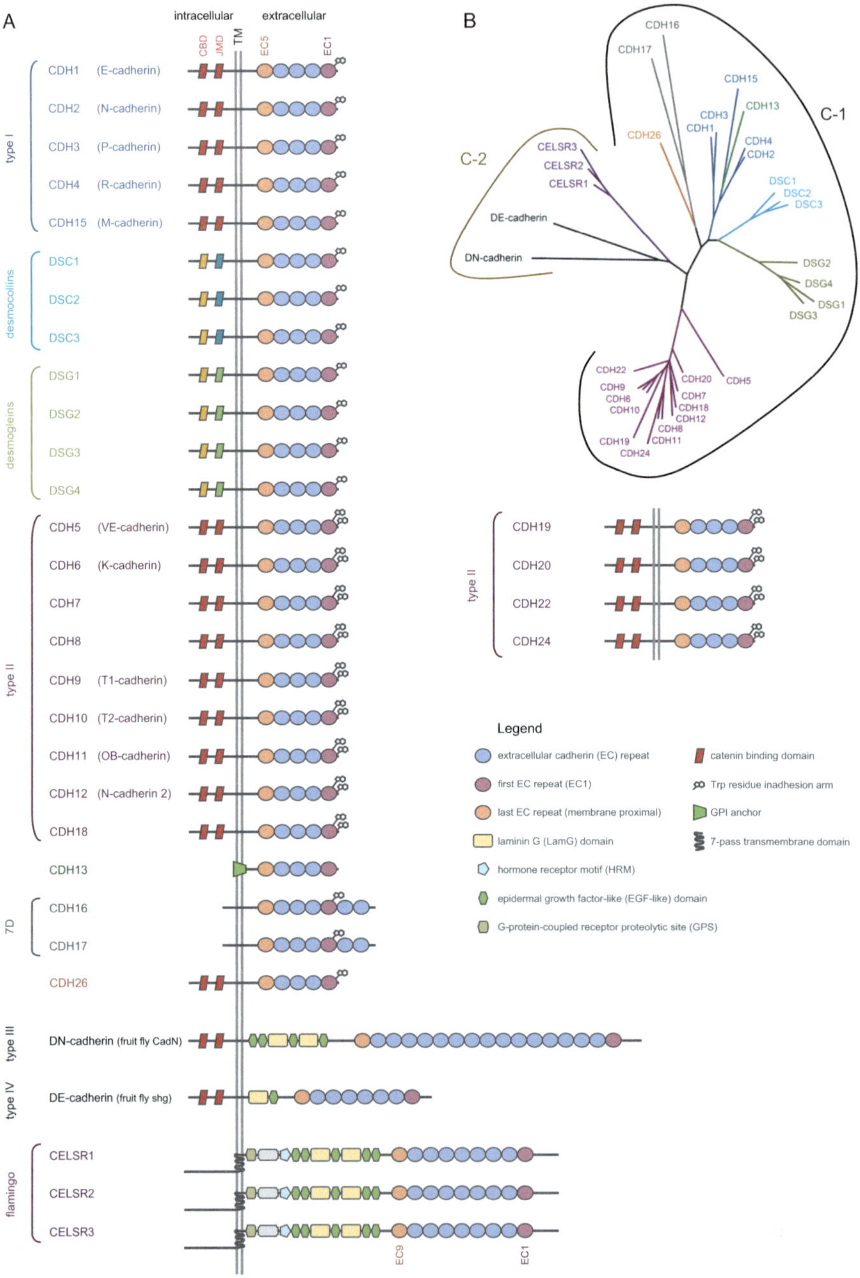

Fig. 2.2 Classification and phylogeny of the cadherin family. (**A**) Schematic representation of the domain organization of human cadherins of the cadherin major branch (CMB). These cadherins can be classified into eight subfamilies: type-I, type-II, type-III, type-IV, and 7D cadherins, desmocollins, and desmogleins, and flamingo/CELSR cadherins. Official gene/protein symbols and widely accepted aliases (shown between the parentheses) are shown on the left. Protein sequences and domain lengths are drawn to scale. See legend on the right for an explanation of symbols. (**B**) Radial phylogram representing the evolutionary relationship of cadherins. First, the

composition of desmosomes is considerably different from that of other junctions in which the other C-1 cadherins are involved. Desmosomes play a crucial role in maintaining the mechanical integrity of tissues exposed to high levels of mechanical stress, and mutations in the desmosomal cadherins lead to skin, hair, and heart diseases (Desai et al. 2009; El-Amraoui and Petit 2013).

CDH13 (T-cadherin/truncated or H-cadherin/heart) and the 7D cadherins CDH16 and CDH17 are solitary members of the C-1 family. These three cadherins do not have a typical Trp at the N-terminus of EC1 (Fig. 2.2A). However, cadherin 16 and 17 might use a single conserved Trp in the beginning of their third EC repeat because the first two domains are duplications. On the other hand, CDH13 does not use a strand-swapping adhesion mechanism but forms an X-shaped dimer through an interface near the first and second EC repeats (EC1-EC2). Moreover, CDH13 is exceptional in that it lacks the typical transmembrane domain present in all other cadherins, and instead is attached to the cell surface by a glycosylphosphatidylinositol (GPI) anchor. Another solitary cadherin is cadherin 26 (CDH26), which is remarkable but poorly studied. Although it is organized like a classical cadherin (Fig. 2.2A), its ectodomain sequence shows limited homology to C-1 cadherins (Fig. 2.2B).

E-cadherin (CDH1) has become a strong paradigm in cadherin research, as illustrated by its impressive prevalence in the literature (Fig. 2.3A). It is essential in development and morphogenesis from the early embryonic stem cell stage to formation and homeostasis of all epithelial tissues, and its role as a strong tumor and invasion suppressor is well known (Pieters and van Roy 2014; van Roy and Berx 2008). E-cadherin is generally replaced by N-cadherin in epithelial–mesenchymal transitions, which play important roles in stem cell development, particular morphogenic processes, and several pathologies, including tumor progression (Pieters and van Roy 2014; De Craene and Berx 2013). In addition, important roles in self-organization of most tissues, in morphogenesis and wiring of the brain, and in either promotion or suppression of tumor malignancy have been ascribed to many other members of the classical C-1 cadherins, although these have generally been studied less intensively than E- and N-cadherin (Fig. 2.3A) (Niessen et al. 2011; Hirano and Takeichi 2012; van Roy 2014). The large increase in the diversity of C-1 cadherins (and catenins, see also below) in vertebrates might have contributed much to the molecular complexity of these more evolved metazoans by amplifying adhesive and signaling functions.

Fig. 2.2 (continued) four N-terminal EC repeats were aligned using MUSCLE (Edgar 2004). Then, the evolutionary history was inferred by using the Bayesian phylogeny employing a mixed model for amino acid substitutions implemented in MrBayes (Ronquist et al. 2012). The phylogenetic branch lengths are to scale and correlate with estimated evolutionary distance

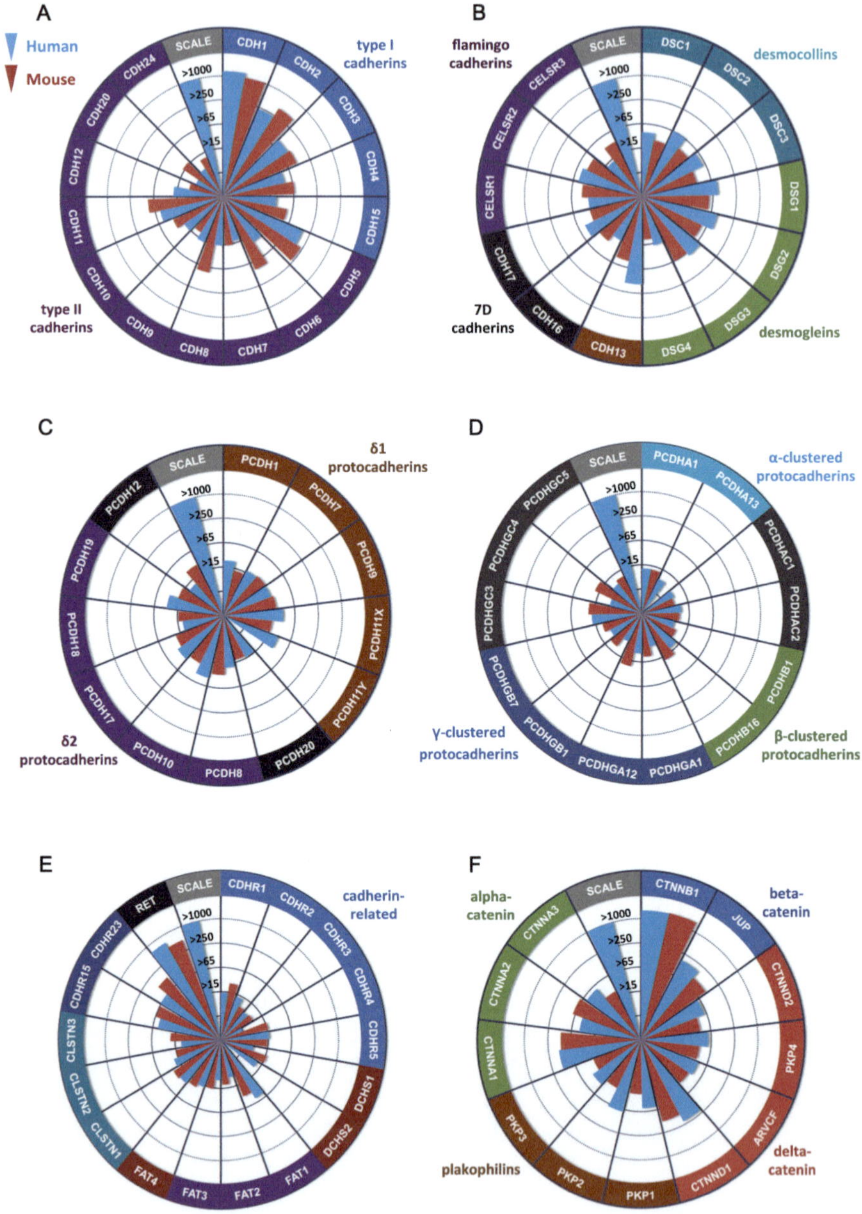

Fig. 2.3 Related articles in *PubMed* for the genes in the cadherin superfamily. Figures (**A–F**) show the related articles in *PubMed* (in March 2015, on a $\log_2$ scale) reported in the Entrez Gene database and being linked with the human (blue wedges) and mouse (red wedges) cadherin/catenin genes. (**A**, **B**) Related articles for members of the CMB. (**C**, **D**) Related articles for nonclustered and clustered protocadherins, respectively. (**E**) Related articles for cadherin-related genes. (**F**) Related articles for catenin genes

2.1.3 The C-2 Sub-Branch of the Cadherin Major Branch (CMB)

The second cadherin sub-branch, C-2, comprises cadherins of type III and type IV, as well as CELSR (cadherin EGF-like LamG-like seven-pass receptors; flamingo or fmi in *Drosophila*) proteins. Several features distinguish these cadherins from the C-1 family: none of the C-2 cadherins have a Trp in their EC1 repeat; they all have many more than five EC repeats in their ectodomain, and in between the EC stretch and the transmembrane domain, several additional domains are present, in particular the so-called laminin-G (LamG) and EGF-like domains (Fig. 2.2A). These extra domains have been collectively called the PCCD, which stands for primitive classic cadherin domain (Oda and Tsukita 1999). Type-III and type-IV cadherins are exemplified, respectively, by fruit fly shotgun (shg, DN-cadherin) and fruit fly E-cadherin (DE-cadherin). They are more ancestral than type-I and type-II cadherins and do not exist in mammals (Fig. 2.1). However, like CELSR proteins, type-III cadherins are present in all nonmammalian bilaterians and also in very basal organisms such as cnidarians and even placozoans. The absence of strand-swapping Trp residues in their EC repeats suggests that cell–cell adhesion or recognition through these C-2 cadherins is mediated by a mechanism that is different from the one used by C-1 family members. Interestingly, the apparent discrepancy between the narrow intercellular space and the long ectodomains of these C-2 cadherins was solved when several calcium-free interdomain linkers were identified in these ectodomains (Jin et al. 2012). These linkers are expected to result in "kinked" molecules that achieve homophilic adhesion by use of internal EC repeats rather than the EC1.

Classical cadherins have evolved spectacularly in composition and in number. Basal metazoans predating the bilaterians, such as *Trichoplax* and *Nematostella* usually have only one huge classical cadherin (an ancestral type-III) consisting of 32 EC repeats and LamG and EGF-like domains proximal to the transmembrane domain (Hulpiau and van Roy 2011). This kind of classical cadherin, which is not present in nonmetazoans, could have been an essential part of the genetic toolkit of the Urmetazoan, the ancestor of all multicellular animals. Such a classical cadherin in the Urmetazoan probably acquired the ability to bind ancestral cytoplasmic armadillo proteins from the beginning (see also below). This is evidenced by the remarkably conserved binding motifs in the cytoplasmic domains of classical cadherins in organisms ranging from placozoans to mammals (Fig. 2.2A). On the basis of comparative analyses of consecutive blocks of EC repeats of classical cadherins throughout evolution, we have reported evidence that the "modern" classical cadherins have arisen from these ancestral cadherins by progressive aminoterminal shortening of the EC stretch and by internal deletion of the additional ancestral motifs in the ectodomain (Hulpiau and van Roy 2011). At the dawn of bilaterians, the number of EC repeats was reduced, as illustrated by "shortened" type-III cadherins found in protostomians and nonvertebrate deuterostomians. In arthropods (protostomians), gene duplication led to another shortened classical-like cadherin represented in the type-IV subfamily. In deuterostomians, the phylum of chordates includes cephalochordates (e.g., lancelet or amphioxus *Branchiostoma floridae*), urochordates (tunicates, e.g., the sea squirt *Ciona intestinalis*) and vertebrates (e.g., mammals, birds, and fish). The first "short" classical cadherins with five ECs and a classical cytoplasmic domain, which belong to the C-1 family, appeared in chordates just before the origin of vertebrates. The two WGD events and multiple individual gene duplications resulted in great expansion and diversification of C-1 cadherins in vertebrates, generating type-I, type-II, and desmosomal cadherins, and in total about 30 C-1 cadherins in mammals. Remarkable is the loss of the type-III cadherin in almost all mammals, both marsupial and placental mammals, but not in platypus (*Ornithorhynchus anatinus*), an egg-laying mammal (monotreme).

CELSR/flamingo proteins show several distinctive features: their ectodomains are less homologous (Fig. 2.2B); they lack the two armadillo–catenin binding sites in their cytoplasmic domain, and, most strikingly, they have a 7-pass TM region, in contrast to the single TM domain of all other cadherins (Fig. 2.2A). On the basis of this 7-pass TM domain, they have also been classified as adhesion GPCR (G-protein coupled receptor) (Krishnan and Schioth 2015), although their coupling to cytoplasmic G proteins is still speculative. Importantly, even in *Nematostella* and *Trichoplax* the CELSR proteins are already present as separate branches of the cadherin superfamily (Hulpiau and van Roy 2011). This indicates that they play essential roles in all metazoans and that these roles differ from those of classical cadherins. Defects in the human or mouse CELSR genes lead to severe defects in brain development, and this has been ascribed to dysfunctions in planar cell polarity, ciliogenesis, and ciliary orientation, affecting correct neuronal migration, axon guidance, and brain homeostasis (Tissir and Goffinet 2013; Boutin et al. 2014).

2.1.4 The Protocadherin Sub-Branch of the Cadherin-Related Major Branch (CrMB)

In addition to the CMB, we can discern a second major branch in the cadherin superfamily: the cadherin-related major branch (CrMB), which includes several branches. The protocadherin (Pcdh) branch is the most densely populated and has more than 60 members in mammals (Figs. 2.1 and 2.4). Pcdhs can be subdivided into clustered and nonclustered protocadherins. In this context, clustering refers to the genomic location of genes and not to a position in the phylogenetic tree or to clustering of the encoded proteins. The nonclustered protocadherins have been named delta-protocadherins (δ-Pcdh). PCDH1, PCDH7, PCDH9, PCDH11 (X- and Y-linked) and PCDH20 have seven EC repeats and form the δ1-protocadherin subfamily (Fig. 2.4A). PCDH8, PCDH10, PCDH12, PCDH17, PCDH18, and PCDH19 have six EC repeats like the clustered protocadherins and are called δ2-

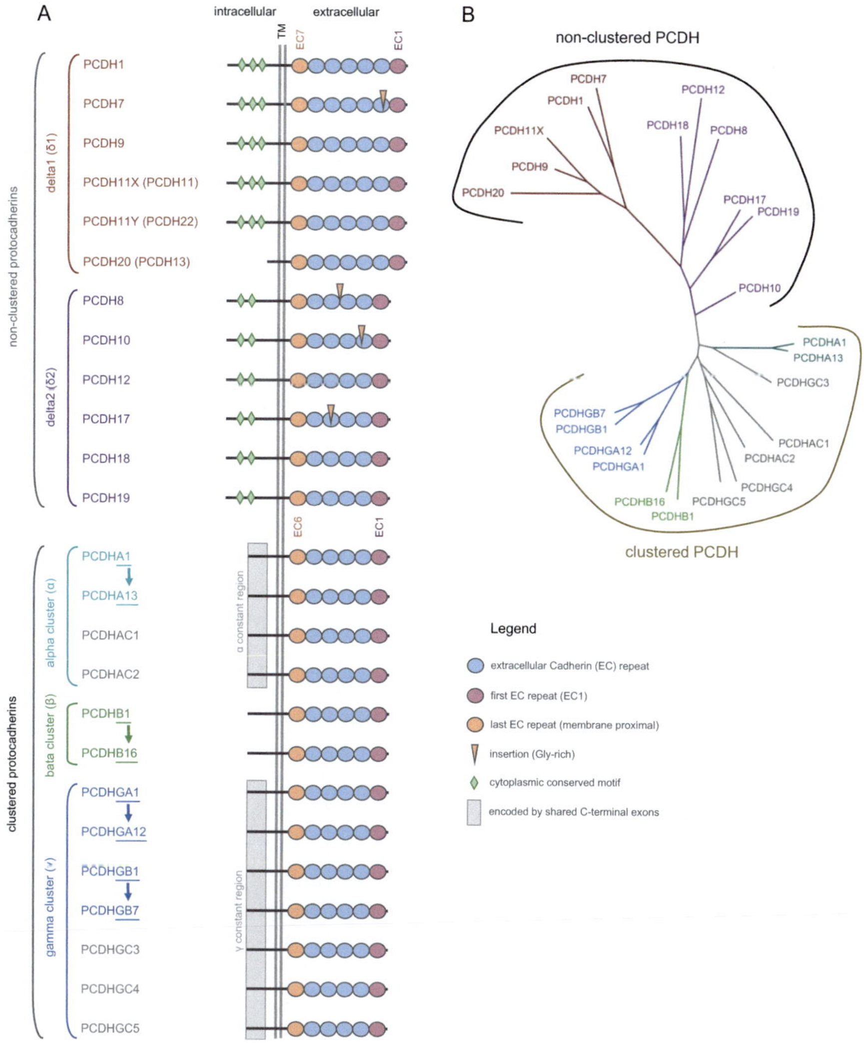

Fig. 2.4 Classification and phylogeny of the protocadherin family. (**A**) Schematic representation of the domain organization of human protocadherins. Protocadherins are classified in two subfamilies, clustered and nonclustered protocadherins. See legend on the right for an explanation of symbols. (**B**) Radial phylogram of protocadherins based on Bayesian inference of phylogeny employing a multiple sequence alignment of all their EC repeats

protocadherins. The δ-Pcdhs also have conserved motifs in their cytoplasmic domains that are not present in clustered Pcdhs: two in δ2-protocadherins and three in δ1-protocadherins (Fig. 2.4A). The cytoplasmic domain of PCDH20 has undergone a C-terminal truncation. This subclassification is largely confirmed when a phylogenetic tree is generated on the basis of the whole region with EC repeats (Fig. 2.4B).

More than 50 Pcdh genes in mammalian genomes are organized in three consecutive gene clusters: Pcdhα, Pcdhβ, and Pcdhγ (Wu and Maniatis 1999; Chen and Maniatis 2013; see also chapter by K. M. Mah and J. A. Weiner). The protocadherins in the Pcdhα and Pcdhγ clusters have one variable exon encoding the entire ectodomain with six EC repeats, the transmembrane domain, and part of the cytoplasmic domain, whereas three constant exons encode the rest of the cytoplasmic domain, which is shared by all members of the same subcluster (Fig. 2.4A). Pcdhβ genes are single-exon genes, and each protocadherin-β has a separate cytoplasmic domain. Each of the cytoplasmic domains of clustered protocadherins differs significantly from those of nonclustered protocadherins, which indicates that downstream intracellular signaling most likely differs much between clustered and nonclustered protocadherins. Also in the case of clustered protocadherins, the subclassification is supported by phylogenetic analysis of the EC1–EC4 region (Fig. 2.4B). From this, it is also obvious that clustered protocadherins have diverged from the clustered ones, whereas PCDH10 has an intermediate position, which might point to its ancestral identity.

As illustrated in Fig. 2.1, the numerous clustered protocadherins arose recently in evolution, as they are found only in vertebrates, whereas nonclustered protocadherins with conserved cytoplasmic motifs are found in Lophotrochozoa and even in Cnidaria (*Nematostella vectensis*), but not in Placozoa (Hulpiau and van Roy 2011). Apparently, the other major group in the Protostomia, that is, the Ecdysozoa comprising the widely used model organisms *Drosophila melanogaster* and *Caenorhabditis elegans*, has lost this ancestral nonclustered protocadherin. As protocadherins are predominantly expressed in neural tissues, one may wonder about their specific functions in lophotrochozoans and cnidarians. On the other hand, their explosive evolution in vertebrates reflects the much more complex neural systems in these more developed animals.

Is the binding mode of protocadherins evolutionarily conserved or has it diverged? Thus far, knowledge of the structures of protocadherin domains is very limited. The conserved Trp residues in the adhesion arm of the EC1 of classical cadherins are absent in protocadherins. Instead, a conserved Tyr residue is present at position 7 of all clustered and nonclustered protocadherins, as well as a Cys-(X) 5-Cys motif that may be involved in both intramolecular and intermolecular interactions (Yagi 2008). For clustered protocadherins, emerging evidence indicates that neurons coexpress multiple isoforms, which associate in combinatorial oligomeric complexes, possibly *cis*-tetramers, at the plasma membrane (Schreiner and Weiner 2010). The strict isoform-specificity of such oligomers, when engaged in homophilic binding across cell–cell contacts (synapses in neurons), provides an extremely rich resource for single-cell identities in the complex vertebrate brain (Hirayama and Yagi 2013; Thu et al. 2014). It has not been reported whether nonclustered protocadherins also form *cis*-oligomers. Although there are only a modest number of publications on nonclustered protocadherins (Fig. 2.3C), several recent findings have increased interest in this PCDH subfamily: cytoplasmic association with the actin-organizing WAVE complex resulting in self-avoidance of neuronal dendrites (Hayashi et al. 2014; Hayashi and Takeichi 2015), involvement in several neurological disorders (Kahr et al. 2013; Hirabayashi and Yagi 2014), and frequent dysregulations in many types of human cancer (Kahr et al. 2013; van Roy 2014; see also chapter by J. D. Jontes).

2.1.5 *Other Cadherins in the Cadherin-Related Major Branch (CrMB)*

Previously, as well as now, many cadherin-related proteins have been named "cadherin"or "protocadherin" although they fit neither classic cadherins nor protocadherins. Therefore, we have proposed to rename them cadherin-related (CDHR) proteins. The relatedness of these proteins to cadherins is based on the presence of at least two consecutive EC repeats in their ectodomains (Hulpiau and van Roy 2009). In addition to this feature, their composition and structure can vary widely as is obvious from Fig. 2.5. This also means that their comparative phylogenetic analysis is challenging and partly speculative. Nonetheless, structural and functional insight into several members within this branch is growing, and some evolutionary inferences can be made.

DCHS1 (Dachsous 1, CDHR6) and FAT4 (CDHR11) constitute a sub-branch of the CDHRs (Fig. 2.5B). The ectodomain of DCHS1 consists of a tandem array of 27 EC repeats (Fig. 2.5A). *Drosophila* fat was the first Fat subfamily member to be identified (Mahoney et al. 1991) and is the orthologue of mammalian FAT4. Their ectodomains have 34 EC repeats followed by a few alternating EGF-like and LamG domains resembling the PCCD of ancestral classical cadherins. In *Drosophila*, the Golgi kinase Four-jointed (Fj) regulates the Fat–Dachsous interaction by phosphorylating their extracellular domains (Fat acts as receptor and Dachsous as ligand), and Fat can activate the Hippo signaling pathway (Ishikawa et al. 2008; Badouel and McNeill 2011). Mutational studies in mice have also shown that the ligand–receptor function of Dchs1 and Fat4 is required for signaling in multiple organs during mouse development (Mao et al. 2011). Recently, the structures of the giant ectodomains of mammalian Dachsous1 and Fat4 were determined (Tsukasaki et al. 2014). These extended protein domains were found to have several hairpin-like self-bending points caused by EC–EC connections deficient in Ca^{2+}-binding, much like what has been described for the extended *Drosophila* classical cadherins (Jin et al. 2012). Moreover, the Dachsous1 ectodomain was found to interact with the Fat4 ectodomain in a Ca^{2+}-dependent heterophilic manner more strongly than with Dachsous1 itself in a homophilic way (Tsukasaki et al. 2014). In view of these specific interactions and downstream signaling, it makes sense that both Dachsous1 and fat/Fat4 are conserved as individual cadherins among all metazoans, including *Trichoplax* (Hulpiau and van Roy 2011).

Mammalian FAT1, FAT2, and FAT3 (CDHR8 to −10) are orthologues of *Drosophila* fat2 (ftl or fat-like) and are positioned in a separate evolutionary branch (Fig. 2.5B; Hulpiau and van Roy 2011; Castillejo-Lopez et al. 2004). This branch

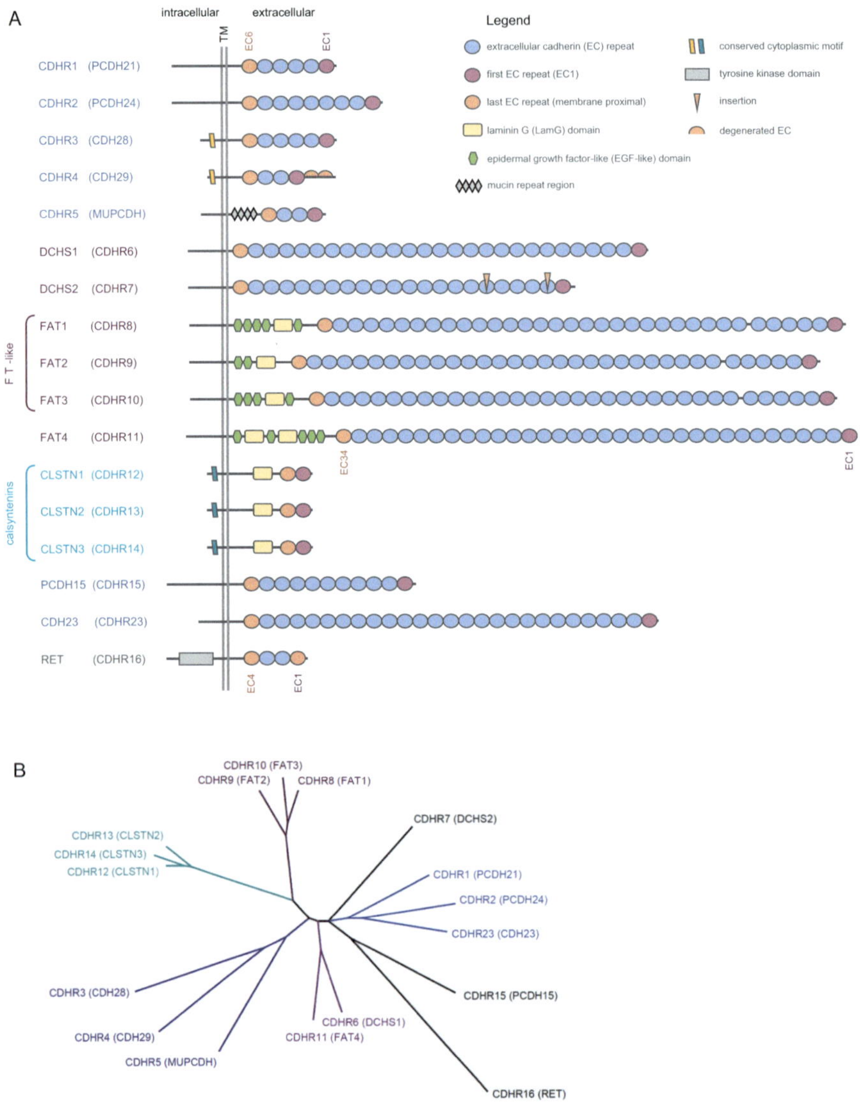

Fig. 2.5 Classification and phylogeny of the cadherin-related family. (**A**) Schematic representation of the domain organization of human cadherin-related family members. Cadherin superfamily members that clearly differ in domain composition and evolutionary history from the cadherin and protocadherin families are classified in the cadherin-related family. See legend on the right for an explanation of symbols. (**B**) Radial phylogram of cadherin-related family members, using up to six N-terminal EC repeats for Bayesian inference of phylogeny. Complex though informative evolutionary relationships are apparent

was named the FAT-like cadherin subfamily, in contrast to *Drosophila* fat and human FAT4 discussed above. Despite significant sequence differences, FAT-like and FAT cadherins have quite similar domain architectures consisting of 34 ECs followed by an ancestral membrane-proximal region consisting of LamG and EGF-like domains (Fig. 2.5A). The clear separation between branches encompassing CDHRs with very long ectodomains, DCHS1, the FAT1 to −3 cluster, and fat/FAT4 dates back to ancient times, as each of them is already present as a discrete molecule in *Trichoplax adhaerens* and *Nematostella vectensis* (Hulpiau and van Roy 2011).

Other sub-branches contain many CDHRs that do not fit into any of the previous groups: CDHR1 (PCDH21), CDHR2 (PCDH24), CDHR3 (CDH28), CHDR4 (CDH29), CDHR5 (MUPCDH), DCHS2 (CDHR7), calsyntenins (CLSTN1 to 3; CDHR12 to -14), RET (CDHR16), CDHR23 (generally called CDH23), and CDHR15 (formally given the confusing name PCDH15; Fig. 2.5).

CDHR5 was originally named MUPCDH or μ-protocadherin because it was discovered as a mucin-like gene with an ectodomain containing four EC repeats and a triple-repeated mucin domain (Goldberg et al. 2002). Although CDHR1, CDHR2, and CDH(R)23 possess 6, 9, and 27 EC repeats, respectively, they contain a peculiar, similarly charged N-terminus that can bind Ca^{2+}. This so-called "Ca0" binding site rules out a trans interaction in strand-swapping mode as seen in classical type-I cadherins (Elledge et al. 2010). This suggests that members of this subfamily have a common new interaction mechanism. Cadherin-related 23 (CDH(R)23) forms tip links between stereocilia in the hair cells of the inner ear by interacting in trans with CDHR15, commonly named PCDH15 and having 11 EC repeats (Kazmierczak et al. 2007; Sotomayor et al. 2012). The aminoterminal ends of these two CDHRs, and specifically EC1 and EC2, engage heterophilically in a mutual "handshake" interaction (Sotomayor et al. 2012). The importance of this interaction is apparent from the inherited deafness and blindness caused by many mutations in these cadherin-related proteins (El-Amraoui and Petit 2013; see also chapter by A. El-Amraoui and C. Petit). This is also reflected in the number of related publications (Fig. 2.3E). Interestingly, orthologous cadherin-related proteins in zebrafish are involved in both hearing and the lateral-line sensory system (Seiler et al. 2005). Also amphioxus has a Cdhr23 protein and even *Nematostella* has a protein with significant homology (Hulpiau and van Roy 2011). In *Nematostella*, the Cdhr23-like protein was localized in hair bundle mechanoreceptors of nematocysts (Watson et al. 2008). CDHR15 (PCDH15) orthologues can be identified in protostomians but not in more basal metazoans.

Calsyntenins (human CLSTN1/CDHR12 to CLSTN3/CDHR14; cals in fruit fly) have only two EC repeats (Fig. 2.5A), and are found in the postsynaptic membrane of excitatory synapses in the central nervous system (Hintsch et al. 2002). They are thought to link extracellular cell adhesion to intracellular calcium signaling and to be important in learning (Ikeda et al. 2008). Finally, RET (CDHR16) is a well-studied receptor tyrosine kinase containing four EC repeats and an intracellular tyrosine kinase domain (Cabrera et al. 2011; Figs. 2.3E and 2.5). RET can associate with and phosphorylate clustered protocadherins upon neuronal differentiation, and this phosphorylation stabilizes the Pcdh/Ret multimeric complex and stimulates downstream intracellular signaling (Schalm et al. 2010).

Of the eight *Trichoplax* cadherins, five are found in nearly all other metazoans, and the other three seem to be specific to placozoans. Similarly, only 9 of the 16 *Nematostella* cadherins are shared with other animals, whereas the others have not been found in more modern metazoan lineages such as vertebrates. One such example is the cadherin Hedgling, which has a Hedge domain (Hh) and a von Willebrand factor type A domain (vWA) preceding the extracellular cadherin repeats. Hedgling is found in *Nematostella* and in the demosponge *Amphimedon queenslandica* (Adamska et al. 2007). Thus, many basic as well as more evolved metazoans have apparently developed their own portfolio of clade-specific cadherin-related proteins. We have indicated this in Fig. 2.1 by putting several question marks for the number of CDHRs in metazoans. Consequently, evolution of the cadherin superfamily might be much more complex than initially thought. The sequencing of more ancestral metazoan genomes might help to resolve current problems in detailed reconstruction of the cadherin superfamily evolution.

2.2 Evolution of Catenins in Metazoans

2.2.1 Structures and Evolution of Metazoan Catenins with Armadillo Domains

Armadillo proteins share the presence of similar imperfect tandem repeats composed of so-called ARM repeats of about 40 amino acids each. ARM repeats were first identified in the *Drosophila* segment polarity protein called armadillo, which is the orthologue of mammalian β-catenin (Riggleman et al. 1989). Although the sequence similarity between individual ARM repeats of a single protein may be very low, all these repeats have conserved three-dimensional structures. Invariably, three helices within each ARM repeat form a compact helical bundle, and the consecutive repeats fold together as an extended curved superhelical structure, which enables versatile high affinity interactions with many proteins (Xing et al. 2008; Striegl et al. 2010).

Armadillo proteins may differ a lot in the number and organization of their ARM repeats, and they are involved in a broad range of biological processes, including cell adhesion, signaling, cytoskeletal regulation, and intracellular transport. Here, we focus on those that can bind to the conserved motifs in the cytoplasmic domains of classical cadherins and can be collectively called armadillo catenins. The name catenin was derived from *catena*, the Latin word for chain, and was proposed by R. Kemler and M. Ozawa (1989). They proposed that the major function of catenins might be to link proteins coding for Ca^{2+}-dependent cell adhesion molecules (CAMs) to cytoskeletal structures. This holds true for β-catenin and plakoglobin, which link classical cadherins indirectly to the actin cytoskeleton, and also for p120 catenin (p120ctn), which links classical cadherins indirectly to both the actomyosin cytoskeleton and the microtubuli (Takeichi 2014).

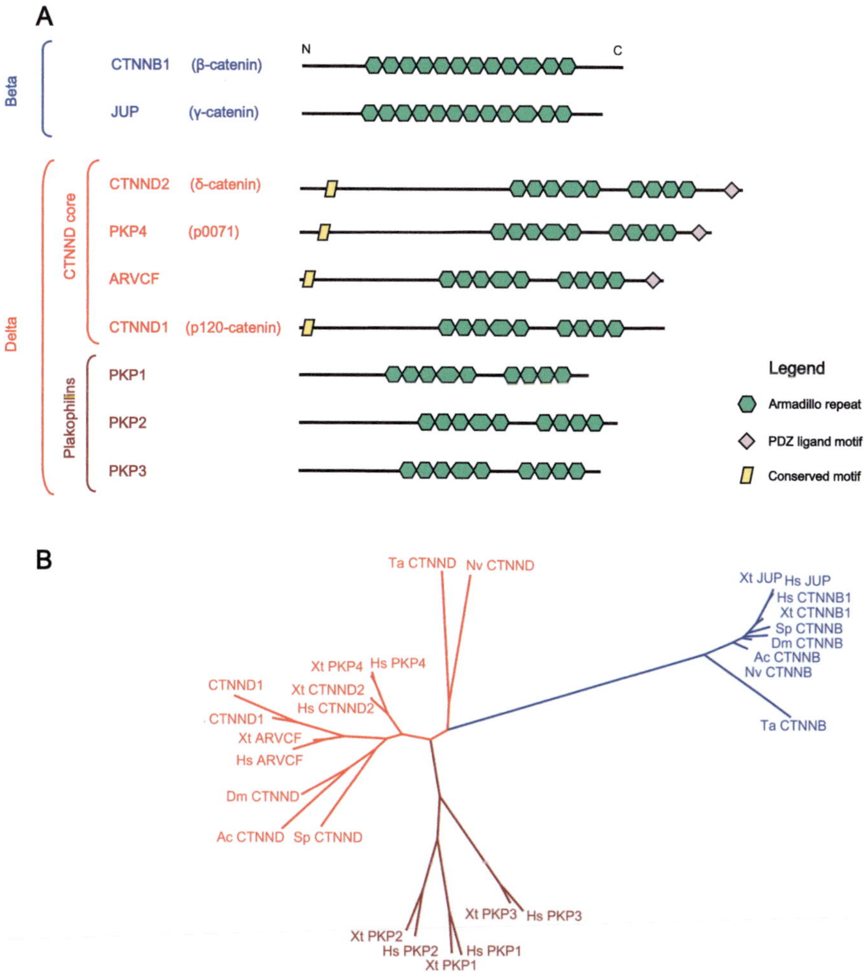

Fig. 2.6 Classification and phylogeny of cadherin-associated armadillo catenins. (**A**) Schematic representation of the domain organization of human armadillo catenins. Armadillo catenins can be classified in two subfamilies: beta-catenins and delta-catenins. Furthermore, the delta-catenin subfamily can be divided into two branches, the p120 or CTNND branch and the plakophilin or PKP branch. The armadillo domains of beta- and delta-catenin subfamily members were annotated here on the basis of published 3D structures (Huber et al. 1997; Choi and Weis 2005; Ishiyama et al. 2010). See explanation of symbols on the right. (**B**) Radial phylogram of armadillo catenins based on Bayesian inference of phylogeny using a multiple sequence alignment of the armadillo repeat regions, which is a common feature of all subfamily members. The protein sequences of all catenins analyzed were retrieved from GenBank and through BLAST. In the case of partial sequences, additional gene prediction analyses were performed by using FGENESH software (Solovyev et al. 2006). Species names are indicated: Hs (*Homo sapiens*, humans), Xt (*Xenopus tropicalis*, Western clawed frog), Sp (*Strongylocentrotus purpuratus*, purple sea urchin), Ac (*Aplysia californica*, California sea hare), Dm (*Drosophila melanogaster*, fruit fly), Nv (*Nematostella vectensis*, starlet sea anemone), and Ta (*Trichoplax adhaerens*, Placozoa). Although the vertebrate gene products are represented by their official gene symbols from GenBank, nonvertebrate gene products are shown as CTNNB and CTNND to indicate the ancestors of β-catenins and δ-catenins, respectively

Based on their sequence homology, the mammalian catenins with ARM repeats can be divided into two subfamilies named after representative members (Fig. 2.6). The β-catenin subfamily consists of β-catenin (encoded by gene *CTNNB1*) and plakoglobin (also called γ-catenin, encoded by *JUP*). The delta-catenin subfamily has seven members and can be divided into two branches, the p120 branch and the plakophilin branch. We propose that the p120 branch be called CTNND core proteins. This branch includes p120ctn (encoded by *CTNND1*), δ-catenin (encoded by *CTNND2*), ARVCF (armadillo repeat gene deleted in velocardiofacial syndrome, encoded by *ARVCF*, although *CTNND3* would be a better gene name), and p0071 (encoded by *PKP4*, although this protein is not a plakophilin, and therefore *CTNND4* would have been more appropriate). The plakophilin branch comprises the real plakophilins: plakophilin-1 (encoded by *PKP1*), -2 (*PKP2*) and -3 (*PKP3*) (McCrea and Gu 2010). Plakophilins show higher sequence and functional similarity to the p120/CTNND core members than to β-catenin (Fig. 2.6B).

β-catenin in particular has been studied extensively (Fig. 2.3F). β-catenin is famous for two reasons: in cell adherens junctions, it bridges the cytoplasmic domain of the classical cadherins to α-catenin (see also below) and indirectly to the actin cytoskeleton (van Roy and Berx 2008; Maiden and Hardin 2011; Pokutta et al. 2014); in the nucleus, it is involved in the activation of Wnt-triggered signaling by activating TCF/Lef family transcription factors (van Es et al. 2003). The related subfamily member plakoglobin functions in both the adherens junctions and the desmosomes. Desmosomal plakoglobin serves as a molecular link between the cytoplasmic domains of desmosomal cadherins (desmogleins, desmocollins) and the aminoterminal ends of desmoplakin (Kowalczyk et al. 1997). The carboxyterminal ends of desmoplakin bind to the intermediate filaments in the desmosomal plaque.

Whereas β-catenin and plakoglobin bind to the conserved motif nearby the C-terminus of classical cadherins, p120ctn binds to the conserved juxtamembrane domain of these cadherins (van Roy and Berx 2008). p120ctn regulates the abundance of classical cadherins by controlling the rate of cadherin turnover (Ireton et al. 2002). Another important function of p120ctn is its regulation of Rho-GTPases, which are essential for the dynamic organization of the actin cytoskeleton (Anastasiadis 2007). The other p120ctn subfamily members, δ-catenin, ARVCF, and p0071, are also reported to be involved in modulation of cadherin stability at cell–cell junctions (McCrea and Park 2007). Moreover, they directly or indirectly interact with a number of Rho-GTPases, which makes intracellular signaling and cytoskeletal control possible. The linkage of intermediate filaments to desmosomal cadherins occurs via plakoglobin (see above) and plakophilins (PKP1 to PKP3; Desai et al. 2009). In contrast to the other armadillo catenins, which use their Arm domain to bind to the cytoplasmic domain of cadherins, plakophilins seem to use their globular aminoterminal domains to associate with desmosomal cadherins (Kowalczyk et al. 1999; Hatzfeld et al. 2014).

Examining the evolutionary history of β-catenin by phylogenetic analysis based on protein sequences showed that it was present far back in premetazoan species, such as the soil-dwelling amoeba *Dictyostelium discoideum* (slime mold)

(Dickinson et al. 2011). However, no β-catenin homologue has been reported in the choanoflagellate *Monosiga brevicollis*, a unicellular organism closely related to metazoans. The ancestral metazoan *Trichoplax adhaerens* has a β-catenin-related ARM protein, designated TaCTNNB in Fig. 2.6B (Hulpiau and van Roy 2011). Later, ancestral β-catenin went through several lineage-specific duplications, in accordance with whole genome duplications and functional diversifications (Zhao et al. 2011). However, only one *CTNNB* gene is present in mammalian genomes. On the other hand, the ancestral *CTNNB* gene underwent a duplication event in the chordate lineage, which gave rise to genes encoding either vertebrate β-catenin or plakoglobin (Fig. 2.1). Like β-catenin, plakoglobin has 12 consecutive ARM repeats (Fig. 2.6A). This gene duplication allowed a neofunctionalization of plakoglobin compared to β-catenin (Swope et al. 2013).

Phylogenctic analysis of the complete ARM repeat region of β-catenin implies strong sequence conservation across lineages, in line with functional conservation. Comparison of individual ARM repeats of the β-catenin homologues in Cnidaria, Arthropoda, Echinodermata, and Chordata showed that each ARM repeat has an individual signature that has been conserved throughout metazoan evolution. Generally, the most conserved residues in the ARM repeat regions are involved in interactions with cadherins, axin, or APC (adenomatous polyposis coli protein) (Schneider et al. 2003). In line with this, we have shown that the β-catenin binding domain in the classical cadherins of the cnidarian *N. vectensis* and the placozoan *T. adhaerens* is also well conserved (Hulpiau and van Roy 2011).

Members of the delta-catenin armadillo family, comprising both p120ctn (CTNND core) and plakophilin (PKP) proteins, are structurally more closely related to each other than to the β-catenin subfamily. Compared to the 12 repeats in β-catenin and plakoglobin, the p120ctn and plakophilin members contain 9 ARM repeats with a long insert between ARM5 and -6, as well as N- and C-terminal regions that diverge substantially from one another (Fig. 2.6A). Although the insert region does not affect the packing of the ARM repeats, it may create a major bend in the ARM repeat region (Choi and Weis 2005).

As with β-catenin and plakoglobin, all seven members of the δ-catenin armadillo family are present in vertebrates (Figs. 2.1 and 2.6B). A recent phylogenetic analysis showed that the ancestor of the p120ctn subfamily is a single δ-catenin/ARVCF-like gene present in metazoans but not in the unicellular premetazoan *M. brevicollis* (Carnahan et al. 2010). Also, ancestral metazoans such as *Trichoplax* and *Nematostella* possess a single subfamily member, designated as CTNND in Fig. 2.6B. In contrast, plakophilins are confined to vertebrates, just as are the desmosomal cadherins with which they associate (Fig. 2.1). From the phylogenetic tree in Fig. 2.6B, one may infer that the p120ctn gene in vertebrates (*CTNND1*) probably arose by duplication of the *ARVCF* gene, whereas the p0071 gene (*PKP4*) arose by duplication of a gene encoding δ-catenin (*CTNND2*). Yet it is not clear which member of the p120ctn subfamily is at the origin of the plakophilins (Zhao et al. 2011).

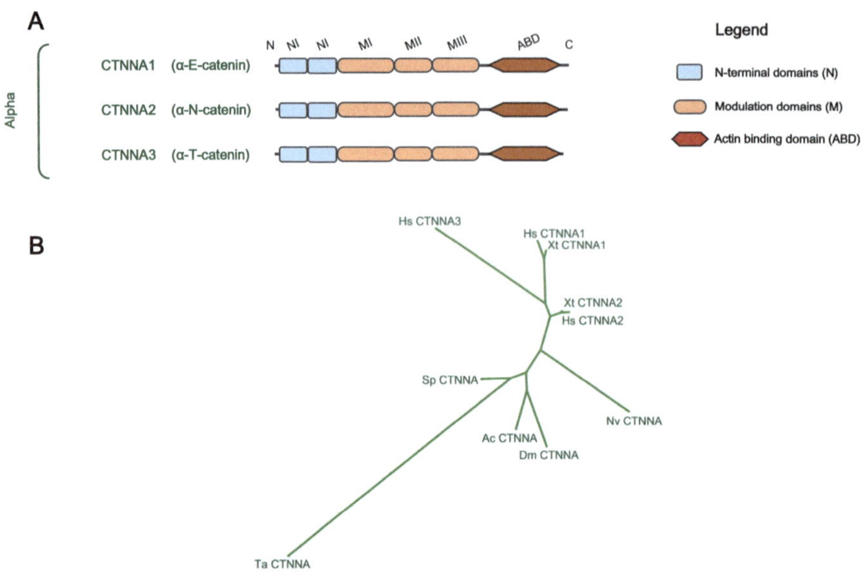

Fig. 2.7 Classification and phylogeny of cadherin-associated alpha-catenins. (**A**) Schematic representation of the domain organization of human α-catenins. Annotation was done on the basis of recently published 3D structures (Ishiyama et al. 2013; Pokutta et al. 2014). (**B**) Radial phylogram of α-catenins based on Bayesian inference of phylogeny using a multiple sequence alignment of their N-terminal domains. The nonvertebrate α-catenins are shown as CTNNA, representing the ancestor of this subfamily

2.2.2 *Structures and Evolution of Alpha-Catenins in Metazoans*

Another catenin family consists of α-catenins and related proteins. Contrary to the armadillo catenins, discussed in the previous section, the α-catenins do not have ARM repeats and do not bind directly to cadherins. Genes *CTNNA1* to *−3* encode αE-, αN-, and αT-catenins, respectively, where E stands for epithelial, N for neural, and T for testis (Janssens et al. 2001; Fig. 2.7). α-Catenins participate in cell–cell adhesion by binding to junctional β-catenin, and they also participate in coordinating actin dynamics in adherens junctions (Kobielak and Fuchs 2004; Maiden and Hardin 2011; Pokutta et al. 2014; Takeichi 2014). Although αN-catenin is a major component of intercellular junctions in the neural system, so far αE-catenin has received particular attention (Fig. 2.3F). In the intercalated discs of cardiomyocytes, αT-catenin is at the basis of the *area composita*, which is a stress-resistant junction consisting of both N-cadherin with associated catenins, and desmosomal cadherins with associated desmosomal components (Goossens et al. 2007; Rampazzo et al. 2014).

In contrast to the armadillo catenins, where just two ancestral molecules, CTNNB and CTNND, have diverged during evolution to nine members in vertebrates, the α-catenins have undergone fewer duplication events in metazoans. Phylogenetic studies showed that αN-catenin is the ancestor of the other two α-catenins (Zhao et al. 2011), being already present in ancestral metazoans (Fig. 2.1 and CTNNA branches in Fig. 2.7B). αN-catenin gave rise to αE-catenin as a result of the vertebrate-specific whole genome duplication, whereas αT-catenin arose as a result of an amniote-specific gene duplication event (Hulpiau et al. 2013). Also in premetazoans such as *Dictyostelium discoideum*, an α-catenin–like protein can form complexes with a β-catenin–like protein. In the social though unicellular protozoan *D. discoideum*, this catenin complex is essential for formation of a peculiar polarized "epithelium" formed despite complete absence of members of the cadherin superfamily (Dickinson et al. 2011). It is currently not well understood how these nonmetazoan "catenins" have evolved into the above-mentioned metazoan proteins with their strongly conserved participation in cadherin-based junctional complexes.

Acknowledgments We thank A. Bredan for critical reading and editing of the manuscript. Research was supported by the Research Foundation - Flanders (FWO) and by the Belgian Science Policy (Interuniversity Attraction Poles - IAP7/07).

References

Adamska M, Matus DQ, Adamski M, Green K, Rokhsar DS, Martindale MQ, Degnan BM (2007) The evolutionary origin of hedgehog proteins. Curr Biol 17(19):R836–R837

Anastasiadis PZ (2007) p120-ctn: a nexus for contextual signaling via Rho GTPases. Biochim Biophys Acta Mol Cell Res 1773(1):34–46

Badouel, C., McNeill, H. (2011) SnapShot: The hippo signaling pathway. Cell 145 (3):484–484.e1.

Bertolotti R, Rutishauser U, Edelman GM (1980) A cell surface molecule involved in aggregation of embryonic liver cells. Proc Natl Acad Sci U S A 77(8):4831–4835

Boutin C, Labedan P, Dimidschstein J, Richard F, Cremer H, Andre P, Yang Y, Montcouquiol M, Goffinet AM, Tissir F (2014) A dual role for planar cell polarity genes in ciliated cells. Proc Natl Acad Sci U S A 111(30):E3129–E3138

Cabrera JR, Bouzas-Rodriguez J, Tauszig-Delamasure S, Mehlen P (2011) RET modulates cell adhesion via its cleavage by caspase in sympathetic neurons. J Biol Chem 286 (16):14628–14638

Carnahan RH, Rokas A, Gaucher EA, Reynolds AB (2010) The molecular evolution of the p120-catenin subfamily and its functional associations. PLoS One 5(12):e15747

Castillejo-Lopez C, Arias WM, Baumgartner S (2004) The fat-like gene of Drosophila is the true orthologue of vertebrate fat cadherins and is involved in the formation of tubular organs. J Biol Chem 279(23):24034–24043

Chen WV, Maniatis T (2013) Clustered protocadherins. Development 140(16):3297–3302

Choi HJ, Weis WI (2005) Structure of the armadillo repeat domain of plakophilin 1. J Mol Biol 346(1):367–376

Damsky CH, Knudsen KA, Dorio RJ, Buck CA (1981) Manipulation of cell-cell and cell-substratum interactions in mouse mammary tumor epithelial cells using broad spectrum antisera. J Cell Biol 89:173–184

De Craene B, Berx G (2013) Regulatory networks defining EMT during cancer initiation and progression. Nat Rev Cancer 13:97–110
Desai BV, Harmon RM, Green KJ (2009) Desmosomes at a glance. J Cell Sci 122(24):4401–4407
Dickinson DJ, Nelson WJ, Weis WI (2011) A polarized epithelium organized by beta- and alpha-catenin predates cadherin and metazoan origins. Science 331(6022):1336–1339
Edgar RC (2004) MUSCLE: multiple sequence alignment with high accuracy and high throughput. Nucleic Acids Res 32(5):1792–1797
El-Amraoui A, Petit C (2013) Cadherin defects in inherited human diseases. Prog Mol Biol Transl Sci 116:361–384
Elledge HM, Kazmierczak P, Clark P, Joseph JS, Kolatkar A, Kuhn P, Muller U (2010) Structure of the N terminus of cadherin 23 reveals a new adhesion mechanism for a subset of cadherin superfamily members. Proc Natl Acad Sci U S A 107(23):10708–10712
Goldberg M, Wei M, Tycko B, Falikovich I, Warburton D (2002) Identification and expression analysis of the human mu-protocadherin gene in fetal and adult kidneys. Am J Physiol Renal Physiol 283(3):F454–F463
Goossens S, Janssens B, Bonné S, De Rycke R, Braet F, van Hengel J, van Roy F (2007) A unique and specific interaction between alpha-T-catenin and plakophilin-2 recruits desmosomal proteins to the adherens junctions of the heart. J Cell Sci 120(12):2126–2136
Harrison OJ, Jin X, Hong S, Bahna F, Ahlsen G, Brasch J, Wu Y, Vendome J, Felsovalyi K, Hampton CM, Troyanovsky RB, Ben-Shaul A, Frank J, Troyanovsky SM, Shapiro L, Honig B (2011) The extracellular architecture of adherens junctions revealed by crystal structures of type I cadherins. Structure 19(2):244–256
Hatzfeld M, Wolf A, Keil R (2014) Plakophilins in desmosomal adhesion and signaling. Cell Commun Adhes 21(1):25–42
Hayashi S, Takeichi M (2015) Emerging roles of protocadherins: from self-avoidance to enhancement of motility. J Cell Sci 128(8):1455–1464
Hayashi S, Inoue Y, Kiyonari H, Abe T, Misaki K, Moriguchi H, Tanaka Y, Takeichi M (2014) Protocadherin-17 mediates collective axon extension by recruiting actin regulator complexes to interaxonal contacts. Dev Cell 30(6):673–687
Hintsch G, Zurlinden A, Meskenaite V, Steuble M, Fink-Widmer K, Kinter J, Sonderegger P (2002) The calsyntenins – a family of postsynaptic membrane proteins with distinct neuronal expression patterns. Mol Cell Neurosci 21(3):393–409
Hirabayashi T, Yagi T (2014) Protocadherins in neurological diseases. Adv Neurobiol 8:293–314
Hirano S, Takeichi M (2012) Cadherins in brain morphogenesis and wiring. Physiol Rev 92 (2):597–634
Hirayama T, Yagi T (2013) Clustered protocadherins and neuronal diversity. Prog Mol Biol Transl Sci 116:145–167
Huber AH, Nelson WJ, Weis WI (1997) Three-dimensional structure of the Armadillo repeat region of beta-catenin. Cell 90:871–882
Hulpiau P, van Roy F (2009) Molecular evolution of the cadherin superfamily. Int J Biochem Cell Biol 41:349–369
Hulpiau P, van Roy F (2011) New insights into the evolution of metazoan cadherins. Mol Biol Evol 28(1):647–657
Hulpiau P, Gul I, van Roy F (2013) New insights into the evolution of metazoan cadherins and catenins. Prog Mol Biol Transl Sci 116:71–94
Ikeda DD, Duan Y, Matsuki M, Kunitomo H, Hutter H, Hedgecock EM, Iino Y (2008) CASY-1, an ortholog of calsyntenins/alcadeins, is essential for learning in *Caenorhabditis elegans*. Proc Natl Acad Sci U S A 105(13):5260–5265
Ireton RC, Davis MA, van Hengel J, Mariner DJ, Barnes K, Thoreson MA, Anastasiadis PZ, Matrisian L, Bundy LM, Sealy L, Gilbert B, van Roy F, Reynolds AB (2002) A novel role for p120 catenin in E-cadherin function. J Cell Biol 159(3):465–476
Ishikawa HO, Takeuchi H, Haltiwanger RS, Irvine KD (2008) Four-jointed is a Golgi kinase that phosphorylates a subset of cadherin domains. Science 321(5887):401–404

Ishiyama N, Lee SH, Liu S, Li GY, Smith MJ, Reichardt LF, Ikura M (2010) Dynamic and static interactions between p120 catenin and E-cadherin regulate the stability of cell-cell adhesion. Cell 141(1):117–128

Ishiyama N, Tanaka N, Abe K, Yang YJ, Abbas YM, Umitsu M, Nagar B, Bueler SA, Rubinstein JL, Takeichi M, Ikura M (2013) An autoinhibited structure of alpha-catenin and its implications for vinculin recruitment to adherens junctions. J Biol Chem 288(22):15913–15925

Janssens B, Goossens S, Staes K, Gilbert B, van Hengel J, Colpaert C, Bruyneel E, Mareel M, van Roy F (2001) alpha-T-Catenin: a novel tissue-specific beta-catenin-binding protein mediating strong cell-cell adhesion. J Cell Sci 114(17):3177–3188

Jin X, Walker MA, Felsovalyi K, Vendome J, Bahna F, Mannepalli S, Cosmanescu F, Ahlsen G, Honig B, Shapiro L (2012) Crystal structures of Drosophila N-cadherin ectodomain regions reveal a widely used class of Ca(2)+-free interdomain linkers. Proc Natl Acad Sci U S A 109 (3):127–134

Kahr I, Vandepoele K, van Roy F (2013) Delta-protocadherins in health and disease. Prog Mol Biol Transl Sci 116:169–192

Kasahara M (2007) The 2R hypothesis: an update. Curr Opin Immunol 19(5):547–552

Kazmierczak P, Sakaguchi H, Tokita J, Wilson-Kubalek EM, Milligan RA, Muller U, Kachar B (2007) Cadherin 23 and protocadherin 15 interact to form tip-link filaments in sensory hair cells. Nature 449(7158):87–91

Kemler R, Ozawa M (1989) Uvomorulin-catenin complex: cytoplasmic achorage of a Ca^{2+} dependent cell adhesion molecule. Bioessays 11:88–91

Kobielak A, Fuchs E (2004) Alpha-catenin: at the junction of intercellular adhesion and actin dynamics. Nat Rev Mol Cell Biol 5(8):614–625

Kowalczyk AP, Bornslaeger EA, Borgwardt JE, Palka HL, Dhaliwal AS, Corcoran CM, Denning MF, Green KJ (1997) The amino-terminal domain of desmoplakin binds to plakoglobin and clusters desmosomal cadherin-plakoglobin complexes. J Cell Biol 139(3):773–784

Kowalczyk AP, Hatzfeld M, Bornslaeger EA, Kopp DS, Borgwardt JE, Corcoran CM, Settler A, Green KJ (1999) The head domain of plakophilin-1 binds to desmoplakin and enhances its recruitment to desmosomes – implications for cutaneous disease. J Biol Chem 274 (26):18145–18148

Krishnan A, Schioth HB (2015) The role of G protein-coupled receptors in the early evolution of neurotransmission and the nervous system. J Exp Biol 218(Pt 4):562–571

Mahoney PA, Weber U, Onofrechuk P, Biessmann H, Bryant PJ, Goodman CS (1991) The *fat* tumor suppressor gene in Drosophila encodes a novel member of the cadherin gene superfamily. Cell 67(5):853–868

Maiden SL, Hardin J (2011) The secret life of alpha-catenin: moonlighting in morphogenesis. J Cell Biol 195(4):543–552

Mao Y, Mulvaney J, Zakaria S, Yu T, Morgan KM, Allen S, Basson MA, Francis-West P, Irvine KD (2011) Characterization of a Dchs1 mutant mouse reveals requirements for Dchs1-Fat4 signaling during mammalian development. Development 138(5):947–957

McCrea PD, Gu D (2010) The catenin family at a glance. J Cell Sci 123(Pt 5):637–642

McCrea PD, Park JI (2007) Developmental functions of the p120-catenin subfamily. Biochim Biophys Acta Mol Cell Res 1773(1):17–33

Nagafuchi A, Shirayoshi Y, Okazaki K, Yasuda K, Takeichi M (1987) Transformation of cell adhesion properties by exogenously introduced E-cadherin cDNA. Nature (London) 329:341–343

Niessen CM, Leckband D, Yap AS (2011) Tissue organization by cadherin adhesion molecules: dynamic molecular and cellular mechanisms of morphogenetic regulation. Physiol Rev 91 (2):691–731

Oda H, Tsukita S (1999) Nonchordate classic cadherins have a structurally and functionally unique domain that is absent from chordate classic cadherins. Dev Biol 216(1):406–422

Patel SD, Ciatto C, Chen CP, Bahna F, Rajebhosale M, Arkus N, Schieren I, Jessell TM, Honig B, Price SR, Shapiro L (2006) Type II cadherin ectodomain structures: implications for classical cadherin specificity. Cell 124(6):1255–1268

Peyrieras N, Hyafil F, Louvard D, Ploegh HL, Jacob F (1983) Uvomorulin: a nonintegral membrane protein of early mouse embryo. Proc Natl Acad Sci U S A 80:6274–6277

Pieters T, van Roy F (2014) Role of cell-cell adhesion complexes in embryonic stem cell biology. J Cell Sci 127:2603–2613

Pokutta S, Choi HJ, Ahlsen G, Hansen SD, Weis WI (2014) Structural and thermodynamic characterization of cadherin beta-catenin alpha-catenin complex formation. J Biol Chem 289 (19):13589–13601

Rampazzo A, Calore M, van Hengel J, van Roy F (2014) Intercalated discs and arrhythmogenic cardiomyopathy. Circ Cardiovasc Genet 7(6):930–940

Riggleman B, Wieschaus E, Schedl P (1989) Molecular analysis of the armadillo locus: uniformly distributed transcripts and a protein with novel internal repeats are associated with Drosophila segment polarity gene. Genes Dev 3:96–113

Ronquist F, Teslenko M, van der Mark P, Ayres DL, Darling A, Hohna S, Larget B, Liu L, Suchard MA, Huelsenbeck JP (2012) MrBayes 3.2: efficient Bayesian phylogenetic inference and model choice across a large model space. Syst Biol 61(3):539–542

Schalm SS, Ballif BA, Buchanan SM, Phillips GR, Maniatis T (2010) Phosphorylation of protocadherin proteins by the receptor tyrosine kinase Ret. Proc Natl Acad Sci U S A 107 (31):13894–13899

Schneider SQ, Finnerty JR, Martindale MQ (2003) Protein evolution: structure-function relationships of the oncogene beta-catenin in the evolution of multicellular animals. J Exp Zool B Mol Dev Evol 295(1):25–44

Schreiner D, Weiner JA (2010) Combinatorial homophilic interaction between {gamma}-protocadherin multimers greatly expands the molecular diversity of cell adhesion. Proc Natl Acad Sci U S A 107(33):14893–14898

Seiler C, FingerBaier KC, Rinner O, Makhankov YV, Schwarz H, Neuhauss SCF, Nicolson T (2005) Duplicated genes with split functions: independent roles of protocadherin15 orthologues in zebraf ish hearing and vision. Development 132(3):615–623

Solovyev V, Kosarev P, Seledsov I, Vorobyev D (2006) Automatic annotation of eukaryotic genes, pseudogenes and promoters. Genome Biol 7 Suppl 1:S10 11–12.

Sotomayor M, Weihofen WA, Gaudet R, Corey DP (2012) Structure of a force-conveying cadherin bond essential for inner-ear mechanotransduction. Nature 492(7427):128–132

Sotomayor M, Gaudet R, Corey DP (2014) Sorting out a promiscuous superfamily: towards cadherin connectomics. Trends Cell Biol 24(9):524–536

Striegl H, Andrade-Navarro MA, Heinemann U (2010) Armadillo motifs involved in vesicular transport. PLoS One 5(2):e8991

Swope D, Li J, Radice GL (2013) Beyond cell adhesion: the role of armadillo proteins in the heart. Cell Signal 25(1):93–100

Takeichi M (2014) Dynamic contacts: rearranging adherens junctions to drive epithelial remodelling. Nat Rev Mol Cell Biol 15(6):397–410

Thu CA, Chen WV, Rubinstein R, Chevee M, Wolcott HN, Felsovalyi KO, Tapia JC, Shapiro L, Honig B, Maniatis T (2014) Single-cell identity generated by combinatorial homophilic interactions between alpha, beta, and gamma protocadherins. Cell 158(5):1045–1059

Tissir F, Goffinet AM (2013) Shaping the nervous system: role of the core planar cell polarity genes. Nat Rev Neurosci 14(8):525–535

Tsukasaki Y, Miyazaki N, Matsumoto A, Nagae S, Yonemura S, Tanoue T, Iwasaki K, Takeichi M (2014) Giant cadherins Fat and Dachsous self-bend to organize properly spaced intercellular junctions. Proc Natl Acad Sci U S A 111(45):16011–16016

van Es JH, Barker N, Clevers H (2003) You Wnt some, you lose some: oncogenes in the Wnt signaling pathway. Curr Opin Genet Dev 13(1):28–33

van Roy F (2014) Beyond E-cadherin: Roles of other cadherin superfamily members in cancer. Nat Rev Cancer 14(2):121–134

van Roy F, Berx G (2008) The cell-cell adhesion molecule E-cadherin. Cell Mol Life Sci 65 (23):3756–3788

Watson GM, Pham L, Graugnard EM, Mire P (2008) Cadherin 23-like polypeptide in hair bundle mechanoreceptors of sea anemones. J Comp Physiol Neuroethol Sens Neural Behav Physiol 194(9):811–820

Wu Q, Maniatis T (1999) A striking organization of a large family of human neural cadherin-like cell adhesion genes. Cell 97(6):779–790

Xing Y, Takemaru K, Liu J, Berndt JD, Zheng JJ, Moon RT, Xu W (2008) Crystal structure of a full-length beta-catenin. Structure 16(3):478–487

Yagi T (2008) Clustered protocadherin family. Dev Growth Differ 50:S131–S140

Yoshida-Noro C, Suzuki N, Takeichi M (1984) Molecular nature of the calcium-dependent cell-cell adhesion system in mouse teratocarcinoma and embryonic cells studied with a monoclonal antibody. Dev Biol 101(1):19–27

Zhao ZM, Reynolds AB, Gaucher EA (2011) The evolutionary history of the catenin gene family during metazoan evolution. BMC Evol Biol 11:198

Part II
Classical Cadherins

Chapter 3
Classical Cadherins

Barry M. Gumbiner

Abstract Classical cadherins mediate a wide range of cellular functions important for morphogenesis, tissue homeostasis and physiology, and disease conditions. Their core adhesive functions, in association with the contractile actomyosin cytoskeleton allow cells to sort into domains and exert forces in order to separate from their neighbors, rearrange relative to their neighbors, or cause a change in tissue shape. The diversity of classical cadherins underlies cell recognition specificity and also provides for more specialized tissue-specific functions and signaling events. Cadherin-mediated adhesion is regulated in a variety of ways to allow them to perform this diversity of physiological and developmental functions in different tissues. Regulatory mechanisms include control of cadherin gene expression and cadherin switching, cadherin membrane trafficking, cytoskeletal remodeling, and the control of the state of the adhesive bond itself. Signaling mechanisms are essential to classical cadherin functions, contributing to both their intrinsic adhesive and force generating functions as well as communicating the state of the tissue to mediate changes in growth and differentiation.

Keywords Cell adhesion • Catenins • Morphogenesis • Actomyosin • Cell sorting • Signal transduction • Barrier function • Regulation • Selectivity • Structure

3.1 Introduction

Classical cadherins have fundamental roles in the formation, maintenance, and physiological functions of tissues of all kinds. One or more classical cadherins is required in virtually all cells that interact physically to form tissues. Nominally classical cadherins mediate cell–cell adhesion, but they do much more than hold cells together. They mediate dynamic interactions between cells (Fig. 3.1), controlling cell movements underlying morphogenesis and changes in cell polarity, cell

B.M. Gumbiner, Ph.D. (✉)
Center for Developmental Biology and Regenerative Medicine, Seattle Children's Research Institute, Seattle, WA, USA

Department of Pediatrics, University of Washington School of Medicine, Seattle, WA, USA
e-mail: gumbiner@uw.edu

S.T. Suzuki, S. Hirano (eds.), *The Cadherin Superfamily*,
DOI 10.1007/978-4-431-56033-3_3

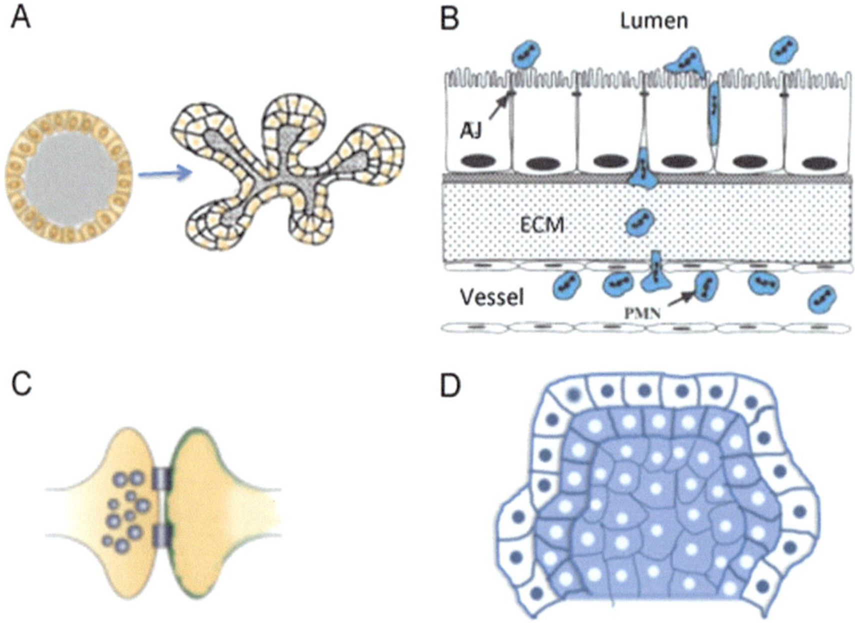

Fig. 3.1 Wide-ranging general functions of cadherins in tissues. (**a**) Morphogenesis in tissue development and regeneration; example shown is branching morphogenesis. (**b**) Physiological regulation of barrier function, in epithelia and endothelia illustration from (Parkos, 1997). (**c**) Physiological control of cell communication, such as synaptic contact and activity. (**d**) Signaling functions in development, growth, regeneration, and cancer; example shown in regulation of the nuclear accumulation of transcription factors by contact inhibition

structure, and tissue architecture. Similarly, they dynamically regulate physiological properties of tissues, including barrier function and synaptic communication. Classical cadherins also mediate numerous intracellular signaling processes to control both cytoplasmic organization and motile behaviors of cells as well as changes in gene expression to control cell differentiation and growth. In this context, it is not surprising that classical cadherins play important roles in diseases involving disturbances in tissue architecture, such as cancer and inflammation.

The classical cadherins constitute a family of highly related proteins that are differentially expressed in tissues and organs (Oda and Takeichi 2011; Takeichi 1991; Hulpiau and van Roy 2009; Gumbiner 2005). They mediate calcium-dependent homophilic binding between adjacent cells to mediate adhesive interactions. In vertebrates they posses five extracellular cadherin (EC) repeat domains, and all classical cadherins interact with catenins via their cytoplasmic domains (Fig. 3.2). Classical cadherins have both common core functions in all types of cells as well as more specialized functions in particular tissues. There have been many hypotheses about the role of cadherin diversity and many suggestions for cadherin-type specific functions. One important early idea is that they mediate cell interaction specificity, cell recognition events, and cell sorting (Takeichi 1991; Nose et al. 1988; discussed in morphogenesis and selectivity Sections 3.1 and 4.1). Other studies have led to suggestions that different cadherins are responsible for different types of cell behaviors. For example, the association of N-cadherin expression with a mesenchymal phenotype during the EMT and with growing neuronal processes has led to the notion that it mediates a more motile type of cell interaction, whereas E-cadherin expression in epithelia mediates more stable adherens junctional interactions (Wheelock et al. 2008; Hazan et al. 2004). However, there are clear exceptions that disprove this rule. N-cadherin can form stable, well-developed adherens junctions in the neural tube and antagonize the EMT of the neural crest (Park and Gumbiner 2010) as well as form (relatively) stable synaptic junctions between neurons that have ceased growing (Takeichi and Abe 2005). Moreover, E-cadherin has been found to be required for border cell migration in the *Drosophila* ovary (Geisbrecht and Montell 2002), a process akin to mesenchymal motility, and similarly found to be required for epithelial tumor cell invasive migration (Shamir et al. 2014). Thus, many of these apparent cadherin specific functions are probably due more to cellular context or variations in their association with interacting proteins or signaling pathways than attributable to the type of classical cadherin expressed. In this context it is important to note that *C. Elegans* development is mediated entirely by a single critical classical cadherin (Pettitt et al. 2003; Hardin et al. 2013), and *Drosophila* development is mediated entirely by E-cadherin and two highly related forms of N-cadherin (Oda and Takeichi 2011; Cavey et al. 2008; Bulgakova et al. 2012). This suggests that in these species either the functions of cadherin diversity have been taken over by other classes of adhesion molecules in these species or, more likely, each of the classical cadherins has the capacity to carry out a diverse range of functions depending on the cellular context.

3.2 Core Functions

Classical cadherins share several structural features related to their functions, but the most universal defining feature is their interactions with catenins via their cytoplasmic domains (Fig. 3.2). The cytoplasmic functions of classical cadherins distinguish this family from other types of cell–cell adhesion proteins, including other members of the cadherin superfamily. Their adhesive functions mediated by the association of β-catenin-α-catenin with the contractile acto-myosin cytoskeleton are important to be able to do work at cell adhesions and junctions, that is, to sense tension and generate forces (Ivanov et al. 2007; Maddugoda et al. 2007; Huveneers et al. 2012). Force generation at adhesions underlies many morphogenetic processes in tissues, including cell rearrangements, tissue buckling, and invagination (Baum and Georgiou 2011; Simoes Sde et al. 2014). It is also important for physiological processes such as wound closure, junction opening during barrier modulation and leukocyte transmigration across epithelial and endothelial cell layers, and synaptic spine modulation (Takeichi and Abe 2005; Ivanov

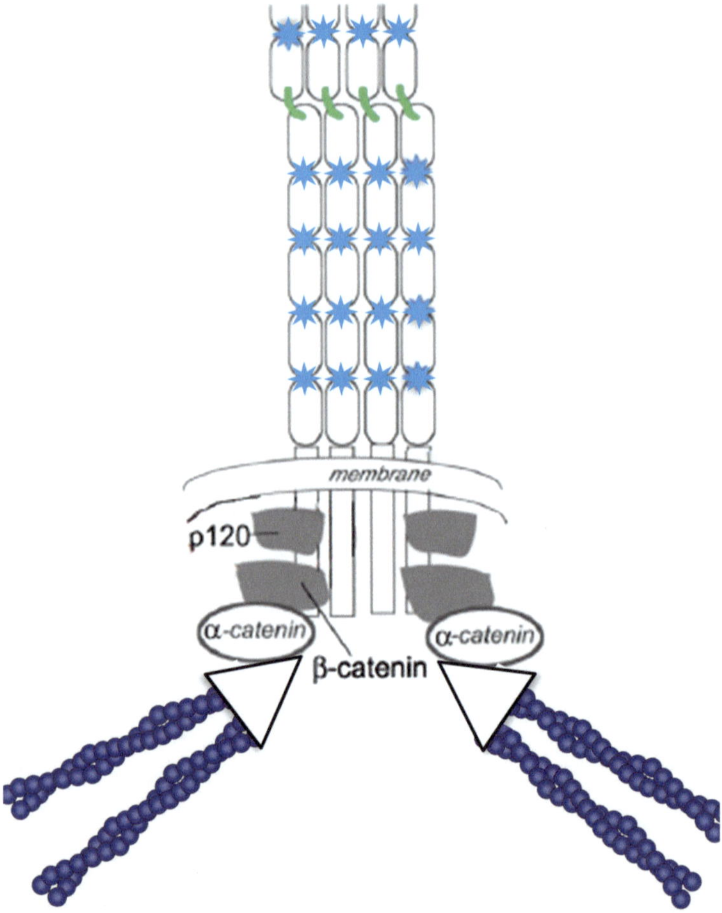

Fig. 3.2 The cadherin–catenin complex. A vertebrate cadherin is shown. The extracellular domain consists of five extracellular cadherin (EC) repeats, with EC domain interfaces mediated by calcium binding sites. The EC1 domain contains key Trp2 residue (or W2W4 residues for type 2 cadherins) that mediate strand exchange and homophilic binding. The cytoplasmic domain binds to catenins, which regulate cadherin function and mediate interactions with the cytoskeleton and signaling molecules

et al. 2007; Ivanov et al. 2010; Franke et al. 2005; Wood et al. 2002; Chazeau et al. 2015). In many ways this association of adhesions with contractile cytoskeleton is analogous to (albeit different in mechanistic detail) the roles of integrins in cell–ECM interactions and cell motility (Hynes 2002).

The interaction of p120-catenin with the juxtamembrane region of the cadherin cytoplasmic domain is involved in cadherin regulation. As has β-catenin, p120-catenin has also been found to interact with various signaling proteins (Reynolds 2007; Smith et al. 2012; Wong et al. 2010). However, p120-catenin has been found to mediate interactions with the microtubule cytoskeleton (Akhmanova et al. 2009;

Shahbazi et al. 2013; Ichii and Takeichi 2007; Meng et al. 2008), in contrast to the prominent interaction of β-catenin and α-catenin with the actomyosin cytoskeleton. p120-catenin and the juxtamembrane domain regulate cadherin internalization and thereby the level of cadherin cell surface expression (Reynolds 2007; Davis et al. 2003; Xiao et al. 2005), but it also regulates the functional activity of cadherins at the cell surface (Petrova et al. 2012). There is still much to learn about the exact role of p120-catenin and its relationship to β-catenin and α-catenin in the various functions of classical cadherins.

The (cytoplasmic) signaling functions of classical cadherins are largely determined by the catenins as well. Most of the cadherin cytoplasmic tail, except for the very C-terminus, is covered by extensive interactions with β-catenin and p120-catenin (Ishiyama et al. 2010), and therefore most of the interactions of the cadherin–catenin complex with cytoplasmic signaling proteins likely occur via the catenins. The signaling functions of cadherin–catenin complexes are quite varied and are discussed in more detail in Section 4.3.

The extracellular domains of vertebrate classical cadherins are also highly similar, and comparisons of their 3D X-ray crystal structures reveals a remarkable similarity in their overall shapes and features (Boggon et al. 2002; Patel et al. 2006; Brasch et al. 2012). All consist of 5 EC cadherin domains linked by calcium-binding sites, which are responsible for the characteristic calcium dependence of this family of adhesion molecules; loss of calcium causes the ectodomain to lose its overall structure. The EC1 domain of most (but not all) type I classical cadherins contains a Trp2 residue that mediates strand exchange to form the homophilic adhesive bond; for most type II cadherins two trp residues, W2W4 mediate strand exchange (Patel et al. 2006; Brasch et al. 2011). (T-cadherin, which is GPI-linked and has no cytoplasmic domain, does not have a trp2 to mediate strand exchange and instead forms an X-dimer (Ciatto et al. 2010), which is thought to be a transient intermediate for other classical cadherins (Harrison et al. 2010)). However, *Drosophila* cadherins are much longer, contain domains in addition to EC cadherin repeats, and do not utilize a strand exchange mechanism via EC1 to form the homophilic bond (Jin et al. 2012). Yet, with regard to both overall function and interactions with catenins, they definitely belong to the classical cadherin family of proteins. This emphasizes that cytoplasmic interactions of cadherins rather than the specific mechanism of homophilic bond formation are the defining functional characteristics of this family.

Despite variations in the mechanism of homophilic binding between classical cadherins, the universal presence of numerous, but variable numbers of, extracellular cadherin repeats linked by calcium-binding sites suggests that these repeats contribute to the overall structure important for cadherin function. X-ray crystallographic studies suggest that these EC repeats serve mainly as spacers to position the EC1 domains to engage in adhesive bond formation (Boggon et al. 2002). The arrangement of the 5 EC domains and their calcium-binding sites produces an elongated curved structure that is thought to orient the EC1 domains to engage in trp2-mediated strand exchange. There is also evidence that specific residues in the other EC domains are important for function (van Roy and Berx 2008; Guilford et al. 1998; Pharoah et al. 2001; Kaurah et al. 2007; Kane et al. 2005), as well as evidence for additional putative dimer interactions observed in the crystal structures (Wu et al. 2010). Our understanding of how cadherins assemble into larger adhesive complexes at the sites of cell–cell interactions is less well developed, and it is possible that the repeat EC structure contributes in ways that are not yet appreciated. This may be best illustrated by the extremely long EC repeat structures found in nonclassical cadherins that make up stereocilia tip links form coiled homodimers with monomers wrapping around each other (Kazmierczak et al. 2007). The conserved curved structure of the classical cadherin repeat could potentially facilitate this sort of coiled interaction. Although it has not been observed in X-ray crystal structures of cadherins, which are dominated interactions favored by crystal packing, a structure of coiled trimmers, which interact via trp2s to form a hexamer, has been reported for an electron microscopic structure of bacterially expressed VE-cadherin ectodomains (Hewat et al. 2007). This structure may have been favored by the lack of glycosylation (Brasch et al. 2011), and aside from the tip link cadherins, it is not known whether such multimeric coiled structures exist in cells.

In many tissues classical cadherins assemble into a higher order structure; the zonula adherens or fascia adherens junctions (Yap et al. 1997; Abe and Takeichi 2008; Li and Radice 2010; Borrmann et al. 2006). It is important to note that these are special forms of cadherin-mediated cell interactions, and that cadherins can mediate robust adhesion in cases where no evidence for discrete adherens junctions exist. For example, strong Ca^{2+}-dependent adhesion between embryonic blastomeres in *Xenopus* is mediated by C-cadherin surrounding the whole cell perimeter (Fagotto and Gumbiner 1994). It is possible in these cases that cadherins form extremely small dispersed clusters distinct from higher order assembly of adherens junctions; indeed AJs may be assembled from smaller units (Cavey et al. 2008; Wu et al. 2015). The AJs are special in having important functional association with the contractile actomyosin cytoskeleton; that is, they are specialized places where the cytoskeleton can do work on localized adhesive structures in the cell. There is evidence for molecular specialization of AJs not found in all areas of cadherin-dependent contact. The Nectin family of Ig adhesion proteins and their associated cytoplasmic afadins, are components of AJs that interact with and modulate classical cadherins (Takai et al. 2008). Nectins and afadin can be even more specifically localized to AJs than E-cadherin in some tissues (Takahashi et al. 1999), suggesting that they may help assemble the specialized structure. Also, a tension-dependent epitope of α-catenin is more highly localized to the AJ than either the bulk E-cadherin–catenin complexes (Yonemura et al. 2010), consistent with the idea that the AJ is especially associated with contractile activity of the actomyosin cytoskeleton. These functional distinctions between AJ-associated cadherins and adhesive nonjunctional cadherins are often obscured or overlooked in the literature, because the term junction is often broadly used to describe any region between cells where cadherins accumulate.

3.3 Physiological and Developmental Roles of Classical Cadherins

3.3.1 Morphogenesis

Classical cadherin functions underlie tissue morphogenesis, both during embryonic development and during homeostasis and regeneration of tissues in organs (Fig. 3.3). They mediate a variety of different, although sometimes overlapping, morphogenetic processes which have been studied separately in different contexts. These include cell sorting, cell condensation/compaction or the reverse: dissociation and EMT, cell rearrangements, tissue contractility (invagination/buckling/wound closure), and cell communication and signaling.

One of the earliest roles of cadherins to be established was the process of cell sorting. Cell sorting was observed very early on as an important feature of embryonic tissues in development whereby cells from different germ layers or embryonic regions could recognize themselves and segregate into their tissue types of origin (Townes and Holtfreter 1955). Indeed, these experimental observations provided the original motivation to identify molecules that mediate cell–cell adhesion. The identification of cadherins as mediators of Ca^{2+}-dependent adhesion with tissue-specific expression provided a molecular explanation for these early observations (Nose et al. 1988). Sorting of cells into different domains within a tissue having distinct gene expression patterns or functional attributes is also quite common, and there are often mechanisms to create distinct boundaries between domains. We now know that multiple mechanisms involving many different types of cell surface molecules participate in cell sorting, segregation into domains, and boundary

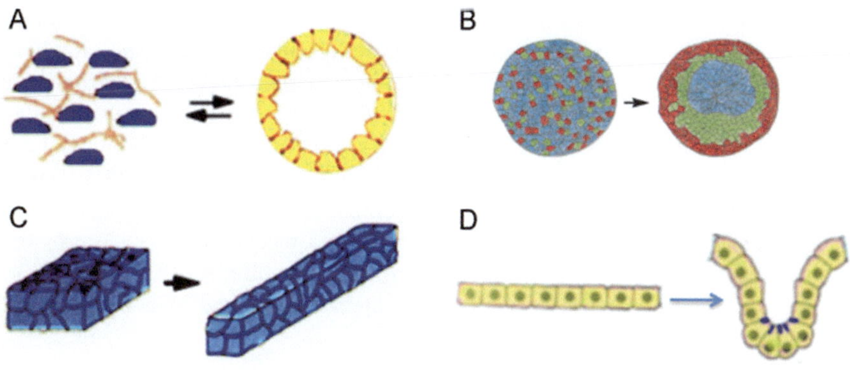

Fig. 3.3 Cadherins mediate tissue morphogenesis. (**a**) Cell compaction or condensation, which is reversible, causing cell dissociation (e.g., EMT). (**b**) Cell sorting into different regions or compartments. Shown is the classic example of the sorting out of the three primary germ layers (endoderm, mesoderm, ectoderm) after cells are aggregated in a mix illustration from figure 21-75 in (Alberts et al. 2002). (**c**) Cell rearrangements, in which cells exchange neighbors, breaking and making new adhesive interactions. Example shown is convergent extension. (**d**) Cytoskeletal-mediated cell and tissue shape changes through forces on the adherens junctions. Apical constriction and invagination are shown

formation. Moreover, the mechanisms by which cadherins can mediate cell sorting events has turned out to be more complex than just changes in cadherin type expression (see Section 4.1).

Classical cadherins mediate gross changes in cell adhesion, such as the compaction of cells into tissue structures as well as the reverse: the dissociation or scattering of cells associated with morphogenetic processes. A striking example of compaction is the early mammalian embryo, whereby E-cadherin–mediated adhesion is responsible for the formation of the trophectoderm, a barrier epithelium, from loosely associated blastomeres (Vestweber et al. 1987; Fleming and Johnson 1988). Similar processes underlie the MET, for example, during kidney formation (Davies 1996), the N-cadherin-mediated condensation of mesenchyme during bone formation (Haas and Tuan 1999), and cadherin-specific condensation of neurons to form brain nuclei (Astick et al. 2014). The reverse process underlies the EMT, for example, the formation of migratory neural crest from the neural tube (Park and Gumbiner 2010), and in tumor cell invasion and metastasis (Thiery et al. 2009; Yang and Weinberg 2008). These processes represent gross changes in Ca^{2+}-dependent cell–cell adhesion mediated by different cadherins, either due to changes in their expression or rapid changes in physiological cadherin activation (see Section 4.2).

Classical cadherins also mediate subtler changes in cell interactions that are very important in morphogenesis, even though they may be difficult to measure experimentally. In particular, they control cell rearrangements that underlie the changes in the shape of tissues. A well-studied example in embryonic development is convergent-extension, whereby cells intercalate to form a narrower and elongated tissue (Keller et al. 2000), but other forms of intercalation are also known (Keller 2002; Keller 2006). In these cases, cells need to break adhesive bonds at one region while creating new adhesive bonds with different cells in another, and the strength of cadherin-mediated adhesion has been shown to be regulated to allow this to happen (Wirtz-Peitz and Zallen 2009; Zhong et al. 1999; Brieher and Gumbiner 1994). Other important examples of cell rearrangements include E-cadherin–mediated branching morphogenesis of tube-like structures such as the airways or kidney tubules (Costantini and Kopan 2010) and VE-cadherin–mediated endothelial cell rearrangements during vasculogenesis and angiogenesis (Dor et al. 2003; Dejana and Vestweber 2013). Directional rearrangements probably require polarized or localized changes in cell adhesion in the cells, and recent studies have provided some evidence supporting this idea (Cai et al. 2014; Peglion et al. 2014). Contractile force generating processes acting on cadherins, discussed below, also play a role in these cell rearrangements.

Cadherin adhesions also create sites of cytoskeletal attachment that allow cells to generate contractile forces that shape tissues. A striking well-known example is the invagination of an epithelial sheet, which involves cytoskeletal attachment to the zonula adherens junctions formed by cadherins and changes in cell shape and tension in a specified region of a tissue (Nishimura and Takeichi 2009; Oda et al. 1998). Closure of a gap or wound in a tissue also involves contractile changes; dorsal closure in *Drosophila* embryos demonstrates the importance of cytoskeletal contraction at the adherens junctions (Franke et al. 2005). In these cases the cadherins may act as stable adhesive structures upon which cytoskeletal elements can exert force to effect changes in cell or tissue shape. In contrast, force generation, associated with cell rearrangements discussed above, appears to work in concert with alterations in the state of adhesion, which must occur for cells to exchange neighbors. Controlling the balance between cadherin bond dissociation to allow neighbor exchange and maintenance of strong stable bonds to hold cells together to withstand forces is an important mechanistic feature of morphogenetic processes (as well as physiological regulation of barrier function; see Section 4.2).

Classical cadherins also participate in morphogenetic processes by cell signaling mechanisms. Signaling is integral to the process of cell adhesion and force generation at cell adhesions, providing feedback mechanisms in response to changes in tension and or state of adhesive bonds (Smith et al. 2012; Piedra et al. 2003; Anastasiadis and Reynolds 2001; Ratheesh et al. 2012; Wheelock and Johnson 2003; Lampugnani et al. 2002). It is therefore not surprising that signaling pathways controlling the actomyosin cytoskeleton, such as Rho GTPases, are tightly coupled to cadherin adhesion processes. Classical cadherins also modulate other signaling pathways that do not intrinsically require cadherins to function, such as signaling by traditional cell surface growth factor receptors (Qian et al. 2004; Rudini et al. 2008; Carmeliet et al. 1999; Curto et al. 2007). An interesting example associated with morphogenesis is the coupling between cadherin expression and BMP signaling during formation of the neural crest from the chick neural tube (Fig. 3.4) (Park and Gumbiner 2010). Cadherin-6B (type II) expressing cells in the neural tube sort out from the bulk of the N-cadherin (type I) expressing neural epithelial cells to form the premigratory neural crest in the dorsal region. Cadherin-6B enhances BMP signaling in these cells, and N-cadherin inhibits it. In a reciprocal feedback loop BMP signaling stimulates cadherin-6B expression and inhibits N-cadherin expression to reinforce this pattern. High levels of dorsally localized BMP signaling causes these cells to delaminate and form the migratory neural crest; in this way the sorting activity of these two cadherins is coupled to the creation of different domains of signaling activity that localizes a morphogenetic event. Classical cadherins also participate in signaling pathways that regulate gene expression in the nucleus to cause changes in cell growth and differentiation (McCrea et al. 2015). Signaling mechanisms are discussed in more detail in Section 4.3.

3.3.2 Barrier Function

Classical cadherins also have very important roles in physiological processes in fully formed and differentiated tissues. In particular they contribute to the formation and regulation of tissue barriers in which cell layers or processes separate physiologically distinct spaces (Fig. 3.1b). A well-known example is the role of E-cadherin in the formation of barriers created by polarized epithelial cell layers lining organs that interface with the external environment, such as lung, digestive

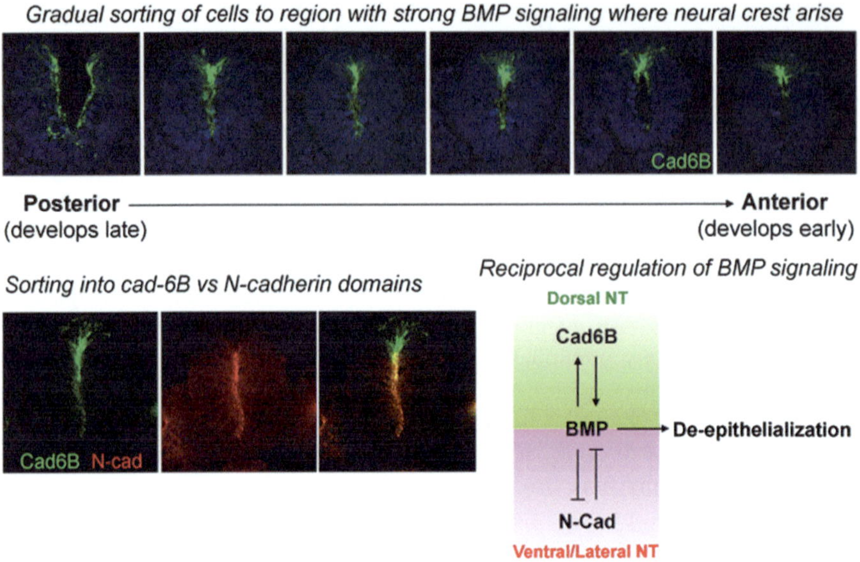

Fig. 3.4 Coupling between selective cell adhesion, cell sorting, and differential cadherin signaling during tissue development. Cadherin-6B (type II) expressing cells in the neural tube sort out from the bulk of the N-cadherin (type I) expressing neural epithelial cells to form the premigratory neural crest in the dorsal region. Cadherin-6B enhances BMP signaling in these cells, whereas N-cadherin inhibits it. In a reciprocal feedback loop BMP signaling stimulates cadherin-6B expression and inhibits N-cadherin expression to reinforce this pattern. High levels of dorsally localized BMP signaling cause these cells to delaminate and form the migratory neural crest; in this way the sorting activity of these two cadherins is coupled to the creation of different domains of signaling activity that localizes a morphogenetic event

tract, and kidneys (Ivanov et al. 2010; Holgate 2010; Nusrat et al. 1997; Ivanov et al. 2004; Shen et al. 2011). Similarly, VE-cadherin mediates the formation of the endothelial barrier in the vasculature of all tissues and organs (Dejana and Vestweber 2013; Vestweber et al. 2010). Cadherins are also associated with other types of tissue barriers, including the sealing of the myelin sheath around axons (Alanne et al. 2009; Fannon et al. 1995), and perhaps other tissue compartments in the nervous system or other organs. In this context the cadherins often function in cooperation with the tight or occluding junctions formed between the cells. The occluding junctions, which form cell interactions via the claudins and occluden, form the actual physiological sealing elements that control the ionic and chemical properties of the barrier; that is, they act as finely tuned sealing gaskets (Anderson and Van Itallie 2009). The classical cadherins form the adjacent zonula adherens junctions and appear to exert grosser effects on the state of the barrier between cells because they govern the overall state of assembly of the occluding junctions. They facilitate occluding junction formation and disruption of the cadherin-mediated ZA, by a variety of potential mechanisms leads to disruption of the barrier. In particular, contractility of the acto-myosin cytoskeleton associated with the ZA and tight junction regulates barrier function in both epithelia and endothelia (Ivanov et al. 2010; Dejana and Vestweber 2013; Nusrat et al. 1997; Vestweber et al. 2010; Samarin et al. 2007; Dejana et al. 2008; Dejana et al. 2009; Cunningham and Turner 2012). Although the ionic properties of occluding junctions and their permeability to small molecules may be governed by direct molecular effects on the claudins themselves (Shen et al. 2011; Tsukita and Furuse 2002; Van Itallie and Anderson 2013), grosser physiological changes in barrier function that control paracellular passage of macromolecules and cells (e.g., leukocytes) involve the functions of the cadherins and ZA junctions.

3.3.3 Cell Communication

Cadherins also facilitate the formation and stabilization of communicating junctions between cells important for a variety of physiological processes in differentiated tissues, including gap junctions (Bauer et al. 2006; Prowse et al. 1997; Frenzel and Johnson 1996). They contribute to synapse formation as well as dendritic spine modulation, and have been implicated in neural plasticity (Takeichi and Abe 2005; Chazeau et al. 2015). Synaptic junctions are quite complex and a large number of proteins mediate the interaction between pre- and postsynaptic membranes. In this context, the role of the cadherins may be similar to their role in barrier formation: controlling grosser physical contact between cells to allow the more specialized synaptic machinery to organize and control the synapse functionally. Cadherins similarly facilitate gap junction formation in tissues to allow direct ionic and small molecule coupling, as well as the formation of other types of intercellular junctions. The broader roles of cadherins in cell signaling are varied and complex and are discussed in more detail in Section 4.3.

3.4 Mechanisms of Classical Cadherin Functions

3.4.1 Selectivity and Sorting

Classical cadherins have long been thought to mediate selective cell interactions, cell sorting out, and formation of tissue layers or compartments (Fig. 3.3b; Takeichi 1991; Nose et al. 1988). The identification of a diverse family of classical cadherins with tissue-specific patterns of expression suggested that the pattern of their expression alone could account for these processes. However, we now know that a variety of different types of cell surface molecules contributes to selective cell recognition events, including ephrins and Eph receptors, nectins, Dscams, and protocadherins (Gumbiner 2005; Takai et al. 2008; Chen and Gumbiner 2006; Chen et al. 2009; Zipursky and Sanes 2010; Cooke and Moens 2002; Kemp et al. 2009). Moreover, the mechanisms by which classical cadherins do mediate cell selectivity are complex and not readily attributable to selective cadherin expression. Indeed, several studies have found that cells expressing different classical cadherins can adhere to one another heterotypically via heterophilic cadherin binding (Niessen and Gumbiner 2002; Steinberg and Takeichi 1994; Volk et al. 1987), and such heterotypic cadherin interactions probably occur in vivo (Straub et al. 2011). The structural features of the binding interactions between different type I classical cadherins observed in crystal structures are remarkably similar across different cadherins, and the binding affinities of heterophilic interactions are within the same range as the homophilic interactions (Patel et al. 2006; Katsamba et al. 2009). However, the type I and type II classical cadherins do exhibit somewhat distinct structures of their binding interfaces, corresponding to greater selectivity of interactions between them.

Nonetheless, classical cadherins can mediate cell sorting events via alternative mechanisms. In one study using controlled cadherin expression in cultured cells, the level of cadherin expression was found to be a more important determinant of cell sorting than the kind of (type I) cadherin (Steinberg and Takeichi 1994). This provides support for the differential adhesion hypothesis described by Steinberg, which invokes quantitative differences in the strengths of adhesion between cells, rather than the specificity of molecular interactions, to determine which cells interact with each other (Steinberg 2007). An important variation on this idea is that the overall strength of adhesion can be determined by the regulation of the activity of the cadherin at the cell surface rather than the level of its expression. Evidence for this was found for C-cadherin in the early *Xenopus* embryo; regulation of its adhesive activity at the surface was found to mediate cell sorting events as well as allow for cell rearrangements during morphogenesis (Zhong et al. 1999; Brieher and Gumbiner 1994). Regulation of cadherin activity and resulting sorting can be triggered either cell nonautonomously by a soluble growth factor, such as activin, or cell autonomously by another surface protein. C-cadherin downregulation and resulting cell sorting in *Xenopus* embryos is triggered by the local expression of paraxial protocadherin (PAPC), which itself does not appear to have very strong cell adhesion activity (Chen and Gumbiner 2006). This suggests an interesting general hypothesis for how other types of cell surface receptors or signaling processes can mediate cell sorting or tissue boundary formation. For example, the formation of boundaries between hindbrain rhombomeres depends on changes in calcium-dependent adhesion (Wizenmann and Lumsden 1997) but ephrins regulate the sorting cells into compartments (Cooke and Moens 2002; Kemp et al. 2009). Similarly, compartment boundaries and cell sorting activities in the *Drosophila* imaginal disk are controlled by hedgehog signaling without any change in E-cadherin expression (Rodriguez and Basler 1997). These signaling pathways probably also cause other adhesion-related cytoskeletal changes, such as repulsion or motility, that may work coordinately with cadherin regulation to effect cell sorting behavior.

Cell sorting may be mediated by changes in the tension of the cell cortex (Maitre et al. 2012; Maitre and Heisenberg 2013). In Zebrafish embryos changes in cortical tension independent of any effect on the affinity of the individual cadherin adhesive bond were found to mediate cell sorting. This may be related to the differential adhesion hypothesis of Steinberg, which involves the generation of differences in tissue tension associated with different strengths of cell adhesion to cause cell sorting. Indeed, at the whole cell level, cortical tension and cell adhesion are inextricably intertwined. The ensemble activity of a great number of adhesive bonds in association with the actomyosin cytoskeleton generates changes in tension at the cell–cell interface, as is nicely illustrated by the process of E-cadherin-mediated cell compaction in the early mouse embryo (Fleming and Johnson 1988). Thus, multiple interrelated changes in cadherin adhesion and associated cytoskeleton probably function together to mediate changes in cell sorting behavior.

3.4.2 Cadherin Regulation

It is quite apparent that cadherin-based adhesion must be very dynamic to mediate its various biological roles. Dynamic changes in cell adhesion occur by a variety of mechanisms suited to the specific process and needs of the tissue (Fig. 3.5).

The simplest mechanism to detect and understand is changes in cadherin expression, usually due to changes in cadherin gene transcription that occurs during many developmental processes and in association with differentiation of different tissue types. Increases in cadherin expression mediate tissue condensation or compaction (Davies 1996; Haas and Tuan 1999; Astick et al. 2014), whereas losses can lead to cell dissociation and scattering, for example, during the EMT (Thiery et al. 2009; Yang and Weinberg 2008). It should be noted, however, that losses of cadherin expression during the EMT are often accompanied by increases in expression of a different cadherin; for transporting epithelia there is often a switch from E-cadherin to N-cadherin (Wheelock et al. 2008) and during formation of the neural crest there is a switch from N-cadherin to cadherin-6 or 6B (Park and Gumbiner 2010). These switches have often been thought to underlie a change from adhesive to motile states, but as mentioned before these states cannot be attributed solely to differences in the intrinsic adhesion properties or strengths of the different cadherins. The EMT is a major change in the differentiation of the cell and numerous other factors can control how the cadherins are utilized. On the other hand there is evidence that cadherins can differentially regulate cell signaling pathways and these differences can account for their different properties/cell behaviors.

A commonly observed mechanism for posttranscriptional regulation of cadherins is internalization from the cell surface by endocytosis (Baum and Georgiou 2011; Gavard and Gutkind 2006; D'Souza-Schorey 2005; Lu et al. 2003; Ogata et al. 2007). At one extreme this is a mechanism for losing cadherin expression. Cadherins have been found to turn over very rapidly even at steady state (e.g., with a half-life of~5 h; Shore and Nelson 1991), so there may be a constitutive and fairly rapid internalization and degradation, and in this context a

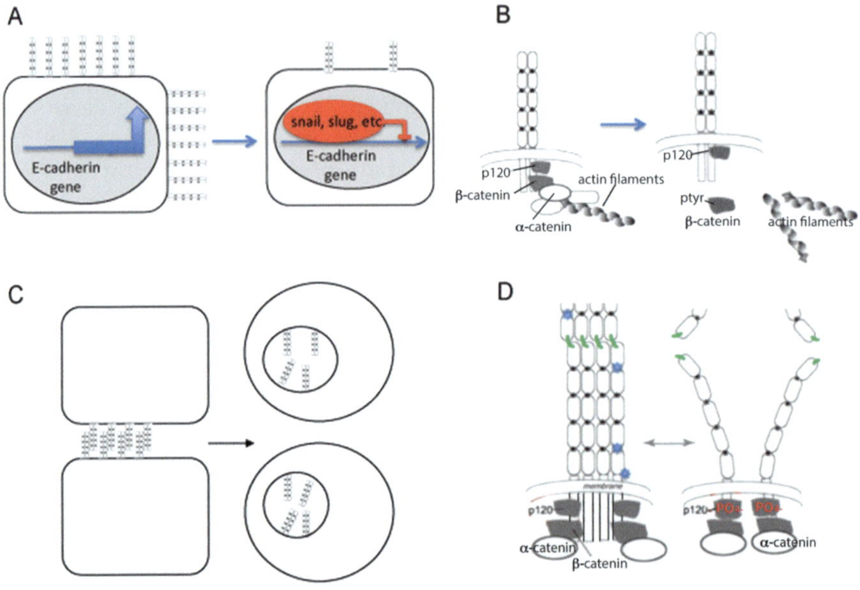

Fig. 3.5 Modes of cadherin regulation. (**a**) Regulation of expression levels, usually via transcriptional mechanisms. (**b**) Signaling induced dissociation of catenins and actin cytoskeleton from the cadherin. (**c**) Removal of cadherins from the cell surface by endocytosis/internalization. (**d**) Regulation of the cadherin homophilic bond itself by bound catenins and changes in the structure or organization of the extracellular domains (perhaps allostery), analogous to inside-out regulation of platelet and leukocyte integrins

loss of gene expression alone could clear the cell of cadherins. Nonetheless, there is evidence that certain signaling events can induce cadherin internalization, perhaps as a way to augment turnover due to decreased gene expression. Alternatively, induced internalization may be a mechanism for transiently causing a partial decrease in cell adhesion needed for morphogenetic changes, such as cell sorting or cell rearrangements in tissues (Classen et al. 2005; Levayer et al. 2011). The p120-catenin has been found to be an important regulator of cadherin internalization. Loss of p120-catenin expression leads to much reduced levels of cadherin protein expression, at the cell surface and overall, apparently due to its control of cadherin endocytosis (Reynolds 2007; Davis et al. 2003; Xiao et al. 2005). p120-catenin may be important for controlling both the constitutive and transiently induced internalization of cadherins. p120-catenin is thought to regulate the interaction of endocytic signals in the cadherin cytoplasmic domain with the endocytic machinery. How the cadherin is removed from the cell junction or adhesive bond to allow internalization is not known; either the endocytic machinery can pull it out with enough force to disrupt the adhesive bond, or else control of the state of the adhesive bond itself is required to allow internalization (see bond regulation below).

Localized cadherin internalization and recycling back to the cell surface, independent of turnover or changes in overall levels, may play a role in regulating adhesive changes during tissue morphogenesis. *Drosophila* cadherins are thought to relocalize from one side of the cell to another (medial to lateral or vice versa) by endocytosis and recycling to change the distribution of adhesive interactions in the tissue (Classen et al. 2005; Levayer et al. 2011). Although not discussed by these authors, these findings have implications for the higher order structure of cadherins in these situations. Given the very small size of the *Drosophila* cells and the lack of any diffusion barrier between lateral and medial surfaces, relocalization of individual cadherin molecules or small oligomers would be very rapidly dissipated by diffusion. For this mechanism to work, the cadherins must form large-scale structures that get disrupted and internalized and then reinserted into the membrane as units, like bricks in a wall. For larger cells over longer distances and/or when there are diffusion barriers between membrane domains, relocalization of individual cadherins by recycling from the cell surface could play a role in redistributing the pattern of cell adhesion.

The state of the cadherin homophilic adhesive bond itself can also be regulated at the cell surface. Mouse embryo compaction at the 8–16 cell stage is triggered by signaling events independent of any changes in E-cadherin expression (Vestweber et al. 1987; Winkel et al. 1990). Changes in the strength of C-cadherin–mediated adhesion in response to the growth factor activin during embryonic tissue morphogenesis in *Xenopus* are not associated with changes in cadherin levels or internalization (Brieher and Gumbiner 1994; Chen et al. 2009). Indeed, a specific set of activating antibodies can bind C-cadherin present at the cell surface and enhance the strength of adhesion (Zhong et al. 1999). Another striking example is the regulation of E-cadherin in the colo205 human colorectal tumor cell line (Petrova et al. 2012; Aono et al. 1999). Under normal culture conditions these cells are completely nonadhesive, even though they have a full complement of cadherin–catenin complexes at the cell surface, but can be triggered to adhere and tightly compact with various agents. Clearly the formation of adhesive bonds by the E-cadherin present at the cell surface is under some sort of control. Several different specific activating mAbs bind to E-cadherin on the cell surface and trigger it to form strong adhesive bonds. Another selective set of mAbs recognizes some sort of change in E-cadherin during activation, binding to it less well than inactive E-cadherin (Petrova et al. 2012). All of these highly selective activating and distinguishing mAbs recognize conformational epitopes at the calcium binding sites that form interfaces between adjacent EC domains, suggesting that the allosteric changes in the E-cadherin extracellular domain may control the state of the adhesive bond.

The molecular mechanisms that control the state of the cadherin adhesive bond are still not well understood, but seem to involve the functions of the catenins. Direct allosteric effects on cadherin binding conformation, changes in cadherin oligomerization or clustering, and interaction with the actomyosin cytoskeleton may all be involved. These mechanisms are analogous to those that have been well established for integrins, where clear conformational changes and alterations in oligomerization underlie inside-out regulation of platelet and leukocyte integrins (Wegener et al. 2007; Hynes 2003; Giancotti 2003). Actomyosin interactions with integrins at focal adhesions are crucial for controlling cell motility (Webb et al. 2002), although less is known about the control of the integrin adhesive bond in this case. For cadherin regulation, one popular and frequently reported model is a catenin dissociation mechanism, whereby tyrosine phosphorylation of β-catenin results in its dissociation from the cadherin cytoplasmic domain (Daugherty and Gottard 2007; Lilien and Balsamo 2005; Chen et al. 2012; Roura et al. 1999), causing the cadherin to disengage from the actomyosin cytoskeleton (Fig. 3.4); presumably loss of the link to the actin cytoskeleton somehow causes cadherin adhesive junctions to fall apart.

There are many clear examples of cadherin regulation that do not involve disruption of their interactions with catenins (Geisbrecht and Montell 2002; Pettitt et al. 2003; Petrova et al. 2012; Zhong et al. 1999; Brieher and Gumbiner 1994; Chen and Gumbiner 2006; Aono et al. 1999; Reynolds et al. 1994; Takeda et al. 1995; Nottebaum et al. 2008; Tamada et al. 2012). In these cases something more subtle and harder to detect with the usual routine assays must be going on. In the case of colo205 cells, dephosphorylation of p120-catenin is associated with cadherin activation, and p120-catenin mutants lacking the N-terminal regulatory domain or lacking ser/thr phosphorylation sites in this domain cause constitutive activation (Petrova et al. 2012; Aono et al. 1999). How p120-catenin phosphorylation might regulate the adhesive state of the E-cadherin extracellular domain is not yet clear. This ser-phosphorylation–dependent function of p120-catenin is separable from its role in regulating cadherin surface levels, as phosphorylation mutants that activate or do not activate adhesion are all similarly able to support high levels of E-cadherin expression (Petrova et al. 2012). Nonetheless, it is intriguing to consider whether these processes may be coordinated; as noted above, it may be necessary to break the adhesive bond between cadherins as a prerequisite for their internalization by the endocytic machinery. Other phosphorylation events are also involved in regulating the adhesive state of cadherins. Direct tyr phosphorylation of the VE-cadherin cytoplasmic domain is associated with control of endothelial barrier function (Dejana et al. 2008; Nottebaum et al. 2008). These phosphorylation events as well as β-catenin tyr phosphorylation could control molecular interactions other than catenin dissociation; for example, they could also recruit regulatory molecules to the cytoplasmic surface of the complex, as is well established for other tyr phosphorylation controlled signaling processes.

It is instructive to distinguish the catenin dissociation model from a model involving the direct regulation of the cadherin homophilic adhesive bond in the context of junctional regulation (Fig. 3.6a). Catenin dissociation leads to the detachment of the actomyosin cytoskeleton from the cadherin in the junction, which is thought to cause junctions somehow to fall apart passively. In contrast, regulation of cadherin homophilic binding could leave the actomyosin cytoskeleton attached to the cadherin and allow the cytoskeleton to do work on the junctions, pulling them apart in a controlled way. Indeed there is evidence for such an actomyosin-mediated contractile mechanism in the rapid transient control of barrier

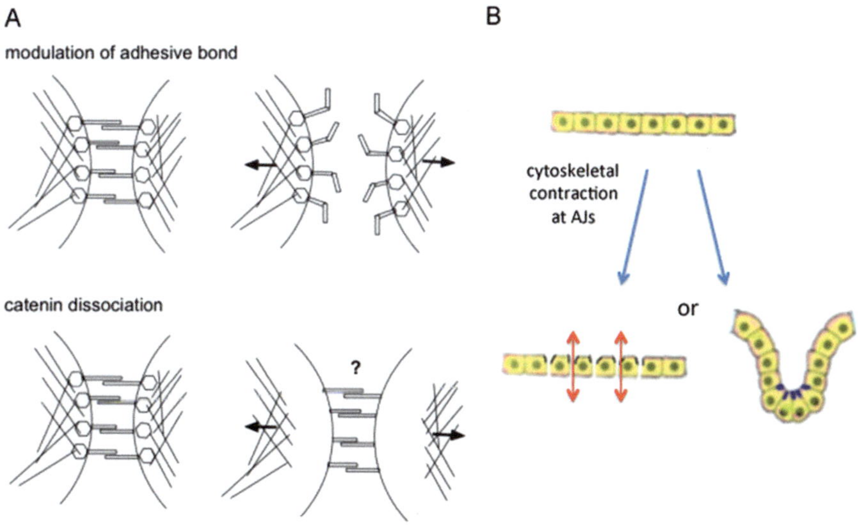

Fig. 3.6 Physiological implications for different models of cadherin regulation. (**a**) Contrast between regulation of cadherin homophilic bond and regulation of catenin/cytoskeletal attachment. Catenin dissociation leads to the detachment of the actomyosin cytoskeleton from the cadherin in the junction, which is thought to cause junctions somehow to fall apart passively. In contrast, regulation of cadherin homophilic binding could leave the actomyosin cytoskeleton attached to the cadherin and allow the cytoskeleton to do work on the junctions, pulling them apart in a controlled way. (**b**) Actomyosin contraction in coordination with bond regulation. If coupled it would control the state of the junctional complex and barrier function; absent coordinate regulation of the bond, the actomyosin cytoskeleton causes apical constriction and tissue invagination

function in both epithelia and endothelia (Ivanov et al. 2010; Dejana and Vestweber 2013; Nusrat et al. 1997; Vestweber et al. 2010; Samarin et al. 2007), especially during inflammation and the transmigration of leukocytes across the cell monolayer. Actomyosin contraction in coordination with bond regulation controls the state of the junctional complex; absent coordinate regulation of the bond, the actomyosin cytoskeleton causes apical constriction and tissue invagination as described above (Fig. 3.6b). Adhesion regulation during cell rearrangements in morphogenesis could involve something in between: a combination of forces acting on junctions with partial disruptions of the bonds allowing rearrangement of adhesive junctions.

3.4.3 Signaling by Classical Cadherins

In addition to controlling the interactions between cells in tissues, cadherins sense the state of the tissue and transduce signals to regulate a plethora of functions. Although there is not always a clean distinction, it is useful to consider two classes of signaling processes. One includes those local signaling events intrinsic to controlling cell adhesion and related morphogenetic changes involving the acto-myosin cytoskeleton and membrane trafficking machinery (Smith et al. 2012; Piedra et al. 2003; Anastasiadis and Reynolds 2001; Ratheesh et al. 2012; Wheelock and Johnson 2003; Lampugnani et al. 2002). Control of these processes is complex and involves numerous signaling steps, such as phosphorylation events or activities of Rho family GTPases that are part and parcel of the overall cellular processes. The second class includes signals that provide instructions to the cell to grow, divide, or differentiate, often involving the transduction of signals to the nucleus where transcription and cell division are controlled (McCrea et al. 2015; Gumbiner and Kim 2014). Of course, there is overlap between these two classes and a given signaling step may transduce both local regulatory mechanisms and nuclear events controlling growth and differentiation. (The focus is mostly on thelatter in the following.)

Classical cadherins interact with and stimulate signaling processes in many different ways (Fig. 3.7). By virtue of being cell–cell adhesion proteins that control the physical interactions between cells, they facilitate signaling through other mechanisms (Fagotto and Gumbiner 1996). In the grossest sense this includes control of cell motility by allowing cells to dissociate from a tissue in order to migrate. They also include facilitating the formation of other cell junctions, including tight junctions and desmosomes and associated generation of cell polarity, which in turn regulate many signaling events in the cell. Also, physical cell

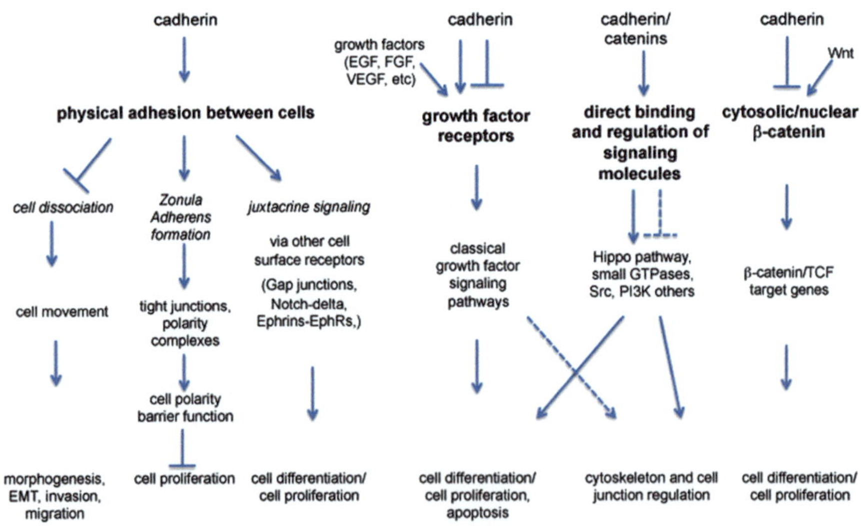

Fig. 3.7 Diversity of ways that cadherins mediate signaling events. *Left*: physical adhesion of cells itself influences cell interactions and signaling in multiple ways. Cadherins also interact with and regulate cell surface growth factor receptors; via catenins they also interact with a number of intracellular signaling molecules; and catenins themselves are intracellular signaling mediators that act in the nucleus

adhesion can facilitate signaling through other juxtacrine cell surface receptors, which require close cell contact in order to engage ligand–receptor pairs. This has been well documented for signaling through gap junctions, which are influenced by the state of cadherin adhesion, but may also be true for other juxtacrine mechanisms, such as notch signaling, ephrins-EphR signaling, or signaling via other membrane-tethered ligands. Facilitation of juxtacrine signaling by cadherin adhesion ensures that cells are closely associated with other cells in a tissue in order to transmit signals, acting as multicellular cohorts for communication.

At the other extreme, classical cadherins interact directly with high affinity with a major nuclear transcriptional effector protein, β-catenin (McCrea et al. 1991; Peifer et al. 1992). The role of cadherins in control of β-catenin nuclear signaling has turned out to be quite complex (McCrea et al. 2015; Nelson and Nusse 2004). The main regulator of β-catenin signaling is the Wnt pathway, in which Wnts act as extracellular ligands via Frizzled and LRP transmembrane receptors to inhibit the degradation of cytosolic and nuclear β-catenin via a specialized destruction complex (Clevers and Nusse 2012). Cadherins are not required for the Wnt pathway, and the levels of cytosolic and nuclear β-catenin can be controlled completely independently of the cadherin. Indeed, cells regulate β-catenin levels in two distinct ways: cytosolic and nuclear β-catenin are stabilized by Wnt signaling and membrane associated β-catenin is stabilized by the level of cadherin expression, by virtue of binding to cadherins. However, several early experiments showed that in certain conditions, cadherins can be made to sequester β-catenin out of the nucleus via direct binding and thus antagonize Wnt signaling, either in embryonic development or in tumor cell proliferation (Heasman et al. 1994; Fagotto et al. 1996; Gottardi et al. 2001; Simcha et al. 2001). Whether, or how often, this antagonism between cadherins and Wnt signaling occurs under physiological conditions in vivo, is not entirely clear. In fact, studies have revealed that β-catenin may be regulated in a way such that it can exist in somewhat distinct pools that mediate either cell adhesion or Wnt signaling (Gottardi and Gumbiner 2004). On the other hand, there is also evidence that cadherins and/or cadherin-dependent cell junctions can either stimulate Wnt–β-catenin nuclear signaling or inhibit β-catenin signaling by directly stimulating the degradation of the cytosolic/nuclear pool (Maher et al. 2009). The many mechanisms by which cadherins affect β-catenin signaling are discussed in a recent comprehensive review on cadherin signaling (McCrea et al. 2015).

Cadherins also interact directly with other proteins known to mediate other classical signaling pathways. Cadherin–catenin complexes have been found to associate with transduction modules, including numerous kinases, phosphatases, GTPases, and GEFs, and so on, that can transduce a variety of different signals (Wheelock and Johnson 2003; McCrea et al. 2015; Dejana et al. 2009; McLachlan and Yap 2007; Radice 2013). Another common mechanism is for cadherins to interact with cell surface growth factor receptors, especially receptor tyrosine kinases as well as TGFβ family receptor kinases. In these cases cadherins either inhibit or enhance signaling by receptors that are stimulated by extracellular ligands. The physiological roles of these interactions are not always clear, but presumably they allow the state of tissue organization to influence the strength of the signals transduced (e.g., Fig. 3.4).

Cadherins have often been found to interact specifically with certain types of cell surface signaling receptors, that is, E-cadherin with EGFR, N-cadherin with FGFR, and VE-cadherin with VEGFR (Qian et al. 2004; Rudini et al. 2008; Carmeliet et al. 1999; Curto et al. 2007; Suyama et al. 2002). As mentioned above, the specificity of these interactions can potentially account for different cellular activities of the cadherins, but in many cases it is not clear whether the apparent different interactions are due to cadherin–receptor specificity or just coexpression in certain types of tissues (e.g., both EGFR and E-cadherin in epithelia). Nonetheless, as described above, cadherin 6B selectively stimulates BMP signaling to induce de-epithelialization of the neural crest whereas N-cadherin inhibits BMP signaling to help maintain the epithelial structure of the neural tube (Park and Gumbiner 2010). Also, N-cadherin has been found to interact specifically with FGFRs via specific regions in the N-cadherin extracellular domain to mediate some of its effects (Williams et al. 2001). Thus, changes in cadherin expression can have effects on cell behavior beyond simple changes in cell adhesion.

Cadherins have long been thought to be important mediators of contact inhibition of motility and proliferation and thereby act as growth/tumor suppressor proteins. Contact inhibition of motility involves local cytoskeletal signaling pathways (Theveneau et al. 2010; Theveneau and Mayor 2012; Weber et al. 2012). Contact inhibition of proliferation is mediated through nuclear changes, and cadherins could potentially mediate contact inhibition through any of the general mechanisms shown in Fig. 3.7, even simply by virtue of mediating physical adhesion and indirect facilitation of other signaling pathways. Nonetheless, there is evidence that they can mediate contact inhibition via their direct interactions with growth factor receptors (Curto et al. 2007; McClatchey and Yap 2012). Also, experiments involving E-cadherin protein coated-bead attachment to isolated cells demonstrated that E-cadherin mediates contact inhibition independent of the formation of any other cell interactions (Perrais et al. 2007). These experiments also showed that cadherin engagement in a homophilic bond is important for contact inhibition signaling, in contrast to the inhibition of β-catenin by sequestering out of the nucleus, which only requires the presence of the cadherin cytoplasmic tail (Gottardi et al. 2001; Simcha et al. 2001). A follow-up study discovered that cadherin engagement causes contact inhibition of proliferation by stimulating the Hippo signaling pathway (Kim et al. 2011). The Hippo pathway is a growth inhibitory pathway mediated by a kinase cascade, which leads to the nuclear exclusion of the growth-promoting transcriptional activators YAP and TAZ (Tumaneng et al. 2012). Hippo signaling is regulated by a variety of upstream effectors, including mechanical sensing, polarity proteins, and growth factor receptors (Gumbiner and Kim 2014; Schroeder and Halder 2012), but more recent studies have confirmed that E-cadherin stimulates the Hippo pathway in other systems, including *Drosophila* imaginal disks and early mouse embryos (Hirate et al. 2013; Yang et al. 2015). The exact mechanism of coupling cadherins to the Hippo pathway is not yet certain, but alpha-catenin is a particularly important regulator of the pathway (Kim et al. 2011;

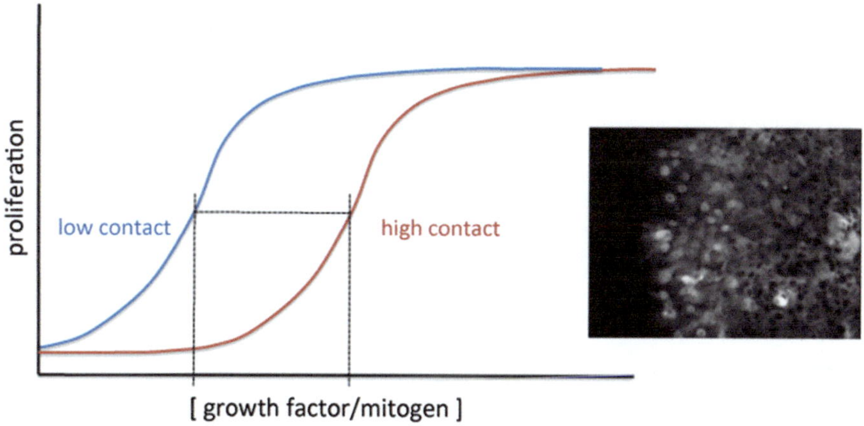

Fig. 3.8 Model for the role of contact inhibition in growth regulation. Contact inhibition to work shifts the dose-dependence of cell proliferation in response to a mitogenic growth factor. Cells at the edge respond to a lower level of a growth factor than those deeper within the tissue. Nuclear localization of TEAD, a Hippo pathway mediator and potent inducer of growth-promoting genes, appears to be regulated in part by this phenomenon

Silvis et al. 2011; Schlegelmilch et al. 2011), and the merlin/NF2 tumor suppressor protein is an upstream regulator of the Hippo pathway that is involved in E-cadherin stimulation of the pathway (Kim et al. 2011; Hirate et al. 2013).

Cadherin-mediated contact inhibition has interesting implications for patterns of tissue growth (Fig. 3.8). Contact inhibition appears to work by shifting the dose-dependence of cell proliferation in response to a mitogenic growth factor such as EGF to a higher dose (Kim et al. 2009). As a result, a cell at the edge of a culture or tissue can respond to a lower level of a growth factor than other cells deeper within the tissue or cell monolayer. Although other mechanisms may explain this phenomenon, it is well known that the nuclear localization of YAP tends to occur at the edge of a cell cluster compared to cells in the middle. Inasmuch as growth factor signaling has been found to stimulate the nuclear localization of YAP (Fan et al. 2013; Reddy and Irvine 2013; Yu et al. 2012), cells at the edge of a colony or a wound in a tissue, or at the edge of a tumor may be more sensitive to growth factors and more likely to proliferate, migrate, or differentiate in a YAP/TAZ-dependent process.

References

Abe K, Takeichi M (2008) EPLIN mediates linkage of the cadherin catenin complex to F-actin and stabilizes the circumferential actin belt. Proc Natl Acad Sci U S A 105(1):13–19

Akhmanova A, Stehbens SJ, Yap AS (2009) Touch, grasp, deliver and control: functional cross-talk between microtubules and cell adhesions. Traffic 10(3):268–274

Alanne MH, Pummi K, Heape AM, Grenman R, Peltonen J, Peltonen S (2009) Tight junction proteins in human Schwann cell autotypic junctions. J Histochem Cytochem 57(6):523–529
Alberts B, Johnson A, Lewis J, et al (2002) Molecular biology of the cell (4th edn). New York: Garland Science
Anastasiadis PZ, Reynolds AB (2001) Regulation of Rho GTPases by p120-catenin. Curr Opin Cell Biol 13(5):604–610
Anderson JM, Van Itallie CM (2009) Physiology and function of the tight junction. Cold Spring Harbor Perspect Biol 1(2):a002584
Aono S, Nakagawa S, Reynolds AB, Takeichi M (1999) p120(ctn) acts as an inhibitory regulator of cadherin function in colon carcinoma cells. J Cell Biol 145(3):551–562
Astick M, Tubby K, Mubarak WM, Guthrie S, Price SR (2014) Central topography of cranial motor nuclei controlled by differential cadherin expression. Curr Biol 24(21):2541–2547
Bauer R, Weimbs A, Lechner H, Hoch M (2006) DE-cadherin, a core component of the adherens junction complex modifies subcellular localization of the Drosophila gap junction protein innexin2. Cell Commun Adhes 13(1–2):103–114
Baum B, Georgiou M (2011) Dynamics of adherens junctions in epithelial establishment, maintenance, and remodeling. J Cell Biol 192(6):907–917
Boggon TJ, Murray J, Chappuis-Flament S, Wong E, Gumbiner BM, Shapiro L (2002) C-cadherin ectodomain structure and implications for cell adhesion mechanisms. Science 296 (5571):1308–1313
Borrmann CM, Grund C, Kuhn C, Hofmann I, Pieperhoff S, Franke WW (2006) The area composita of adhering junctions connecting heart muscle cells of vertebrates. II. Colocalizations of desmosomal and fascia adhaerens molecules in the intercalated disk. Eur J Cell Biol 85(6):469–485
Brasch J, Harrison OJ, Ahlsen G, Carnally SM, Henderson RM, Honig B, Shapiro L (2011) Structure and binding mechanism of vascular endothelial cadherin: a divergent classical cadherin. J Mol Biol 408(1):57–73
Brasch J, Harrison OJ, Honig B, Shapiro L (2012) Thinking outside the cell: how cadherins drive adhesion. Trends Cell Biol 22(6):299–310
Brieher WM, Gumbiner BM (1994) Regulation of C-cadherin function during activin induced morphogenesis of Xenopus animal caps. J Cell Biol 126(2):519–527
Bulgakova NA, Klapholz B, Brown NH (2012) Cell adhesion in Drosophila: versatility of cadherin and integrin complexes during development. Curr Opin Cell Biol 24(5):702–712
Cai D, Chen SC, Prasad M, He L, Wang X, Choesmel-Cadamuro V, Sawyer JK, Danuser G, Montell DJ (2014) Mechanical feedback through E-cadherin promotes direction sensing during collective cell migration. Cell 157(5):1146–1159
Carmeliet P, Lampugnani MG, Moons L, Breviario F, Compernolle V, Bono F, Balconi G, Spagnuolo R, Oosthuyse B, Dewerchin M, Zanetti A, Angellilo A, Mattot V, Nuyens D, Lutgens E et al (1999) Targeted deficiency or cytosolic truncation of the VE-cadherin gene in mice impairs VEGF-mediated endothelial survival and angiogenesis. Cell 98(2):147–157
Cavey M, Rauzi M, Lenne PF, Lecuit T (2008) A two-tiered mechanism for stabilization and immobilization of E-cadherin. Nature 453(7196):751–756
Chazeau A, Garcia M, Czondor K, Perrais D, Tessier B, Giannone G, Thoumine O (2015) Mechanical coupling between transsynaptic N-cadherin adhesions and actin flow stabilizes dendritic spines. Mol Biol Cell 26(5):859–873
Chen X, Gumbiner BM (2006) Paraxial protocadherin mediates cell sorting and tissue morphogenesis by regulating C-cadherin adhesion activity. J Cell Biol 174(2):301–313
Chen X, Koh E, Yoder M, Gumbiner BM (2009) A protocadherin-cadherin-FLRT3 complex controls cell adhesion and morphogenesis. PLoS One 4(12):e8411
Chen XL, Nam JO, Jean C, Lawson C, Walsh CT, Goka E, Lim ST, Tomar A, Tancioni I, Uryu S, Guan JL, Acevedo LM, Weis SM, Cheresh DA, Schlaepfer DD (2012) VEGF-induced vascular permeability is mediated by FAK. Dev Cell 22(1):146–157

Ciatto C, Bahna F, Zampieri N, VanSteenhouse HC, Katsamba PS, Ahlsen G, Harrison OJ, Brasch J, Jin X, Posy S, Vendome J, Ranscht B, Jessell TM, Honig B, Shapiro L (2010) T-cadherin structures reveal a novel adhesive binding mechanism. Nat Struct Mol Biol 17 (3):339–347

Classen AK, Anderson KI, Marois E, Eaton S (2005) Hexagonal packing of Drosophila wing epithelial cells by the planar cell polarity pathway. Dev Cell 9(6):805–817

Clevers H, Nusse R (2012) Wnt/beta-catenin signaling and disease. Cell 149(6):1192–1205

Cooke JE, Moens CB (2002) Boundary formation in the hindbrain: Eph only it were simple. Trends Neurosci 25(5):260–267

Costantini F, Kopan R (2010) Patterning a complex organ: branching morphogenesis and nephron segmentation in kidney development. Dev Cell 18(5):698–712

Cunningham KE, Turner JR (2012) Myosin light chain kinase: pulling the strings of epithelial tight junction function. Ann N Y Acad Sci 1258:34–42

Curto M, Cole BK, Lallemand D, Liu CH, McClatchey AI (2007) Contact-dependent inhibition of EGFR signaling by Nf2/Merlin. J Cell Biol 177(5):893–903

D'Souza-Schorey C (2005) Disassembling adherens junctions: breaking up is hard to do. Trends Cell Biol 15(1):19–26

Daugherty RL, Gottard CJ (2007) Phospho-regulation of Beta-catenin adhesion and signaling functions. Physiology (Bethesda) 22:303–309

Davies JA (1996) Mesenchyme to epithelium transition during development of the mammalian kidney tubule. Acta Anat 156(3):187–201

Davis MA, Ireton RC, Reynolds AB (2003) A core function for p120-catenin in cadherin turnover. J Cell Biol 163(3):525–534

Dejana E, Vestweber D (2013) The role of VE-cadherin in vascular morphogenesis and permeability control. Prog Mol Biol Transl Sci 116:119–144

Dejana E, Orsenigo F, Lampugnani MG (2008) The role of adherens junctions and VE-cadherin in the control of vascular permeability. J Cell Sci 121(Pt 13):2115–2122

Dejana E, Tournier-Lasserve E, Weinstein BM (2009) The control of vascular integrity by endothelial cell junctions: molecular basis and pathological implications. Dev Cell 16 (2):209–221

Dor Y, Djonov V, Keshet E (2003) Making vascular networks in the adult: branching morphogenesis without a roadmap. Trends Cell Biol 13(3):131–136

Fagotto F, Gumbiner BM (1994) Beta-catenin localization during Xenopus embryogenesis: accumulation at tissue and somite boundaries. Development 120(12):3667–3679

Fagotto F, Gumbiner BM (1996) Cell contact-dependent signaling. Dev Biol 180(2):445–454

Fagotto F, Funayama N, Gluck U, Gumbiner BM (1996) Binding to cadherins antagonizes the signaling activity of beta-catenin during axis formation in Xenopus. J Cell Biol 132 (6):1105–1114

Fan R, Kim NG, Gumbiner BM (2013) Regulation of Hippo pathway by mitogenic growth factors via phosphoinositide 3-kinase and phosphoinositide-dependent kinase-1. Proc Natl Acad Sci U S A 110(7):2569–2574

Fannon AM, Sherman DL, Ilyina-Gragerova G, Brophy PJ, Friedrich VL Jr, Colman DR (1995) Novel E-cadherin-mediated adhesion in peripheral nerve: Schwann cell architecture is stabilized by autotypic adherens junctions. J Cell Biol 129(1):189–202

Fleming TP, Johnson MH (1988) From egg to epithelium. Ann Rev Cell Biol 4:459–485

Franke JD, Montague RA, Kiehart DP (2005) Nonmuscle myosin II generates forces that transmit tension and drive contraction in multiple tissues during dorsal closure. Curr Biol 15 (24):2208–2221

Frenzel EM, Johnson RG (1996) Gap junction formation between cultured embryonic lens cells is inhibited by antibody to N-cadherin. Dev Biol 179(1):1–16

Gavard J, Gutkind JS (2006) VEGF controls endothelial-cell permeability by promoting the beta-arrestin-dependent endocytosis of VE-cadherin. Nat Cell Biol 8(11):1223–1234

Geisbrecht ER, Montell DJ (2002) Myosin VI is required for E-cadherin-mediated border cell migration. Nat Cell Biol 4(8):616–620

Giancotti FG (2003) A structural view of integrin activation and signaling. Dev Cell 4(2):149–151

Gottardi CJ, Gumbiner BM (2004) Distinct molecular forms of beta-catenin are targeted to adhesive or transcriptional complexes. J Cell Biol 167(2):339–349

Gottardi CJ, Wong E, Gumbiner BM (2001) E-cadherin suppresses cellular transformation by inhibiting beta-catenin signaling in an adhesion-independent manner. J Cell Biol 153 (5):1049–1060

Guilford P, Hopkins J, Harraway J, McLeod M, McLeod N, Harawira P, Taite H, Scoular R, Miller A, Reeve AE (1998) E-cadherin germline mutations in familial gastric cancer. Nature 392(6674):402–405

Gumbiner BM (2005) Regulation of cadherin-mediated adhesion in morphogenesis. Nat Rev Mol Cell Biol 6(8):622–634

Gumbiner BM, Kim NG (2014) The Hippo-YAP signaling pathway and contact inhibition of growth. J Cell Sci 127(Pt 4):709–717

Haas AR, Tuan RS (1999) Chondrogenic differentiation of murine C3H10T1/2 multipotential mesenchymal cells: II. Stimulation by bone morphogenetic protein-2 requires modulation of N-cadherin expression and function. Differentiation 64(2):77–89

Hardin J, Lynch A, Loveless T, Pettitt J (2013) Cadherins and their partners in the nematode worm *Caenorhabditis elegans*. Prog Mol Biol Transl Sci 116:239–262

Harrison OJ, Bahna F, Katsamba PS, Jin X, Brasch J, Vendome J, Ahlsen G, Carroll KJ, Price SR, Honig B, Shapiro L (2010) Two-step adhesive binding by classical cadherins. Nat Struct Mol Biol 17(3):348–357

Hazan RB, Qiao R, Keren R, Badano I, Suyama K (2004) Cadherin switch in tumor progression. Ann N Y Acad Sci 1014:155–163

Heasman J, Crawford A, Goldstone K, Garner-Hamrick P, Gumbiner B, McCrea P, Kintner C, Noro CY, Wylie C (1994) Overexpression of cadherins and underexpression of beta-catenin inhibit dorsal mesoderm induction in early Xenopus embryos. Cell 79(5):791–803

Hewat EA, Durmort C, Jacquamet L, Concord E, Gulino-Debrac D (2007) Architecture of the VE-cadherin hexamer. J Mol Biol 365(3):744–751

Hirate Y, Hirahara S, Inoue K, Suzuki A, Alarcon VB, Akimoto K, Hirai T, Hara T, Adachi M, Chida K, Ohno S, Marikawa Y, Nakao K, Shimono A, Sasaki H (2013) Polarity-dependent distribution of angiomotin localizes hippo signaling in preimplantation embryos. Curr Biol 23 (13):1181–1194

Holgate ST (2010) Has the time come to rethink the pathogenesis of asthma? Curr Opin Allergy Clin Immunol 10(1):48–53

Hulpiau P, van Roy F (2009) Molecular evolution of the cadherin superfamily. Int J Biochem Cell Biol 41(2):349–369

Huveneers S, Oldenburg J, Spanjaard E, van der Krogt G, Grigoriev I, Akhmanova A, Rehmann H, de Rooij J (2012) Vinculin associates with endothelial VE-cadherin junctions to control force-dependent remodeling. J Cell Biol 196(5):641–652

Hynes RO (2002) Integrins: bidirectional, allosteric signaling machines. Cell 110(6):673–687

Hynes RO (2003) Structural biology. Changing Partners Sci 300(5620):755–756

Ichii T, Takeichi M (2007) p120-catenin regulates microtubule dynamics and cell migration in a cadherin-independent manner. Genes Cells 12(7):827–839

Ishiyama N, Lee SH, Liu S, Li GY, Smith MJ, Reichardt LF, Ikura M (2010) Dynamic and static interactions between p120 catenin and E-cadherin regulate the stability of cell-cell adhesion. Cell 141(1):117–128

Ivanov AI, Nusrat A, Parkos CA (2004) The epithelium in inflammatory bowel disease: potential role of endocytosis of junctional proteins in barrier disruption. Novartis Found Symp 263:115–124; discussion 124–32, 211–8

Ivanov AI, Bachar M, Babbin BA, Adelstein RS, Nusrat A, Parkos CA (2007) A unique role for nonmuscle myosin heavy chain IIA in regulation of epithelial apical junctions. PLoS One 2(7): e658

Ivanov AI, Parkos CA, Nusrat A (2010) Cytoskeletal regulation of epithelial barrier function during inflammation. Am J Pathol 177(2):512–524

Jin X, Walker MA, Felsovalyi K, Vendome J, Bahna F, Mannepalli S, Cosmanescu F, Ahlsen G, Honig B, Shapiro L (2012) Crystal structures of Drosophila N-cadherin ectodomain regions reveal a widely used class of Ca(2)+-free interdomain linkers. Proc Natl Acad Sci U S A 109 (3):E127–E134

Kane DA, McFarland KN, Warga RM (2005) Mutations in half baked/E-cadherin block cell behaviors that are necessary for teleost epiboly. Development 132(5):1105–1116

Katsamba P, Carroll K, Ahlsen G, Bahna F, Vendome J, Posy S, Rajebhosale M, Price S, Jessell TM, Ben-Shaul A, Shapiro L, Honig BH (2009) Linking molecular affinity and cellular specificity in cadherin-mediated adhesion. Proc Natl Acad Sci U S A 106(28):11594–11599

Kaurah P, MacMillan A, Boyd N, Senz J, De Luca A, Chun N, Suriano G, Zaor S, Van Manen L, Gilpin C, Nikkel S, Connolly-Wilson M, Weissman S, Rubinstein WS, Sebold C et al (2007) Founder and recurrent CDH1 mutations in families with hereditary diffuse gastric cancer. JAMA 297(21):2360–2372

Kazmierczak P, Sakaguchi H, Tokita J, Wilson-Kubalek EM, Milligan RA, Muller U, Kachar B (2007) Cadherin 23 and protocadherin 15 interact to form tip-link filaments in sensory hair cells. Nature 449(7158):87–91

Keller R (2002) Shaping the vertebrate body plan by polarized embryonic cell movements. Science 298(5600):1950–1954

Keller R (2006) Mechanisms of elongation in embryogenesis. Development 133(12):2291–2302

Keller R, Davidson L, Edlund A, Elul T, Ezin M, Shook D, Skoglund P (2000) Mechanisms of convergence and extension by cell intercalation. Philos Trans R Soc Lond Ser B Biol Sci 355 (1399):897–922

Kemp HA, Cooke JE, Moens CB (2009) EphA4 and EfnB2a maintain rhombomere coherence by independently regulating intercalation of progenitor cells in the zebrafish neural keel. Dev Biol 327(2):313–326

Kim JH, Kushiro K, Graham NA, Asthagiri AR (2009) Tunable interplay between epidermal growth factor and cell-cell contact governs the spatial dynamics of epithelial growth. Proc Natl Acad Sci U S A 106(27):11149–11153

Kim NG, Koh E, Chen X, Gumbiner BM (2011) E-cadherin mediates contact inhibition of proliferation through Hippo signaling-pathway components. Proc Natl Acad Sci U S A 108 (29):11930–11935

Lampugnani MG, Zanetti A, Breviario F, Balconi G, Orsenigo F, Corada M, Spagnuolo R, Betson M, Braga V, Dejana E (2002) VE-cadherin regulates endothelial actin activating Rac and increasing membrane association of Tiam. Mol Biol Cell 13(4):1175–1189

Levayer R, Pelissier-Monier A, Lecuit T (2011) Spatial regulation of Dia and Myosin-II by RhoGEF2 controls initiation of E-cadherin endocytosis during epithelial morphogenesis. Nat Cell Biol 13(5):529–540

Li J, Radice GL (2010) A new perspective on intercalated disc organization: implications for heart disease. Dermatol Res Pract 2010:207835

Lilien J, Balsamo J (2005) The regulation of cadherin-mediated adhesion by tyrosine phosphorylation/dephosphorylation of beta-catenin. Curr Opin Cell Biol 17(5):459–465

Lu Z, Ghosh S, Wang Z, Hunter T (2003) Downregulation of caveolin-1 function by EGF leads to the loss of E-cadherin, increased transcriptional activity of beta-catenin, and enhanced tumor cell invasion. Cancer Cell 4(6):499–515

Maddugoda MP, Crampton MS, Shewan AM, Yap AS (2007) Myosin VI and vinculin cooperate during the morphogenesis of cadherin cell cell contacts in mammalian epithelial cells. J Cell Biol 178(3):529–540

Maher MT, Flozak AS, Stocker AM, Chenn A, Gottardi CJ (2009) Activity of the beta-catenin phosphodestruction complex at cell-cell contacts is enhanced by cadherin-based adhesion. J Cell Biol 186(2):219–228

Maitre JL, Heisenberg CP (2013) Three functions of cadherins in cell adhesion. Curr Biol 23(14): R626–R633

Maitre JL, Berthoumieux H, Krens SF, Salbreux G, Julicher F, Paluch E, Heisenberg CP (2012) Adhesion functions in cell sorting by mechanically coupling the cortices of adhering cells. Science 338(6104):253–256

McClatchey AI, Yap AS (2012) Contact inhibition (of proliferation) redux. Curr Opin Cell Biol 24 (5):685–694

McCrea PD, Turck CW, Gumbiner B (1991) A homolog of the armadillo protein in Drosophila (plakoglobin) associated with E-cadherin. Science 254(5036):1359–1361

McCrea PD, Maher MT, Gottardi CJ (2015) Nuclear signaling from cadherin adhesion complexes. Curr Top Dev Biol 112:129–196

McLachlan RW, Yap AS (2007) Not so simple: the complexity of phosphotyrosine signaling at cadherin adhesive contacts. J Mol Med 85(6):545–554

Meng W, Mushika Y, Ichii T, Takeichi M (2008) Anchorage of microtubule minus ends to adherens junctions regulates epithelial cell-cell contacts. Cell 135(5):948–959

Nelson WJ, Nusse R (2004) Convergence of Wnt, beta-catenin, and cadherin pathways. Science 303(5663):1483–1487

Niessen CM, Gumbiner BM (2002) Cadherin-mediated cell sorting not determined by binding or adhesion specificity. J Cell Biol 156(2):389–399

Nishimura T, Takeichi M (2009) Remodeling of the adherens junctions during morphogenesis. Curr Top Dev Biol 89:33–54

Nose A, Nagafuchi A, Takeichi M (1988) Expressed recombinant cadherins mediate cell sorting in model systems. Cell 54:993–1001

Nottebaum AF, Cagna G, Winderlich M, Gamp AC, Linnepe R, Polaschegg C, Filippova K, Lyck R, Engelhardt B, Kamenyeva O, Bixel MG, Butz S, Vestweber D (2008) VE-PTP maintains the endothelial barrier via plakoglobin and becomes dissociated from VE-cadherin by leukocytes and by VEGF. J Exp Med 205(12):2929–2945

Nusrat A, Parkos CA, Liang TW, Carnes DK, Madara JL (1997) Neutrophil migration across model intestinal epithelia: monolayer disruption and subsequent events in epithelial repair. Gastroenterology 113(5):1489–1500

Oda H, Takeichi M (2011) Evolution: structural and functional diversity of cadherin at the adherens junction. J Cell Biol 193(7):1137–1146

Oda H, Tsukita S, Takeichi M (1998) Dynamic behavior of the cadherin-based cell-cell adhesion system during Drosophila gastrulation. Dev Biol 203(2):435–450

Ogata S, Morokuma J, Hayata T, Kolle G, Niehrs C, Ueno N, Cho KW (2007) TGF-beta signaling-mediated morphogenesis: modulation of cell adhesion via cadherin endocytosis. Genes Dev 21 (14):1817–1831

Park KS, Gumbiner BM (2010) Cadherin 6B induces BMP signaling and de-epithelialization during the epithelial mesenchymal transition of the neural crest. Development 137 (16):2691–2701

Parkos CA (1997) Molecular events in neutrophil transepithelial migration. BioEssays 19 (10):865–873

Patel SD, Ciatto C, Chen CP, Bahna F, Rajebhosale M, Arkus N, Schieren I, Jessell TM, Honig B, Price SR, Shapiro L (2006) Type II cadherin ectodomain structures: implications for classical cadherin specificity. Cell 124(6):1255–1268

Peglion F, Llense F, Etienne-Manneville S (2014) Adherens junction treadmilling during collective migration. Nat Cell Biol 16(7):639–651

Peifer M, McCrea PD, Green KJ, Wieschaus E, Gumbiner BM (1992) The vertebrate adhesive junction proteins beta-catenin and plakoglobin and the Drosophila segment polarity gene armadillo form a multigene family with similar properties. J Cell Biol 118(3):681–691

Perrais M, Chen X, Perez-Moreno M, Gumbiner BM (2007) E-cadherin homophilic ligation inhibits cell growth and epidermal growth factor receptor signaling independently of other cell interactions. Mol Biol Cell 18(6):2013–2025

Petrova YI, Spano MM, Gumbiner BM (2012) Conformational epitopes at cadherin calcium-binding sites and p120-catenin phosphorylation regulate cell adhesion. Mol Biol Cell 23 (11):2092–2108

Pettitt J, Cox EA, Broadbent ID, Flett A, Hardin J (2003) The Caenorhabditis elegans p120 catenin homologue, JAC-1, modulates cadherin-catenin function during epidermal morphogenesis. J Cell Biol 162(1):15–22

Pharoah PD, Guilford P, Caldas C (2001) Incidence of gastric cancer and breast cancer in CDH1 (E-cadherin) mutation carriers from hereditary diffuse gastric cancer families. Gastroenterology 121(6):1348–1353

Piedra J, Miravet S, Castano J, Palmer HG, Heisterkamp N, Garcia de Herreros A, Dunach M (2003) p120 Catenin-associated Fer and Fyn tyrosine kinases regulate beta-catenin Tyr-142 phosphorylation and beta-catenin-alpha-catenin Interaction. Mol Cell Biol 23(7):2287–2297

Prowse DM, Cadwallader GP, Pitts JD (1997) E-cadherin expression can alter the specificity of gap junction formation. Cell Biol Int 21(12):833–843

Qian X, Karpova T, Sheppard AM, McNally J, Lowy DR (2004) E-cadherin-mediated adhesion inhibits ligand-dependent activation of diverse receptor tyrosine kinases. EMBO J 23 (8):1739–1748

Radice GL (2013) N-cadherin-mediated adhesion and signaling from development to disease: lessons from mice. Prog Mol Biol Transl Sci 116:263–289

Ratheesh A, Gomez GA, Priya R, Verma S, Kovacs EM, Jiang K, Brown NH, Akhmanova A, Stehbens SJ, Yap AS (2012) Centralspindlin and alpha-catenin regulate Rho signalling at the epithelial zonula adherens. Nat Cell Biol 14(8):818–828

Reddy BV, Irvine KD (2013) Regulation of Hippo signaling by EGFR-MAPK signaling through Ajuba family proteins. Dev Cell 24(5):459–471

Reynolds AB (2007) p120-catenin: past and present. Biochim Biophys Acta 1773(1):2–7

Reynolds AB, Daniel J, McCrea PD, Wheelock MJ, Wu J, Zhang Z (1994) Identification of a new catenin: the tyrosine kinase substrate p120cas associates with E-cadherin complexes. Mol Cell Biol 14(12):8333–8342

Rodriguez I, Basler K (1997) Control of compartmental affinity boundaries by hedgehog. Nature 389(6651):614–618

Roura S, Miravet S, Piedra J, Garcia A (1999) de Herreros, and M. Dunach. Regulation of E-cadherin/Catenin association by tyrosine phosphorylation. J Biol Chem 274 (51):36734–36740

Rudini N, Felici A, Giampietro C, Lampugnani M, Corada M, Swirsding K, Garre M, Liebner S, Letarte M, ten Dijke P, Dejana E (2008) VE-cadherin is a critical endothelial regulator of TGF-beta signalling. EMBO J 27(7):993–1004

Samarin SN, Ivanov AI, Flatau G, Parkos CA, Nusrat A (2007) Rho/Rho-associated kinase-II signaling mediates disassembly of epithelial apical junctions. Mol Biol Cell 18(9):3429–3439

Schlegelmilch K, Mohseni M, Kirak O, Pruszak J, Rodriguez JR, Zhou D, Kreger BT, Vasioukhin V, Avruch J, Brummelkamp TR, Camargo FD (2011) Yap1 acts downstream of alpha-catenin to control epidermal proliferation. Cell 144(5):782–795

Schroeder MC, Halder G (2012) Regulation of the Hippo pathway by cell architecture and mechanical signals. Semin Cell Dev Biol 23(7):803–811

Shahbazi MN, Megias D, Epifano C, Akhmanova A, Gundersen GG, Fuchs E, Perez-Moreno M (2013) CLASP2 interacts with p120-catenin and governs microtubule dynamics at adherens junctions. J Cell Biol 203(6):1043–1061

Shamir ER, Pappalardo E, Jorgens DM, Coutinho K, Tsai WT, Aziz K, Auer M, Tran PT, Bader JS, Ewald AJ (2014) Twist1-induced dissemination preserves epithelial identity and requires E-cadherin. J Cell Biol 204(5):839–856

Shen L, Weber CR, Raleigh DR, Yu D, Turner JR (2011) Tight junction pore and leak pathways: a dynamic duo. Annu Rev Physiol 73:283–309

Shore EM, Nelson WJ (1991) Biosynthesis of the cell adhesion molecule uvomorulin (E-cadherin) in Madin-Darby canine kidney epithelial cells. J Biol Chem 266(29):19672–19680

Silvis MR, Kreger BT, Lien WH, Klezovitch O, Rudakova GM, Camargo FD, Lantz DM, Seykora JT, Vasioukhin V (2011) alpha-catenin is a tumor suppressor that controls cell accumulation by regulating the localization and activity of the transcriptional coactivator Yap1. Sci Signal 4 (174):ra33

Simcha I, Kirkpatrick C, Sadot E, Shtutman M, Polevoy G, Geiger B, Peifer M, Ben-Ze'ev A (2001) Cadherin sequences that inhibit beta-catenin signaling: a study in yeast and mammalian cells. Mol Biol Cell 12(4):1177–1188

Simoes Sde M, Mainieri A, Zallen JA (2014) Rho GTPase and Shroom direct planar polarized actomyosin contractility during convergent extension. J Cell Biol 204(4):575–589

Smith AL, Dohn MR, Brown MV, Reynolds AB (2012) Association of Rho-associated protein kinase 1 with E-cadherin complexes is mediated by p120-catenin. Mol Biol Cell 23(1):99–110

Steinberg MS (2007) Differential adhesion in morphogenesis: a modern view. Curr Opin Genet Dev 17(4):281–286

Steinberg MS, Takeichi M (1994) Experimental specification of cell sorting, tissue spreading, and specific spatial patterning by quantitative differences in cadherin expression. Proc Natl Acad Sci U S A 91(1):206–209

Straub BK, Rickelt S, Zimbelmann R, Grund C, Kuhn C, Iken M, Ott M, Schirmacher P, Franke WW (2011) E-N-cadherin heterodimers define novel adherens junctions connecting endoderm-derived cells. J Cell Biol 195(5):873–887

Suyama K, Shapiro I, Guttman M, Hazan RB (2002) A signaling pathway leading to metastasis is controlled by N-cadherin and the FGF receptor. Cancer Cell 2(4):301–314

Takahashi K, Nakanishi H, Miyahara M, Mandai K, Satoh K, Satoh A, Nishioka H, Aoki J, Nomoto A, Mizoguchi A, Takai Y (1999) Nectin/PRR: an immunoglobulin-like cell adhesion molecule recruited to cadherin-based adherens junctions through interaction with Afadin, a PDZ domain-containing protein. J Cell Biol 145(3):539–549

Takai Y, Miyoshi J, Ikeda W, Ogita H (2008) Nectins and nectin-like molecules: roles in contact inhibition of cell movement and proliferation. Nat Rev Mol Cell Biol 9(8):603–615

Takeda H, Nagafuchi A, Yonemura S, Tsukita S, Behrens J, Birchmeier W (1995) V-src kinase shifts the cadherin-based cell adhesion from the strong to the weak state and beta catenin is not required for the shift. J Cell Biol 131(6 Pt 2):1839–1847

Takeichi M (1991) Cadherin cell adhesion receptors as a morphogenetic regulator. Science 251 (5000):1451–1455

Takeichi M, Abe K (2005) Synaptic contact dynamics controlled by cadherin and catenins. Trends Cell Biol 15(4):216–221

Tamada M, Farrell DL, Zallen JA (2012) Abl regulates planar polarized junctional dynamics through beta-catenin tyrosine phosphorylation. Dev Cell 22(2):309–319

Theveneau E, Mayor R (2012) Cadherins in collective cell migration of mesenchymal cells. Curr Opin Cell Biol 24(5):677–684

Theveneau E, Marchant L, Kuriyama S, Gull M, Moepps B, Parsons M, Mayor R (2010) Collective chemotaxis requires contact-dependent cell polarity. Dev Cell 19(1):39–53

Thiery JP, Acloque H, Huang RY, Nieto MA (2009) Epithelial-mesenchymal transitions in development and disease. Cell 139(5):871–890

Townes PL, Holtfreter J (1955) Directed movements and selective adhesion of embryonic amphibian cells. J Exp Zool 128:53–120

Tsukita S, Furuse M (2002) Claudin-based barrier in simple and stratified cellular sheets. Curr Opin Cell Biol 14(5):531–536

Tumaneng K, Russell RC, Guan KL (2012) Organ size control by Hippo and TOR pathways. Curr Biol 22(9):R368–R379

Van Itallie CM, Anderson JM (2013) Claudin interactions in and out of the tight junction. Tissue Barriers 1(3):e25247

van Roy F, Berx G (2008) The cell-cell adhesion molecule E-cadherin. Cell Mol Life Sci 65 (23):3756–3788
Vestweber D, Gossler A, Boller K, Kemler R (1987) Expression and distribution of cell adhesion molecule uvomorulin in mouse preimplantation embryos. Dev Biol 124:451–456
Vestweber D, Broermann A, Schulte D (2010) Control of endothelial barrier function by regulating vascular endothelial-cadherin. Curr Opin Hematol 17(3):230–236
Volk T, Cohen O, Geiger B (1987) Formation of heterotypic adherens-type junctions between L-CAM-containing liver cells and A-CAM-containing lens cells. Cell 50(6):987–994
Webb DJ, Parsons JT, Horwitz AF (2002) Adhesion assembly, disassembly and turnover in migrating cells -- over and over and over again. Nat Cell Biol 4(4):E97–E100
Weber GF, Bjerke MA, DeSimone DW (2012) A mechanoresponsive cadherin-keratin complex directs polarized protrusive behavior and collective cell migration. Dev Cell 22(1):104–115
Wegener KL, Partridge AW, Han J, Pickford AR, Liddington RC, Ginsberg MH, Campbell ID (2007) Structural basis of integrin activation by talin. Cell 128(1):171–182
Wheelock MJ, Johnson KR (2003) Cadherin-mediated cellular signaling. Curr Opin Cell Biol 15 (5):509–514
Wheelock MJ, Shintani Y, Maeda M, Fukumoto Y, Johnson KR (2008) Cadherin switching. J Cell Sci 121(Pt 6):727–735
Williams EJ, Williams G, Howell FV, Skaper SD, Walsh FS, Doherty P (2001) Identification of an N-cadherin motif that can interact with the fibroblast growth factor receptor and is required for axonal growth. J Biol Chem 276(47):43879–43886
Winkel GK, Ferguson JE, Takeichi M, Nuccitelli R (1990) Activation of protein kinase C triggers premature compaction in the four-cell stage mouse embryo. Dev Biol 138(1):1–15
Wirtz-Peitz F, Zallen JA (2009) Junctional trafficking and epithelial morphogenesis. Curr Opin Genet Dev 19(4):350–356
Wizenmann A, Lumsden A (1997) Segregation of rhombomeres by differential chemoaffinity. Mol Cell Neurosci 9(5–6):448–459
Wong LE, Reynolds AB, Dissanayaka NT, Minden A (2010) p120-catenin is a binding partner and substrate for Group B Pak kinases. J Cell Biochem 110(5):1244–1254
Wood W, Jacinto A, Grose R, Woolner S, Gale J, Wilson C, Martin P (2002) Wound healing recapitulates morphogenesis in Drosophila embryos. Nat Cell Biol 4(11):907–912
Wu Y, Jin X, Harrison O, Shapiro L, Honig BH, Ben-Shaul A (2010) Cooperativity between trans and cis interactions in cadherin mediated junction formation. Proc Natl Acad Sci U S A 107 (41):17592–17597
Wu Y, Kanchanawong P, Zaidel-Bar R (2015) Actin-delimited adhesion-independent clustering of E-cadherin forms the nanoscale building blocks of adherens junctions. Dev Cell 32(2):139–154
Xiao K, Garner J, Buckley KM, Vincent PA, Chiasson CM, Dejana E, Faundez V, Kowalczyk AP (2005) p120-Catenin regulates clathrin-dependent endocytosis of VE-cadherin. Mol Biol Cell 16(11):5141–5151
Yang J, Weinberg RA (2008) Epithelial-mesenchymal transition: at the crossroads of development and tumor metastasis. Dev Cell 14(6):818–829
Yang CC, Graves HK, Moya IM, Tao C, Hamaratoglu F, Gladden AB, Halder G (2015) Differential regulation of the Hippo pathway by adherens junctions and apical-basal cell polarity modules. Proc Natl Acad Sci U S A 112(6):1785–1790
Yap AS, Brieher WM, Gumbiner BM (1997) Molecular and functional analysis of cadherin-based adherens junctions. Annu Rev Cell Dev Biol 13:119–146
Yonemura S, Wada Y, Watanabe T, Nagafuchi A, Shibata M (2010) alpha-Catenin as a tension transducer that induces adherens junction development. Nat Cell Biol 12(6):533–542
Yu FX, Zhao B, Panupinthu N, Jewell JL, Lian I, Wang LH, Zhao J, Yuan H, Tumaneng K, Li H, Fu XD, Mills GB, Guan KL (2012) Regulation of the Hippo-YAP pathway by G-protein-coupled receptor signaling. Cell 150(4):780–791
Zhong Y, Brieher WM, Gumbiner BM (1999) Analysis of C-cadherin regulation during tissue morphogenesis with an activating antibody. J Cell Biol 144(2):351–359
Zipursky SL, Sanes JR (2010) Chemoaffinity revisited: dscams, protocadherins, and neural circuit assembly. Cell 143(3):343–353

Chapter 4
Structure and Function of Cadherin Extracellular Regions

Lawrence Shapiro

Abstract Cell-surface glycoproteins of the cadherin superfamily are defined by the presence of extracellular cadherin (EC) β-sandwich domains in their extracellular regions. EC domains adopt a fold similar to immunoglobulin domains, but most EC domains ligate calcium through stereotyped sites positioned between successive domains; Ca^{2+}-binding at these sites rigidifies cadherin extracellular regions. Although the superfamily is highly diverse and may serve numerous functions, the best-characterized members are the vertebrate "classical" cadherins, which mediate cell–cell adhesion via homodimerization between their membrane-distal EC1 domains. Nonclassical and invertebrate cadherins have evolved distinct mechanisms for cell recognition and adhesion, and are only now beginning to be understood.

Keywords Cadherin • Classical cadherin • Extracellular cell adhesion • Crystal structure crystallography • Cell adhesion • Adherens junctions

4.1 Introduction

Cadherins embody a large family of cell surface proteins, the best characterized of which function in cell–cell recognition and adhesion (Nagafuchi et al. 1987; Ogou et al. 1983; Takeichi 1990, 1991). In mediating this function, cadherins bind between cells through their extracellular regions, the structure and function of which are the topic of this review. Cadherin extracellular regions are diverse in structure (Fig. 4.1) (Nollet et al. 2000; Shapiro and Weis 2009) and, as described here, serve varied remarkable functions. Extracellular regions of cadherins are characterized by the presence of distinctive protein domains of ~100 amino acids called extracellular cadherin (EC) domains (Hatta et al. 1988; Overduin et al. 1995; Shapiro et al. 1995a, b). The number of EC domains in the extracellular regions of various cadherins varies widely, however, distinctive EC domain sequences can be

L. Shapiro (✉)
Department of Biochemistry and Molecular Biophysics, Columbia University, New York, NY 10032, USA
e-mail: shapiro@convex.hhmi.columbia.edu

S.T. Suzuki, S. Hirano (eds.), *The Cadherin Superfamily*,
DOI 10.1007/978-4-431-56033-3_4

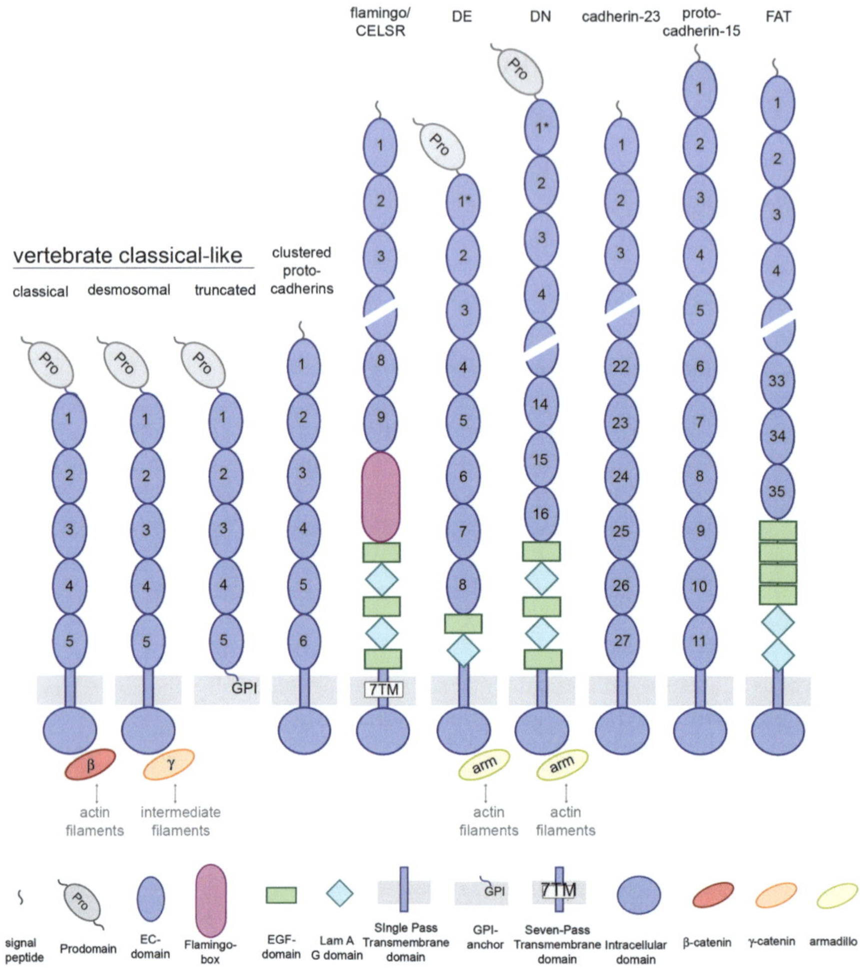

Fig. 4.1 Schematic diagram of domain arrangements in numerous cadherin subclasses. EC domains are numbered, and domains of other types are shown in the legends

detected in cadherins from widely varying species, including vertebrates, invertebrates, and some single-celled animals (Nollet et al. 2000; Posy et al. 2008b).

EC domains have topology similar to immunoglobulin domains (Overduin et al. 1995; Shapiro et al. 1995a), although arrangements of their hydrophobic core residues are distinct (Shapiro et al. 1995b). Two β -sheets are formed by seven strands; the immunoglobulin strand-topology nomenclature has been adopted for cadherin domains, with one sheet formed from strands ACFG, and the other by strands BED. The N-terminal A strand enters at the "top" of the domain, whereas the C-terminus of the final G strand exits through the "bottom", with the long-axis of the domain running parallel to the β-strands. Thus, EC domains can be efficiently assembled as contiguous repeats. For most, but not all EC domains, binding sites for

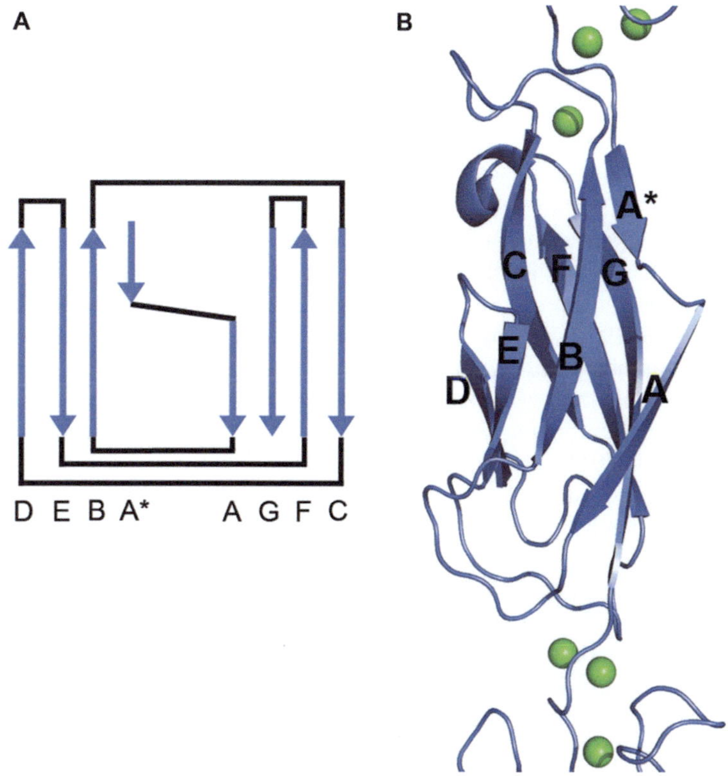

Fig. 4.2 Folding topology and role of Ca^{2+} binding by EC domains. The folding topology of an EC domain is shown schematically in (**a**). The topology is identical to immunoglobulin domains (in which the *A* strand can also associate with either sheet in different cases), and the immunoglobulin β-strand nomenclature is used. Panel (**b**) shows a ribbon diagram of a "middle" (EC2) cadherin domain, showing the classical role of Ca^{2+} ligation in rigidifying the connections between successive EC domains

Ca^{2+} ions are situated at each end of the domain (Boggon et al. 2002). Because Ca^{2+} ligands are donated from both preceding and following EC domains, Ca^{2+} ligation serves to rigidify the overall ectodomain structure (Boggon et al. 2002; Harrison et al. 2011; Fig. 4.2). Thus, cadherin EC domains provide a platform from which loops can be elaborated, as for immunoglobulin domains, and Ca^{2+} binding provides a mechanism to rigidify an overall superstructure of tandemly repeated EC domains.

Cadherins of vertebrates have been the most extensively studied. Numerous vertebrate cadherin subfamilies can be identified by phylogenetic analysis (Nollet et al. 2000; Posy et al. 2008b), including the classical cadherins, which appear in two distinctive sequence clusters referred to as type I and type II classical cadherins. Type I cadherins, including N-, E-, P-, and R-cadherins are broadly expressed, and mediate Ca^{2+}-dependent adhesion with primarily (but not exclusively) homophilic specificities (Katsamba et al. 2009). Their homophilic specificity makes cadherins ideal for formation of cell layers composed of a single cell type. The distinctive specificities of different type I classical cadherins are also thought to provide a driving force for the separation of cell layers (Hatta et al. 1987; Hatta and Takeichi 1986; Hirano et al. 1987; Nagafuchi et al. 1987; Duguay et al. 2003). Thus, type I cadherins commonly mediate homophilic adhesion between cells of various layers, and also play a role in mediating the relationships between layers, which can involve either separation or heterophilic adhesion.

Type I cadherins provide the transmembrane intercellular adhesive components of actin-attached adherens junctions (Harris and Tepass 2010; Meng and Takeichi 2009; Yap et al. 1997; Yonemura 2011). The formation of such junctions is among the most remarkable functions of cadherin ectodomains, where the combination of trans interactions (between cadherins on apposing cells) and cis interactions (between cadherins on the same cell) underlies an ordered junction structure (Harrison et al. 2011). Remarkably, the structure of these ordered assemblies appears to be encoded in the extracellular region itself, as type I cadherins lacking their cytoplasmic domain can spontaneously assemble junction-like structures between cadherins presented on juxtaposed membrane surfaces (Harrison et al. 2011; Taveau et al. 2008).

Another subfamily of vertebrate cadherins, the desmosomal cadherins (Delva et al. 2009; Garrod 2010; Holthofer et al. 2007; Koeser et al. 2003; Lewis et al. 1994), also form ordered intercellular attachments at desmosome junctions. Desmosomal cadherins, which include two subgroups, desmocollins and desmogleins, have structures (as inferred by sequence analysis) expected to be highly similar to type I classical cadherins (Posy et al. 2008a). However, structures of functional desmosmal cadherin ectodomains have not yet been obtained. Desmosomes are extremely dense and form junction superstructures that appear well-ordered in the extracellular space, suggesting the presence of lateral interactions between cadherin ectodomains (He et al. 2003; Al-Amoudi et al. 2007; Al-Amoudi and Frangakis 2008). Although sequence conservation suggests that their adhesive binding will likely be similar to that of type I cadherins (Posy et al. 2008a), the nature of potential lateral interactions among desmosomal cadherins remains unknown. Furthermore, as desmosome junctions contain both desmocollins and desmogleins (Chitaev and Troyanovsky 1997; Delva et al. 2009; Franke et al. 1994; Nollet et al. 2000), the roles and interactions of members of either subfamily remain unknown; thus the specific architecture of desmosomes remains largely to be determined.

Another family of vertebrate cadherins that function in cell adhesion is the type II family of classical cadherins (Patel et al. 2006; Shimoyama et al. 1999). Their structures are similar overall to type I cadherins (Patel et al. 2006), but their expression patterns differ. In general, type II cadherins, encoded by 13 different genes conserved in vertebrates (Nollet et al. 2000), are coexpressed in subsets together (Marthiens et al. 2005; Price et al. 2002). Thus, although a given cell type usually expresses a single type I cadherin, such as N- or E-cadherin (Hatta and Takeichi 1986; Hirano et al. 1987; Nakagawa and Takeichi 1998; Takeichi 1991,

1995), most type II cadherin-expressing cells produce a small set of type II cadherins. A well-studied example can be found in the spinal cord, where motor neuron cell bodies are grouped together in structures called motor pools; each motor pool innervates an individual muscle. Each motor pool expresses a distinct combination of type II cadherins, and these distinct sets of type II cadherins function to hold the cells of each motor pool together through homophilic adhesion, while separating them from cells of other motor pools (Patel et al. 2006; Price et al. 2002). Misexpression of type II cadherins in this system disrupts motor pool organization (Patel et al. 2006; Price et al. 2002). Overall, the coexpression of type II cadherins has complicated their functional analyses. As a result, less is known about their function than for type I cadherins.

Numerous other branches of the cadherin family are also found in vertebrate genomes, and have been functionally characterized to varying degrees. A group of proteins referred to, sometimes confusingly, as protocadherins is also represented (Nollet et al. 2000). Many protocadherins appear to function in cell–cell adhesion, but there are two notable classes of outliers. First, is a group of extremely large cadherins that help to form filamentous signaling structures (Ahmed et al. 2006; Kazmierczak et al. 2007). In a key example, the hair cells of the inner ear produce stereocilia, which are linked at their apex by a thin helical structure called the "tip-link". This structure, which is required for hearing, is composed of a complex between the very large proteins cadherin-23 (27 EC domains) and protocadherin-15 (11 EC domains) (Ahmed et al. 2006; Kazmierczak et al. 2007; Elledge et al. 2010; Sotomayor et al. 2010). There is binding at the tips between these two large cadherins that extend from adjacent stereocilia (Sotomayor et al. 2012), however, the cable-like structure they produce is tuned to sensing vibration and transmitting it to the hair cells via associated ion channels to produce neural representations of sound (Kazmierczak et al. 2007). Similar proteins are found in other sensory systems as well, but their functions remain unknown (Seiler et al. 2005).

Another important outlying example is found in a distinctive set of genes arranged in three clusters that encode the α, β, and γ "clustered" protocadherins, highly specialized proteins that help to mediate neural self-recognition and self-avoidance (Hayashi and Takeichi 2015; Thu et al. 2014; Chen and Maniatis 2013; Yagi 2012). These proteins are functionally distinct from the majority of cadherins, still mediating cell–cell recognition, but primarily to activate avoidance between neurites emanating from the same neuron, while allowing interaction between neurites of different neurons (Lefebvre et al. 2012; Chen et al. 2012). Thus, cell–cell recognition by cadherins can mediate avoidance as well as adhesion. The mechanism by which protocadherins enable self-recognition by processes from the same neuron remains unclear but, as described below, preliminary structure/function studies give clues about their remarkable function.

4.2 Vertebrate Classical Cadherins

Extracellular regions of classical cadherins protrude from opposing cell surfaces and form trans adhesive homodimers through their membrane-distal EC1 domains (Fig. 4.3). The cadherin/cadherin interface that mediates this interaction has been characterized in detail from atomic resolution structures of numerous classical cadherins (Harrison et al. 2011; Shapiro et al. 1995a; Boggon et al. 2002; Haussinger et al. 2004; Parisini et al. 2007). All classical cadherins share a common binding mechanism in which the most N-terminal portion of the β strand A, called

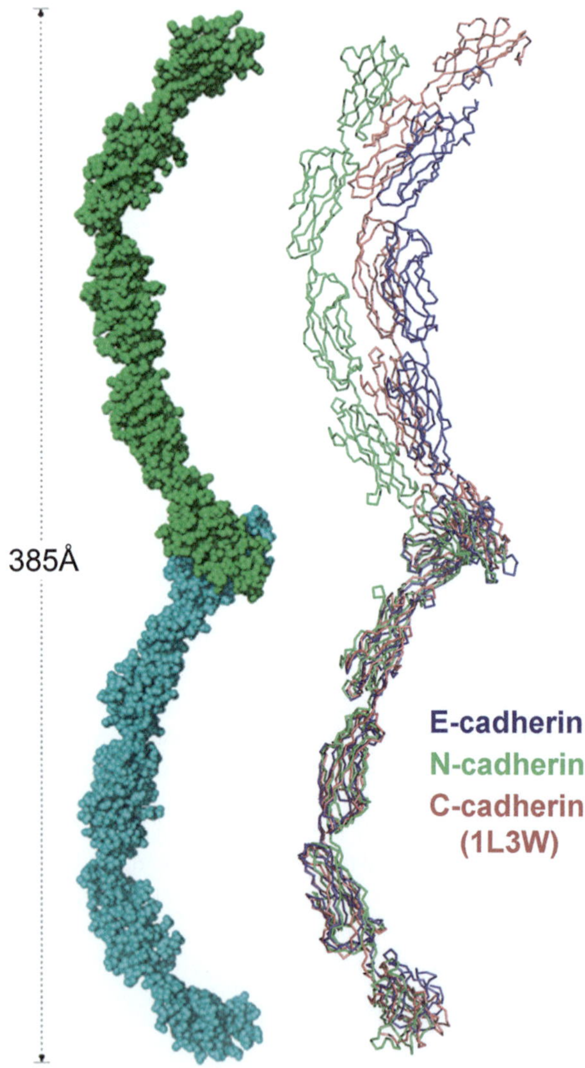

Fig. 4.3 Crystal structures of the full adhesive ectodomains in the adhesive dimer conformation for N-, E-, and C-cadherin structures. *Left*, C-cadherin adhesive dimer structure in space-filling representation. *Right*, Superposition of all three dimer structures, showing their variability. Only the lower molecule is used in the superposition, and the variation of the upper molecules represents mainly angular differences at the adhesive interface

the A* strand, swaps between EC1 domains of the adhesive partner protomers, a form of 3D domain swapping (Bennett et al. 1995). Conserved hydrophobic anchor residues located on the A* strand—Trp2 for type I cadherins and Trp2 and Trp4 in type II cadherins—dock into a conserved hydrophobic pocket in the body of the adhesive partner EC1 domain in trans (from the apposed cell). These conserved anchoring Trp residues are necessary for cell adhesion, and point mutation at these residues provides a convenient knockout for classical cadherin function (Harrison et al. 2005; Meng and Takeichi 2009; Patel et al. 2003; Shapiro and Weis 2009; Troyanovsky et al. 2003; Patel et al. 2003).

4.3 Strand-Swap Binding

The exchange of β-strands between interacting classical cadherins provides one of the clearest biological examples of the protein interaction mode referred to as "3D domain swapping" (Bennett et al. 1995). 3D domain swapping is defined by the presence of two alternative conformations, the unbound and bound states, which differ in that the "swapping" region is self-bound in the monomer, but interacts identically with its partner domain in the bound state. In the case of vertebrate classical cadherins, the swapping "domain" (the A*-strand) can dock into a binding pocket in the body of its own protomer to form a "closed"monomer (Fig. 4.4, left panel), or can dock into the pocket of a partner EC1 domain to form a strand-swapped dimer (Fig. 4.4, right panel). A necessary step in the transition of the closed monomer to the swapped dimer is rearrangement of the monomer state such that the swapping domain, the A* strand, undocks, allowing dimers to form between two open monomers. As is characteristic of 3D-domain swap binding, the swapping domain (the A* strand) is found in nearly identical residue environments in the "closed" monomer and in the swapped dimer. The closed monomer state can thus be thought of as a competitive inhibitor for the swapped dimer. This competition generally weakens interactions for 3D domain-swapped interfaces, and underlies the relatively weak binding affinities of classical cadherins (~10–100μM) (Chen et al. 2005). This property also requires that structural differences exist that stabilize the dimer and/or destabilize the monomer so that adhesive dimers are favored at points of cell–cell contact. Several factors that favor the formation of strand-swapped dimers (Posy et al. 2008b) have been identified, including a shortened A-strand, which strains to self-dock, but which is free to dock in the less geometrically constrained setting of a dimer.

Binding sites for three Ca^{2+} ions are found at interdomain linkers between each set of successive EC domains (Fig. 4.2). Glu11, a residue conserved at the base of EC1 A strands, coordinates Ca^{2+} in all classical cadherins. Anchoring of the A strand at both ends—at the base by Ca^{2+} binding to Glu11 and at the N terminus by Trp2 docking—induces strain in the shortened A strand. This strain destabilizes the closed monomer and thus favors strand-swapped dimer formation (Vendome et al. 2011; Vunnam and Pedigo 2011a, b).

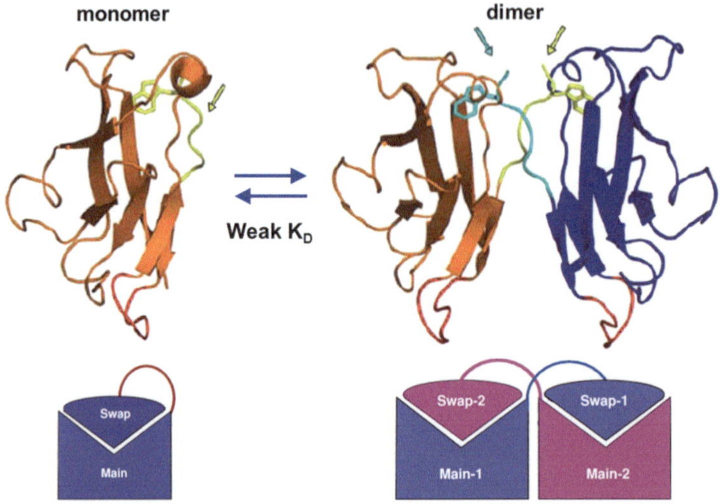

Fig. 4.4 3D domain swapping in classical cadherins. *Left*, a ribbon representation of the closed form of the EC1 domain of E-cadherin (*top*), and schematic representation of a closed monomer in 3D domain swapping. The swap domain in the *lower panel* corresponds to the swapping A* strand in the *upper panel*. In the dimer (*right*), the atomic environment of the swapped strand is nearly identical, but in the intermolecular rather than the intramolecular context shown at *left*. The similarity between the monomer structure at *left* and the dimer structure at *right* leads to small energy differences, and hence weak binding is weak. Arrows indicate the swapping A* strands

All vertebrate classical cadherins form cell-adhesive dimers through a similar strand-swapping mechanism. However, there are characteristic differences between members of the type I and type II cadherin subfamilies (Fig. 4.5). Adhesive interfaces of type I cadherins are restricted to the Trp2-acceptor pocket region near the N-terminal apex of EC1 and the partner A* strand region that includes Trp2. By contrast, in type II cadherins, two tryptophan residues, Trp2 and Trp4, anchor the swapped strand. The dimer interfaces of type II family members also extend beyond the swapping region, along the entire face of the EC1 domain. These extended interface regions mediate interactions involving conserved hydrophobic residues at positions 8, 10, and 13 (Patel et al. 2006). VE- cadherin, a divergent classical cadherin and the primary adhesion protein of the vascular endothelium (Harris and Nelson 2010), blurs the definition between type I and type II cadherins. As do type II cadherins, VE-cadherin docks Trp2 and Trp4 into the hydrophobic pocket of its partner, but as do type I cadherins it lacks hydrophobic interactions along the rest of its EC1 domain and thus has an overall dimer arrangement more similar to that of type I cadherins (Brasch et al. 2011).

The specificity of classical cadherin adhesive binding is controlled by the EC1 domain, as demonstrated by experiments in which EC1 domains were shuffled between different cadherins with different specificities (Patel et al. 2006; Klingelhofer et al. 2000; Nose et al. 1990; Shan et al. 2000, 2004). Type I cadherins do not bind to type II cadherins (Shapiro and Weis 2009; Patel et al. 2006;

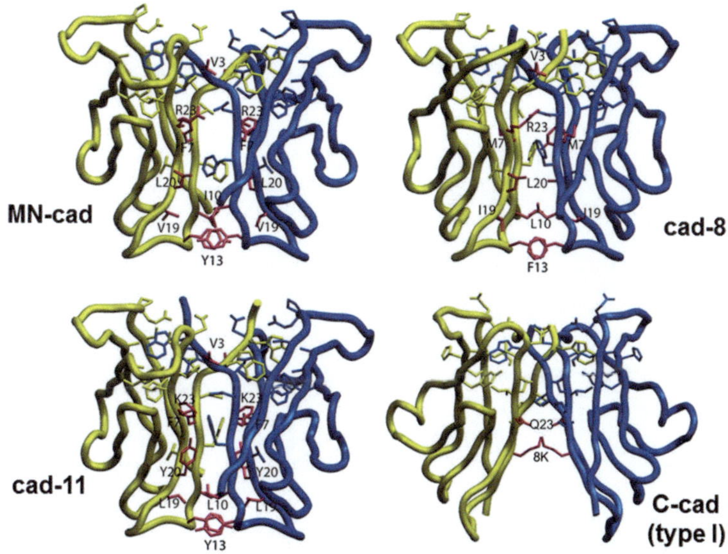

Fig. 4.5 Comparison of type I and type II cadherin adhesive interfaces. Worm diagrams are shown for three type II cadherin EC1 domain adhesive dimers: MN-cadherin, and cadherins -8 and -11. The adhesive EC1 dimer is also shown for one type I cadherin, C-cadherin. Note that the adhesive interface for type I cadherins involves only the strand-swap region, whereas type II cadherins have an extra hydrophobic interface toward the base, which zips up over the whole length of the EC1 domain

Katsamba et al. 2009; Shimoyama et al. 1999), consistent with the differences between the adhesive interface structures of these cadherin subfamilies. They are often described as having homophilic specificities, however, classical cadherins interact promiscuously within subfamilies (Katsamba et al. 2009; Patel et al. 2006; Shimoyama et al. 1999; Shimoyama et al. 2000). Type I cadherins are usually expressed singly, but type II cadherins are in general expressed in combinations; the biological effect of type II cadherin coexpression remains to be fully understood.

4.4 X-Dimers Facilitate Strand Swapping in Type I and Type II Classical Cadherins

To form strand-swapped cell-adhesive dimers requires that each partner classical cadherin protomer refold to transition from the "closed" monomer form (Fig. 4.4, left panel) to the "open" dimer form (Fig. 4.4, right panel). This conformational change can present a kinetic barrier. Indeed, in other examples of 3D domain swapping, this process can occur over very long time periods (Bennett et al. 1995). Results of single-molecule fluorescence resonance energy transfer (FRET) experiments have provided evidence for an encounter complex

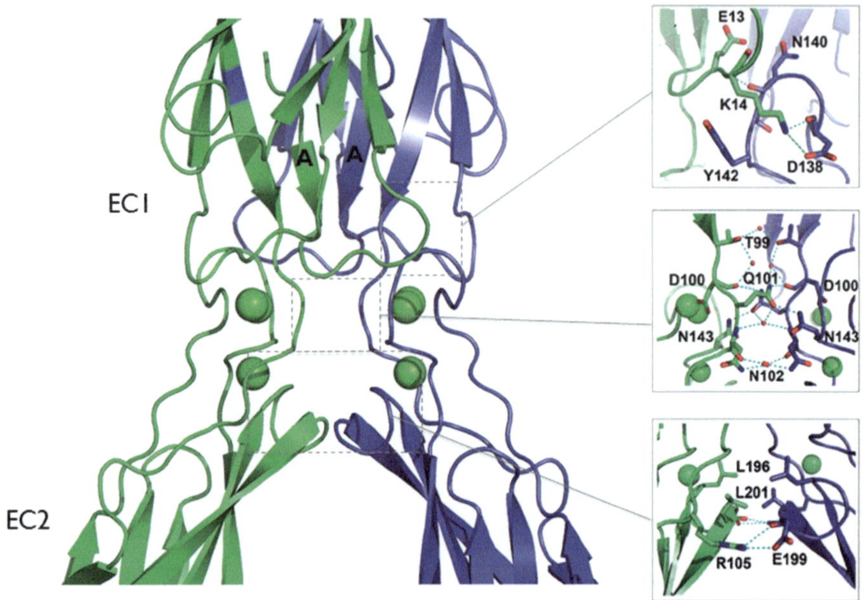

Fig. 4.6 X-dimer structure. The overall structure of the X-dimer interface is shown in ribbon representation for the E-cadherin strand swapping-incompetent mutant E89A. All strand-swapping–incompetent mutants of type I and type II cadherins determined thus far are in the X-dimer conformation. The X-interface includes the bottom of EC1, the EC1-EC2 linker, and the top of EC2; these three regions are shown in expanded view at right

intermediate. When strand swapping was ablated by a Trp2 to Ala mutation, dimers still formed between EC1 domains, with FRET distances slightly altered as compared with swapped dimers, suggesting the existence of a nonswapped dimer form. Additionally, atomic force microscopy (AFM) experiments showed the nonswapped mutant dimers to be weaker than strand-swapped, wild-type dimers, energetically consistent with a role as a binding intermediate (Sivasankar et al. 2009).

Structural studies of strand swap-impaired classical cadherin mutants have revealed the molecular details of this encounter complex (Harrison et al. 2010). Numerous strand swap-impaired mutants adopt a similar conformation: a dimer with its interface centered around the EC1–EC2 interdomain linker (Fig. 4.6). These dimeric structures are now called "X-dimers" due to their X-like overall shape. X-dimers interact through surface residues, thus requiring no refolding for interaction. X-dimers can therefore form with fast-binding kinetics. Most importantly, in the X-dimer the A strands of each protomer are positioned in parallel to each other in close proximity, as if poised to swap (Harrison et al. 2010). Thus, X-dimers form quickly and position the EC1 domains of interacting cadherins, holding them in place to enable refolding to adopt the strand-swapped conformation. Type II cadherins (Harrison et al. 2010). The role of the X-dimer conformation as a kinetic intermediate has been confirmed by cell biological and biophysical observations (Sivasankar et al. 2009). Mutations designed to prevent X-dimer formation, but leave strand swapping intact, fail to mediate cell adhesion (Harrison et al. 2010). The association rates of type I E-cadherin and type II cadherin-6 are dramatically slowed in such mutants such that no dimerization could be observed in short (~1 min) SPR experiments. However, in sedimentation equilibrium analytical ultracentrifugation experiments (~48 h equilibration time) wild-type binding affinities are observed (Harrison et al. 2010).

Interestingly, T-cadherin, a divergent vertebrate classical-like cadherin which is GPI-anchored and lacks a cytoplasmic region, does not interact by strand swapping. Rather, the adhesive state of T-cadherin represents an X-dimer formed between T-cadherin ectodomains from juxtaposed cells (Ciatto et al. 2010). Mutations targeting the X-dimer interface in T-cadherin abolish its function in neurite outgrowth regulation, whereas mutations targeted to the region involved in strand swapping for other classical cadherins had no effect on T-cadherin function or homodimerization (Ciatto et al. 2010). The close phylogenetic relation to type I classical cadherins suggests that T-cadherin represents a classical cadherin that has lost its ability to bind through strand swapping.

4.5 Cis Interactions, Adherens Junctions, and Desmosomes

In mature tissues cadherins localize primarily to intercellular structures with defined morphology called junctions. There are two primary junction types: adherens junctions, which are formed by classical cadherins and are linked to the actin cytoskeleton, and desmosome junctions formed by specialized desmosomal cadherins with members of two distinct subfamilies, the desmocollins and desmogleins, discussed below. Cadherin ectodomains appear to play a critical role in junction assembly. Experiments with purified classical cadherin ectodomains show that, when bound to liposomes, ectodomains alone self-assemble into structures closely resembling adherens junctions (Harrison et al. 2011; Taveau et al. 2008). Mutations at the crystallographically identified cis interface destroy these junction-like structures, suggesting a structural basis for self-assembly of adherens junctions through these cis and trans interfaces.

For classical cadherins, the lateral *cis*-interaction site shows a conservation signal above background among type vertebrate type I classical cadherins, and critically has been observed in all crystal structures of full-length type I cadherin ectodomains (Boggon et al. 2002; Harrison et al. 2011). In addition to the adhesive strand-swap interface, this lateral cis interface, formed between the base of the EC1 domain of one protomer and a region near the apex of EC2, is found in the structures of all three currently available structures of full-length cadherins, N-, E-, and C-cadherins (Fig. 4.7; Boggon et al. 2002; Harrison et al. 2011). The combination of cis and trans interactions for each cadherin molecule creates similar molecular layers within each crystal form, which likely to correspond to the fully

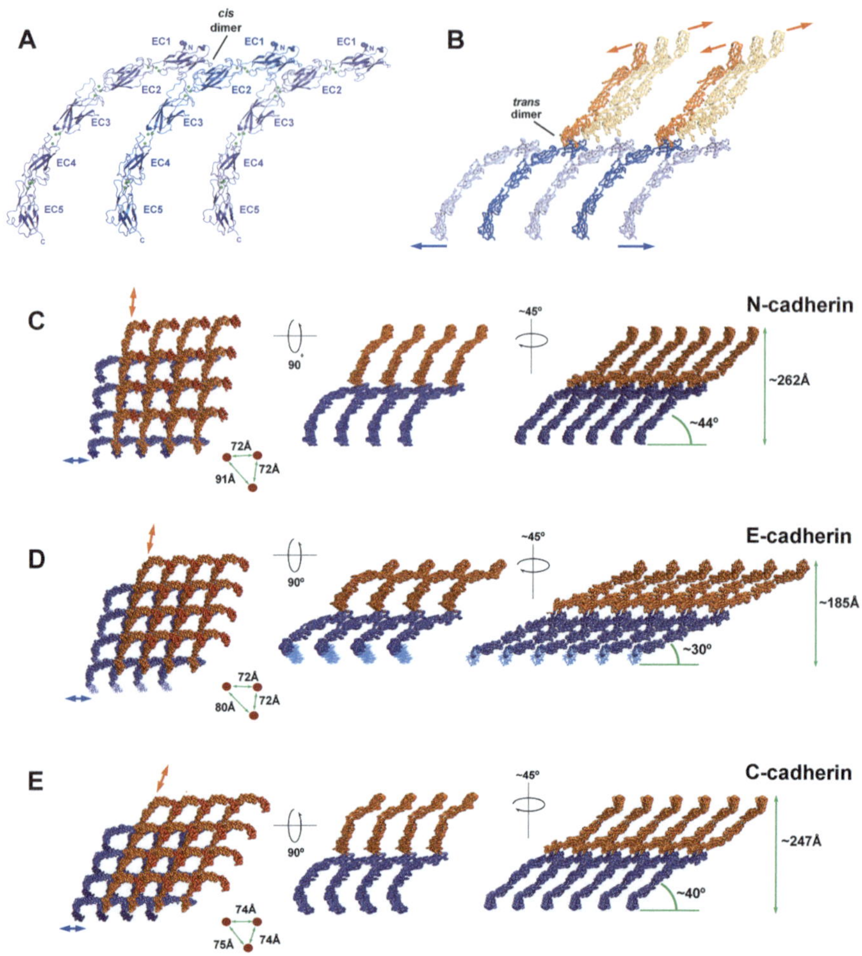

Fig. 4.7 The likely extracellular structure of adherens junctions. (**a**) The cis interface, similar among all type I cadherins investigated, is shown as it appears in crystals of N-cadherin. (**b**) The orange molecules at top, which also partake in cis interactions, attach to the blue cis interface-polymer via adhesive EC1 interactions. (**c**, **d**, and **f**) These interfaces combine to form similar lattices in unrelated crystals of N-, E-, and C-cadherins, respectively

bound state of cadherin ectodomains in adherens junctions (Boggon et al. 2002; Harrison et al. 2011). The region of EC1 involved in this cis interface is opposite to the strand-swapping site, so that cis and trans interactions can form simultaneously resulting in a continuous two-dimensional lattice with dimensions near to those expected for adherens junctions (Fig. 4.7). No sequence conservation above background level is observed for this region in other cadherin subfamilies, many of which lack elements of the interface through residue deletions. These three proteins share identities of 58 %(C/E), 58 %/(E/N), and 39 %(C/N), and it would be highly unlikely for all three proteins to form a nearly identical interface (the cis interface) in all three unrelated crystals, arguing for its biological function.

In an artificial system in which purified His-tagged cadherin ectodomains are bound to the surface of controlled-size (~200nm) Ni^{2+}-chelating liposomes, time-dependent Ca^{2+}-dependent liposome aggregation was observed, and cryo-EM analysis revealed ordered junction-like structures that resemble the layer of molecules, composed of cadherins arranged through cis and trans interfaces, observed in the unrelated crystal lattices of C- (Boggon et al. 2002), E-, and N-cadherin (Harrison et al. 2011). This liposome system, and assays using transfected cells have been employed to test the idea that the cis interface underlies lateral assembly of adherens junctions comprised of type I cadherins. Mutants compromised for binding at the cis interface of E- cadherin, but wild-type for trans strand-swap binding, showed adhesion between liposomes, but at a reduced level. In transfected cells, these cis interface-mutant cadherins a dominant phenotype in which junctions incorporating them became unstable and transient. In cells lacking wild-type cadherins, cis mutant proteins showed both diffuse localization as well as some degree of concentration at sites of cell contact, but this degree of concentration was significantly less than observed for wild-type cadherins (Harrison et al. 2011). Finally, cryo-EM analysis of adherent liposomes produced with purified cis interface-mutant cadherins showed that the ordered tooth-like structure characteristic of the wild-type reconstituted junctions was absent in the cis interface mutant junctions (Harrison et al. 2011) . Taken together, these data strongly implicate the cis interface identified in crystallographic studies in the lateral assembly of cadherin trans dimers in adherens junctions.

Remarkably, cis interactions among classical type I cadherins are too weak to be detected by analytical ultracentrifugation (detection limited to K_Ds <1 mM) or other typical solution-binding experiments, despite its apparent biological role in junction assembly revealed in mutagenesis studies with cellular or cryo-EM readouts. This apparent paradox is likely due to the significant differences expected for protein–protein interactions in solution and in the context of a membrane, where positional and rotational freedom are limited (Wu et al. 2011). Indeed, *in silico* simulation experiments suggest that when type I cadherin ectodomain dimers form in trans, their motional freedom is dramatically reduced because they are attached to one other through the adhesive interface while tethered at each end to one of the apposed cell membranes (Wu et al. 2011). Thus trans interactions between cadherins are expected to lower the entropic penalty associated with cis dimer formation (Wu et al. 2010; Wu et al. 2011), triggering a cooperative junction assembly process when two cells come into contact. The observation that cadherins do not cluster on the cell surface in the absence of an apposed cadherin-expressing cell (Gumbiner 2005; Hajra and Fearon 2002) can be at least partially explained by this model.

The self-assembly of junctions, at least in part through specific interactions of cadherin extracellular regions, could have multiple functions. Such lateral intercadherin interactions are likely to increase the mechanical stability of intercellular adhesion, and the concentration of proteins at intercellular appositions could in principle play a critical role in signaling, although the nature of signaling at adherens junctions remains poorly understood overall. When proteins that bind one another are presented on apposing cell surfaces, their binding creates a "diffusion trap" mechanism whereby the proteins will concentrate at sites of cell membrane apposition. Cadherin assembly, however, yields a degree of concentration at junctions that is higher than can be achieved by a diffusion trap alone. The cis interface mutants described above show that adhesive binding alone, in the absence of self-assembly mediated by the cis interface, yields a substantially lower concentration of cadherin at cell–cell contacts than observed for wild-type cadherins (Harrison et al. 2011). This experiment demonstrates that cis interactions play a critical role in enhancing the localization of classical cadherins to adherens junctions, likely through assembly of small punctate junctions with bound structures produced by the cis and trans interfaces similar overall to those observed in liposome-reconstituted junctions and the N-, E-, and C-cadherin crystal structures. Large cellular adherens junctions, as observed in fluorescence microscopy, are likely to be assembled from numerous subdomains with the lattice-like structure described above. The lattice structure is directional such that any two subdomains would have to meet with an appropriate orientation to merge. Although the vertebrate type II classical cadherins are highly similar to type I cadherins and have the same adhesive mechanism as type I cadherins, they do not appear to partake in self-assembly through the cis interface described above (Brasch et al. 2011; Harrison et al. 2010; Patel et al. 2006) .

Desmosome junctions, which are extremely dense and stable structures, also assemble differently from adherens junctions formed by type I cadherins. Analyses of sequence conservation between desmosomal and type I cadherins (Thomason et al. 2010) suggest that they also adhere through a strand-swap binding mechanism—although whether the X-dimer kinetic intermediate is also used cannot be inferred by sequence comparison—but their lateral interactions are likely to differ. Both subfamilies of desmosomal cadherins, desmocollins and desmogleins, conserve the strand swap-anchoring Trp residue conserved at position 2, and hydrophobic residues corresponding to the Trp binding pocket in classical cadherins (Posy et al. 2008b; Thomason et al. 2010). Also as for type I and type II classical cadherins, mutation of Trp2 or its acceptor pocket abolishes trans binding of desmocollin 2 in cross-linking experiments (Nie et al. 2011). The structure of an EC1-domain fragment of human desmoglein-2, determined by NMR spectroscopy (pdb-ID: 2YQG) (NMR) shows a domain fold similar to that of vertebrate type I cadherins. This structure is monomeric with Trp2 self-docked, perhaps due to the inclusion of 10 residues preceding the native N terminus from a cloning artifact. Extensions of this type have been shown to prevent strand-swap dimerization in classical cadherins (Harrison et al. 2010; Haussinger et al. 2004) . Two groups have independently produced high-resolution electron microscopy tomograms of in situ desmosomes. The first of these, from the Stokes group (He et al. 2003), examined desmosomes from mouse skin embedded in plastic and sectioned. This reconstruction revealed a dense network of interacting desmosomal cadherin ectodomains, but the arrangement of ectodomains appeared far less ordered than expected from two-dimensional EM images of desmosomes. This apparent lack of order could have arisen as an artifact of the sectioning procedure. A second more recent desmosome reconstruction was produced by the Frangakis group. This study was based on cryo-electron tomography of vitreous sections from human epidermis, and revealed a regular array of curve-shaped densities resembling classical cadherin ectodomain structures spaced at ~70Å intervals along the midline (Al-Amoudi et al. 2007). Despite this seemingly clear result, both this study and the earlier one were unable to define a molecular model for ectodomain interactions in desmosomes. A significant part of this uncertainty arises from uncertainty about the composition of the desmosomes studied. It is thought that a given desmosome will contain both desmocollins and desmogleins, but the protein compositions of the desmosomes studied by both the Frangakis and Stokes groups were unknown, and no attempt was made in either work to distinguish desmocollins from desmogliens.

4.6 "Giant" Cadherins

Both vertebrate and invertebrate genomes encode numerous proteins containing large numbers of tandem EC domains, so-called "Giant" cadherins. Relatively little is known about their structure/function relations, but early insights are exciting. The Giant cadherins protocadherin-15 (11 EC domains) and cadherin-23 (27 EC domains) provide a remarkable example. These proteins, each involved in inherited deafness, link adjacent stereocilia of sound-sensing hair cells by formation of a cable-like structure known as the tip-link. Scanning transmission electron microscopy images suggest that the tip-link is comprised of a double helix formed by two cadherin-23 molecules emanating from one stereocilium interacting at the tip with the tip of a double helix formed by two protocadherin-15 molecules emanating from adjacent stereocilium (Kazmierczak et al. 2007). Atomic resolution structures of an N-terminal EC1–EC2 fragment from cadherin-23, and its complex formed by interaction with an EC1–EC2 fragment from protocadherin-15, yield significant insights into how this head-to-tail oriented complex forms an extended handshake interaction involving both EC1 and EC2 domains. Interestingly, Pcdh 15 has an elongated N-terminus which extends as a helix beyond the body of EC1; this helix forms much of the interface with Pcdh 23 EC2. Unlike classical cadherins, there is no strand-swap interaction. The authors use molecular dynamics simulations to highlight ways in which the cadherein 23/Pcdh 15 interface is optimized to resist force in transducing vibrational signals.

Another well-studied pair of interacting giant cadherins are Fat and Dachsous, which regulate cell polarity and proliferation (Ishiuchi et al. 2009; Tanoue and Takeichi 2005). Fat is the largest cadherin, with 34 EC repeats, and binds to Dachsous, another Giant cadherin with 27 EC domains. Despite their large sizes, the mammalian proteins Fat4 and Dachsous1 are detected in intercellular spaces contiguous with adherens junctions, raising the question of how such large molecules can be accommodated in a relatively small space, one which classical cadherins are known to traverse with only five EC domains from each adherent cell surface. A recent study of purified Fat4 and Dachsous1 ectodomains reveals that each molecule is made up of elongated sets of contiguous domains, with hairpin bends distributed at specific interdomain linkers (Tsukasaki et al. 2014). These hairpin bends appear to be associated with interdomain linker regions that lack the canonical Ca^{2+} binding sites, which normally help rigidify linkages between EC domains. Consistently, earlier work had shown that a four-domain fragment of DN-cadherin had a hairpin bend at just such a Ca^{2+}-free linker (Jin et al. 2012). Thus, Fat and Dachsous appear to bind tip to tip in the intercellular space, and have long, folded-up multi-EC ectodomains that could in principle traverse the intercellular space multiple times (Tsukasaki et al. 2014).

4.7 The Clustered Protocadherins and Neurite Self-Avoidance

The clustered protocadherins are a family of highly related vertebrate cadherin-like proteins encoded in three novel contiguous gene clusters (α, β, and γ) and are predominantly expressed in the nervous system. Protocadherins help to establish single-neuron identity to establish specific self-avoidance between neurites emanating from the same neuron. In mouse there are 58 Pcdh proteins, and each neuron expresses a defined subset of these (up to about 15) via a mechanism involving stochastic promoter choice. Neurites from the same neuron express the same Pcdhs, and thus recognize one another and repel; neurites of different neurons have different sets of Pcdhs, and hence are free to interact (no repulsion is signaled). Structure/function relationships in protocadherins remain largely obscure. Although a number of single-domain structures have been determined (pdb IDs 2EE0, 2YST, 1WYJ, 1WUZ; Morishita et al. 2006), none reveal functional recognition sites. Aggregation assays with transfected cells have shown that singly expressed Pcdhs have homophilic binding specificities, but how these specificities relate to self-avoidance in the case where many Pcdh isoforms are expressed remains unclear (Schreiner and Weiner 2010). Domain shuffling experiments suggest that Pcdh domains EC1–EC3 are crucial for trans adhesion, with domains EC2 and EC3 appearing to control protocadherin specificity in cell aggregation assays (Schreiner and Weiner 2010) . Domains EC2 and EC3 show the highest sequence diversity among individual protocadherin isoforms, consistent with the possibility of their contribution to specificity (Schreiner and Weiner 2010).

4.8 Concluding Remarks

Vertebrate classical cadherin ectodomains and vertebrate desmosomal cadherins—close relatives of classical cadherins—contain sequence elements indicative of strand-swap binding. However, other superfamily members, including protocadherins and all invertebrate cadherins, are likely to use distinct mechanisms, and these will become clear only with further structure/function studies.

References

Ahmed ZM, Goodyear R, Riazuddin S, Lagziel A, Legan PK, Behra M, Burgess SM, Lilley KS, Wilcox ER, Griffith AJ et al (2006) The tip-link antigen, a protein associated with the transduction complex of sensory hair cells, is protocadherin-15. J Neurosci 26:7022–7034

Al-Amoudi A, Frangakis AS (2008) Structural studies on desmosomes. Biochem Soc Trans 36:181–187

Al-Amoudi A, Diez DC, Betts MJ, Frangakis AS (2007) The molecular architecture of cadherins in native epidermal desmosomes. Nature 450:832–837

Bennett MJ, Schlunegger MP, Eisenberg D (1995) 3D domain swapping: a mechanism for oligomer assembly. Protein Sci 4:2455–2468

Boggon TJ, Murray J, Chappuis-Flament S, Wong E, Gumbiner BM, Shapiro L (2002) C-cadherin ectodomain structure and implications for cell adhesion mechanisms. Science 296:1308–1313

Brasch J, Harrison OJ, Ahlsen G, Carnally SM, Henderson RM, Honig B, Shapiro L (2011) Structure and binding mechanism of vascular endothelial cadherin: a divergent classical cadherin. J Mol Biol 408:57–73

Chen WV, Maniatis T (2013) Clustered protocadherins. Development 140:3297–3302

Chen CP, Posy S, Ben-Shaul A, Shapiro L, Honig BH (2005) Specificity of cell-cell adhesion by classical cadherins: critical role for low-affinity dimerization through beta-strand swapping. Proc Natl Acad Sci U S A 102:8531–8536

Chen WV, Alvarez FJ, Lefebvre JL, Friedman B, Nwakeze C, Geiman E, Smith C, Thu CA, Tapia JC, Tasic B et al (2012) Functional significance of isoform diversification in the protocadherin gamma gene cluster. Neuron 75:402–409

Chitaev NA, Troyanovsky SM (1997) Direct Ca^{2+}-dependent heterophilic interaction between desmosomal cadherins, desmoglein and desmocollin, contributes to cell-cell adhesion. J Cell Biol 138:193–201

Ciatto C, Bahna F, Zampieri N, VanSteenhouse HC, Katsamba PS, Ahlsen G, Harrison OJ, Brasch J, Jin X, Posy S et al (2010) T-cadherin structures reveal a novel adhesive binding mechanism. Nat Struct Mol Biol 17:339–347

Delva E, Tucker DK, Kowalczyk AP (2009) The desmosome. Cold Spring Harb Perspect Biol 1: a002543

Duguay D, Foty R, Steinberg M (2003) Cadherin-mediated cell adhesion and tissue segregation: qualitative and quantitative determinants. Dev Biol 253:309–323

Elledge HM, Kazmierczak P, Clark P, Joseph JS, Kolatkar A, Kuhn P, Muller U (2010) Structure of the N terminus of cadherin 23 reveals a new adhesion mechanism for a subset of cadherin superfamily members. Proc Natl Acad Sci U S A 107:10708–10712

Franke WW, Koch PJ, Schafer S, Heid HW, Troyanovsky SM, Moll I, Moll R (1994) The desmosome and the syndesmos: cell junctions in normal development and in malignancy. Princess Takamatsu Symp 24:14–27

Garrod D (2010) Desmosomes in vivo. Dermatol Res Pract 2010:212439

Gumbiner BM (2005) Regulation of cadherin-mediated adhesion in morphogenesis. Nat Rev Mol Cell Biol 6:622–634

Hajra KM, Fearon ER (2002) Cadherin and catenin alterations in human cancer. Genes Chromosome Cancer 34:255–268

Harris ES, Nelson WJ (2010) VE-cadherin: at the front, center, and sides of endothelial cell organization and function. Curr Opin Cell Biol 22:651–658

Harris TJ, Tepass U (2010) Adherens junctions: from molecules to morphogenesis. Nat Rev Mol Cell Biol 11:502–514

Harrison OJ, Corps EM, Berge T, Kilshaw PJ (2005) The mechanism of cell adhesion by classical cadherins: the role of domain 1. J Cell Sci 118:711–721

Harrison OJ, Bahna F, Katsamba PS, Jin X, Brasch J, Vendome J, Ahlsen G, Carroll KJ, Price SR, Honig B et al (2010) Two-step adhesive binding by classical cadherins. Nat Struct Mol Biol 17:348–357

Harrison OJ, Jin X, Hong S, Bahna F, Ahlsen G, Brasch J, Wu Y, Vendome J, Felsovalyi K, Hampton CM et al (2011) The extracellular architecture of adherens junctions revealed by crystal structures of type I cadherins. Structure 19:244–256

Hatta K, Takeichi M (1986) Expression of N-cadherin adhesion molecules associated with early morphogenetic events in chick development. Nature 320:447–449

Hatta K, Takagi S, Fujisawa H, Takeichi M (1987) Spatial and temporal expression pattern of N-cadherin cell adhesion molecules correlated with morphogenetic processes of chicken embryos. Dev Biol 120:215–227

Hatta K, Nose A, Nagafuchi A, Takeichi M (1988) Cloning and expression of cDNA encoding a neural calcium-dependent cell adhesion molecule: its identity in the cadherin gene family. J Cell Biol 106:873–881

Haussinger D, Ahrens T, Aberle T, Engel J, Stetefeld J, Grzesiek S (2004) Proteolytic E-cadherin activation followed by solution NMR and X-ray crystallography. EMBO J 23:1699–1708

Hayashi S, Takeichi M (2015) Emerging roles of protocadherins: from self-avoidance to enhancement of motility. J Cell Sci 128:1455–1464

He W, Cowin P, Stokes D (2003) Untangling desmosomal knots with electron tomography. Science 302:109–113

Hirano S, Nose A, Hatta K, Kawakami A, Takeichi M (1987) Calcium-dependent cell-cell adhesion molecules (cadherins): subclass specificities and possible involvement of actin bundles. J Cell Biol 105:2501–2510

Holthofer B, Windoffer R, Troyanovsky S, Leube RE (2007) Structure and function of desmosomes. Int Rev Cytol 264:65–163

Ishiuchi T, Misaki K, Yonemura S, Takeichi M, Tanoue T (2009) Mammalian Fat and Dachsous cadherins regulate apical membrane organization in the embryonic cerebral cortex. J Cell Biol 185:959–967

Jin X, Walker MA, Felsovalyi K, Vendome J, Bahna F, Mannepalli S, Cosmanescu F, Ahlsen G, Honig B, Shapiro L (2012) Crystal structures of Drosophila N-cadherin ectodomain regions reveal a widely used class of Ca^{2+}-free interdomain linkers. Proc Natl Acad Sci U S A 109: E127–E134

Katsamba P, Carroll K, Ahlsen G, Bahna F, Vendome J, Posy S, Rajebhosale M, Price S, Jessell TM, Ben-Shaul A et al (2009) Linking molecular affinity and cellular specificity in cadherin-mediated adhesion. Proc Natl Acad Sci U S A 106(28):11594–11599

Kazmierczak P, Sakaguchi H, Tokita J, Wilson-Kubalek EM, Milligan RA, Muller U, Kachar B (2007) Cadherin 23 and protocadherin 15 interact to form tip-link filaments in sensory hair cells. Nature 449:87–91

Klingelhofer J, Troyanovsky RB, Laur OY, Troyanovsky S (2000) Amino-terminal domain of classic cadherins determines the specificity of the adhesive interactions. J Cell Sci 113 (Pt 16):2829–2836

Koeser J, Troyanovsky SM, Grund C, Franke WW (2003) De novo formation of desmosomes in cultured cells upon transfection of genes encoding specific desmosomal components. Exp Cell Res 285:114–130

Lefebvre JL, Kostadinov D, Chen WV, Maniatis T, Sanes JR (2012) Protocadherins mediate dendritic self-avoidance in the mammalian nervous system. Nature 488:517–521

Lewis JE, Jensen PJ, Wheelock MJ (1994) Cadherin function is required for human keratinocytes to assemble desmosomes and stratify in response to calcium. J Invest Dermatol 102:870–877

Marthiens V, Gavard J, Padilla F, Monnet C, Castellani V, Lambert M, Mege RM (2005) A novel function for cadherin-11 in the regulation of motor axon elongation and fasciculation. Mol Cell Neurosci 28:715–726

Meng W, Takeichi M (2009) Adherens junction: molecular architecture and regulation. Cold Spring Harb Perspect Biol 1:a002899

Morishita H, Umitsu M, Murata Y, Shibata N, Udaka K, Higuchi Y, Akutsu H, Yamaguchi T, Yagi T, Ikegami T (2006) Structure of the cadherin-related neuronal receptor/protocadherin-alpha first extracellular cadherin domain reveals diversity across cadherin families. J Biol Chem 281:33650–33663

Nagafuchi A, Shirayoshi Y, Okazaki K, Yasuda K, Takeichi M (1987) Transformation of cell adhesion properties by exogenously introduced E-cadherin cDNA. Nature 329:341–343

Nakagawa S, Takeichi M (1998) Neural crest emigration from the neural tube depends on regulated cadherin expression. Development 125:2963–2971

Nie Z, Merritt A, Rouhi-Parkouhi M, Tabernero L, Garrod D (2011) Membrane-impermeable cross-linking provides evidence for homophilic, isoform-specific binding of desmosomal cadherins in epithelial cells. J Biol Chem 286:2143–2154

Nollet F, Kools P, van Roy F (2000) Phylogenetic analysis of the cadherin superfamily allows identification of six major subfamilies besides several solitary members. J Mol Biol 299:551–572

Nose A, Tsuji K, Takeichi M (1990) Localization of specificity determining sites in cadherin cell adhesion molecules. Cell 61:147–155

Ogou SI, Yoshida-Noro C, Takeichi M (1983) Calcium-dependent cell-cell adhesion molecules common to hepatocytes and teratocarcinoma stem cells. J Cell Biol 97:944–948

Overduin M, Harvey T, Bagby S, Tong K, Yau P, Takeichi M, Ikura M (1995) Solution structure of the epithelial cadherin domain responsible for selective cell adhesion. Science 267:386–389

Parisini E, Higgins JM, Liu JH, Brenner MB, Wang JH (2007) The Crystal Structure of Human E-cadherin Domains 1 and 2, and Comparison with other Cadherins in the Context of Adhesion Mechanism. J Mol Biol 373:401–411

Patel SD, Chen CP, Bahna F, Honig B, Shapiro L (2003) Cadherin-mediated cell-cell adhesion: sticking together as a family. Curr Opin Struct Biol 13:690–698

Patel SD, Ciatto C, Chen CP, Bahna F, Rajebhosale M, Arkus N, Schieren I, Jessell TM, Honig B, Price SR et al (2006) Type II cadherin ectodomain structures: implications for classical cadherin specificity. Cell 124:1255–1268

Posy S, Shapiro L, Honig B (2008a) Sequence and structural determinants of strand swapping in cadherin domains: do all cadherins bind through the same adhesive interface? J Mol Biol 378:954–968

Posy S, Shapiro L, Honig B (2008b) Sequence and structural determinants of strand swapping in cadherin domains: do all cadherins bind through the same adhesive interface? J Mol Biol 378:952–966

Price SR, De Marco Garcia NV, Ranscht B, Jessell TM (2002) Regulation of motor neuron pool sorting by differential expression of type II cadherins. Cell 109:205–216

Schreiner D, Weiner JA (2010) Combinatorial homophilic interaction between gamma-protocadherin multimers greatly expands the molecular diversity of cell adhesion. Proc Natl Acad Sci U S A 107:14893–14898

Seiler C, Finger-Baier KC, Rinner O, Makhankov YV, Schwarz H, Neuhauss SC, Nicolson T (2005) Duplicated genes with split functions: independent roles of protocadherin15 orthologues in zebrafish hearing and vision. Development 132:615–623
Shan W, Tanaka H, Phillips G, Arndt K, Yoshida M, Colman D, Shapiro L (2000) Functional cis-heterodimers of N- and R-cadherins. J Cell Biol 148:579–590
Shan W, Yagita Y, Wang Z, Koch A, Svenningsen A, Gruzglin E, Pedraza L, Colman D (2004) The minimal essential unit for cadherin-mediated intercellular adhesion comprises extracellular domains 1 and 2. J Biol Chem 279:55914–55923
Shapiro L, Weis WI (2009) Structure and biochemistry of cadherins and catenins. Cold Spring Harb Perspect Biol 1:a003053
Shapiro L, Fannon AM, Kwong PD, Thompson A, Lehmann MS, Grubel G, Legrand JF, Als-Nielsen J, Colman DR, Hendrickson WA (1995a) Structural basis of cell-cell adhesion by cadherins. Nature 374:327–337
Shapiro L, Kwong PD, Fannon AM, Colman DR, Hendrickson WA (1995b) Considerations on the folding topology and evolutionary origin of cadherin domains. Proc Natl Acad Sci U S A 92:6793–6797
Shimoyama Y, Takeda H, Yoshihara S, Kitajima M, Hirohashi S (1999) Biochemical characterization and functional analysis of two type II classic cadherins, cadherin-6 and -14, and comparison with E-cadherin. J Biol Chem 274:11987–11994
Shimoyama Y, Tsujimoto G, Kitajima M, Natori M (2000) Identification of three human type-II classic cadherins and frequent heterophilic interactions between different subclasses of type-II classic cadherins. Biochem J 349:159–167
Sivasankar S, Zhang Y, Nelson WJ, Chu S (2009) Characterizing the initial encounter complex in cadherin adhesion. Structure 17:1075–1081
Sotomayor M, Weihofen WA, Gaudet R, Corey DP (2010) Structural determinants of cadherin-23 function in hearing and deafness. Neuron 66:85–100
Sotomayor M, Weihofen WA, Gaudet R, Corey DP (2012) Structure of a force-conveying cadherin bond essential for inner-ear mechanotransduction. Nature 492:128–132
Takeichi M (1990) Cadherins: a molecular family important in selective cell-cell adhesion. Annu Rev Biochem 59:237–252
Takeichi M (1991) Cadherin cell adhesion receptors as a morphogenetic regulator. Science 251:1451–1455
Takeichi M (1995) Morphogenetic roles of classic cadherins. Curr Opin Cell Biol 7:619–627
Tanoue T, Takeichi M (2005) New insights into Fat cadherins. J Cell Sci 118:2347–2353
Taveau JC, Dubois M, Le Bihan O, Trepout S, Almagro S, Hewat E, Durmort C, Heyraud S, Gulino-Debrac D, Lambert O (2008) Structure of artificial and natural VE-cadherin-based adherens junctions. Biochem Soc Trans 36:189–193
Thomason HA, Scothern A, McHarg S, Garrod DR (2010) Desmosomes: adhesive strength and signalling in health and disease. Biochem J 429:419–433
Thu CA, Chen WV, Rubinstein R, Chevee M, Wolcott HN, Felsovalyi KO, Tapia JC, Shapiro L, Honig B, Maniatis T (2014) Single-cell identity generated by combinatorial homophilic interactions between alpha, beta, and gamma protocadherins. Cell 158:1045–1059
Troyanovsky RB, Sokolov E, Troyanovsky SM (2003) Adhesive and lateral E-cadherin dimers are mediated by the same interface. Mol Cell Biol 23:7965–7972
Tsukasaki Y, Miyazaki N, Matsumoto A, Nagae S, Yonemura S, Tanoue T, Iwasaki K, Takeichi M (2014) Giant cadherins Fat and Dachsous self-bend to organize properly spaced intercellular junctions. Proc Natl Acad Sci U S A 111:16011–16016
Vendome J, Posy S, Jin X, Bahna F, Ahlsen G, Shapiro L, Honig B (2011) Molecular design principles underlying beta-strand swapping in the adhesive dimerization of cadherins. Nat Struct Mol Biol 18:693–700
Vunnam N, Pedigo S (2011a) Calcium-induced strain in the monomer promotes dimerization in neural cadherin. Biochemistry 50:8437–8444

Vunnam N, Pedigo S (2011b) Prolines in betaA-sheet of neural cadherin act as a switch to control the dynamics of the equilibrium between monomer and dimer. Biochemistry 50:6959–6965

Wu Y, Jin X, Harrison O, Shapiro L, Honig BH, Ben-Shaul A (2010) Cooperativity between trans and cis interactions in cadherin-mediated junction formation. Proc Natl Acad Sci U S A 107:17592–17597

Wu Y, Vendome J, Shapiro L, Ben-Shaul A, Honig B (2011) Transforming binding affinities from three dimensions to two with application to cadherin clustering. Nature 475:510–513

Yagi T (2012) Molecular codes for neuronal individuality and cell assembly in the brain. Front Mol Neurosci 5:45

Yap AS, Brieher WM, Gumbiner BM (1997) Molecular and functional analysis of cadherin-based adherens junctions. Annu Rev Cell Dev Biol 13:119–146

Yonemura S (2011) Cadherin-actin interactions at adherens junctions. Curr Opin Cell Biol 23:515–522

Chapter 5
Regulation of Cadherin–Catenin Biology by Mechanical Force and Phosphorylation

Jiongyi Tan, Blair W. Benham-Pyle, William I. Weis, and W. James Nelson

Abstract In the adherens junction (AJ), cadherin and catenin proteins form a cell–cell adhesion complex that is indispensable for tissue morphogenesis and homeostasis. The complex mechanically couples neighboring cells through intercellular binding by cadherins, and actin binding and regulation by the cytoplasmic catenins. In addition, the cadherin–catenin complex participates in signaling pathways that direct cellular organization, proliferation, and motility. Some of these signaling pathways can be regulated by mechanical stimulation or posttranslational modification of the components of the AJ. In light of these findings, we discuss our current understanding of how AJ signaling and mechanical functions are regulated by phosphorylation and force, and speculate on the mechanisms underlying the coordination between these two types of modifications.

Keywords Cadherin • Catenin • Force • Tension • Phosphorylation • Proliferation • Actin • Kinase • Phosphatase

---

J. Tan and B.W. Benham-Pyle contributed equally with all other contributors.

J. Tan
Biophysics Program, Stanford University, Stanford, CA 94305, USA

B.W. Benham-Pyle
Cancer Biology Program, Stanford University, Stanford, CA 94305, USA

W.I. Weis
Biophysics Program, Stanford University, Stanford, CA 94305, USA

Cancer Biology Program, Stanford University, Stanford, CA 94305, USA

Department of Structural Biology, Stanford University, Stanford, CA 94305, USA

Department of Molecular and Cell Physiology, Stanford University, Stanford, CA 94305, USA

W.J. Nelson (✉)
Cancer Biology Program, Stanford University, Stanford, CA 94305, USA

Department of Molecular and Cell Physiology, Stanford University, Stanford, CA 94305, USA

Department of Biology, Stanford University, Stanford, CA 94305, USA
e-mail: wjnelson@stanford.edu

S.T. Suzuki, S. Hirano (eds.), *The Cadherin Superfamily*,
DOI 10.1007/978-4-431-56033-3_5

5.1 Cadherin Extracellular Domain Interactions

The adherens junction (AJ) contains classical cadherins, which are single-pass transmembrane proteins with five extracellular cadherin (EC) repeat domains that form a rigid curved structure stabilized by Ca^{++} (Shapiro and Weis 2009; Pokutta et al. 1994). Cell–cell adhesion is established through trans binding between the N-terminal EC1 domain of cadherins on opposing cells, and X-ray crystal structures have revealed two kinds of interfaces between these interacting EC1 domains (Manibog et al. 2014; Rakshit et al. 2012; Brasch et al. 2012; Harrison et al. 2010). In the first, the N-terminal β-strands of each domain exchange to form part of a β-sheet in the partner molecule (strand-swap dimer). The second interface involves association of the base of EC1 and the Ca^{++}-binding site between it and EC2 to form an X-dimer. Kinetic and equilibrium measurements, as well as atomic force microscopy assays and steered molecular dynamics simulations, indicate that the X-dimer is an intermediate in the formation of the more stable strand-swap dimer (Manibog et al. 2014; Rakshit et al. 2012; Brasch et al. 2012; Harrison et al. 2010). The strand-swap dimer is formed by molecular interactions very similar to those in the unbound monomer, and involves the kinetically unfavorable refolding of the interacting EC1s to accommodate the partner β-strand. Thus, the X-dimer may be a low energy "encounter complex" intermediate that overcomes the kinetic barrier to the strand swap (Fig. 5.1).

The extracellular region of E-cadherins of the same cell can form *cis* interactions, which appear to contribute to the stability of cell–cell contacts. The existence of cis clusters has been inferred from crystal structures (Harrison et al. 2011), fusion constructs (Pertz et al. 1999), and chemical cross-linking (Takeda et al. 1999), but the interactions are apparently too weak to be detected in solution or in single-molecule assays, suggesting that rates of association and, thus, binding probabilities are low (Zhang et al. 2009). Combined atomic force microscopy and FRET measurements found that even though two cadherin extracellular domains do not bind in cis as single molecules, their proximity increases the probability of establishing a trans interaction (Zhang et al. 2009). Mutational disruption of E-cadherin cis interactions inferred from crystal structures prevented recruitment of endogenous E-cadherin to cell–cell junctions, indicating that cis interactions are required for AJ maturation (Harrison et al. 2011). Moreover, disruption of either trans or cis E-cadherin interactions by site-specific mutagenesis demonstrated that trans interactions in turn stabilize *cis*-mediated clusters of E-cadherin lacking the cytoplasmic domain, and that anchoring E-cadherin to the actin cytoskeleton guides the assembly of these clusters (Hong et al. 2013). Together, these studies indicate that trans and cis cadherin binding may cooperate during formation of cell–cell contacts.

Mechanical force may also have a role in stabilizing the cadherin adhesive interaction (Fig. 5.1). Notably, the two *trans*-dimer configurations have different unbinding kinetics in response to applied tension (that is, an opposing mechanical force): the X-dimer forms a catch bond, whose lifetime increases with tension,

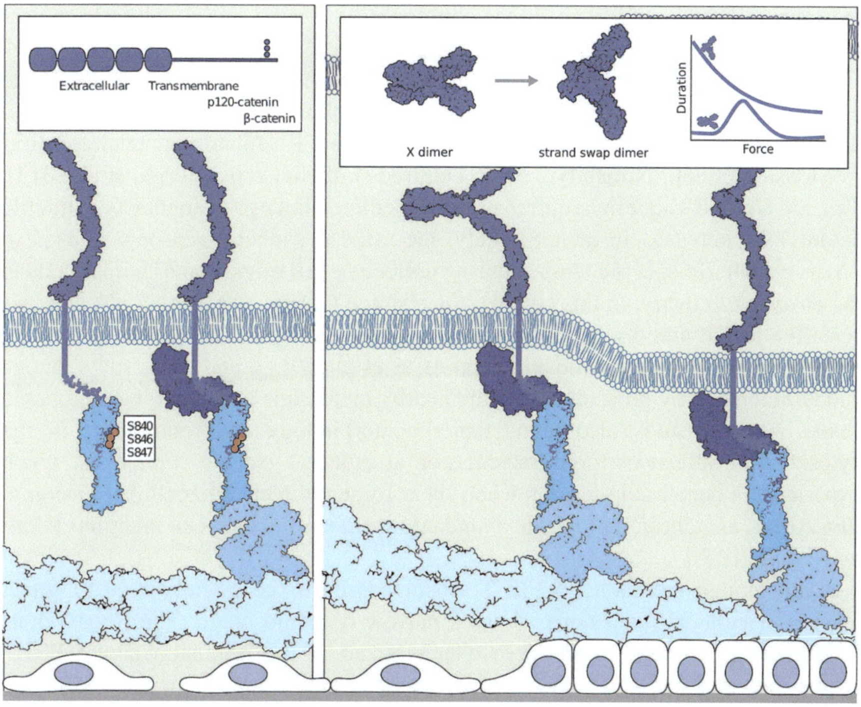

Fig. 5.1 Regulation of E-cadherin interactions by phosphorylation and force. Cadherin-mediated adhesion regulates the maturation of initial cell–cell recognition to loosely adherent cell clusters, to compacted groups of cells in colonies. E-cadherin is constitutively phosphorylated at S840, S846, and S847, facilitating binding to β-catenin and shuttling to the plasma membrane. E-cadherin is under constitutive tension after being incorporated into the plasma membrane and association with the actin cytoskeleton. E-cadherin trans X-dimer bonds are stabilized by force and may precede formation of stable strand-swap dimers, E-cadherin cis interactions, and the mature Adherens junction

whereas the strand-swap dimer behaves as a slip bond, whose duration decreases monotonically with respect to applied tension (Rakshit et al. 2012). It is unclear if cadherin catch bonds have a significant role in vivo. Formation of E-cadherin strand-swap dimers does not seem to require tension. In vitro spectroscopy experiments indicate that most extracellular dimers can form strand-swap slip bonds after an unloaded (no tension) contact time of 3 s (Rakshit et al. 2012). Moreover, mutations that compromise the affinity of the X-dimer slow, but still permit, the formation of the strand-swap dimer (Harrison et al. 2010). The X-dimer bond is most stable at ~30 pN, a level that is unlikely to be reached by a single myosin motor (Norstrom et al. 2010) coupled to the cadherin–catenin/actin complex. It is possible that catch bond behavior enables lower levels of tension to extend the lifetime of the X-dimer bond and thereby increase the probability of transition into the more robust strand-swap dimer conformation during initial cell–cell contact formation.

Cadherins are under tension in mature cell–cell contacts. A Forster resonance energy transfer (FRET)-based tension sensor (Grashoff et al. 2010) introduced into the cytoplasmic domain of E-cadherin indicated that E-cadherin is under constitutive tension of approximately 2 pN in cultured epithelial cells (Borghi et al. 2012). Tension along E-cadherin required catenin-mediated linkage to an intact contractile actomyosin network. In another study, the same E-cadherin sensor was used to observe cadherin-specific tension during collective cell migration of border cells in the *Drosophila* ovary. In this context, the average tension was also ~2 pN, and was sensitive to the activity of the small Rho family GTPase Rac. Rac regulates the nucleation of branched actin filaments (Cai et al. 2014), and these may change cadherin tension by protruding into the nearby membrane and changing membrane shape. Another study also found that the morphology and contractility of the cytoskeleton influences force transmission at cadherin–catenin complexes, which experience a decrease in tension when shear force redirects intercellular tension to PECAM-1, an adhesion molecule abundant in endothelial cell–cell junctions (Conway et al. 2013).

Even though the cadherin FRET sensor has been successfully used to detect tension at cell–cell adhesions, it has a narrow dynamic range. The force versus FRET efficiency calibration curve characterized in the original vinculin FRET sensor showed that FRET indices at forces greater than 7.5 pN are indistinguishable from the background signal (Grashoff et al. 2010). Due to this limitation, the cadherin force sensor cannot be used to test if intercellular cadherin bonds in cells are ever subject to 30-pN forces, which stabilize the bonds in the X-dimer conformation. Since the inception of the vinculin tension sensor, several FRET-based genetically encoded and synthetic tension sensors have been developed (Cost et al. 2015). Unfortunately, these sensors are subject to their own unique limitations, and further techniques will need to be developed to chart a comprehensive map of forces at cell–cell junctions.

At the cellular scale, forces at cell–cell junctions have been inferred using traction force microscopy based on the principle of mechanical equilibrium (Maruthamuthu et al. 2011; Ng et al. 2014; Sim et al. 2015). In these experiments, cells are plated on a compliant substrate functionalized with extracellular matrix (ECM), whose deformation can be used to calculate stresses at the cell–substrate interface. Cells typically do not move substantially during the timescale of substrate deformation, so the cells are assumed to be under mechanical equilibrium in which cell–cell forces balance cell–ECM forces. Using this strategy, cell–cell junctions were found to be subject to hundreds of nN of tension. However, this tension is not confined to the AJ, as epithelial cells also form intermediate filament-bound desmosomes and an actin filament-bound tight junction. A study combining traction force microscopy and the cadherin FRET tension sensor found that average tension along cadherin molecules was constant in spite of significant changes in cell–ECM and cell–cell forces (Sim et al. 2015). Moreover, increased forces at cell–cell contacts did not result in changes of total cadherin levels at cell–cell junctions. Instead, cadherin was found to be locally enriched at the edges of the contacts as cell–cell forces increased. These findings suggest that cells may maintain molecular-level mechanical homeostasis at the AJ by modulating the localization of cadherin-based complexes.

5.2 Cadherin Intracellular Domain Interactions: p120-Catenin

Interactions between cadherin, catenin proteins, and the actin cytoskeleton are tightly regulated to coordinate AJ assembly and disassembly in response to external or internal cues. In epithelial tissues, the cadherin–catenin complex is composed of E-cadherin and its associated cytoplasmic catenins: p120-catenin, β-catenin, and αE-catenin. β-Catenin binds the cytoplasmic domain of E-cadherin upon synthesis in the endoplasmic reticulum. After delivery of the heterodimer to the plasma membrane, the complex is stabilized by intercellular trans E-cadherin interactions (see above) and p120-catenin binding to the cadherin juxtamembrane domain. Finally, αE-catenin mechanically integrates the cytoskeletons of adjacent cells by binding to β-catenin and linking actin filaments to the complex (Ozawa et al. 1990; Hinck et al. 1994).

p120-Catenin regulates the rate of cadherin endocytosis, and the dynamics of the actin cytoskeleton through interactions with Rho family GTPases (Fig. 5.2, left). p120-Catenin was first identified as a Src kinase substrate in a study designed to screen for genes related to transformation (Reynolds et al. 1994), but subsequent studies demonstrated that in nontransformed cells, direct binding between p120-catenin and cadherin stabilizes cadherin at the plasma membrane at the onset of strong cell–cell adhesion (Thoreson et al. 2000; Yap et al. 1998; Davis and Reynolds 2006). Moreover, internalization assays demonstrated that p120-catenin binding prevents cadherin endocytosis by blocking binding of Hakai, an E3 ligase that ubiquitylates the E-cadherin cytoplasmic domain, targeting the complex to the endocytic machinery (Hartsock and Nelson 2012; Xiao et al. 2005). Src phosphorylation of p120-catenin at Y217 and Y228 increases p120-catenin affinity for E-cadherin and RhoA GTPase (Roura et al. 1999). Similarly, Fyn/Fer kinases phosphorylate p120-catenin and increase its affinity for E-cadherin (Rosato et al. 1998). However, Src and Fer/Fyn kinases also phosphorylate β-catenin Y654 and Y142, respectively, leading to dissociation from E-cadherin and αE-catenin and subsequent deterioration of cell–cell adhesion (Roura et al. 1999; Piedra et al. 2003). Together, these results raise the question: why is the affinity of p120-catenin to cadherin increased by kinases that also destabilize cadherin's interactions with the other catenins?

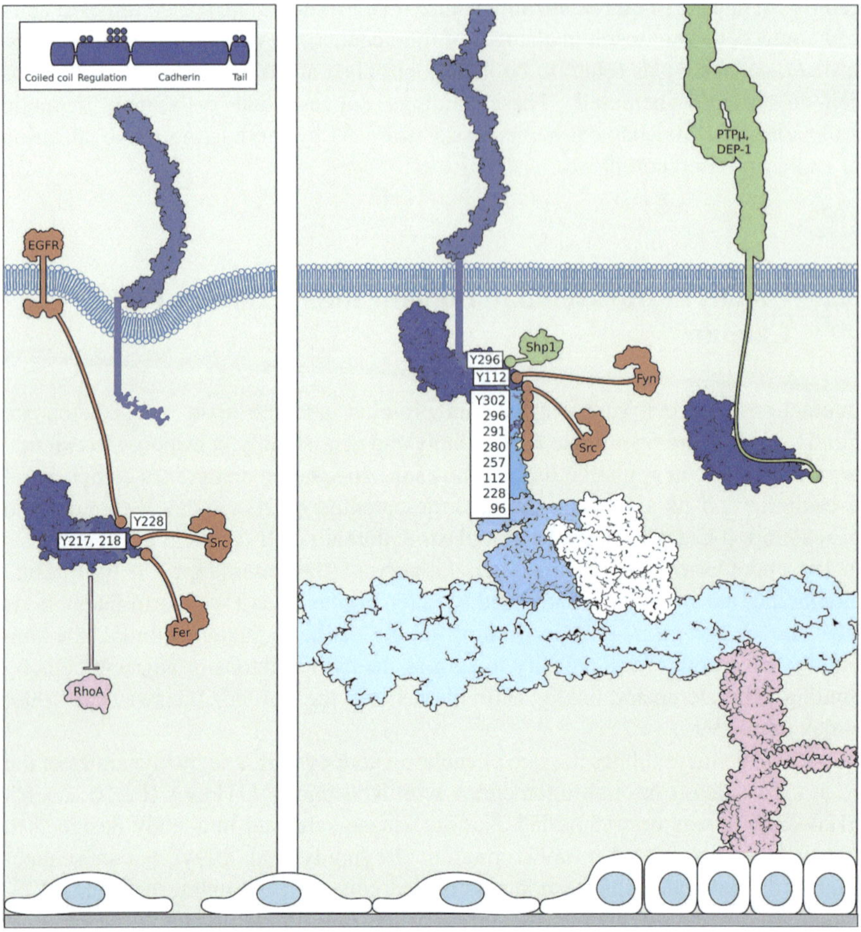

Fig. 5.2 p120-Catenin-mediated regulation of actin dynamics, E-cadherin endocytosis, and phosphatase activity at the AJ. Cytosolic p120-catenin tyrosine-phosphorylated by growth factor cascades and/or Src and Fer kinases (*orange*) downregulates RhoA GTPase activity. Without p120-catenin binding, E-cadherin is targeted for endocytosis. Upon recruitment to the AJ at high cell densities, p120-catenin and its associated kinases can activate phosphatases (*green*) that counteract tyrosine-phosphorylation of β-catenin and αE-catenin, stabilizing the cadherin–catenin complex at the AJ

Phosphorylation-mediated disruption of the cadherin–catenin complex may be opposed by p120-catenin (Fig. 5.2, right). In addition to binding to E-cadherin, p120-catenin associates with several tyrosine phosphatases, including the receptor-type tyrosine phosphatases PTPμ (Zondag et al. 2000) and DEP-1 (Holsinger et al. 2002), and the cytosolic tyrosine phosphatase SHP-1 (Reynolds et al. 1994). The receptor-type tyrosine phosphatases are upregulated at high cell density (Ostman et al. 1994) and could counteract Src and Fer/Fyn phosphorylation of the cadherin–catenin complex during cell–cell junction maturation. In addition, p120-catenin recruits Fer to cell–cell adhesions, promoting activation of PTP1B, a cytosolic tyrosine phosphatase that counteracts phosphorylation of β-catenin Y142 and Y654 (El Sayegh et al. 2005; Xu et al. 2004). Thus, p120-catenin may play a critical role in maintaining the balance of kinase and phosphatase activity in the context of cell–cell adhesion.

p120-catenin also regulates actin dynamics through its interactions with Rho family GTPases (Grosheva et al. 2001; Noren et al. 2000; Anastasiadis et al. 2000). Actin dynamics regulate the architecture of the cytoskeleton, thus p120-catenin likely affects how the cadherin–catenin complex transmits mechanical stimuli. In one study, for example, overexpression of p120-catenin inhibited RhoA activity, resulting in formation of branch-like actin protrusions and destabilization of stress fibers (Reynolds et al. 1996). These findings indicate that p120-catenin, when dissociated from E-cadherin, induces a more migratory phenotype driven by branch-like actin protrusions (Noren et al. 2000; Reynolds et al. 1996). This phenotype is evident in nascent cell–cell contacts (Toret et al. 2014; Yamada and Nelson 2007), but is suppressed by RhoA activity as cell–cell contacts expand (Yamada and Nelson 2007). In addition, studies show that RhoA activity can be mechanically activated in a variety of cell types (Zhao et al. 2007a; Abiko et al. 2015) and is correlated with local levels of stress (Reffay et al. 2014). Whether p120-catenin plays a role in this pathway remains to be determined. p120-Catenin can modulate GTPase activity by acting as a guanine nucleotide dissociation inhibitor (Anastasiadis et al. 2000) or associating with guanine nucleotide exchange factors such as p190RhoGAP (Wildenberg et al. 2006) and Vav2 (Fukuyama et al. 2006). How these interactions are affected by mechanical perturbation of cell–cell contacts has not been investigated.

5.3 Cadherin Intracellular Interactions: β-Catenin

β-Catenin, an armadillo repeat protein (Huber et al. 1997a), binds to the cytoplasmic domain of E-cadherin distal to the juxtamembrane domain and the p120-catenin binding site. In turn, β-catenin binds the actin binding protein αE-catenin (Huber et al. 1997b). Binding of β-catenin confers structure to the cytoplasmic domain of E-cadherin, which protects cadherin from proteolysis (Huber et al. 2001) and reduces the turnover rate of the E-cadherin/β-catenin heterodimer at the plasma membrane. Calorimetry and mutagenesis studies indicate that the affinity of E-cadherin/β-catenin is increased when S840, S846, and S847 in the E-cadherin cytoplasmic domain are phosphorylated (Lickert et al. 2000; Serres et al. 2000; Choi et al. 2006; Fig. 5.1). These phosphorylation events occur constitutively (McEwen et al. 2014) and may stabilize the cadherin–catenin complex.

There are many posttranslational modifications that regulate the turnover of β-catenin in the cadherin–catenin complex (Fig. 5.3, top left). Phosphorylation of Y654 by Src or Abl, both cytoplasmic kinases, disrupts a hydrogen bond between the β-catenin Y654 phenolic hydroxyl group and a cadherin aspartate residue

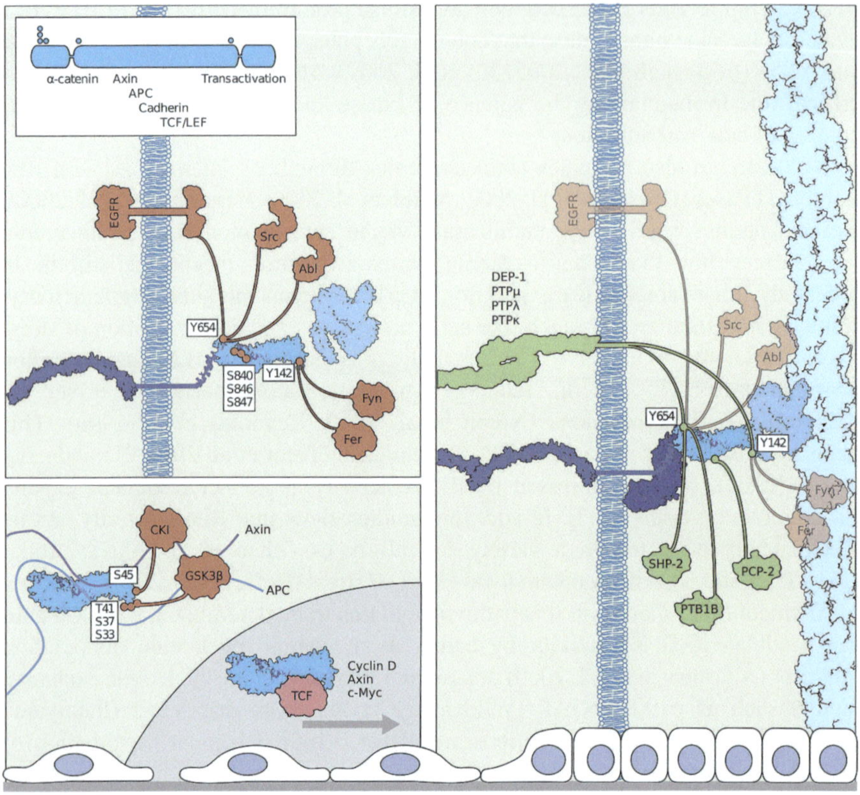

Fig. 5.3 Regulation of β-catenin localization, stability, and transcriptional activity by cell density and the balance of tyrosine kinase and phosphatase activities. Interactions of β-catenin with E-cadherin and αE-catenin are negatively regulated by phosphorylation of β-catenin by receptor and cytoplasmic tyrosine kinases EGFR, Src, Abl, Fer, and Fyn (*red/orange components*), which phosphorylate Y654 and Y142 residues in β-catenin. In contrast, β-catenin interactions with E-cadherin and αE-catenin are positively regulated by serine/threonine phosphorylation of E-cadherin (S840, S846, and S847) and β-catenin dephosphorylation (Y654 and Y142) by protein tyrosine phosphatases that bind p120 and β-catenin (*green components*). Degradation of cytoplasmic β-catenin is driven by phosphorylation by CKI and GSK3β and scaffolding by the tumor suppressors Axin and APC. The localization and phosphorylation state of β-catenin are associated with changes in cell density and affect cell–cell adhesion, cell migration, and the level of transcriptionally active β-catenin

(Huber and Weis 2001), resulting in at least a fifteen-fold reduction in affinity (Roura et al. 1999; Catimel et al. 2006). Another means of perturbing this interaction is via Src-mediated phosphorylation of N-cadherin Y860, as found in endothelial cells (Qi et al. 2005). Although Src disrupts E-cadherin/β-catenin heterodimerization, the p120-catenin-associated cytoplasmic kinases Fer and Fyn (Kim and Wong 1995) disrupt β-catenin/αE-catenin interactions through tyrosine-phosphorylation of β-catenin (Rosato et al. 1998) Y142 (Piedra et al. 2003), which is located in the β-catenin/αE-catenin binding interface (Pokutta and Weis 2000).

Fer and Fyn are examples of kinases that regulate cadherin–catenin complex stability downstream of signaling pathways mediated by receptor tyrosine kinases (RTKs). One of the most studied RTKs known to regulate the cadherin–catenin complex is the epidermal growth factor (EGF) receptor. EGF receptor activation induces dissociation of cell aggregates, cell rounding, and membrane ruffling (Fujii et al. 1996). The EGF receptor can bind directly to β-catenin (Hoschuetzky et al. 1994) and phosphorylate Y654 (Hazan and Norton 1998), weakening β-catenin affinity for E-cadherin. Without the cadherin–β-catenin interaction, αE-catenin cannot link the actin cytoskeletons of neighboring cells, resulting in reduced cell–cell adhesion and transition to a migratory phenotype. There is also evidence for the intersection of Src kinase and EGFR activation pathways, as inhibition of Src kinase blocks EGF-stimulated DNA synthesis and subsequent proliferation (Bromann et al. 2004). Activation of MET tyrosine kinase, another RTK, by hepatocyte growth factor (HGF) also results in β-catenin phosphorylation and subsequent nuclear accumulation (Monga et al. 2002).

When not associated with E-cadherin, β-catenin can participate in Wnt-dependent and -independent proliferation pathways (Fig. 5.3, bottom left). These require the translocation of β-catenin to the nucleus (McCrea et al. 1991; Nelson and Nusse 2004), where it associates with TCF/LEF transcription factors and induces specific gene transcription (He et al. 1998; Korinek et al. 1997; Morin et al. 1997). The amount of cytoplasmic β-catenin and thereby its transcriptional function can be regulated by a proteasome-targeted destruction complex (Aberle et al. 1997) comprising the tumor suppressors Adenomatous Polyposis Coli (APC) (Rubinfeld et al. 1993; Su et al. 1993) and axin (Zeng et al. 1997), the serine and threonine kinases GSK-3 (Dominguez et al. 1995; He et al. 1995; Kimelman and Pierce 1996) and CK1 (Liu et al. 2002; Amit et al. 2002), protein phosphatase 2A (Seeling et al. 1999), and the E3-ligase β-TrCP (Winston et al. 1999). Axin scaffolds the phosphorylation of β-catenin S45 by CKI (Amit et al. 2002; Sakanaka 2002), and then T41, S37, and S33 by GSK3 (Liu et al. 2002; Sadot et al. 2002); phosphorylation of S33 and S37 leads to ubiquitylation by β-TrCP and destruction in the proteasome. Canonical Wnt signaling promotes cell proliferation by inhibiting the activity of the β-catenin destruction complex, and these pathways are dysfunctional in many cancers (Fodde and Brabletz 2007).

Mechanical strain activates the transcriptional function of β-catenin independently of the Wnt signaling pathway during gastrulation in *Danio rerio* and *Drosophila melanogaster* (Brunet et al. 2013; Desprat et al. 2008). During gastrulation, the blastula, a spherical sheet of cells, folds inwards to create the gastrula, a structure comprising the three germ layers that give rise to specific organs during embryonic development. Folding of the blastula requires actomyosin contractility and correlates with Src-mediated phosphorylation of β-catenin Y654. In the absence of endogenous actomyosin contractility, β-catenin phosphorylation could be rescued by exogenous compression of the blastula using magnetic beads (Brunet et al. 2013). Mechanical strain across a contact-inhibited epithelial monolayer in vitro also results in increased β-catenin nuclear signaling, and cell-cycle progression (Benham-Pyle et al. 2015). This increase in signaling requires cadherin-mediated cell–cell adhesion, as expression of a truncated E-cadherin lacking the extracellular domain blocked activation of β-catenin and cell-cycle progression following mechanical strain.

At present, it is unclear how mechanical strain is transduced to Src or β-catenin activation. Because Src phosphorylation was rescued using nonspecific magnetic compression of tissues (Desprat et al. 2008), Src may be subject to mechanical regulation independently of the cadherin–catenin complex at sites of cell–cell adhesion. Abl kinase, which affects cell–cell adhesion similarly to Src, possesses an actin binding domain (Van Etten et al. 1994), and myristoylation anchors the kinase to the plasma membrane (Hantschel et al. 2003). Interestingly, combined actin binding and myristoylation inhibit Abl activity (Hantschel et al. 2003; Woodring et al. 2001), which is lower in stable cell–cell contacts (Bays et al. 2014). These results suggest that mechanical stimuli could activate Abl at cell–cell contacts by dissociating it from the actin cytoskeleton.

Wnt-independent nuclear localization of β-catenin also depends on cell density (Dietrich et al. 2002). As cell density increases, β-catenin shifts from a nuclear pool to a junctional pool, and confluent cells stop proliferating due to contact inhibition. Cell density changes are accompanied by dramatic changes in cell morphology, and these changes may affect force generation and transmission at the AJ. Thus, it is possible that changes in mechanical strain and cell density modulate β-catenin junctional stability and transcriptional activity in similar ways (Brunet et al. 2013; Desprat et al. 2008; Benham-Pyle et al. 2015).

Phosphorylation of β-catenin and hence its transcriptional activity can be inhibited by several protein tyrosine phosphatases (PTPs) at cell–cell junctions (Fig. 5.3, right). PTPκ binds β-catenin in vitro and dephosphorylates tyrosine-phosphorylated β-catenin from cell lysates (Fuchs et al. 1996), and PTPλ similarly associates with β-catenin (Cheng et al. 1997). Several protein tyrosine phosphatases such as the cytosolic PTP-PCP2 also dephosphorylate β-catenin that had been phosphorylated downstream of growth factor signaling pathways (Yan et al. 2002). In high-density cultures, phosphatases localize to cell–cell junctions (Rijksen et al. 1993) and cadherin–catenin complexes may be directly involved in their recruitment (Piedra et al. 2003).

5.4 Cadherin–Catenin Intracellular Interactions: αE-Catenin

αE-catenin, which binds to cadherin through β-catenin, anchors the AJ to the actin cytoskeleton directly or indirectly through different actin-binding partners. The amino terminus of αE-catenin comprises a β-catenin binding domain and, in the mammalian homologue, an overlapping homodimerization domain (Pokutta and Weis 2000). The N-terminus is followed by a modulation domain that binds several actin-binding proteins including vinculin (Hazan et al. 1997; Choi et al. 2012),

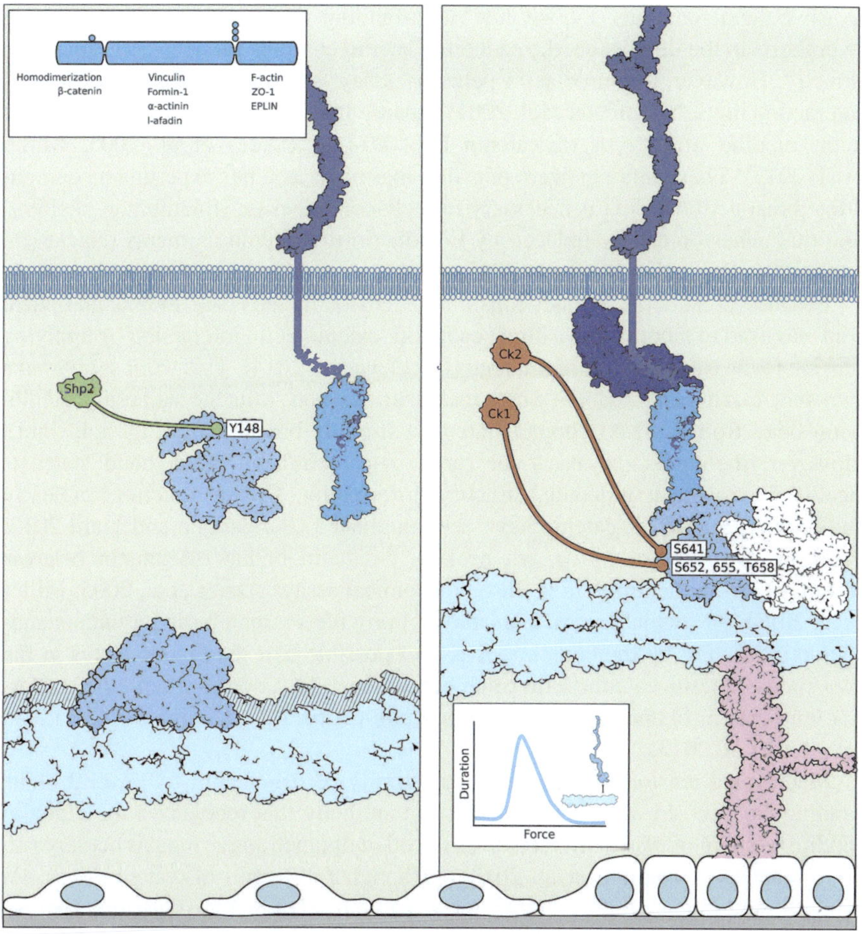

Fig. 5.4 Regulation of cytosolic and junctional αE-catenin. Cytosolic αE-catenin can be dephosphorylated by Shp2 phosphatase (*green*), and can also form homodimers that have a higher affinity for actin filaments and inhibit Arp2/3-mediated branching. Junctional αE-catenin is subject to phosphorylation by CKI/II (*orange*), and acto-myosin generated tension which increases the actin binding affinity of the cadherin–catenin complex by modulating transitions between weakly and strongly bound catch bond states. Under tension, αE-catenin acquires an open conformational state associated with vinculin recruitment (*dark purple*), and possibly other actin binding proteins (see domain organization)

l-afadin (Pokutta et al. 2002), formin-1 (Kobielak et al. 2004), and α-actinin (Knudsen et al. 1995); the C-terminal domain also binds ZO-1 (Itoh et al. 1997) and EPLIN (Abe and Takeichi 2008). Thus, the cadherin–catenin complex can bind the actin cytoskeleton and regulate its nucleation (Kobielak et al. 2004; Tang and Brieher 2012) and morphology (Abe and Takeichi 2008) through multiple actin binding partners. The C-terminal domain of αE-catenin binds directly to actin filaments (Pokutta et al. 2002; Rimm et al. 1995; Fig. 5.4, left).

αE-Catenin contains a bone fide actin-binding domain and a long-standing hypothesis in the field is that the cadherin–catenin complex binds to actin filaments directly. However, a simple actin pelleting assay was unable to reconstitute this interaction in vitro (Yamada et al. 2005) because binding to β-catenin decreases the actin binding affinity of αE-catenin by >20-fold (Drees et al. 2005; Miller et al. 2013). These findings were puzzling inasmuch as other experiments demonstrated that actin binding is necessary for cell–cell adhesion (Imamura et al. 1999) and that adhesion can be induced by E-cadherin-αE-catenin chimeras (Nagafuchi et al. 1994; Pacquelet and Rørth 2005).

Because E-cadherin is under constitutive tension in cells (see above), an optical trap was used to reconstitute a direct cadherin–catenin/actin interaction by applying tension to the αE-catenin/F-actin bond (Buckley et al. 2014). This work supported a two-state catch bond model in which increasing tension shifts the cadherin–catenin/actin bond from a weakly bound state to a strongly bound state (Fig. 5.4, right). However, the molecular basis for cadherin–catenin/actin catch bond states is unclear due to a lack of detailed structural information. Crystal structures of nearly full-length dimeric αE-catenin have been published (Rangarajan and Izard 2012; Rangarajan and Izard 2013), but because β-catenin–bound αE-catenin behaves differently from the dimer in in vitro biochemical assays (Drees et al. 2005; Miller et al. 2013), the available structures may not provide a strong basis for understanding actin binding by the complex. It is also possible that the kinetic states in the two-state cadherin–catenin/actin catch bond model are associated with the conformation of actin filaments, which change upon cooperative binding of αE-catenin (Hansen et al. 2013).

Mechanical tension may regulate the affinity of αE-catenin for several of its binding partners. In cell culture models, an antibody that recognizes the vinculin binding domain of αE-catenin localizes to cell–cell junctions as long as actomyosin is contractile (Yonemura et al. 2010). Although full-length αE-catenin does not bind full-length vinculin in solution, the vinculin head domain readily binds part of the modulation domain of αE-catenin, and the affinity decreases as flanking domains of αE-catenin are included (Choi et al. 2012). Significantly, stretching of αE-catenin using magnetic tweezers promotes vinculin head domain binding (Yao et al. 2014), but whether this force-mediated structural change is sufficient to recruit full-length vinculin to the cadherin–catenin complex at the AJ is unclear. Pulling on cadherin-coated magnetic beads attached to cells recruits full-length vinculin to cadherin-mediated attachment sites (le Duc et al. 2010), and this recruitment requires Src/Abl phosphorylation of vinculin Y822 in the head domain (Bays et al. 2014). Abl phosphorylates the vinculin head domain in vitro, but it may not phosphorylate full-length vinculin due to autoinhibitory interactions between the actin-binding domain and the rest of the molecule. Together, these data indicate that the actin-binding activity of vinculin at cell–cell junctions may be coactivated by force-induced conformational changes of αE-catenin and phosphorylation by Abl. It is possible that phosphorylation regulates αE-catenin interactions as well; for example, the linker that connects the αE-catenin modulation and actin binding domains is constitutively phosphorylated by CKI and CKII (Fig. 5.4, right).

However, these particular modifications do not seem to affect binding of actin (Drees et al. 2005) or vinculin and other actin binding partners (Drees et al. 2005; Escobar et al. 2015).

αE-catenin may also mediate crosstalk between the cadherin–catenin complex and other adhesion complexes at cell–cell junctions. A prominent example is the nectin family of Ig superfamily adhesion proteins (Takai et al. 2008), which can affect the spatial localization of cadherin–catenin complexes during assembly by recruiting them to nascent cell–cell junctions. The recruitment may occur through afadin, an actin-binding protein that can bind directly to nectins (Takai et al. 2008) and the cadherin–catenin complex through αE-catenin (Pokutta et al. 2002). Ponsin and vinculin may also mediate interactions between afadin and the cadherin–catenin complex (Tachibana et al. 2000; Mandai et al. 1999). However, ponsin does not bind afadin and vinculin simultaneously in vitro (Mandai et al. 1999), but vinculin coimmunoprecipitates with ponsin when αE-catenin is present (Peng et al. 2012), suggesting it may be necessary to reconstitute a ponsin/afadin/vinculin complex.

Recent work has shown that αE-catenin also regulates the Hippo pathway protein YAP1, implicating the AJ in another cell proliferation pathway. Initially discovered in *Drosophila*, the Hippo pathway is a serine/threonine kinase cascade comprising Hippo (Harvey et al. 2003), Warts (Xu et al. 1995), Salvador (Pantalacci et al. 2003), and Mats (Lai et al. 2005). To control organ size during development, the Hippo–Salvador complex activates the Warts–Mats complex, which phosphorylates *Drosophila* YAP1, deactivates YAP1 transcriptional activity, and excludes it from the nucleus (Dong et al. 2007; Oh and Irvine 2008; Zhao et al. 2007b). Recent studies indicate that αE-catenin acts as a suppressor of the transcriptional activity of YAP1 (Schlegelmilch et al. 2011; Silvis et al. 2011). This function of αE-catenin is cell-density dependent and requires an interaction with the scaffolding protein 14-3-3 to sequester YAP1 at the AJ and in the cytosol. Expression of a truncated E-cadherin lacking the extracellular domain disrupts YAP1 sequestration in the cytoplasm (Benham-Pyle et al. 2015), suggesting that trans interactions between E-cadherin and mechanical coupling between cells may be required for sequestration of YAP1 in the cytoplasm or interaction with the cadherin–catenin complex. As does β-catenin, YAP1 becomes localized to the nucleus and transcriptionally active upon mechanical strain of contact-inhibited epithelial cells, but the molecular mechanism of activation is unknown (Benham-Pyle et al. 2015). Because YAP1 activation is sensitive to the morphology and contractile state of the actin cytoskeleton (Dupont et al. 2011; Wada et al. 2011), it is possible that YAP1 is mechanically activated through interactions with cytosolic αE-catenin. Cytosolic αE-catenin forms homodimers that bind and bundle actin filaments in the absence of tension and inhibit Arp2/3-mediated actin polymerization (Drees et al. 2005; Benjamin et al. 2010). Finally, both YAP1 and αE-catenin have been linked to the β-catenin destruction complex (Brunet et al. 2013), indicating that YAP1 phosphorylation independent of the Hippo pathway may disrupt Yap1 interactions with the 14-3-3 scaffold.

5.5 Moving Forward

Several themes emerge from the large body of work seeking to understand how cadherin and catenin proteins are regulated by phosphorylation and mechanical force. Cell biology, biochemistry, and genetic data indicate that a balance of cell density-dependent phosphorylation and dephosphorylation events regulates cadherin-mediated adhesion. At low cell densities, high tyrosine kinase activity, some of which is downstream of growth factor signaling pathways, upregulates the motility and proliferation machinery necessary to develop a dense multicellular organization. As cell density increases, tyrosine phosphatase activity increases, perhaps to the point of counteracting kinase activity, resulting in the stabilization of the cadherin–catenin complex at cell–cell junctions while turning off motility and proliferation signals. Interestingly, many receptor protein tyrosine phosphatases possess extracellular domains similar to those found in cell adhesion molecules (Stoker 2005), and thus these phosphatases may be recruited and activated by cadherin-mediated adhesion via mechanisms similar to those reconstituted on lipid bilayers (Hui and Vale 2014; Greene et al. 2014; Lin et al. 2014).

Recent biophysical and bioengineering methods have uncovered evidence that cadherin and catenin biology is regulated mechanically. A salient finding is that cadherin and αE-catenin form catch bonds between *trans*-interacting E-cadherin extracellular domains (Manibog et al. 2014) and F-actin (Buckley et al. 2014), respectively. However, additional experiments are needed to determine whether E-cadherin and αE-catenin catch bonds contribute to signaling in a cellular environment. The rate of tension loaded in force spectroscopy experiments is much faster than that generated by molecular motors associated with the cytoskeleton in the cytoplasm (Finer et al. 1994). If this rate is too low, then bonds dissociate before experiencing levels of tension that slow down unbinding (Dudko et al. 2008). Thus, it is not clear if cadherin and αE-catenin "feel" sufficient force in vivo to display catch bond behavior. To date, the best evidence of a catch bond operating in physiological conditions comes from studies of neutrophils detaching from selectin-binding surfaces under shear flow (Schmidtke and Diamond 2000; Yago et al. 2004). Gathering additional evidence for this type of cadherin/cadherin or αE-catenin/F-actin bond in vivo will likely require a combination of FRET-based force measurements and single-molecule tracking.

It seems increasingly likely that mechanical force not only alters the structure and molecular composition of the AJ, but also contributes to signaling from the AJ to regulate growth, invasion, and cell division. Mechanical strain across contact-inhibited epithelial monolayers induces cell-cycle entry and DNA synthesis, which require trans interactions between neighboring cells (Benham-Pyle et al. 2015). Density-dependent mechanical properties regulate the exclusion of transcription factors (YAP1, β-catenin) from the nucleus. Moreover, mechanical perturbations of the AJ can result in numerous phosphorylation events, triggering remodeling and release of previously sequestered signaling molecules (Brunet et al. 2013; Desprat et al. 2008; Benham-Pyle et al. 2015). It remains unknown how mechanical force at the AJ triggers increased kinase activity or release of sequestered transcription factors, and this will be an important topic for future work.

The morphology of the actin cytoskeleton regulates how force is generated and transmitted at the AJ. Actin networks can adopt distinct architectures: a highly branched network that is nucleated downstream of Rac1 and Cdc42, and an unbranched contractile network downstream of RhoA (Ridley 2006). These types of networks have different mechanical properties. Branched networks can tolerate compressive forces better than linear networks because network-level forces dissipate at the nodes connecting actin branches, and the high spatial density of these nodes generates short branches that buckle at larger compressive forces (Pujol et al. 2012). Thus, a branched network is better suited for generating protrusive forces, such as those found at the leading edge of migrating cells. At the AJ, these protrusive forces may move the plasma membrane locally and associated cadherin–catenin complexes. In addition, these complexes could experience an increase in tension if they are anchored to actin filament bundles that do not move with respect to the branching network. In contrast to branched networks, contractile networks are comprised of actin filaments bundled by myosin motors or other actin bundling proteins such as α-actinin and cytosolic αE-catenin. In these networks, motors generate contractile forces, and the bundled filaments transmit tension without undergoing much deformation (strain) given the Young's modulus of individual actin filaments (~50 pN/nm; Kojima et al. 1994). Due to this mechanical resilience, a contractile actin network can efficiently induce mechanical strain on associated protein scaffolds (Claessens et al. 2006). In turn, the strain on these components can manifest as changes in conformation and dissociation rates. This myosin-dependent process drives morphogenetic changes such as planar cell intercalation, where AJs perpendicular to the axis of elongation disassemble to give rise to aligned AJs (Bertet et al. 2004).

The combination of phosphorylation and mechanical studies of the cadherin–catenin complex generate a model in which a stable E-cadherin/β-catenin/α-catenin complex is buttressed on either end by force-dependent interactions with E-cadherin molecules on neighboring cells and F-actin in the cytoplasm. The stability of E-cadherin/β-catenin/α-catenin interactions can then be tightly regulated by kinases and phosphatases to quickly dissociate the complex when needed, for example, in response to tissue wounding or other morphogenetic signals. It is likely that the combination of mechanical and biochemical modifications facilitates switches between different functions of cadherin and catenin proteins. As such, it will be important to address how mechanical forces contribute to phosphatase and kinase activities at the AJ, and how these modifications then contribute to the regulation of cell migration and growth.

Acknowledgments This work was supported by Predoctoral Fellowships from the NSF (JT, BB-P), a Stanford Bio-X Pre-doctoral Fellowship (JT), NSF EFRI Award (1136790) to WJN and WIW, and NIH GM35527 (WJN).

References

Abe K, Takeichi M (2008) EPLIN mediates linkage of the cadherin catenin complex to F-actin and stabilizes the circumferential actin belt. Proc Natl Acad Sci U S A 105:13–19

Aberle H, Bauer A, Stappert J, Kispert A, Kemler R (1997) beta-catenin is a target for the ubiquitin-proteasome pathway. EMBO J 16:3797–3804

Abiko H et al (2015) Rho guanine nucleotide exchange factors involved in cyclic-stretch-induced reorientation of vascular endothelial cells. J Cell Sci 128:1683–1695

Amit S et al (2002) Axin-mediated CKI phosphorylation of beta-catenin at Ser 45: a molecular switch for the Wnt pathway. Genes Dev 16:1066–1076

Anastasiadis PZ et al (2000) Inhibition of RhoA by p120 catenin. Nat Cell Biol 2:637–644

Bays JL et al (2014) Vinculin phosphorylation differentially regulates mechanotransduction at cell-cell and cell-matrix adhesions. J Cell Biol 205:251–263

Benham-Pyle BW, Pruitt BL, Nelson WJ (2015) Cell adhesion. Mechanical strain induces E-cadherin-dependent Yap1 and β-catenin activation to drive cell cycle entry. Science 348:1024–1027

Benjamin JM et al (2010) AlphaE-catenin regulates actin dynamics independently of cadherin-mediated cell-cell adhesion. J Cell Biol 189:339–352

Bertet C, Sulak L, Lecuit T (2004) Myosin-dependent junction remodelling controls planar cell intercalation and axis elongation. Nature 429:667–671

Borghi N et al (2012) E-cadherin is under constitutive actomyosin-generated tension that is increased at cell-cell contacts upon externally applied stretch. Proc Natl Acad Sci U S A 109:12568–12573

Brasch J, Harrison OJ, Honig B, Shapiro L (2012) Thinking outside the cell: how cadherins drive adhesion. Trends Cell Biol 22:299–310

Bromann PA, Korkaya H, Courtneidge SA (2004) The interplay between Src family kinases and receptor tyrosine kinases. Oncogene 23:7957–7968

Brunet T et al (2013) Evolutionary conservation of early mesoderm specification by mechanotransduction in Bilateria. Nat Commun 4:2821

Buckley CD et al (2014) Cell adhesion. The minimal cadherin-catenin complex binds to actin filaments under force. Science 346:1254211

Cai D et al (2014) Mechanical feedback through E-cadherin promotes direction sensing during collective cell migration. Cell 157:1146–1159

Catimel B et al (2006) In situ phosphorylation of immobilized receptors on biosensor surfaces: application to E-cadherin/beta-catenin interactions. Anal Biochem 357:277–288

Cheng J et al (1997) A novel protein-tyrosine phosphatase related to the homotypically adhering kappa and mu receptors. J Biol Chem 272:7264–7277

Choi HJ, Huber AH, Weis WI (2006) Thermodynamics of beta-catenin-ligand interactions: the roles of the N- and C-terminal tails in modulating binding affinity. J Biol Chem 281:1027–1038

Choi HJ et al (2012) alphaE-catenin is an autoinhibited molecule that coactivates vinculin. Proc Natl Acad Sci U S A 109:8576–8581

Claessens MM, Bathe M, Frey E, Bausch AR (2006) Actin-binding proteins sensitively mediate F-actin bundle stiffness. Nat Mater 5:748–753

Conway DE et al (2013) Fluid shear stress on endothelial cells modulates mechanical tension across VE-cadherin and PECAM-1. Curr Biol 23:1024–1030

Cost AL, Ringer P, Chrostek-Grashoff A, Grashoff C (2015) How to measure molecular forces in cells: a guide to evaluating genetically-encoded fret-based tension sensors. Cell Mol Bioeng 8:96–105

Davis MA, Reynolds AB (2006) Blocked acinar development, E-cadherin reduction, and intraepithelial neoplasia upon ablation of p120-catenin in the mouse salivary gland. Dev Cell 10:21–31

Desprat N, Supatto W, Pouille PA, Beaurepaire E, Farge E (2008) Tissue deformation modulates twist expression to determine anterior midgut differentiation in Drosophila embryos. Dev Cell 15:470–477

Dietrich C, Scherwat J, Faust D, Oesch F (2002) Subcellular localization of beta-catenin is regulated by cell density. Biochem Biophys Res Commun 292:195–199

Dominguez I, Itoh K, Sokol SY (1995) Role of glycogen synthase kinase 3 beta as a negative regulator of dorsoventral axis formation in Xenopus embryos. Proc Natl Acad Sci U S A 92:8498–8502

Dong J et al (2007) Elucidation of a universal size-control mechanism in Drosophila and mammals. Cell 130:1120–1133

Drees F, Pokutta S, Yamada S, Nelson WJ, Weis WI (2005) Alpha-catenin is a molecular switch that binds E-cadherin-beta-catenin and regulates actin-filament assembly. Cell 123:903–915

Dudko OK, Hummer G, Szabo A (2008) Theory, analysis, and interpretation of single-molecule force spectroscopy experiments. Proc Natl Acad Sci U S A 105:15755–15760

Dupont S et al (2011) Role of YAP/TAZ in mechanotransduction. Nature 474:179–183

El Sayegh TY et al (2005) Phosphorylation of N-cadherin-associated cortactin by Fer kinase regulates N-cadherin mobility and intercellular adhesion strength. Mol Biol Cell 16:5514–5527

Escobar DJ et al (2015) α-Catenin phosphorylation promotes intercellular adhesion through a dual-kinase mechanism. J Cell Sci 128:1150–1165

Finer JT, Simmons RM, Spudich JA (1994) Single myosin molecule mechanics: piconewton forces and nanometre steps. Nature 368:113–119

Fodde R, Brabletz T (2007) Wnt/beta-catenin signaling in cancer stemness and malignant behavior. Curr Opin Cell Biol 19:150–158

Fuchs M, Müller T, Lerch MM, Ullrich A (1996) Association of human protein-tyrosine phosphatase kappa with members of the armadillo family. J Biol Chem 271:16712–16719

Fujii K, Furukawa F, Matsuyoshi N (1996) Ligand activation of overexpressed epidermal growth factor receptor results in colony dissociation and disturbed E-cadherin function in HSC-1 human cutaneous squamous carcinoma cells. Exp Cell Res 223:50–62

Fukuyama T, Ogita H, Kawakatsu T, Inagaki M, Takai Y (2006) Activation of Rac by cadherin through the c-Src-Rap1-phosphatidylinositol 3-kinase-Vav2 pathway. Oncogene 25:8–19

Grashoff C et al (2010) Measuring mechanical tension across vinculin reveals regulation of focal adhesion dynamics. Nature 466:263–266

Greene AC et al (2014) Spatial organization of EphA2 at the cell-cell interface modulates trans-endocytosis of ephrinA1. Biophys J 106:2196–2205

Grosheva I, Shtutman M, Elbaum M, Bershadsky AD (2001) p120 catenin affects cell motility via modulation of activity of Rho-family GTPases: a link between cell-cell contact formation and regulation of cell locomotion. J Cell Sci 114:695–707

Hansen SD et al (2013) alphaE-catenin actin-binding domain alters actin filament conformation and regulates binding of nucleation and disassembly factors. Mol Biol Cell 24:3710–3720

Hantschel O et al (2003) A myristoyl/phosphotyrosine switch regulates c-Abl. Cell 112:845–857

Harrison OJ et al (2010) Two-step adhesive binding by classical cadherins. Nat Struct Mol Biol 17:348–357

Harrison OJ et al (2011) The extracellular architecture of adherens junctions revealed by crystal structures of type I cadherins. Structure 19:244–256

Hartsock A, Nelson WJ (2012) Competitive regulation of E-cadherin juxtamembrane domain degradation by p120-catenin binding and Hakai-mediated ubiquitination. PLoS One 7:e37476

Harvey KF, Pfleger CM, Hariharan IK (2003) The Drosophila Mst ortholog, hippo, restricts growth and cell proliferation and promotes apoptosis. Cell 114:457–467

Hazan RB, Norton L (1998) The epidermal growth factor receptor modulates the interaction of E-cadherin with the actin cytoskeleton. J Biol Chem 273:9078–9084

Hazan RB, Kang L, Roe S, Borgen PI, Rimm DL (1997) Vinculin is associated with the E-cadherin adhesion complex. J Biol Chem 272:32448–32453

He X, Saint-Jeannet JP, Woodgett JR, Varmus HE, Dawid IB (1995) Glycogen synthase kinase-3 and dorsoventral patterning in Xenopus embryos. Nature 374:617–622
He TC et al (1998) Identification of c-MYC as a target of the APC pathway. Science 281:1509–1512
Hinck L, Näthke IS, Papkoff J, Nelson WJ (1994) Dynamics of cadherin/catenin complex formation: novel protein interactions and pathways of complex assembly. J Cell Biol 125:1327–1340
Holsinger LJ, Ward K, Duffield B, Zachwieja J, Jallal B (2002) The transmembrane receptor protein tyrosine phosphatase DEP1 interacts with p120(ctn). Oncogene 21:7067–7076
Hong S, Troyanovsky RB, Troyanovsky SM (2013) Binding to F-actin guides cadherin cluster assembly, stability, and movement. J Cell Biol 201:131–143
Hoschuetzky H, Aberle H, Kemler R (1994) Beta-catenin mediates the interaction of the cadherin-catenin complex with epidermal growth factor receptor. J Cell Biol 127:1375–1380
Huber AH, Weis WI (2001) The structure of the beta-catenin/E-cadherin complex and the molecular basis of diverse ligand recognition by beta-catenin. Cell 105:391–402
Huber AH, Nelson WJ, Weis WI (1997a) Three-dimensional structure of the armadillo repeat region of beta-catenin. Cell 90:871–882
Huber O, Krohn M, Kemler R (1997b) A specific domain in alpha-catenin mediates binding to beta-catenin or plakoglobin. J Cell Sci 110(Pt 15):1759–1765
Huber AH, Stewart DB, Laurents DV, Nelson WJ, Weis WI (2001) The cadherin cytoplasmic domain is unstructured in the absence of beta-catenin. A possible mechanism for regulating cadherin turnover. J Biol Chem 276:12301–12309
Hui E, Vale RD (2014) In vitro membrane reconstitution of the T-cell receptor proximal signaling network. Nat Struct Mol Biol 21:133–142
Imamura Y, Itoh M, Maeno Y, Tsukita S, Nagafuchi A (1999) Functional domains of alpha-catenin required for the strong state of cadherin-based cell adhesion. J Cell Biol 144:1311–1322
Itoh M, Nagafuchi A, Moroi S, Tsukita S (1997) Involvement of ZO-1 in cadherin-based cell adhesion through its direct binding to alpha catenin and actin filaments. J Cell Biol 138:181–192
Kim L, Wong TW (1995) The cytoplasmic tyrosine kinase FER is associated with the catenin-like substrate pp 120 and is activated by growth factors. Mol Cell Biol 15:4553–4561
Kimelman D, Pierce SB (1996) Regulation of dorsal-ventral axis formation in Xenopus by intercellular and intracellular signalling. Biochem Soc Symp 62:13–23
Knudsen KA, Soler AP, Johnson KR, Wheelock MJ (1995) Interaction of alpha-actinin with the cadherin/catenin cell-cell adhesion complex via alpha-catenin. J Cell Biol 130:67–77
Kobielak A, Pasolli HA, Fuchs E (2004) Mammalian formin-1 participates in adherens junctions and polymerization of linear actin cables. Nat Cell Biol 6:21–30
Kojima H, Ishijima A, Yanagida T (1994) Direct measurement of stiffness of single actin filaments with and without tropomyosin by in vitro nanomanipulation. Proc Natl Acad Sci U S A 91:12962–12966
Korinek V et al (1997) Constitutive transcriptional activation by a beta-catenin-Tcf complex in APC-/- colon carcinoma. Science 275:1784–1787
Lai ZC et al (2005) Control of cell proliferation and apoptosis by mob as tumor suppressor, mats. Cell 120:675–685
le Duc Q et al (2010) Vinculin potentiates E-cadherin mechanosensing and is recruited to actin-anchored sites within adherens junctions in a myosin II-dependent manner. J Cell Biol 189:1107–1115
Lickert H, Bauer A, Kemler R, Stappert J (2000) Casein kinase II phosphorylation of E-cadherin increases E-cadherin/beta-catenin interaction and strengthens cell-cell adhesion. J Biol Chem 275:5090–5095
Lin WC et al (2014) H-Ras forms dimers on membrane surfaces via a protein-protein interface. Proc Natl Acad Sci U S A 111:2996–3001

Liu C et al (2002) Control of beta-catenin phosphorylation/degradation by a dual-kinase mechanism. Cell 108:837–847

Mandai K et al (1999) Ponsin/SH3P12: an l-afadin- and vinculin-binding protein localized at cell-cell and cell-matrix adherens junctions. J Cell Biol 144:1001–1017

Manibog K, Li H, Rakshit S, Sivasankar S (2014) Resolving the molecular mechanism of cadherin catch bond formation. Nat Commun 5:3941

Maruthamuthu V, Sabass B, Schwarz US, Gardel ML (2011) Cell-ECM traction force modulates endogenous tension at cell-cell contacts. Proc Natl Acad Sci U S A 108:4708–4713

McCrea PD, Turck CW, Gumbiner B (1991) A homolog of the armadillo protein in Drosophila (plakoglobin) associated with E-cadherin. Science 254:1359–1361

McEwen AE, Maher MT, Mo R, Gottardi CJ (2014) E-cadherin phosphorylation occurs during its biosynthesis to promote its cell surface stability and adhesion. Mol Biol Cell 25:2365–2374

Miller PW et al (2013) Danio rerio alphaE-catenin is a monomeric F-actin binding protein with distinct properties from Mus musculus alphaE-catenin. J Biol Chem 288:22324–22332

Monga SP et al (2002) Hepatocyte growth factor induces Wnt-independent nuclear translocation of beta-catenin after Met-beta-catenin dissociation in hepatocytes. Cancer Res 62:2064–2071

Morin PJ et al (1997) Activation of beta-catenin-Tcf signaling in colon cancer by mutations in beta-catenin or APC. Science 275:1787–1790

Nagafuchi A, Ishihara S, Tsukita S (1994) The roles of catenins in the cadherin-mediated cell adhesion: functional analysis of E-cadherin-alpha catenin fusion molecules. J Cell Biol 127:235–245

Nelson WJ, Nusse R (2004) Convergence of Wnt, beta-catenin, and cadherin pathways. Science 303:1483–1487

Ng MR, Besser A, Brugge JS, Danuser G (2014) Mapping the dynamics of force transduction at cell-cell junctions of epithelial clusters. Elife 3:e03282

Noren NK, Liu BP, Burridge K, Kreft B (2000) p120 catenin regulates the actin cytoskeleton via Rho family GTPases. J Cell Biol 150:567–580

Norstrom MF, Smithback PA, Rock RS (2010) Unconventional processive mechanics of non-muscle myosin IIB. J Biol Chem 285:26326–26334

Oh H, Irvine KD (2008) In vivo regulation of Yorkie phosphorylation and localization. Development 135:1081–1088

Ostman A, Yang Q, Tonks NK (1994) Expression of DEP-1, a receptor-like protein-tyrosine-phosphatase, is enhanced with increasing cell density. Proc Natl Acad Sci U S A 91:9680–9684

Ozawa M, Ringwald M, Kemler R (1990) Uvomorulin-catenin complex formation is regulated by a specific domain in the cytoplasmic region of the cell adhesion molecule. Proc Natl Acad Sci U S A 87:4246–4250

Pacquelet A, Rørth P (2005) Regulatory mechanisms required for DE-cadherin function in cell migration and other types of adhesion. J Cell Biol 170:803–812

Pantalacci S, Tapon N, Léopold P (2003) The Salvador partner Hippo promotes apoptosis and cell-cycle exit in Drosophila. Nat Cell Biol 5:921–927

Peng X, Maiers JL, Choudhury D, Craig SW, DeMali KA (2012) α-Catenin uses a novel mechanism to activate vinculin. J Biol Chem 287:7728–7737

Pertz O et al (1999) A new crystal structure, Ca2+ dependence and mutational analysis reveal molecular details of E-cadherin homoassociation. EMBO J 18:1738–1747

Piedra J et al (2003) p120 Catenin-associated Fer and Fyn tyrosine kinases regulate beta-catenin Tyr-142 phosphorylation and beta-catenin-alpha-catenin interaction. Mol Cell Biol 23:2287–2297

Pokutta S, Weis WI (2000) Structure of the dimerization and beta-catenin-binding region of alpha-catenin. Mol Cell 5:533–543

Pokutta S, Herrenknecht K, Kemler R, Engel J (1994) Conformational changes of the recombinant extracellular domain of E-cadherin upon calcium binding. Eur J Biochem 223:1019–1026

Pokutta S, Drees F, Takai Y, Nelson WJ, Weis WI (2002) Biochemical and structural definition of the l-afadin- and actin-binding sites of alpha-catenin. J Biol Chem 277:18868–18874

Pujol T, du Roure O, Fermigier M, Heuvingh J (2012) Impact of branching on the elasticity of actin networks. Proc Natl Acad Sci U S A 109:10364–10369
Qi J, Chen N, Wang J, Siu CH (2005) Transendothelial migration of melanoma cells involves N-cadherin-mediated adhesion and activation of the beta-catenin signaling pathway. Mol Biol Cell 16:4386–4397
Rakshit S, Zhang Y, Manibog K, Shafraz O, Sivasankar S (2012) Ideal, catch, and slip bonds in cadherin adhesion. Proc Natl Acad Sci U S A 109:18815–18820
Rangarajan ES, Izard T (2012) The cytoskeletal protein alpha-catenin unfurls upon binding to vinculin. J Biol Chem 287:18492–18499
Rangarajan ES, Izard T (2013) Dimer asymmetry defines alpha-catenin interactions. Nat Struct Mol Biol 20:188–193
Reffay M et al (2014) Interplay of RhoA and mechanical forces in collective cell migration driven by leader cells. Nat Cell Biol 16:217–223
Reynolds AB et al (1994) Identification of a new catenin: the tyrosine kinase substrate p120cas associates with E-cadherin complexes. Mol Cell Biol 14:8333–8342
Reynolds AB, Daniel JM, Mo YY, Wu J, Zhang Z (1996) The novel catenin p120cas binds classical cadherins and induces an unusual morphological phenotype in NIH3T3 fibroblasts. Exp Cell Res 225:328–337
Ridley AJ (2006) Rho GTPases and actin dynamics in membrane protrusions and vesicle trafficking. Trends Cell Biol 16:522–529
Rijksen G, Völler MC, van Zoelen EJ (1993) The role of protein tyrosine phosphatases in density-dependent growth control of normal rat kidney cells. FEBS Lett 322:83–87
Rimm DL, Koslov ER, Kebriaei P, Cianci CD, Morrow JS (1995) Alpha 1(E)-catenin is an actin-binding and -bundling protein mediating the attachment of F-actin to the membrane adhesion complex. Proc Natl Acad Sci U S A 92:8813–8817
Rosato R, Veltmaat JM, Groffen J, Heisterkamp N (1998) Involvement of the tyrosine kinase fer in cell adhesion. Mol Cell Biol 18:5762–5770
Roura S, Miravet S, Piedra J, García de Herreros A, Duñach M (1999) Regulation of E-cadherin/Catenin association by tyrosine phosphorylation. J Biol Chem 274:36734–36740
Rubinfeld B et al (1993) Association of the APC gene product with beta-catenin. Science 262:1731–1734
Sadot E et al (2002) Regulation of S33/S37 phosphorylated beta-catenin in normal and transformed cells. J Cell Sci 115:2771–2780
Sakanaka C (2002) Phosphorylation and regulation of beta-catenin by casein kinase I epsilon. J Biochem 132:697–703
Schlegelmilch K et al (2011) Yap1 acts downstream of α-catenin to control epidermal proliferation. Cell 144:782–795
Schmidtke DW, Diamond SL (2000) Direct observation of membrane tethers formed during neutrophil attachment to platelets or P-selectin under physiological flow. J Cell Biol 149:719–730
Seeling JM et al (1999) Regulation of beta-catenin signaling by the B56 subunit of protein phosphatase 2A. Science 283:2089–2091
Serres M et al (2000) The disruption of adherens junctions is associated with a decrease of E-cadherin phosphorylation by protein kinase CK2. Exp Cell Res 257:255–264
Shapiro L, Weis WI (2009) Structure and biochemistry of cadherins and catenins. Cold Spring Harb Perspect Biol 1:a003053
Silvis MR et al (2011) α-catenin is a tumor suppressor that controls cell accumulation by regulating the localization and activity of the transcriptional coactivator Yap1. Sci Signal 4:ra33
Sim JY et al (2015) Apatial distribution of cell-cell and cell-ecm adhesions regulates force balance while maintaining E-cadherin molecular tension in cell pairs. Mol Biol Cell 26(13):2456–2465
Stoker AW (2005) Protein tyrosine phosphatases and signalling. J Endocrinol 185:19–33

Su LK, Vogelstein B, Kinzler KW (1993) Association of the APC tumor suppressor protein with catenins. Science 262:1734–1737

Tachibana K et al (2000) Two cell adhesion molecules, nectin and cadherin, interact through their cytoplasmic domain-associated proteins. J Cell Biol 150:1161–1175

Takai Y, Miyoshi J, Ikeda W, Ogita H (2008) Nectins and nectin-like molecules: roles in contact inhibition of cell movement and proliferation. Nat Rev Mol Cell Biol 9:603–615

Takeda H, Shimoyama Y, Nagafuchi A, Hirohashi S (1999) E-cadherin functions as a cis-dimer at the cell-cell adhesive interface in vivo. Nat Struct Biol 6:310–312

Tang VW, Brieher WM (2012) α-Actinin-4/FSGS1 is required for Arp2/3-dependent actin assembly at the adherens junction. J Cell Biol 196:115–130

Thoreson MA et al (2000) Selective uncoupling of p120(ctn) from E-cadherin disrupts strong adhesion. J Cell Biol 148:189–202

Toret CP, Collins C, Nelson WJ (2014) An Elmo-Dock complex locally controls Rho GTPases and actin remodeling during cadherin-mediated adhesion. J Cell Biol 207:577–587

Van Etten RA et al (1994) The COOH terminus of the c-Abl tyrosine kinase contains distinct F- and G-actin binding domains with bundling activity. J Cell Biol 124:325–340

Wada K, Itoga K, Okano T, Yonemura S, Sasaki H (2011) Hippo pathway regulation by cell morphology and stress fibers. Development 138:3907–3914

Wildenberg GA et al (2006) p120-catenin and p190RhoGAP regulate cell-cell adhesion by coordinating antagonism between Rac and Rho. Cell 127:1027–1039

Winston JT et al (1999) The SCFbeta-TRCP-ubiquitin ligase complex associates specifically with phosphorylated destruction motifs in IkappaBalpha and beta-catenin and stimulates IkappaBalpha ubiquitination in vitro. Genes Dev 13:270–283

Woodring PJ, Hunter T, Wang JY (2001) Inhibition of c-Abl tyrosine kinase activity by filamentous actin. J Biol Chem 276:27104–27110

Xiao K et al (2005) p120-Catenin regulates clathrin-dependent endocytosis of VE-cadherin. Mol Biol Cell 16:5141–5151

Xu T, Wang W, Zhang S, Stewart RA, Yu W (1995) Identifying tumor suppressors in genetic mosaics: the Drosophila lats gene encodes a putative protein kinase. Development 121:1053–1063

Xu G et al (2004) Continuous association of cadherin with beta-catenin requires the non-receptor tyrosine-kinase Fer. J Cell Sci 117:3207–3219

Yago T et al (2004) Catch bonds govern adhesion through L-selectin at threshold shear. J Cell Biol 166:913–923

Yamada S, Nelson WJ (2007) Localized zones of Rho and Rac activities drive initiation and expansion of epithelial cell-cell adhesion. J Cell Biol 178:517–527

Yamada S, Pokutta S, Drees F, Weis WI, Nelson WJ (2005) Deconstructing the cadherin-catenin-actin complex. Cell 123:889–901

Yan HX et al (2002) Physical and functional interaction between receptor-like protein tyrosine phosphatase PCP-2 and beta-catenin. Biochemistry 41:15854–15860

Yao M et al (2014) Force-dependent conformational switch of alpha-catenin controls vinculin binding. Nat Commun 5:4525

Yap AS, Niessen CM, Gumbiner BM (1998) The juxtamembrane region of the cadherin cytoplasmic tail supports lateral clustering, adhesive strengthening, and interaction with p120ctn. J Cell Biol 141:779–789

Yonemura S, Wada Y, Watanabe T, Nagafuchi A, Shibata M (2010) alpha-Catenin as a tension transducer that induces adherens junction development. Nat Cell Biol 12:533–542

Zeng L et al (1997) The mouse Fused locus encodes Axin, an inhibitor of the Wnt signaling pathway that regulates embryonic axis formation. Cell 90:181–192

Zhang Y, Sivasankar S, Nelson WJ, Chu S (2009) Resolving cadherin interactions and binding cooperativity at the single-molecule level. Proc Natl Acad Sci U S A 106:109–114

Zhao XH et al (2007a) Force activates smooth muscle alpha-actin promoter activity through the Rho signaling pathway. J Cell Sci 120:1801–1809

Zhao B et al (2007b) Inactivation of YAP oncoprotein by the Hippo pathway is involved in cell contact inhibition and tissue growth control. Genes Dev 21:2747–2761

Zondag GC, Reynolds AB, Moolenaar WH (2000) Receptor protein-tyrosine phosphatase RPTPmu binds to and dephosphorylates the catenin p120(ctn). J Biol Chem 275:11264–11269

Chapter 6
Cooperative Roles of Nectins with Cadherins in Physiological and Pathological Processes

Takeshi Fujiwara, Akira Mizoguchi, and Yoshimi Takai

Abstract Mammalian tissues and organs are composed of cells of different types and these cells adhere to one another to form societies of cells including the mesenchyme and the epithelium. Cell–cell adhesion in the mesenchyme is weak whereas cells are organized into closely adherent barrier-forming sheets by cell–cell adhesions in the epithelium. Cell–cell adhesion is organized to allow the release or incorporation of individual cells during various physiological or pathological processes, for example, organ development and growth, maintenance and repair of tissues, or tumorigenesis. One prominent cell–cell adhesion apparatus is adherens junctions (AJs). The central structural components of AJs are transmembrane cell–cell adhesion molecules, cadherins and nectins. Nectins recruit cadherins to the cell–cell contact sites and mediate cell–cell adhesions that control local membrane dynamics for cell polarization and coordinate shape change at the cellular level. Nectins are engaged in calcium-independent homophilic and heterophilic *trans*-interactions between opposing cells in numerous tissues and cell types to form homotypic and heterotypic cell–cell adhesion, whereas cadherins are primarily involved in calcium-dependent homophilic *trans*-interactions between opposing cells to form homotypic cell–cell adhesion. Nectins function cooperatively with or independently of cadherins to control physiological processes and cooperative ones include the formation of AJs and apico-basal polarity, apical constriction, contact inhibition of cell movement and proliferation, formation of synapse, formation of checkerboard-like cell arrangement, ciliary epithelium and lens, and

T. Fujiwara • A. Mizoguchi
Department of Neural Regeneration and Cell Communication, Mie University Graduate School of Medicine, 2-174 Edobashi, Tsu, Mie 514-8507, Japan

CREST, Japan Science and Technology Agency, Kobe, Japan
e-mail: ts-fujiwara@doc.medic.mie-u.ac.jp; amizomie@doc.medic.mie-u.ac.jp

Y. Takai (✉)
Division of Pathogenetic Signaling, Department of Biochemistry and Molecular Biology, Kobe University Graduate School of Medicine, 1-5-6 Minatojima-minamimachi, Chuo-ku, Kobe, Hyogo 650-0047, Japan

CREST, Japan Science and Technology Agency, Kobe, Japan
e-mail: ytakai@med.kobe-u.ac.jp

S.T. Suzuki, S. Hirano (eds.), *The Cadherin Superfamily*,
DOI 10.1007/978-4-431-56033-3_6

neuronal cell adhesion and migration. Moreover, nectins are involved in pathological processes including virus infection and human inherited disorder.

Keywords Nectin • Cadherin • Afadin • Cell adhesion • Cell polarity • Synapse formation • Apical constriction • Cell migration • Disease

6.1 Introduction

In multicellular organisms, tissues and organs are composed of different types of cells that adhere to each other to form cell societies of different characteristics including the mesenchyme and the epithelium. Adhesion between mesenchymal cells is relatively weak and cells are easily detached from one another, whereas adhesion between epithelial cells is relatively strong to form barrier-forming sheets in the epithelium. Cell–cell adhesion plays an important role in the developmental processes and the maintenance and regeneration of tissues and organs. Examples are: the process of a single fertilized egg developing into a multicellular blastula that consists of a layer of blastomeres; the assembly and disassembly of coherent barrier-forming sheets of epithelial and endothelial cells that line the inner and outer surfaces of the organ including those of intestine, skin, and blood vessels; the maintenance of the barrier function by holding cells together by turnovers and remodeling of tissues to allow extrusion of old cells and incorporation of newly generated cells derived from progenitors and/or stem cells as observed in intestine; and the process of wound healing, in which cells move and proliferate to form new cell–cell adhesions between encountered cells from the opposing sides of the wound.

Mammalian tissues and organs are composed of two or more cell types that form cell–cell adhesion homotypically, which is interaction between cells of the same type, or heterotypically, which is interaction between cells of different types (Fig. 6.1a). Homotypic cell–cell adhesion consists of symmetric and asymmetric adhesions. Symmetric homotypic cell–cell adhesion is formed between cells of the same type observed in intestinal absorptive epithelial cell junctions, vascular endothelial cell junctions, and between fibroblasts. Asymmetric homotypic cell–cell adhesion is formed between cells of the same type observed in axons and dendrites of neurons and in wing cells of *Drosophila* containing bristles at the apical surface. Heterotypic cell–cell adhesion is formed between cells of different types observed in pigment and nonpigment epithelial cells in the eye, auditory hair cells and supporting cells in the auditory epithelium of the inner ear, and neurons and glia cells in the central and peripheral nervous system.

One prominent cell–cell adhesion apparatus is adherens junctions (AJs), which is formed by the main cell–cell adhesion molecules (CAMs), cadherins and nectins (Takai et al. 2008a, b; Takeichi 1991; Takeichi 2014). Cadherins are transmembrane calcium-dependent CAMs and the cadherin protein superfamily comprises more than 100 members including the subfamilies of classical cadherins,

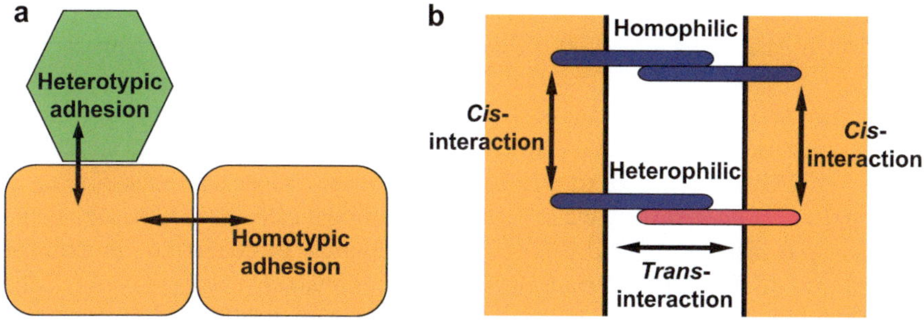

Fig. 6.1 Types of cell–cell adhesion and cell adhesion molecule-dependent interactions. (**a**). Cell–cell adhesion occurs in a homotypic or heterotypic fashion. (**b**) Cell adhesion molecules interact homophilically (i.e., between same molecules) or heterophilically (i.e., between two different molecules). Cell adhesion molecules interact in a cis (interaction between molecules on the same cell surface) or in a trans (interaction between molecules on the opposing cell surface) manner

desmogleins, desmocollins, protocadherins, flamingo/Celsrs, and fat-related cadherins, according to their sequence similarities (Yagi and Takeichi 2000). Cadherins primarily *trans*-interact between the same cadherin members expressed on the opposing cell surface almost exclusively homophilically (Fig. 6.1b) (Zhang et al. 2009). Classical cadherins that form the core of AJs show distinct tissue distribution patterns and the nomenclature is determined from the tissue where they are predominantly expressed: E-cadherin in the epithelium, N-cadherin in the nervous system, and VE-cadherin in the vascular endothelium. Classical cadherins share a common extracellular domain of five cadherin repeats, a single plasma membrane-spanning region, and a cytoplasmic domain. The cytoplasmic domain interacts with regulatory proteins including $p120^{ctn}$ and β-catenin. These catenins interact with various proteins including α-catenin and promote the cadherin-mediated cell–cell adhesion along with the activation of intracellular signaling pathways (Niessen et al. 2011; Petrova et al. 2012).

Nectins, which were originally identified as virus receptors, are another family of CAMs at AJs. They are calcium-independent immunoglobulin-like CAMs (IgCAMs) of which family comprises four protein members, nectin-1, nectin-2, nectin-3, and nectin-4, encoded by *Pvrl1*, *Pvrl2*, *Pvrl3*, and *Pvrl4* genes, respectively. Nectins are capable of interacting with the same and different nectin family members to form homophilic and heterophilic *trans*-interactions (Fig. 6.1b). Nectins show higher binding affinities with their heterophilic *trans*-interactions than with their homophilic *trans*-interactions. All nectins, except for nectin-1γ, possess an extracellular domain containing three Ig-like loops, a single plasma membrane-spanning region, and a cytoplasmic domain. The cytoplasmic domain is associated with the filamentous actin (F-actin)-binding protein afadin (Takahashi et al. 1999). The association between the nectin and cadherin systems is physically mediated by afadin, α-catenin, and their binding proteins, such as ponsin, LMO7, ADIP, vinculin, and α-actinin leading to the formation of symmetric homotypic cell–cell adhesion (Fig. 6.2) (Mandai et al. 1997; Mandai et al. 1999; Reymond et al. 2001; Tachibana et al. 2000; Takai et al. 2008a). The nectin system cooperatively functions with the cadherin system and also mediates asymmetric homotypic and heterotypic cell–cell adhesions, which are observed in various tissues and cell types, such as asymmetric homotypic adhesion in puncta adherentia junctions (PAJs) of synapses formed between the mossy fiber terminals of granule cells in the dentate gyrus and the dendrites of pyramidal cells in the CA3 area in the hippocampus (Mizoguchi et al. 2002), and heterotypic adhesion in the formation of a checkerboard-like cellular arrangement of auditory hair cells and supporting cells in the auditory epithelium of the inner ear (Togashi et al. 2011). On the other hand,

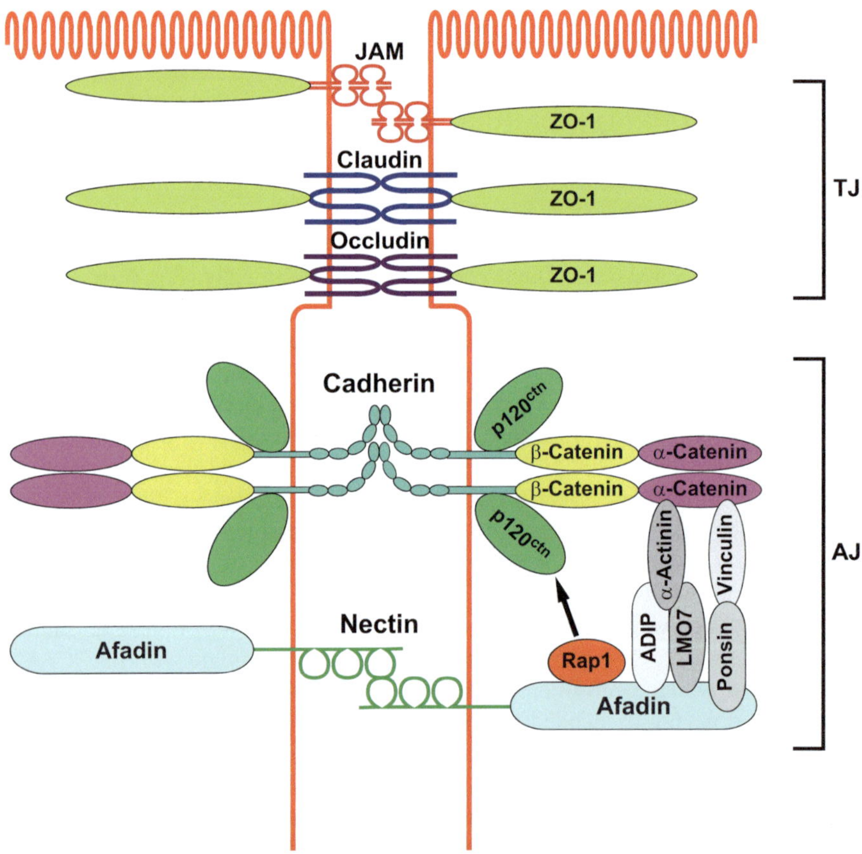

Fig. 6.2 The nectin–afadin and cadherin–catenin complexes at epithelial cell–cell junctions. The nectin–afadin complex is localized at AJs and is linked with the cadherin−catenin complex through the complexes including the ponsin–vinculin, α-actinin–ADIP, and α-actinin–LMO7 complexes, which are associated with F-actin. Activated Rap1 induces the association between afadin with $p120^{ctn}$, which increases the activity of the *trans*-interaction between cadherins and leads to the stabilization of AJ. Afadin is required for the formation of AJs and TJs. F-actin is not depicted

the nectin system functions independently of the cadherin system and mediates homotypic and heterotypic cell–cell adhesions, which are observed in various tissues and cell types including: homotypic adhesion between lateral dendrites and between primary and lateral dendrites of mitral cells in the developing olfactory bulb (Inoue et al. 2015), heterotypic adhesion between lateral dendrites of mitral cells and granule cell dendritic spine necks in the developing olfactory bulb (Inoue et al. 2015), between luminal and basal epithelial cells in the mammary gland (Takai Y et al. unpublished observations), between commissural axons and floor plate cells in the neural tube (Okabe et al. 2004b), between Sertoli cells and germ cells in the testis (Inagaki et al. 2006; Mueller et al. 2003; Ozaki-Kuroda et al. 2002), between ameloblasts and stratum intermedium in the developing tooth (Yoshida et al. 2010), and between T lymphocytes and endothelial cells during *trans*-endothelial migration of T lymphocytes (Devilard et al. 2013). Nectins cooperatively or independently with cadherins are associated with various human diseases, such as virus infection, hereditary ectodermal dysplasia, cataract, Alzheimer's disease, hydrocephalus and subcortical band heterotopia, contextual fear memory, stress-induced cognitive dysfunction, and cancer.

In this chapter, we first introduce the general properties of nectins, then the roles of nectins in cooperation with cadherins in physiological processes, and finally roles of nectins in pathological processes. Due to limited space and our focus on the cooperative roles of nectins with cadherins, the roles of nectins independent of cadherins in physiological processes are summarized elsewhere (Irie et al. 2004; Mandai et al. 2015; Miyoshi and Takai 2005, 2007; Mori et al. 2014; Nakanishi and Takai 2004; Ogita et al. 2008, 2010; Ogita and Takai 2006, 2008; Rikitake et al. 2012; Rikitake and Takai 2008; Sakisaka et al. 2007; Sakisaka and Takai 2004; Shimizu and Takai 2003; Shimono et al. 2012; Takai et al. 2003a, b, 2008a, b; Takai and Nakanishi 2003).

6.2 General Properties of Nectins

The structures of cadherins and their binding proteins in relation to their adhesive activity and localization are fully introduced in other chapters. Thus, in this section, we introduce the structures of nectins and their interacting proteins, the adhesive properties of nectins, and the regulation of the adhesive activity and localization of cadherins by nectins.

6.2.1 Structures of Nectins and Their Interacting Proteins

Nectins comprise four protein family members, each member of which consists of two or three splice variants. Nectins, except for nectin-1γ, contain an extracellular domain with three Ig-like loops that comprise one V type and two C2 type from the

N-terminus, a single plasma membrane-spanning region, and a cytoplasmic domain which binds to the proteins involved in the cytoskeletal reorganization and intracellular signaling (Figs. 6.2 and 6.3a). The V-type Ig-like loop in the extracellular domain is required for the *trans*-dimer formation (Reymond et al. 2001). By crystallographic analysis, nectin-1 forms a V-shaped homophilic *cis*-dimer through the V-type Ig-like loop (Narita et al. 2011). Structure-based site-directed mutagenesis of this domain identifies four essential amino acid residues that are involved in the homophilic *cis*-dimer formation, suggesting that the V-type Ig-like loop is important for nectins to form *trans*- and *cis*-dimers. However, recent crystallographic study shows that the homodimers formed between the V-type Ig-like loop of nectins are likely to be in a trans manner (Harrison et al. 2012). Nectins also *trans*-interact with other Ig-like transmembrane proteins, including nectin-like molecules (Necls or CADMs) (Ikeda et al. 2003; Kakunaga et al. 2004), tactile (CD96), DNAX accessory molecule 1 (DNAM1 or CD226) (Bottino et al. 2003), and T cell immunoreceptor with Ig and ITIM domains (TIGIT) (Stanietsky et al. 2009), expressed on the opposing cell surface (Fig. 6.3a, b). Particularly, nectin-1 interacts with nectin-3, nectin-4, Necl-1 (CADM3), and tactile; nectin-2 interacts with nectin-3, DNAM1, and TIGIT; nectin-3 interacts with Necl-1, Necl-2 (CADM1/SynCAM1), and Necl-5 (PVR, Tage4, or CD155) (Fig. 6.3b). In addition, nectin-2 interacts with N-cadherin between their extracellular domains (Morita et al. 2010), but whether this interaction is in a trans or in a cis manner remains elusive.

Nectins directly bind afadin at their cytoplasmic tails (Figs. 6.2 and 6.3a; Mandai et al. 1997; Satoh-Horikawa et al. 2000; Takahashi et al. 1999). Afadin has two major isoforms, l- and s-afadins. l-Afadin is a longer version that binds F-actin at its C-terminus and s-afadin is a shorter version which lacks the F-actin–binding domain at its C-terminus (Mandai et al. 1997). s-Afadin has insertions of peptide sequences at its C-terminal region that are not present in l-afadin (Kobayashi et al. 2014). The two afadin isoforms display distinct tissue expression patterns as l-afadin is ubiquitous whereas s-afadin is brain specific (Mandai et al. 1997). With the exception of nectin-1β, nectin-3γ, and nectin-4, all nectin family members possess a conserved typical motif (E/A–x–Y–V with x representing any amino acid) at their C-termini, which serves as a binding motif for the PDZ (PSD95–DLG1–ZO-1) domain of afadin (Fig. 6.3a). Moreover, a recent crystal structural study provided evidence that the last three amino acids of nectins (x-Y-V) are sufficient for the binding between nectins and the PDZ domain of afadin (Fujiwara et al. 2015). One exception is that nectin-4, which does not contain the typical conserved motif, interacts with the PDZ domain of afadin at its C-terminus. s-Afadin binds more preferentially to nectins than l-afadin and the deletion of the F-actin–binding domain at the C-terminus does not affect the binding ability of l-afadin to nectins, which indicates that the s-afadin–specific insertions of peptide sequences at the C-terminal region are involved in this preference between l-afadin and s-afadin (Kobayashi et al. 2014). In addition to the different binding properties of l-afadin and s-afadin to nectins, they show distinctive functions in cultured neurons and epithelial cells. In neurons, l-afadin binds to R-Ras small

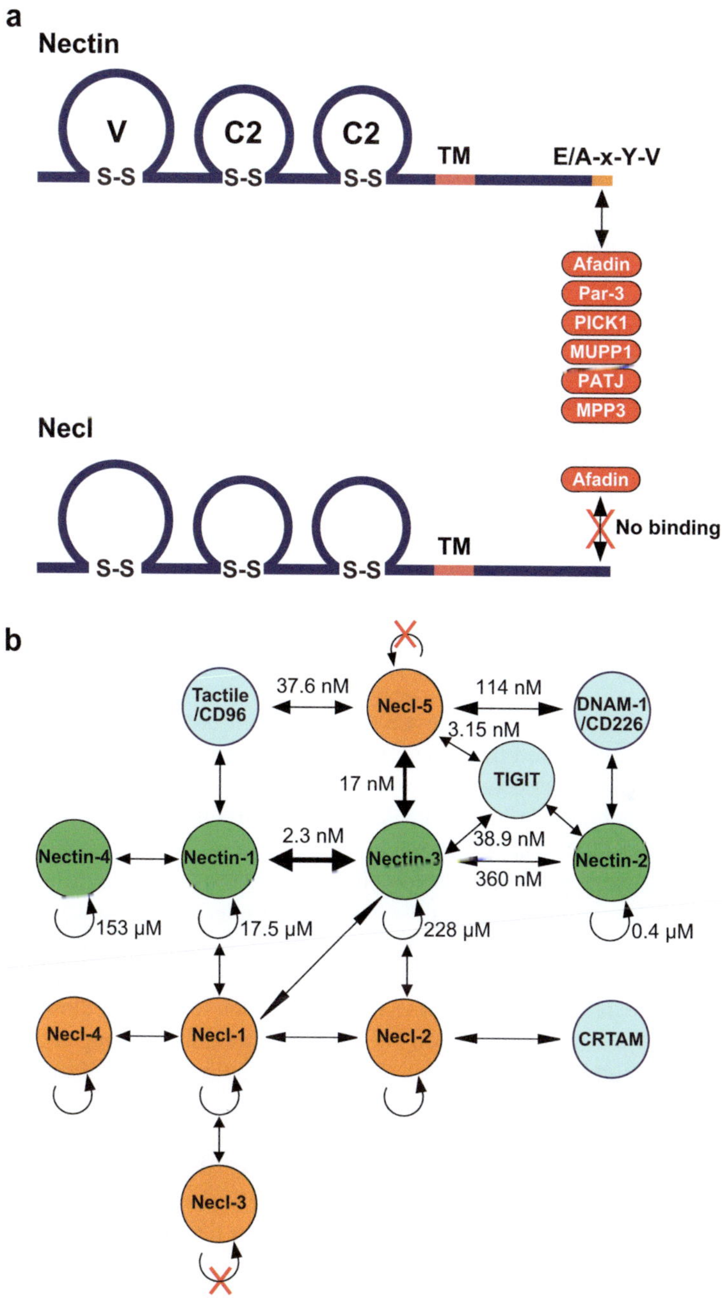

Fig. 6.3 Molecular structures of nectins and Necls and adhesive properties between the nectin family members. (**a**) Nectins and Necls share three Ig-like loops extracellular domain, a single plasma membrane-spanning region (TM), and a cytoplasmic domain. The C-terminus of nectins contains an interaction motif (E/A-x-Y-V) that allows the interactions with afadin, Par-3, PICK1, MUPP1, PATJ, and MPP3. Necls do not interact with afadin. (**b**) The *trans*-interactions between nectins, Necls, and other Ig-like molecules. The homophilic (*looped arrows*) and heterophilic

GTP-binding protein (G protein) and is translocated to the plasma membrane and promotes axon branching through F-actin reorganization (Iwasawa et al. 2012). s-Afadin blocks the binding of l-afadin to R-Ras and inhibits the translocation of l-afadin to the plasma membrane, which results in suppression of axon branching (Umeda et al. 2015). Thus, l-afadin and s-afadin have opposing functions in axon branching. In epithelial cells, l-afadin stabilizes the formation of nectin–l-afadin and E-cadherin–β-catenin–based AJs due to the interaction with F-actin whereas s-afadin is less capable of forming AJs due to the lack of the F-actin–binding domain at its C-terminus (Lorger and Moelling 2006). From now on, unless stated, l-afadin is described as afadin.

Some of the nectin family members bind to cytoplasmic proteins other than afadin (Fig. 6.3a). Those include partitioning defective 3 homologue (Par-3 or PARD3 in mammals; Takekuni et al. 2003), protein interacting with PRKCA1 (PICK1; Reymond et al. 2005), multiple PDZ domain protein (MUPP1 or MPDZ), Pals1-associated tight junction protein (PATJ; Adachi et al. 2009), membrane palmitoylated protein 3 (MPP3; Dudak et al. 2011), zyxin (Gregory Call et al. 2011), and willin (Ishiuchi and Takeichi 2012).

6.2.2 *Adhesive Properties of Nectins*

The adhesive properties of nectins are different from those of cadherins. Nectins form *cis*-homodimers, and then undergo lateral *cis*-clustering on the cell surface, and further *trans*-interact between *cis*-clustering nectins expressed on the opposing cell surface to achieve the nectin-mediated cell–cell adhesion (Fig. 6.4a) (Narita et al. 2011; Reymond et al. 2001; Satoh-Horikawa et al. 2000). Among the *trans*-interactions between nectins, the heterophilic interactions between different nectin family members are stronger than the homophilic interactions between the same nectin family members (Fig. 6.3b). The dissociation constants for the heterophilic *trans*-interactions of nectin-1–nectin-3 and nectin-2–nectin-3 are 2.3 and 360 nM, respectively (Ikeda et al. 2003). These values are much lower than those measured for the homophilic *trans*-interaction between E-cadherins that ranges from 80 to 270 μM (Häussinger et al. 2004; Koch et al. 1997). On the contrary, the dissociation constants for the homophilic *trans*-interactions of nectin-1–nectin-1, nectin-2–nectin-2, nectin-3–nectin-3, and nectin-4–nectin-4 are 17.5, 0.4, 228, and 153 μM, respectively (Harrison et al. 2012; Takai et al. 2008a). By using semi-quantitative surface plasmon resonance and analytical ultracentrifugation technique, the heterophilic interactions are stronger than the homophilic interactions

Fig. 6.3 (continued) (*double-headed arrows*) *trans*-interactions are depicted. The heterophilic *trans*-interaction between nectin-1 and nectin-3 is the strongest followed by those between nectin-3 and Necl-5 and between nectin-2 and nectin-3. Red crossing bars indicate absence of the homophilic *trans*-interaction. Values beside arrows are known dissociation constants (*K*d)

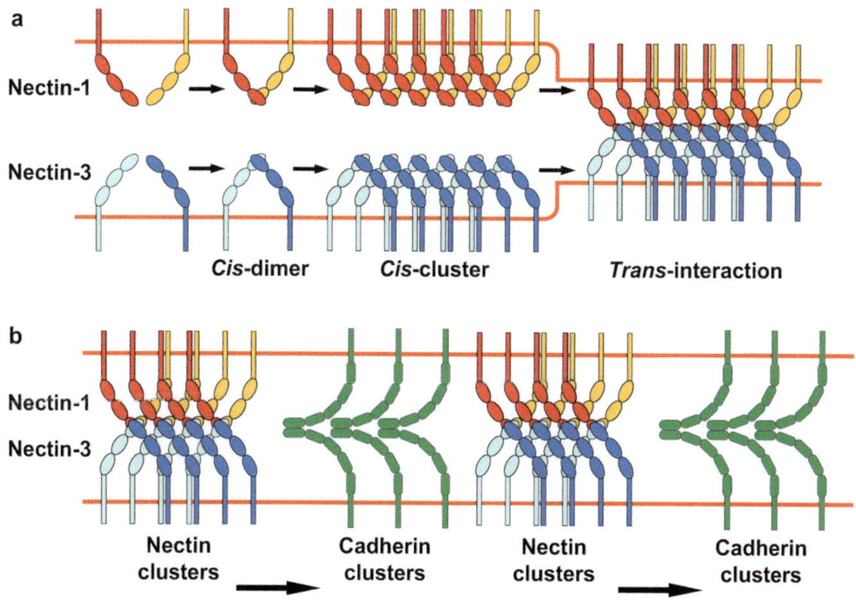

Fig. 6.4 Formation of the nectin-based *trans*-interactions and the alignment of nectin and cadherin clusters in AJs. (**a**) Nectins form *cis*- and *trans*-interactions through the first Ig-like loop (V type). The *cis*-dimers of nectins undergo the lateral cluster formation on the cell surface and these clusters are suggested to interact in a trans manner with the clusters on the opposing cell surface. (**b**) The clusters of the nectin-based and cadherin-based *trans*-interactions are spatially distinct from each other, resulting in a mosaic localization pattern in AJs. The recruitment and stabilization of the nascent cadherin clusters at the nectin-based cell–cell adhesion sites are facilitated by afadin (*arrows*)

between nectins and the stoichiometry of the heterophilic interactions in solution is 1:1 between nectin-1 and nectin-3, and between nectin-1 and nectin-4 (Harrison et al. 2012). This binding property of nectins is further confirmed by the measurement of separation force of cell doublets that express nectin-1 or nectin-3 and that the homophilic interaction between cadherins is stronger than the homophilic and heterophilic interactions of nectins (Martinez-Rico et al. 2005). Moreover, by using intermolecular force microscopy, *trans*-interaction between nectins contains three different bound states that suits the rapid binding reaction of nectins whereas *trans*-interaction between cadherins contains four different bound states that suits the slow and stable binding reaction of cadherins (Tsukasaki et al. 2007). Thus, compared with the homophilic *trans*-interaction between E-cadherins, the binding strength of the heterophilic *trans*-interaction between the nectin family members is weaker, which suits the formation of transient nectin-based cell–cell adhesions and repeated turnovers of these adhesions. The strength of *trans*-interactions depends at least on the plasmalemmal concentration of CAMs and their affinity for *trans*-interaction. The affinity for the homophilic *trans*-interaction between cadherins expressed on the opposing cell surface is low and shows slow lateral diffusion on the cell surface. By contrast, the affinity for the heterophilic *trans*-interactions between nectins expressed on the opposing cell surface is high and shows rapid lateral diffusion on the cell surface.

The crystallographic structures of all four nectins are resolved. Nectins form a homodimer in the crystal lattice, which is consistent with the behavior in solution (Harrison et al. 2012; Narita et al. 2011; Samanta et al. 2012). The V-type Ig-like loop in the extracellular domain exhibits a two-layered β-sheet sandwich topology which is present in other V-type Ig-like loop structures, and contain the front and back sheets composed of 5 and 4 strands, respectively. The dimer interface is formed by nearly orthogonal association of the front 5-stranded β-sheets of two engaging V-type Ig-like loops to organize the nectin homodimers similar to the quaternary structure observed in a number of other physiologically relevant dimers in the IgSF. In mouse nectin-2, F136 residue resides in the V-type Ig-like loop and is buried at the dimer interface, which is important for the interaction between nectin-2 (Harrison et al. 2012; Miyahara et al. 2000). Point mutation, F136D, in the V-type Ig-like loop of nectin-2 inactivates homodimerization and severely attenuates recruitment of nectin-2 to the cell–cell contact sites, indicating that the dimerization interface observed in the crystal structures is required for the recruitment of nectins to the cell–cell contact sites. Targeted cross-linking experiments with nectin-2 suggest that the crystallographically observed dimers common to all nectins are in a trans manner, implicating that the V-type Ig-like loop is the adhesive domain of nectins to achieve both *cis*- and *trans*-interactions.

6.3 Cooperative Roles of Nectins with Cadherins in Physiological Processes

Regulation of physiological processes by the nectin and cadherin systems depends on their adhesive properties and interacting proteins at their cytoplasmic domains. In this section, we introduce the cooperative roles of nectins with cadherins in physiological processes grouped by the types of cell–cell adhesion: symmetric homotypic adhesion in the formation of belt-like structures and apico-basal polarity of epithelial cells, apical constriction of neuroepithelial cells during the formation of the neural tube, and contact inhibition of cell movement and proliferation; asymmetric homotypic adhesion in the formation of the synapse; and heterotypic adhesion in the formation of checkerboard-like cell arrangement, the formation of ciliary epithelium, the formation of the lens, and neuronal cell adhesion and migration.

6.3.1 Symmetric Adhesion

6.3.1.1 Symmetric Homotypic Adhesion: Belt-Like Cell–Cell Adhesion Structures and Apico-Basal Polarity

Epithelial cells adhere to one another through AJs, tight junctions (TJs), and desmosomes to form sheets of cells called the epithelium. AJs have F-actin undercoats that form a circumferential and continuous belt-like structure characteristic to epithelial cells, but noncircumferential and discontinuous in fibroblasts (Yonemura et al. 1995). Epithelial cells have apico-basal polarity with the apical membrane facing the lumen of internal cavities and the basolateral membrane positioning away from the lumen by AJs and TJs (Fig. 6.2; Tsukita et al. 2008). For the establishment of apico-basal polarity, AJs are first formed and subsequently TJs are formed on the apical side of AJs at the epithelial cell–cell adhesion sites. Epithelial cells also show planar cell polarity (PCP) defined by formation of specialized structures in cells within the plane of the epithelium.

The establishment of an epithelial belt-like structure in mammals is dependent on the formation of AJs that mediate the interactions between adjacent epithelial cells within the epithelium. AJs are formed by the nectin and E-cadherin systems (Takai et al. 2008a; Takeichi 2014). In the absence of cell–cell interaction, epithelial cells show flattened morphology with migratory protrusions. However, in groups of cells with cell–cell interactions, epithelial cells show cobblestone-like morphology with little motile activity. Nectins and E-cadherin appear very early at the initial cell–cell contact sites in the course of the formation of AJs and are present at primordial spot-like AJs or puncta which are formed at the tips of actin-rich lamellipodial or filopodial protrusions of two contacting cells (Asakura et al. 1999; Nejsum and Nelson 2009; Perez-Moreno et al. 2003; Takai et al. 2008a). Following the initial cell–cell contact, AJs are maturated by the *trans*-interaction between nectins expressed on the opposing cell surface followed by the *trans*-interaction between E-cadherins (Komura et al. 2008; Martinez-Rico et al. 2005; Sato et al. 2006). The E-cadherin system is recruited to the nectin-based cell–cell contact sites mainly via the interaction between nectin-bound afadin and α-catenin associating with the E-cadherin system (Tachibana et al. 2000; Takai et al. 2008a).

As afadin, a nectin- and F-actin-binding scaffolding molecule (Mandai et al. 1997), directly binds to α-catenin, which directly binds to β-catenin, the recruitment of E-cadherin to the nectin-based cell–cell contact sites is achieved at least by the afadin–α-catenin interaction (Pokutta et al. 2002; Tachibana et al. 2000). The nectin–afadin system is spatially localized in distinct and independent domains from those of the E-cadherin–β-catenin system in AJs, which results in a mosaic pattern of the clusters of the nectin–afadin and E-cadherin–-β-catenin systems in AJs (Fig. 6.4b; Indra et al. 2013). To achieve this mosaic pattern localization of the nectin–afadin and E-cadherin–β-catenin systems in AJs, the interaction between these two systems occurs predominantly on the periphery or outside of the clusters of the E-cadherin–β-catenin system for the recruitment of the

E-cadherin–β-catenin system to the nectin–afadin-based cell–cell contact sites. Afadin facilitates this process by stabilizing the nascent clusters of the E-cadherin–-β-catenin system at the nectin–afadin-based cell–cell contact sites (Indra et al. 2014).

The *trans*-interactions between nectins result in the recruitment and activation of c-Src, followed by the activation of two guanine-nucleotide exchange factors, FRG (FGD-related Cdc42 GEF) and Vav2, which are specific for the activation of Cdc42 and Rac1 small G proteins, respectively (Fukuhara et al. 2003; Kawakatsu et al. 2002, Kawakatsu et al. 2005). The cytoplasmic domain of E-cadherin forms a multiprotein complex consisting of $p120^{ctn}$ and β-catenin, which are the protein family members containing the armadillo repeat domain. The adhesive activity and localization of E-cadherin at AJs are dependent on β-catenin and $p120^{ctn}$, respectively (Perez-Moreno et al. 2003; Pokutta and Weis 2007). Conversely, the stability of β-catenin to circumvent ubiquitin-proteasome–dependent degradation and $p120^{ctn}$ localization at cell–cell contact sites are dependent on E-cadherin (Nelson and Nusse 2004). $p120^{ctn}$ also has a function in regulating the activity of Rho small G proteins as $p120^{ctn}$ interacts with Vav2 and RhoA inhibitor p190RhoGAP to achieve the activation of Rac1 and the inhibition of RhoA at the early formation of cell–cell contact sites (Noren et al. 2000; Wildenberg et al. 2006). The activation of Cdc42 and Rac1 serves to increase the cell surface membranes by the formation of filopodia and lamellipodia, respectively, and the inhibition of RhoA activity attenuates stress fiber formation and cell migration to facilitate new cell–cell adhesion (Perez-Moreno and Fuchs 2006).

As described above, the nectin and E-cadherin systems induce the activation of Cdc42 and Rac1 for the formation of AJs and regulate the reorganization of the actin cytoskeleton at the cell–cell adhesion sites. In terms of the physical link of the nectin–afadin or E-cadherin–β-catenin system to F-actin, the link of the nectin–afadin complex to F-actin is robust. Afadin directly binds to nectins through its PDZ domain and binds to F-actin at its C-terminus, indicating that afadin directly links the actin cytoskeleton to the nectin-based cell–cell adhesion sites (Mandai et al. 1997; Takahashi et al. 1999). In addition to the binding to F-actin, afadin is associated with the F-actin–binding protein vinculin through its interaction with ponsin/SH3P12/sorbs1 (Mandai et al. 1999), and is associated with the F-actin–binding protein α-actinin through its interaction with ADIP (afadin dilute domain-interacting protein) and LMO7 (lim domain only 7; Fig. 6.2; Asada et al. 2003; Ooshio et al. 2004). Thus, the nectin–afadin and E-cadherin–β-catenin systems are physically linked to each other at AJs through multiple protein interactions to influence their activity and localization (Sakisaka et al. 2007).

The nectin–afadin system is important not only for the recruitment of cadherins to the cell–cell contact sites for the formation of AJs, but also for the formation of TJs to establish apico-basal polarity. Reduced afadin expression inhibits the formation of AJs and TJs, eventually impairing apico-basal polarity (Ikeda et al. 1999; Ooshio et al. 2010). The C-terminal conserved four amino acid motif, E/A–x–Y–V, of nectin-1 and nectin-3 binds Par-3, a cell polarity protein that forms a complex with Par-6 (PARD6A in mammals) and atypical protein kinase C (aPKC; Takekuni et al. 2003). The Par-3–Par-6–aPKC complex regulates the formation of TJs and Par-3 contributes to the establishment of apico-basal polarity in concert with afadin (Ooshio et al. 2007). Thus, the nectin–afadin system is required for the establishment of epithelial apico-basal polarity by recruiting the E-cadherin–β-catenin system to the sites of cell–cell contacts for the formation of AJs and further promoting the formation of TJs through the Par-3–Par-6–aPKC complex and afadin.

Recent findings elucidate the molecular mechanism for the alignment of TJs at the apical side of AJs. Nectin-based and junctional adhesion molecule (JAM)-based cell–cell adhesions are critical for the alignment of TJs at the apical side of AJs in epithelial cells (Fig. 6.2), and this is achieved by cooperative functions of cell polarity proteins including Par-3 (Yamada et al. 2013). By exogenous introduction of CAMs including nectin-3, JAM-A, E-cadherin, and claudin-1 into fibroblasts, in which those CAMs are endogenously expressed at negligible levels, the AJ-like nectin-3–based cell–cell adhesion sites are formed at the apical side of the TJ-like JAM-A–based cell–cell adhesion sites showing inverse alignment of the AJ-like and TJ-like cell–cell adhesion sites compared with that in epithelial cells. E-Cadherin and claudin-1 are recruited to the AJ-like nectin-3–based and the TJ-like JAM-A–based cell–cell adhesion sites, respectively. This inverse alignment of the AJ-like and the TJ-like cell–cell adhesion sites is converted to the normal alignment of TJ-like JAM-A–based cell–cell adhesion sites at the apical side of the AJ-like nectin-3–based cell–cell adhesion sites by complementary expression of cell polarity molecules including Par-3, Par-6, aPKC, Crb3, Pals1, and PATJ.

6.3.1.2 Symmetric Homotypic Adhesion: Apical Constriction

Neural tube formation is a dynamic morphogenetic process in vertebrate development. It involves dynamic rearrangements of cells including apical constriction (Colas and Schoenwolf 2001; Davidson and Keller 1999). It is a change in morphology of neuroepithelial cells by acute contraction in their apical cell surface coordinated with the elongation of cells in the apico-basal axis to achieve a wedge-like cell shape and neural folding (Fig. 6.5a). Apical constriction is also observed in invertebrates including invagination of mesoderms and dorsal closure in *Drosophila melanogaster* (Lecuit and Lenne 2007). Apical accumulation and bundling of F-actin are vital for the process of apical constriction. The process of apical constriction is regulated by molecular pathways involving small G proteins including Rap1 for vertebrates (Haigo et al. 2003), and Rho GEF, Rap1, and Rho1 for invertebrates (Barrett et al. 1997; Häcker and Perrimon 1998; Sawyer et al. 2009). These small G proteins activate Rho kinases that in turn activate the mechanical force of the apically localized actomyosin belt to initiate the narrowing of the apical epithelial cell surface (Dawes-Hoang et al. 2005; Nishimura and Takeichi 2008; Wei et al. 2001). Actin-binding molecules, such as the Shroom family molecules, are implicated in facilitating apical accumulation of F-actin and apical constriction in cultured cells and *Xenopus* embyos (Fairbank et al. 2006; Haigo et al. 2003;

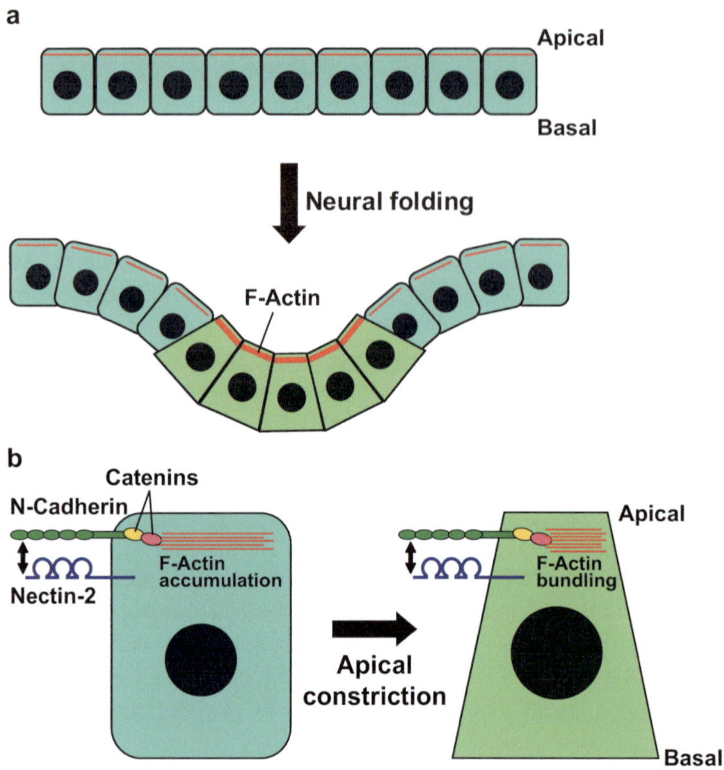

Fig. 6.5 Apical constriction regulated by nectin-2 and N-cadherin. Change in cell shape of neuroepithelium during neural folding. (**a**) The neuroepithelial cells become wedge-shaped (*light green*) in the center at which neuroepithelium bends to induce neural folding. Accumulation of apical F-actin (*red*) is observed in the wedge-shaped neuroepithelial cells. (**b**) The interaction between apically localized nectin-2 and N-cadherin through their extracellular domains facilitates the apical accumulation of F-actin (*left*) and induces the F-actin bundling apical constriction in the neuroepithelial cells (*right*) which leads to the neural folding and the formation of the neural tube

Hildebrand 2005; Lee et al. 2007; Nishimura and Takeichi 2008). However, the detailed mechanism of apical constriction remained poorly understood.

Nectin-2 regulates the process of apical constriction in *Xenopus* neuroepithelial cells by facilitating the apical accumulation of F-actin in cooperation with N-cadherin during neural tube formation (Fig. 6.5b; Morita et al. 2010). Nectin-2 is expressed at the superficial layer of neuroepithelium and depletion of nectin-2 by morpholino oligos causes impaired apical constriction in neuroepithelial cells and defective neural tube closure. Nectin-2 overexpression in nonneural ectoderm, which does not undergo apical constriction, induces ectopic apical accumulation of F-actin and apical constriction by locating nectin-2 to the apical cell membrane (Okabe et al. 2004a; Takai and Nakanishi 2003). Thus, nectin-2 is sufficient to induce apical constriction. The nectin-2-induced apical constriction is dependent on the extracellular domain of nectin-2. N-Cadherin is expressed predominantly in the neural tube and required for neural tube folding during *Xenopus* neurulation (Detrick et al. 1990; Fujimori et al. 1990; Nandadasa et al. 2009). In nonneural ectoderm, E-cadherin and C-cadherin are expressed (Choi and Gumbiner 1989; Ginsberg et al. 1991; Levi et al. 1991). By biochemical analysis, the extracellular domain of nectin-2 preferentially interacts with the extracellular domain of N-cadherin, suggesting a heterophilic interaction between the extracellular domains of nectin-2 and N-cadherin (Fig. 6.5b). Moreover, introduction of *nectin-2* and *N-cadherin* mRNAs into nonneural ectoderm at low dosage induces ectopic apical accumulation of F-actin and apical constriction. Thus, nectin-2 induces apical constriction in neuroepithelial cells in cooperation with N-cadherin.

6.3.1.3 Contact Inhibition of Cell Movement and Proliferation

Contact inhibition of cell movement and proliferation is critical for proper development of organs, and maintenance and remodeling of tissues during regeneration (Mayor and Carmona-Fontaine 2010; McClatchey and Yap 2012). Dysregulation of this process contributes to tumor development. Contact inhibition is observed between two moving and proliferating cells in culture. They collide with one another and in most cases these cells terminate their movement at the direction of the collision and proliferation (Mayor and Carmona-Fontaine 2010; McClatchey and Yap 2012). Molecules involved in this process include integrins and growth factor receptors in cell movement, and cadherins in contact inhibition. In addition to these molecules, nectins and Necls play roles in these processes (Takai et al. 2008b). Necl-5 interacts with both integrin $\alpha_V\beta_3$ and the platelet-derived growth factor (PDGF) receptor and stimulates directional cell movement by both integrin $\alpha_V\beta_3$- and PDGF receptor-induced signaling pathways, such as the activation of Rac, which regulates the formation of peripheral ruffles and focal complexes (Amano et al. 2008; Ikeda et al. 2004; Minami et al. 2007; Miyata et al. 2009). Moreover, Necl-5 attracts growing microtubules to the plasma membrane of the leading edge in moving cells (Minami et al. 2010). Necl-5 is also involved in contact inhibition of cell movement. When two moving cells collide with one another, Necl-5 expressed on the surface of one cell heterophilically interacts in a trans manner with nectin-3, which is diffusely distributed along the surface of another cell, to initiate the formation of cell–cell junctions (Fig. 6.6a) (Ikeda et al. 2003). Then Necl-5 is downregulated by clathrin-dependent endocytosis from the plasma membrane, leading to the reduction of cell movement (Fig. 6.6b; Fujito et al. 2005). On the other hand, nectin-3 dissociated from Necl-5 is retained on the plasma membrane and subsequently interacts in a trans manner with nectin-1 expressed on the surface of the opposing cell (Fig. 6.6b). This nectin-1–nectin-3 *trans*-interaction recruits the cadherin system to the nectin-based cell adhesion sites for the formation of AJs (Fig. 6.2; Takai et al. 2008a). Thus, the cell–cell contact-induced *trans*-interaction between nectin-3 and Necl-5 and subsequent downregulation of Necl-5 on the cell surface are at least one of the mechanisms

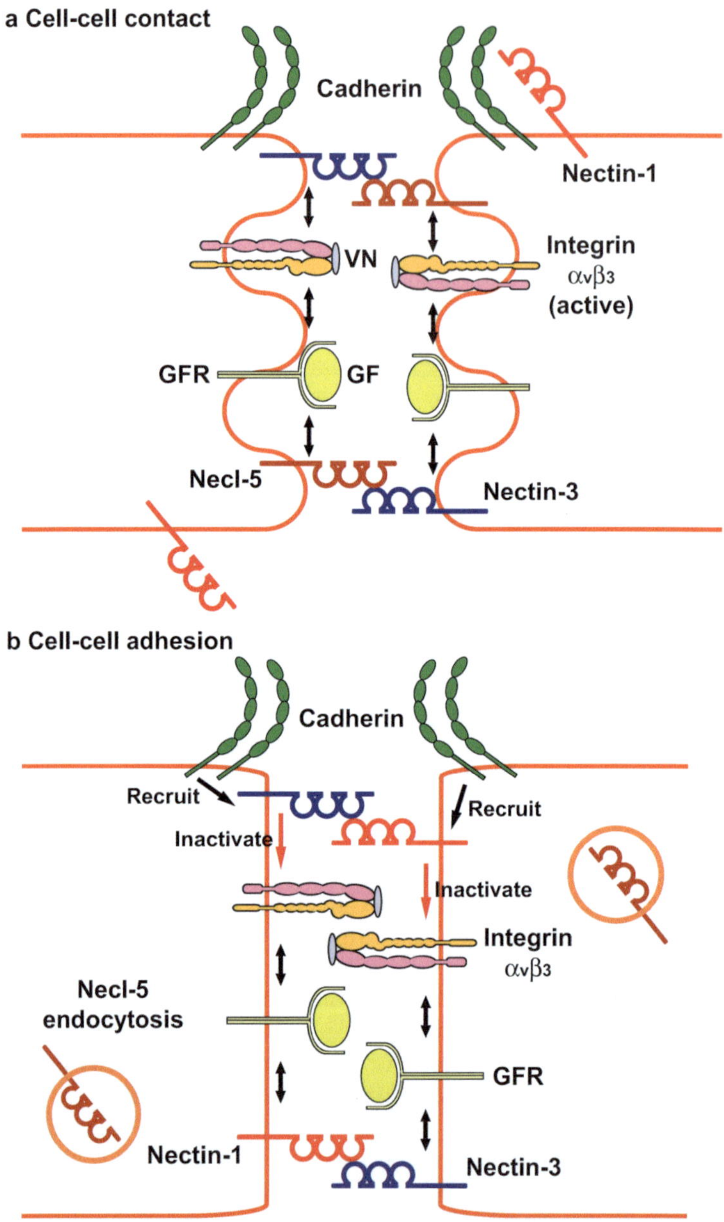

Fig. 6.6 Contact inhibition of cell movement and proliferation by the downregulation of Necl-5 and the inactivation of integrin $\alpha_v\beta_3$ by nectins. (**a**) The initial cell–cell contact is formed at the leading edge of a moving cell by the *trans*-interaction between nectin-3 and Necl-5. Nectins and cadherins are sparsely distributed on the cell surface. (**b**) The *trans*-interaction between nectin-3 and Necl-5 is transient as Necl-5 is subsequently downregulated on the cell surface by endocytosis. This leads to the *trans*-interaction between nectin-1 and nectin-3 which in turn initiates the recruitment of cadherins to the cell–cell adhesion sites (*black arrows*) and the homophilic *trans*-interaction between cadherins as shown in Fig. 6.2. The *trans*-interaction between nectin-1 and nectin-3 induces the inactivation of integrin $\alpha_v\beta_3$ (*red arrows*). Double-headed arrows, physical

for the initiation of contact inhibition of cell movement (Fujito et al. 2005). After the formation of AJs, integrin $\alpha_V\beta_3$ is inactivated but retains its colocalization with nectins at AJs (Sakamoto et al. 2008). All the nectin family members can potentially interact with PTPμ phosphatase through their extracellular regions, and the *trans*-interactions between nectins enhance the phosphatase activity of PTPμ, which leads to the inactivation of integrin $\alpha_V\beta_3$ (Sakamoto et al. 2008). This indicates that in addition to the contact inhibition of cell movement by the *trans*-interaction between nectin-3 and Necl-5 and the downregulation of Necl-5, the inactivation of integrin $\alpha_V\beta_3$ by nectins at AJs provides an additional mechanism for contact inhibition of cell movement.

The nectin-3–Necl-5 *trans*-interaction at cell–cell adhesion sites and the subsequent downregulation of Necl-5 on the cell surface have important roles also in contact inhibition of cell proliferation. Cell proliferation is regulated by growth factors that promote the entry into the cell cycle. Necls are involved in the interface between the growth factor-induced signaling and the cell cycle. Necl-5 enhances the PDGF-induced cell proliferation with a shortened period of the G(0)/G(1) phase of the cell cycle (Kakunaga et al. 2004). Necl-5 also enhances the PDGF-induced activation of the Ras–Raf–MEK–ERK pathway and consequently upregulates cyclins D2 and E, and downregulates p27Kip1 (Hirota et al. 2005). Moreover, Necl-5 regulates vascular endothelial growth factor (VEGF)-induced angiogenesis by controlling the interaction of the VEGF receptor 2 with integrin $\alpha_V\beta_3$ and the VEGF receptor 2-mediated activation of the downstream proangiogenic and survival signals, including Rap1, Akt, and endothelial nitric oxide synthase (Kinugasa et al. 2012). Furthermore, Necl-5 interacts with sprouty2 to prevent it from being tyrosine-phosphorylated by c-Src (Kajita et al. 2007). Sprouty is a negative regulator of growth factor-induced signaling for cell proliferation (Christofori 2003; Kim and Bar-Sagi 2004). When sprouty is tyrosine-phosphorylated by c-Src in response to growth factors, it blocks the growth factor-induced activation of the Ras–Raf–MEK–ERK pathway (Kim and Bar-Sagi 2004). When Necl-5 is downregulated on the cell surface by the *trans*-interaction with nectin-3 at cell–cell adhesion sites, sprouty2 is released from Necl-5, is phosphorylated by c-Src, and inhibits the PDGF-induced activation of Ras. This inhibition further suppresses the de novo synthesis of Necl-5. Thus, this system is at least one of the mechanisms underlying contact inhibition of cell proliferation.

A recent finding shows a novel mechanism which involves Necl-4 in the regulation of contact inhibition of cell movement and proliferation (Yamana et al. 2015). In confluently cultured cells, Necl-4 is upregulated and *cis*-interacts with the VEGF receptor 2 at the cell–cell contact sites. This interaction inhibits the tyrosine phosphorylation of the VEGF receptor 2 through a protein tyrosine phosphatase, nonreceptor type 13 (PTPN13), and reduces cell movement and

Fig. 6.6 (continued) interactions; GF, growth factor; GFR, growth factor receptor; VN, vitronectin. F-actin is not depicted

proliferation. On the contrary, in sparsely cultured cells, Necl-4 is downregulated but is accumulated at leading edges where it inhibits the activation of Rho kinase through PTPN13, which leads to the VEGF-induced activation of Rac1 and enhancement of cell movement. Necl-4 further facilitates the activation of ERK1/2 and enhances cell proliferation. Thus, Necl-4 regulates contact inhibition of cell movement and proliferation cooperatively with the VEGF receptor 2 and PTPN13.

6.3.2 Asymmetric Homotypic Adhesion: Formation of Synapse

The synapse is the site for neurotransmission between the axons and dendrites of neurons, a site of asymmetric homotypic cell–cell adhesion. At the synapses between mossy-fiber terminals of dentate gyrus granule cells and the dendrites of CA3 pyramidal cells (MF synapses) at the stratum lucidum of the hippocampus, two types of junctions between axons and dendrites are highly differentiated and remodeled in an activity-dependent manner: synaptic junctions (SJs), which function as the sites for neurotransmission, and PAJs, which function as the sites for mechanical adhesion between axon terminals and their targets (Fig. 6.7; Amaral and Dent 1981; Spacek and Lieberman 1974). The active zone is the site at the presynaptic side of SJs where neurotransmitter release takes place, and the postsynaptic density (PSD), where the neurotransmitter receptors are accumulated, resides beneath the postsynaptic membrane. At the postnatal developmental stage, the MF synapses are gradually differentiated and remodeled into SJs and PAJs (Amaral and Dent 1981). PAJs are highly developed at the CA3 region of the hippocampus as well as at the calyx of Held of the auditory nucleus in the brainstem. The active remodeling of MF synapses is implicated in the synaptic plasticity that underlies learning and memory processes (Yuste and Bonhoeffer

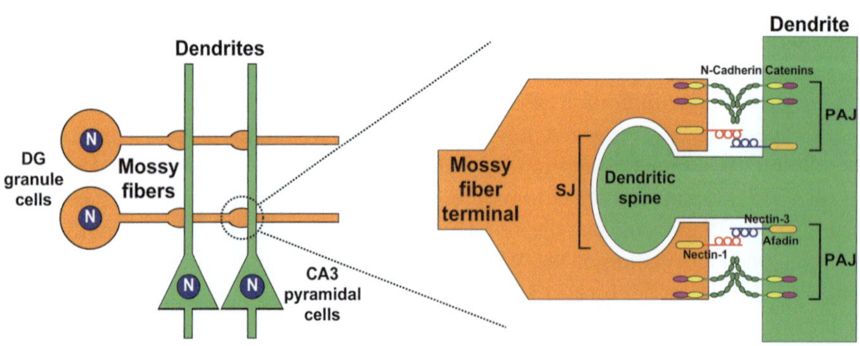

Fig. 6.7 The formation of synapse. Localization and role of nectin-1, nectin-3, afadin, N-cadherin, and catenins at PAJs of synapse between mossy fiber terminals of dentate gyrus (DG) granule cells and dendrites of CA3 pyramidal cells in the hippocampus. F-actin is not depicted

2001). The structure and function of SJs are asymmetric compared with those of AJs being symmetric. At the CA3 region of the mouse hippocampus, N-cadherin is localized symmetrically at both sides of PAJs (Uchida et al. 1996). Nectin-1 and nectin-3 are asymmetrically localized at the presynaptic and postsynaptic membranes, respectively, and afadin is localized symmetrically at both sides of PAJs in the mouse hippocampus (Fig. 6.7; Mizoguchi et al. 2002; Togashi et al. 2006). Thus, SJs are surrounded by nectin-based and N-cadherin–based cell–cell adhesions. At the stratum lucidum of the CA3 region in the *nectin-1*–knockout and *nectin-3*–knockout hippocampi, the number of PAJs at the MF synapses is reduced, indicating that nectins are involved in the formation of PAJs (Honda et al. 2006). Afadin is required for the formation of synapses (Beaudoin et al. 2012; Toyoshima et al. 2014; Xie et al. 2005). Recent findings show that afadin regulates the synaptic localization of N-cadherin and β-catenin at the stratum lucidum of the CA3 region in the mouse hippocampus and the synaptic transmission in cultured hippocampal neurons (Toyoshima et al. 2014). The asymmetric localization of nectin-1 and nectin-3 is important for the specificity of interactions between axons and dendrites for the formation of synapses, and crucial for the synaptic localization of N-cadherin and β-catenin (Honda et al. 2006). The importance of afadin for the formation of MF synapses is further emphasized by the visual three-dimensional reconstruction of MF synapses in the *afadin*-knockout hippocampus using serial block-face scanning electron microscopy. Afadin is required for the maintenance of the structure and number of PAJs and PSDs, the robust coverage of dendritic spines by a MF bouton, and the maintenance of the number of the readily releasable pool synaptic vesicles. These afadin-dependent structural properties are consistent with the afadin-dependent synaptic electrophysiological properties (Takai Y et al. unpublished observations).

6.3.3 Heterotypic Adhesion

6.3.3.1 Heterotypic Adhesion: Formation of Checkerboard-Like Cell Arrangement

In the organ of Corti, which is the auditory epithelium of the snail-shaped cochlea in the inner ear (Fig. 6.8a), sensory hair cells (HCs) are interdigitated with nonsensory supporting cells (SCs) to form a checkerboard-like cell arrangement (Kelley 2006). HCs have a uniform orientation of stereociliary bundles on their apical surfaces that are essential for mechanotransduction (Fig. 6.8b; Montcouquiol and Kelley 2003). This checkerboard-like cell arrangement by HCs and SCs is essential for appropriate perception of sound (Fig. 6.8c; Yoshida and Liberman 1999). The stereociliary bundles on HCs are V-shaped actin-based structures aligned on the apices of HCs pointing toward the outer-lateral boarder of the cochlear duct. The V-shape patterning of stereociliary bundles involves a tubulin-based primary cilium, called the kinocilium (Fig. 6.8c). Recent findings show that nectin-1 and nectin-3 expressed in

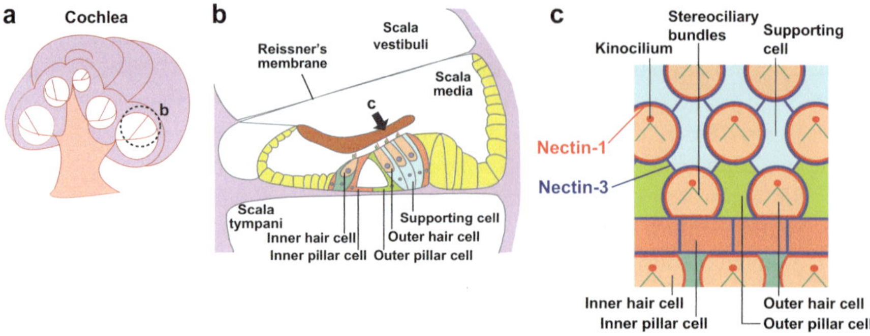

Fig. 6.8 The formation of checkerboard-like cell arrangement. (**a**). Schematic illustration of the cochlea. The area of encircled dashed dots is enlarged in **b**. (**b**) Cross-sectional illustration of the organ of Corti. The cochlea includes three chambers and the organ of Corti is located in the scala media. A single row of inner hair cells is located on the medial side of the epithelium, whereas three rows of outer hair cells are located more laterally. The regions of inner and outer hair cells are separated by the tunnel of Corti, which is surrounded by single rows of inner and outer pillar cells. An arrow indicates the direction of view shown in **c**. (**c**). Luminal surface illustration of the auditory epithelium in the organ of Corti. Inner hair cells, outer hair cells, and various supporting cells are arranged in a checkerboard-like cell arrangement in the auditory epithelium. Nectin-1 and nectin-3 are expressed exclusively by hair cells and supporting cells, respectively. The *trans*-interaction between nectin-1 and nectin-3 occurs at the boundaries between these cells

the auditory epithelium regulate the checkerboard-like cell arrangement by HCs and SCs (Fukuda et al. 2014; Togashi et al. 2011). Nectin-1 and nectin-3 are differentially expressed in HCs and SCs, respectively (Fig. 6.8c), whereas nectin-2 is expressed in both HCs and SCs. The expression pattern of nectin-1 and nectin-3 in HCs and SCs, respectively, is not observed in the vestibular epithelium of the saccule, which indicates a unique mosaic expression pattern of nectin-1 and nectin-3 in the cochlear epithelium. In the cochlear epithelium, the heterophilic *trans*-interaction between nectin-1 and nectin-3 mediates the heterotypic cell–cell adhesion between HCs and SCs and contributes to the checkerboard-like cell arrangement. Molecules involved in PCP and ciliary molecules, including frizzled (Fz), disheveled, Celsr, vang-like (Vangl), prickle, Kif3a, and Lis1, regulate the polarity of kinocilium and stereociliary bundles in HCs (Goodrich and Strutt 2011; Gray et al. 2011; McKenzie et al. 2004; Montcouquiol et al. 2003; Sipe et al. 2013; Sipe and Lu 2011). In the *nectin-3*–knockout mice, abnormal morphology of stereociliary bundles and positioning of the kinocilium in the auditory epithelium are observed in addition to the aberrant attachments between HCs formed by the homophilic *trans*-interaction of nectin-1 leading to the disruption of the checkerboard-like cell arrangement (Fukuda et al. 2014). *Nectin-3* knockout generates a spectrum of abnormalities in the cochlear auditory epithelium, but not in the vestibular epithelium of the saccule: disordered nonuniform orientation and aberrant non-V-shaped flat stereociliary bundles or split bundles that contain several clumps; abnormal localization of the kinocilium located between and near the boundary of aberrantly attached HCs; and mislocalization of the basal body of the kinocilium near the aberrantly attached sites of HCs. In the *nectin-3*–knockout auditory epithelium, molecules involved in PCP, such as Vangl1 and Fz6, maintain their normal localization along the boundary between HCs and SCs. These results suggest that the abnormal phenotypes of stereociliary bundles and the positioning of the kinocilium in HCs by *nectin-3* knockout are independent of the PCP pathway. Abnormal morphology of stereociliary bundles in HCs by *nectin-3* knockout are also observed in the *nectin-1*–knockout auditory epithelium: aberrant attachment between HCs with disordered nonuniform orientation and aberrant morphology of stereociliary bundles. These common phenotypes observed in the *nectin-1*– and *nectin-3*–knockout cochlear auditory epithelium indicate a noncell autonomous effect of the heterophilic *trans*-interaction between nectin-1 and nectin-3 expressed in HCs and SCs, respectively. Components of AJs, such as nectin-2, afadin, E-cadherin, and β-catenin, are concentrated at the boundaries between HCs and SCs and between neighboring SCs whereas nectin-1 is concentrated specifically at the boundaries between HCs and SCs. At the boundaries, nectin-1, nectin-2, and afadin are concentrated at the apical junctional complex which includes AJs and TJs, whereas E-cadherin and β-catenin are extended along the apico-basal axis compared with nectins and afadin. In the *nectin-3*–knockout auditory epithelium, nectin-1, afadin, E-cadherin, and β-catenin are markedly localized at adhesion sites between the aberrantly attached HCs at the apical junctional complex with ectopic extension toward the basal side, but are hardly localized at the boundaries between HCs and SCs and between neighboring SCs. These findings suggest that the heterophilic *trans*-interaction between nectin-1 and nectin-3 regulates the localization of the E-cadherin–β-catenin system at the boundaries between HCs and SCs in the cochlear auditory epithelium.

6.3.3.2 Heterotypic Adhesion: Formation of Ciliary Epithelium

The eye is derived from the neural tube and the optic vesicle develops into an optic cup, which then develops into the retina, ciliary marginal zone, the ciliary epithelium, the pigmented epithelium, and the iris. Nectin-1, nectin-3, and P-cadherin are localized at the apex–apex junctions between the pigment and nonpigment cell layers of the ciliary epithelia (Fig. 6.9a). In the *nectin-1*– and/or *nectin-3*–knockout ciliary epithelia, microphthalmia and separation of apex–apex adhesion between the pigment and nonpigment epithelia of the ciliary body are observed (Inagaki et al. 2005). Localization of P-cadherin at the apex–apex junctions between the pigment and nonpigment cell layers is observed later than those of nectin-1 and nectin-3 during ocular development, suggesting that P-cadherin is recruited to the apex–apex junctions between the pigment and nonpigment cell layers during ocular development.

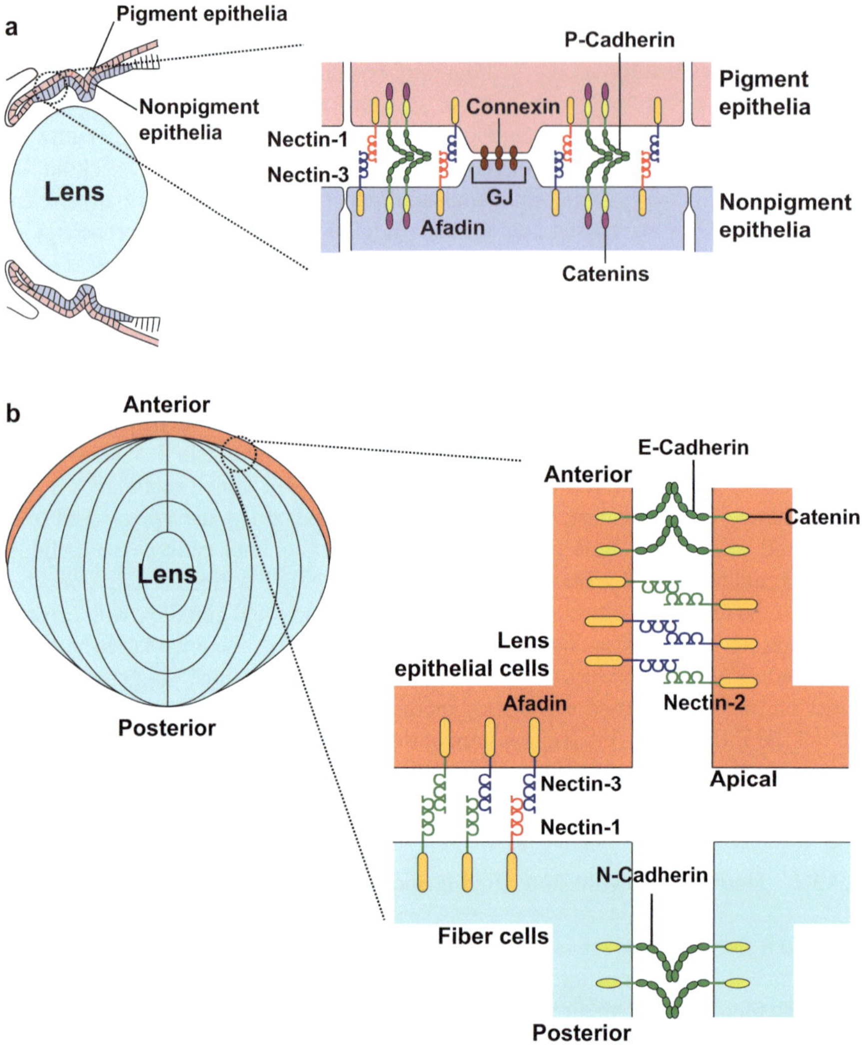

Fig. 6.9 The formation of ciliary epithelium and lens. (**a**) Localization of nectin-1, nectin-3, afadin, P-cadherin, and catenins at the cell–cell adhesion sites between the pigment and nonpigment cell layers of the ciliary epithelium in the eye. GJ, gap junction. (**b**) Localization of nectin-1, -2, and -3, afadin, E- and N-cadherins, and catenin at the cell–cell adhesion sites between the lens epithelial cells, the fiber cells, and between the lens epithelial and fiber cells in the developing lens. F-actin is not depicted

6.3.3.3 Heterotypic Adhesion: Formation of Lens

The shape and morphological development of the ocular lens depend on the elongation of the lens epithelial cells and their differentiation into the fiber cells, followed by the migration and symmetric packing arrangement of the fiber cells

(Chow and Lang 2001; McAvoy 1980). All four nectin genes, *Pvrl1*, *Pvrl2*, *Pvrl3*, and *Pvrl4* are expressed in the developing mouse lens (Lachke et al. 2012; Okabe et al. 2004a). Among them, nectin-1 is localized at the cell surface of the fiber cells adjacent to the lens epithelial cells (Lachke et al. 2012; Maddala et al. 2011), nectin-2 is distributed in both the lens epithelial and fiber cells (Okabe et al. 2004a), and nectin-3 is localized at the apical surface of the lens epithelial cells, which face the fiber cells (Fig. 6.9b; Lachke et al. 2012). On the other hand, E- and N-cadherins are expressed in the developing mouse lens (Cain et al. 2008; Lachke et al. 2012; Maddala et al. 2011), where E-cadherin is localized at the cell–cell junction of the lens epithelial cells (Lachke et al. 2012), and N-cadherin at the cell–cell junction of the fiber cells (Fig. 6.9b; Maddala et al. 2011). In the *nectin-3*–knockout lens, small slit-like separations in the fiber cells adjacent to the lens epithelial cells are observed (Lachke et al. 2012), suggesting that the *trans*-interactions of nectin-1 and -3 and nectin-2 and -3 between the lens epithelial and fiber cells are crucial for lens development.

6.3.3.4 Heterotypic Adhesion: Neuronal Cell Adhesion and Migration

The mammalian cerebral cortex has a six-layered structure and those layers consist of the late-born neurons locating more superficially than the early-born neurons. Those neurons are derived after cell divisions of radial glial cells (RGCs) and intermediate progenitor cells at the ventricular zone and the subventricular zone, respectively, and become projection neurons that migrate through the intermediate zone and the cortical plate along radial glial fibers exerting glia-dependent migration (Nadarajah et al. 2001; Tabata et al. 2009; Tabata and Nakajima 2003; Rakic 1972). When the leading processes of projection neurons reach the most superficial layer of the developing cerebral cortex or the marginal zone, the cell body detaches from radial glial fibers with their leading processes attached to the marginal zone, and move toward the surface of the cortex to complete their migration beneath the marginal zone (Fig. 6.10; Nadarajah et al. 2001; Sekine et al. 2011).

In addition to the glia-dependent migration of projection neurons, the migration of projection neurons independent of radial glial fibers is required for the development of the cerebral cortex (Nadarajah et al. 2001; Tabata and Nakajima 2003). The glia-independent migration of projection neurons largely depends on Cajal–Retzius (CR) cells located in the marginal zone of the cerebral cortex (Soriano and Del Rio 2005). CR cells secrete a glycoprotein, reelin, which regulates the glia-independent migration of projection neurons through N-cadherin, which is expressed in RGCs, projection neurons, and CR cells (Franco et al. 2011; Jossin and Cooper 2011; Sekine et al. 2011). At the ventricular zone, N-cadherin forms homophilic cell–cell adhesions at AJs between RGCs (Fig. 6.10). N-Cadherin is required for cell–cell adhesion between migrating projection neurons and RGCs during glia-dependent migration and the attachment between projection neurons and CR cells in the marginal zone during glia-independent migration (Gil-Sanz et al. 2013; Kawauchi

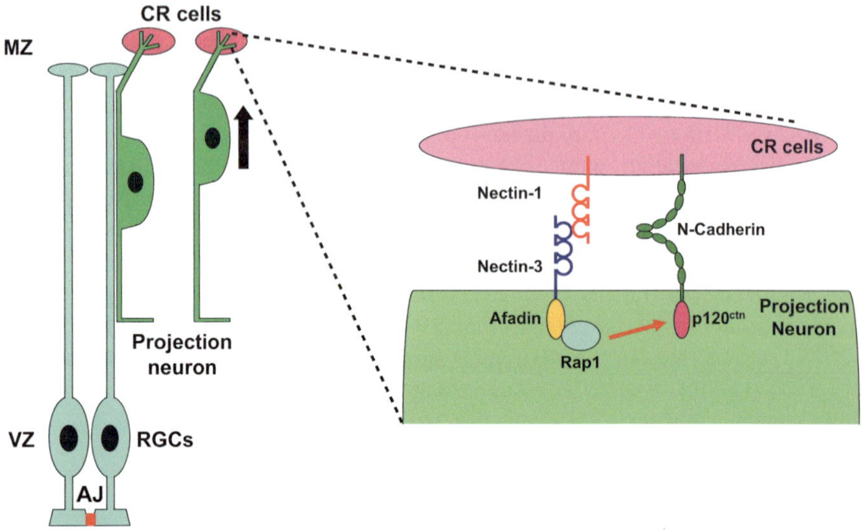

Fig. 6.10 Neuronal cell adhesion and migration. RGCs adhere to one another by AJs in the ventricular zone (VZ) of the cerebral cortex (*left*). Projection neurons derived from RGCs in the VZ migrate and extend their leading processes toward the cortical surface while the trailing processes become axons. When projection neurons reach the cortical surface, the leading processes extend to the marginal zone (MZ), adhere to CR cells accompanied by local branching, and detach from RGC fibers to achieve glia-independent migration (*black arrow*). The heterophilic *trans*-interaction between nectin-1 and nectin-3 expressed in the surface of CR cells and projection neurons, respectively, promotes the homophilic *trans*-interaction between N-cadherin molecules and glia-independent migration of projection neurons by inducing a molecular hierarchy of afadin, Rap1, $p120^{ctn}$, and N-cadherin in projection neurons (*right, red arrow*)

et al. 2010). However, how N-cadherin acquires heterotypic cell–cell adhesions between CR cells and projection neurons remained unclear.

A recent finding shows that nectin-1 and nectin-3 play a central role in the glia-independent migration of projection neurons (Gil-Sanz et al. 2013). Nectin-1 and nectin-3 show the highest affinity among the heterophilic *trans*-interactions between the nectin family members (Ikeda et al. 2003; Takai et al. 2008a). In the developing cerebral cortex of E13.5–14.5, nectin-1 is predominantly expressed in CR cells in the marginal zone and the cortical hem. By contrast, nectin-3 and afadin are expressed throughout the neocortical wall. Nectin-1 and nectin-3 are localized at the cell–cell adhesion sites between CR cells and the leading processes of migrating projection neurons, respectively, and the *trans*-interaction between nectin-1 and nectin-3 mediate the glia-independent migration of projection neurons (Fig. 6.10). The defective form of nectin-1 in CR cells does not affect the migration of CR cells within the cortical marginal zone, but alters the glia-independent migration of projection neurons showing reduced branching of the leading processes in the marginal zone. The defective form of N-cadherin in CR cells phenocopies the defective form of nectin-1 in CR cells. On the other hand, nectin-3 and afadin expressed in projection neurons are important for anchorage and arborization of the extended leading processes in the marginal zone. The glia-independent migration defect of projection neurons caused by RNA interference of nectin-3 or afadin is rescued by the overexpression of N-cadherin, indicating that nectin-1, nectin-3, and afadin function upstream of N-cadherin in projection neurons. Nectin-1 and nectin-3 are colocalized with N-cadherin at the cell–cell adhesion sites between CR cells and the leading processes of projection neurons. In epithelial cells, AJ is formed by the molecular link between the nectin–afadin and E-cadherin–β-catenin systems mediated by afadin- and α-catenin–binding proteins, including ponsin, ADIP, LMO7, and α-actinin (Fig. 6.2). The molecular link between nectins and N-cadherin in the glia-independent migration of projection neurons depends on $p120^{ctn}$ and Rap1 small G protein, which binds to afadin and $p120^{ctn}$ (Fig. 6.2; Hoshino et al. 2005; Sato et al. 2006). Rap1 functions downstream of the nectin-3–afadin complex, and upstream of the N-cadherin–$p120^{ctn}$ complex in projection neurons. Thus, there is a molecular hierarchy of nectin-3–afadin–Rap1–$p120^{ctn}$–N-cadherin for the regulation of cell–cell adhesion between CR cells and projection neurons in the marginal zone and the glia-independent migration of projection neurons (Fig. 6.10).

In cultured neurons, the heterophilic *trans*-interaction between dish-coated nectin-1 and nectin-3 expressed in neurons recruits N-cadherins to the *trans*-interaction sites between nectin-1 and nectin-3 in an afadin-dependent manner (Gil-Sanz et al. 2013). Treatment with reelin enhances this recruitment via the enhanced interaction between afadin and $p120^{ctn}$ and results in the homophilic *trans*-interaction between dish-coated N-cadherin and N-cadherin expressed in neurons. Taken together, reelin initiates the signaling of the molecular hierarchy of nectin-3–afadin–Rap1–$p120^{ctn}$–N-cadherin and recruits N-cadherin to the nectin-based cell–cell contact sites between CR cells and the leading processes of projection neurons to achieve cell–cell adhesion and promotes the glia-independent migration of projection neurons.

6.4 Roles of Nectins in Pathological Processes

In this section, we introduce the roles of nectins in the onset and/or progression of human diseases including mouse models of disease. The functional relationship between nectins and cadherins in the onset and/or progression of human diseases is dependent on molecular and cellular aspects, including different expression patterns of nectins and cadherins, and cooperative or independent functions of nectins and cadherins in cells. Nectins function cooperatively with cadherins for the intercellular spreading of virus and the onset and/or progression of hereditary ectodermal dysplasia and hydrocephalus and subcortical band heterotopias. On the other hand, although the functional relationship with cadherins is unclear, nectins function in the onset and/or progression of Alzheimer's disease, contextual fear memory, and stress-induced cognitive dysfunction. The expression of nectins is altered in various carcinoma tissues, but the roles of nectins cooperatively with or independently of cadherins in the formation of carcinomas remain elusive due to differing expression profiles of nectins and cadherins in various carcinoma tissues.

6.4.1 Virus Infection

Viral infection is achieved by adhering of virus to its specific receptor protein expressed on the cell surface. Thus, some viruses display distinct tissue tropism. For example, poliovirus and herpes viruses show a tropism for neurons. *Nectin-1* and *nectin-2* are isolated as the human poliovirus receptor (PVR)-related genes and named *PRR1* and *PRR2*, respectively (Eberle et al. 1995; Lopez et al. 1995). However, both PRR1 and PRR2 do not serve as a PVR but serve as α-herpes virus receptors to facilitate their entry and intercellular spreading, and are renamed HveC and HveB, respectively (Geraghty et al. 1998; Spear et al. 2000). Human nectin-1 allows entry of all α-herpes viruses thus far tested, including herpes simplex viruses (HSV) types 1 and 2, pseudorabies virus, and bovine herpes virus type 1 (Geraghty et al. 1998). In HSV disease, the intercellular spreading of the virus significantly contributes to the pathogenesis. The nectin-1–afadin and E-cadherin–catenin systems increase the efficacy of the intercellular spreading of HSV-1 (Sakisaka et al. 2001). In addition, the E-cadherin–catenin system interacts with nectin-1α and increases the efficacy of the entry of HSV-1 (Sakisaka et al. 2001). A further study using the *nectin-1*–knockout mice shows that nectin-1 is required for infection by HSV-2 of the vaginal epithelium and viral spread to the nervous system (Taylor et al. 2007). Nectin-4 is the epithelial cell receptor for measles virus, canine distemper virus, and peste des petits ruminants virus (Birch et al. 2013; Muhlebach et al. 2011; Noyce et al. 2011; Noyce et al. 2013; Pratakpiriya et al. 2012), which are members of the Morbillivirus genus. Nectin-4 interacts with the viral attachment protein with high affinity through its membrane-distal domain. As for Necls, human Necl-5 is originally isolated as a receptor for poliovirus and named PVR (Koike et al. 1990; Mendelsohn et al. 1989). Poliovirus is the causative factor of an acute disease of the central nervous system, poliomyelitis. Poliovirus is a neurotropic virus that produces severe lesions selectively in the central nervous system, particularly in the motor neurons. It seems that there are Necl-5/PVR-dependent and -independent poliovirus trafficking pathways toward the central nervous system (Ohka et al. 2012). Thus, nectins and Necls are not only CAMs but also viral receptors that play a critical role in the pathogenesis of neurotrophic viral infections. As for nectin-3, and Necl-1, Necl-2, Necl-3, and Necl-4, they are not identified as viral receptors.

6.4.2 Hereditary Ectodermal Dysplasia

Mutations in the *nectin-1* gene have been shown to be responsible for an autosomal recessive cleft lip/palate ectodermal dysplasia syndrome, also called Zlotogora–Ogur syndrome and Margarita Island ectodermal dysplasia (Sozen et al. 2001; Suzuki et al. 2000). The clinical characteristics include cleft lip/palate, hidrotic ectodermal dysplasia, developmental defects in the hands, and intellectual disability in some cases (Suzuki et al. 1998). Mutations are positioned in the *nectin-1* gene corresponding to the W185 in the second C2-type Ig-like loop and G323 in the third C2-type Ig-like loop that result in a nonsense mutation or a frame shift in the gene. However, the *nectin-1*–knockout mice do not develop cleft lip/palate phenotype (Barron et al. 2008; Yoshida et al. 2012; Yoshida et al. 2010). Thus, the exact role for nectin-1 in the formation of lip and palate remains elusive. On the other hand, the cleft palate phenotype is observed in the *Tgf-β3*–mutant mice and the fusion of palatal shelves is rescued by exogenous administration of Tgf-β3 into cultures of the mutant palate (Taya et al. 1999). Nectin-1 is expressed in the medial edge epithelium, but this expression is lost in the *Tgf-β3*–mutant mice, suggesting that nectin-1 is involved in the initial adhesion between the opposing medial edge epithelium (Martinez-Sanz et al. 2008). Thus, nectin-1 is not essential for but involved in the fusion of palatal shelves. Mutations in the human *nectin-4* gene also cause an ectodermal dysplasia−syndactyly syndrome-1 (EDSS1) that is characterized by the combination of hair and tooth abnormalities, alopecia, and cutaneous syndactyly (Brancati et al. 2010). Mutations are located in the *nectin-4* gene corresponding to the T185M in the second C2-type Ig-like loop and R284 and P304 in the third C2-type Ig-like loop that result in abnormal splicing and a frameshift in the gene, respectively. Other mutations, including P212R and V242M in the second C2-type Ig-like loop, are also reported (Fortugno et al. 2014; Jelani et al. 2011). A recent finding identifies a mechanism leading to the onset of EDSS1 by the *nectin-4* gene mutation, which is the defect in the formation of AJs in EDSS1 (Fortugno et al. 2014). Nectin-4 with T185M or V242M mutation perturbs the nectin-1 clustering at the cell–cell contact sites and causes a significant delay in the recruitment of E-cadherin to the nectin-based cell–cell contact sites and the formation of AJs in cultured primary keratynocytes derived from skin biopsies of EDSS1 patients carrying these mutations. Thus, the attenuation of the recruitment of E-cadherin to the cell–cell contact sites by nectins plays an important role on the onset of hereditary ectodermal dysplasia.

6.4.3 Cataract

Among ocular defects that affect humans, cataract, which is a clinical term of opacity in the ocular lens, is classified as congenital or age-related and is the leading cause of blindness (Graw 2009). Congenital cataract is responsible for approximately 10 % of all childhood blindness and occurs in association with or independent of other defects of ocular development (Bermejo and Martinez-Frias 1998; Hejtmancik 2008; Jensen and Goldschmidt 1971). There is a report of a patient with a balanced translocation, 46,XY,t(1;3)(q31.3;q13.13), which contains a breakpoint residing 515 kb upstream of the *Pvrl3* gene in chromosome 3 (Lachke et al. 2012). As a result, the expression of nectin-3 is reduced to approximately 40% compared with control levels (Lachke et al. 2012). The patient has severe bilateral congenital cataracts, an increased head circumference, prominent extra-axial cerebrospinal fluid spaces, and mild developmental delay (Lachke et al. 2012). Thus, nectin-3 associates with congenital ocular defects.

6.4.4 Alzheimer's Disease

Alzheimer's disease is a common form of disease with dementia. Recent genome-wide association studies show significant association of single nucleotide polymorphisms (SNPs) in the *nectin-2/pvrl2* locus with Alzheimer's disease (Harold et al. 2009; Logue et al. 2011; Takei et al. 2009). First, SNPs in a genomic region encompassing *pvrl2*, *tomm40*, *apoe*, and *apocl* are associated with late onset Alzheimer's disease around the *apoe* (apolipoprotein E) locus in Japanese. Second, a SNP in the 3'UTR region of *nectin-2* is one of the thirteen genome-wide significant SNPs at the *apoe* locus, which are associated with Alzheimer's disease. Third, one of SNPs in the *nectin-2* gene in the region of the *apoe* locus is significantly associated with Alzheimer's disease after adjusting for the confounding effects of the *apoe* genotype in African Americans. In addition, nectin-3 is depleted from the stratum lacunosum moleculare in the CA1 area of the hippocampus, in transgenic mice that express wild-type or mutant human Tau protein, in a chronical or subacute condition (Maurin et al. 2013). It is not known whether the expression of nectin-3 is decreased in the brain of Alzheimer's disease patients, but the decreased expression of nectin-3 in the stratum lacunosum moleculare may serve as an early marker of impaired transport, and eventual synaptic problems, caused by tauopathy.

6.4.5 Hydrocephalus and Subcortical Band Heterotopia

Congenital hydrocephalus is a developmental brain disorder and the neuroepithelial and ependymal cells lining the ventricular and aqueductal walls of the developing brain play an important role in the onset of this disorder (Wagner et al. 2003). The nectin–afadin system maintains the formation of AJs between RGCs in the midbrain and the dorsal telencephalon and between ependymal cells in the midbrain in concert with N-cadherin for the formation of the layer structure in the developing cerebral cortex (Gil-Sanz et al. 2014; Yamamoto et al. 2013, 2015). In the midbrain, loss of afadin causes hydrocephalus by mislocalization of RGCs in the ventricular and intermediate zones, mislocalization of neurons at the surface of the cerebral aqueduct, and loss of ependymal cells from the ventricular and aqueductal surfaces, accompanied by the reduced localization of nectin-1 and N-cadherin and the disruption of AJs. In the dorsal telencephalon, loss of afadin leads to a double cortex phenotype in which heterotopic gray matter is interposed between zones of white matter. This subcortical band heterotopia-like phenotype is due to the disruption of AJs between RGCs (Fig. 6.10). Loss of afadin causes mislocalization of N-cadherin and β-catenin on the cell surface of RGCs. Loss of N-cadherin phenocopies loss of afadin in the dorsal telencephalon. Thus, the nectin–afadin system cooperatively functions with the N-cadherin–β-catenin system to regulate the maintenance of AJs between RGCs.

6.4.6 Contextual Fear Memory

Plasticity of hippocampal synapse is implicated in learning paradigms, including contextual fear conditioning (Garcia et al. 1998; Motanis and Maroun 2010; Restivo et al. 2009). Contextual fear conditioning takes place when a neutral context is associated with an aversive unconditioned stimulus which leads to the induction of emotional memory dependent on the hippocampal and amygdale functions. From recent evidence using rodent models of contextual fear memory, nectin-1 plays a key role in this process (Fantin et al. 2013). After contextual fear conditioning, nectin-1 expression is upregulated exclusively in the ventral hippocampus leading to the induction of contextual fear memory. This induction of contextual fear memory is attenuated by infusion of an antibody against nectin 1 in the ventral hippocampus. Thus, recruitment of nectin-1 to the perisynapse in the ventral hippocampus plays an important role in the formation of contextual fear memory.

6.4.7 Stress-Induced Cognitive Dysfunction

Cognitive dysfunction and dysregulation of social behaviors, including social exploration, social memory, and aggressive behaviors, are induced by acute or chronic exposure to stress (de Kloet et al. 2005; McEwen 2003; Sandi and Richter-Levin 2009). The hippocampus is one of the brain regions that display vulnerability to these stresses. Evidence using rodent models shows that nectin-3 is involved in this process. The expression of nectin-3 in the CA3 area of the hippocampus is downregulated by chronic stress in a corticotropin-releasing hormone receptor 1-dependent manner and is associated with spatial memory and dendritic complexity (Wang et al. 2011). Moreover, nectin-3 in the hippocampus is required for the effects of acute stress on memory and structural plasticity (Wang et al. 2013). On the other hand, a recent finding using rodent models shows that the expression of nectin-3 in the CA1 area of the hippocampus is downregulated by chronic stress in a matrix metalloproteinase 9-dependent manner and is associated with social behaviors and CA1-mediated cognition (van der Kooij et al. 2014). Thus, the corticotropin-releasing hormone–corticotropin-releasing hormone receptor 1 signaling pathway and the activity of matrix metalloproteinase 9 functionally interacts with the nectin–afadin system and mediates the stress-induced effects on memory and structural plasticity.

6.4.8 Cancer

Accumulating evidence shows that the expression of nectins is altered, serves as a prognostic marker, and presents a marker for cancer therapy in various types of carcinoma tissues. The expression of nectin-1 in AJs is downregulated whereas that of the cadherin–catenin is preserved at the early stages of malignant transformation of keratinocytes, such as basal and squamous cell carcinomas (Matsushima et al. 2003). The expression of nectin-1 is also downregulated in the advancing edge of invasive squamous carcinomas of the human uterine cervix (Guzman et al. 2006). On the other hand, the expression of nectin-1 is upregulated in the serum of patients with castration-resistant prostate cancer and colorectal endometriotic tissues which are benign endometrial glands observed outside the uterus which potentially leads to tumor formation (Ballester et al. 2012; Kälin et al. 2011). In addition, high expressions of nectin-1 in head and neck squamous carcinoma cells, thyroid cancer cells, and invasive murine skin squamous carcinoma cells serve as a target for HSV-1 oncolytic therapy, which promotes infection-mediated lysis of cancer cells (Huang et al. 2007; Yu et al. 2007; Yu et al. 2005). The expression of nectin-2 is upregulated in hepatocellular carcinoma tissues (Kurokawa et al. 2006), squamous cell/adenosquamous carcinomas and adenocarcinoma tissues of gallbladder (Miao et al. 2013), and breast and ovarian carcinoma tissues (Oshima et al. 2013). The expression of nectin-3 is downregulated in metastatic breast cancer tissues (Martin et al. 2013). On the other hand, the expression of nectin-3 is upregulated in lung adenocarcinoma (Maniwa et al. 2012) and ovarian, colorectal, and peritoneal endometriotic tissues (Ballester et al. 2012). The expression of nectin-4 is upregulated in breast carcinoma tissues (Athanassiadou et al. 2011; Fabre-Lafay et al. 2005; Pavlova et al. 2013), non-small–cell lung cancer tissues (Takano et al. 2009), ovarian cancer tissues (Derycke et al. 2010), pancreatic cancer tissues (Izumi et al. 2015; Nishiwada et al. 2015), and eutopic endometrium tissues with endometriosis (Ballester et al. 2012). Nectin-2 is involved in cancer immunity by modulating the signaling cascade in certain T cells and NK cells (Pende et al. 2005; Stanietsky et al. 2009). The interaction between nectin-2 expressed in target cells and DNAM-1 expressed in NK cells enhances the NK-mediated lysis of tumor cells. On the other hand, the interaction between nectin-2 expressed in target cells and TIGIT expressed in human NK cells inhibits NK-mediated cytotoxicity. AF-6/afadin also provides a potential marker of metastasis and a prognostic predicator of human breast and colon cancer as the expression of AF-6/afadin is reduced in human breast and colon cancer tissues, which is correlated with poor prognosis of breast and colon cancer patients (Letessier et al. 2007; Sun et al. 2014). Loss of AF-6/afadin aggravates the malignant phenotype of cancer cells by inducing cell migration, invasion, and tumor growth through the activation of ERK in breast and colon cancer cells (Fournier et al. 2011; Sun et al. 2014).

6.5 Conclusions and Perspectives

As described in this chapter, it is clear that the nectin–afadin system plays pivotal roles in various developmental processes including the formation of cell–cell adhesion, and the control of cell migration and proliferation. The remaining important questions concerning the roles of the nectin–afadin system include (1) the remodeling and reorganization of once-formed cell–cell adhesion, (2) the maintenance of remodeled and reorganized cell–cell adhesion, and (3) how the cadherin system is involved in these processes. A suitable cellular system for testing these hypotheses is the activity-dependent remodeling, reorganization, and maintenance of synapses in the central nervous system. The broad distribution of expression of nectins and afadin in mammalian tissues underscores the importance of the roles of the nectin–afadin system independently of the cadherin system in various developmental processes, aging, and in the onset and progression of human disease. The physiological and pathological roles of the nectin–afadin system either cooperatively with or independently of the cadherin system will be continuously revealed in the future by precise and detailed dissection of their molecular mechanisms. In addition, various mouse models including conditional mutant and knock-in mouse lines of nectins, afadin, cadherins, and functionally related genes will be powerful tools for elucidating the molecular impact on various tissues and organs. Moreover, unbiased high-throughput analyses and human genetic studies to identify disease-linked variants in genes of the nectin–afadin and cadherin–catenin systems will greatly contribute to our understanding of in vivo functions of these systems and translational research of these and related molecules.

Acknowledgments We thank our colleagues and collaborators for their enormous contributions and outstanding achievements, Drs. Kenji Mandai, Yoshiyuki Rikitake, Kiyohito Mizutani, and Tomohiko Maruo for their helpful advice and discussions. This work was supported by Grants-in-Aid for Scientific Research (S) (21227005 to Y.T.) and (B) (23300284 to A.M.) from the Japan Society for the Promotion of Science and Grant-in-Aid for Scientific Research on Innovative Areas (26114007 to Y.T.) from the Ministry of Education, Culture, Sports, Science and Technology, Japan.

References

Adachi M, Hamazaki Y, Kobayashi Y et al (2009) Similar and distinct properties of MUPP1 and Patj, two homologous PDZ domain-containing tight-junction proteins. Mol Cell Biol 29:2372–2389

Amano H, Ikeda W, Kawano S et al (2008) Interaction and localization of Necl-5 and PDGF receptor beta at the leading edges of moving NIH3T3 cells: Implications for directional cell movement. Genes Cells 13:269–284

Amaral DG, Dent JA (1981) Development of the mossy fibers of the dentate gyrus: I. A light and electron microscopic study of the mossy fibers and their expansions. J Comp Neurol 195:51–86

Asada M, Irie K, Morimoto K et al (2003) ADIP, a novel Afadin- and alpha-actinin-binding protein localized at cell-cell adherens junctions. J Biol Chem 278:4103–4111

Asakura T, Nakanishi H, Sakisaka T et al (1999) Similar and differential behaviour between the nectin-afadin-ponsin and cadherin-catenin systems during the formation and disruption of the polarized junctional alignment in epithelial cells. Genes Cells 4:573–581

Athanassiadou AM, Patsouris E, Tsipis A et al (2011) The significance of Survivin and Nectin-4 expression in the prognosis of breast carcinoma. Folia Histochem Cytobiol 49:26–33

Ballester M, Gonin J, Rodenas A et al (2012) Eutopic endometrium and peritoneal, ovarian and colorectal endometriotic tissues express a different profile of nectin-1, -3, -4 and nectin-like molecule 2. Hum Reprod 27:3179–3186

Barrett K, Leptin M, Settleman J (1997) The Rho GTPase and a putative RhoGEF mediate a signaling pathway for the cell shape changes in Drosophila gastrulation. Cell 91:905–915

Barron MJ, Brookes SJ, Draper CE et al (2008) The cell adhesion molecule nectin-1 is critical for normal enamel formation in mice. Hum Mol Genet 17:3509–3520

Beaudoin GM 3rd, Schofield CM, Nuwal T et al (2012) Afadin, a Ras/Rap effector that controls cadherin function, promotes spine and excitatory synapse density in the hippocampus. J Neurosci 32:99–110

Bermejo E, Martinez-Frias ML (1998) Congenital eye malformations: clinical-epidemiological analysis of 1,124,654 consecutive births in Spain. Am J Med Genet 75:497–504

Birch J, Juleff N, Heaton MP et al (2013) Characterization of ovine Nectin-4, a novel peste des petits ruminants virus receptor. J Virol 87:4756–4761

Bottino C, Castriconi R, Pende D et al (2003) Identification of PVR (CD155) and Nectin-2 (CD112) as cell surface ligands for the human DNAM-1 (CD226) activating molecule. J Exp Med 198:557–567

Brancati F, Fortugno P, Bottillo I et al (2010) Mutations in PVRL4, encoding cell adhesion molecule nectin-4, cause ectodermal dysplasia-syndactyly syndrome. Am J Hum Genet 87:265–273

Cain S, Martinez G, Kokkinos MI et al (2008) Differential requirement for β-catenin in epithelial and fiber cells during lens development. Dev Biol 321:420–433

Choi YS, Gumbiner B (1989) Expression of cell adhesion molecule E-cadherin in Xenopus embryos begins at gastrulation and predominates in the ectoderm. J Cell Biol 108:2449–2458

Chow RL, Lang RA (2001) Early eye development in vertebrates. Annu Rev Cell Dev Biol 17:255–296

Christofori G (2003) Split personalities: the agonistic antagonist Sprouty. Nat Cell Biol 5:377–379

Colas JF, Schoenwolf GC (2001) Towards a cellular and molecular understanding of neurulation. Dev Dyn 221:117–145

Davidson LA, Keller RE (1999) Neural tube closure in Xenopus laevis involves medial migration, directed protrusive activity, cell intercalation and convergent extension. Development 126:4547–4556

Dawes-Hoang RE, Parmar KM, Christiansen AE et al (2005) folded gastrulation, cell shape change and the control of myosin localization. Development 132:4165–4178

de Kloet ER, Joels M, Holsboer F (2005) Stress and the brain: from adaptation to disease. Nat Rev Neurosci 6:463–475

Derycke MS, Pambuccian SE, Gilks CB et al (2010) Nectin 4 overexpression in ovarian cancer tissues and serum: potential role as a serum biomarker. Am J Clin Pathol 134:835–845

Detrick RJ, Dickey D, Kintner CR (1990) The effects of N-cadherin misexpression on morphogenesis in Xenopus embryos. Neuron 4:493–506

Devilard E, Xerri L, Dubreuil P et al (2013) Nectin-3 (CD113) interacts with Nectin-2 (CD112) to promote lymphocyte transendothelial migration. PLoS One 8:e77424

Dudak A, Kim J, Cheong B et al (2011) Membrane palmitoylated proteins regulate trafficking and processing of nectins. Eur J Cell Biol 90:365–375

Eberle F, Dubreuil P, Mattei MG et al (1995) The human PRR2 gene, related to the human poliovirus receptor gene (PVR), is the true homolog of the murine MPH gene. Gene 159:267–272

Fabre-Lafay S, Garrido-Urbani S, Reymond N et al (2005) Nectin-4, a new serological breast cancer marker, is a substrate for tumor necrosis factor-alpha-converting enzyme (TACE)/ADAM-17. J Biol Chem 280:19543–19550

Fairbank PD, Lee C, Ellis A et al (2006) Shroom2 (APXL) regulates melanosome biogenesis and localization in the retinal pigment epithelium. Development 133:4109–4118

Fantin M, van der Kooij MA, Grosse J et al (2013) A key role for nectin-1 in the ventral hippocampus in contextual fear memory. PLoS One 8:e56897

Fortugno P, Josselin E, Tsiakas K et al (2014) Nectin-4 mutations causing ectodermal dysplasia with syndactyly perturb the rac1 pathway and the kinetics of adherens junction formation. J Invest Dermatol 134:2146–2153

Fournier G, Cabaud O, Josselin E et al (2011) Loss of Af6/afadin, a marker of poor outcome in breast cancer, induces cell migration, invasiveness and tumor growth. Oncogene 30:3862–3874

Franco SJ, Martinez-Garay I, Gil-Sanz C et al (2011) Reelin regulates cadherin function via Dab1/Rap1 to control neuronal migration and lamination in the neocortex. Neuron 69:482–497

Fujimori T, Miyatani S, Takeichi M (1990) Ectopic expression of N-cadherin perturbs histogenesis in Xenopus embryos. Development 110:97–104

Fujito T, Ikeda W, Kakunaga S et al (2005) Inhibition of cell movement and proliferation by cell-cell contact-induced interaction of Necl-5 and nectin-3. J Cell Biol 171:165–173

Fujiwara Y, Goda N, Tamashiro T et al (2015) Crystal structure of afadin PDZ domain–nectin-3 complex shows the structural plasticity of the ligand-binding site. Protein Sci 24:376–385

Fukuda T, Kominami K, Wang S et al (2014) Aberrant cochlear hair cell attachments caused by Nectin-3 deficiency result in hair bundle abnormalities. Development 141:399–409

Fukuhara A, Shimizu K, Kawakatsu T et al (2003) Involvement of nectin-activated Cdc42 small G protein in organization of adherens and tight junctions in Madin-Darby canine kidney cells. J Biol Chem 278:51885–51893

Garcia R, Tocco G, Baudry M et al (1998) Exposure to a conditioned aversive environment interferes with long-term potentiation induction in the fimbria-CA3 pathway. Neuroscience 82:139–145

Geraghty RJ, Krummenacher C, Cohen GH et al (1998) Entry of alphaherpesviruses mediated by poliovirus receptor-related protein 1 and poliovirus receptor. Science 280:1618–1620

Gil-Sanz C, Franco SJ, Martinez-Garay I et al (2013) Cajal-Retzius cells instruct neuronal migration by coincidence signaling between secreted and contact-dependent guidance cues. Neuron 79:461–477

Gil-Sanz C, Landeira B, Ramos C et al (2014) Proliferative defects and formation of a double cortex in mice lacking Mltt4 and Cdh2 in the dorsal telencephalon. J Neurosci 34:10475–10487

Ginsberg D, DeSimone D, Geiger B (1991) Expression of a novel cadherin (EP-cadherin) in unfertilized eggs and early Xenopus embryos. Development 111:315–325

Goodrich LV, Strutt D (2011) Principles of planar polarity in animal development. Development 138:1877–1892

Graw J (2009) Mouse models of cataract. J Genet 88:469–486

Gray RS, Roszko I, Solnica-Krezel L (2011) Planar cell polarity: coordinating morphogenetic cell behaviors with embryonic polarity. Dev Cell 21:120–133

Gregory Call S, Brereton D, Bullard JT et al (2011) A zyxin-nectin interaction facilitates zyxin localization to cell-cell adhesions. Biochem Biophys Res Commun 415:485–489

Guzman G, Oh S, Shukla D et al (2006) Nectin-1 expression in the normal and neoplastic human uterine cervix. Arch Pathol Lab Med 130:1193–1195

Häcker U, Perrimon ND (1998) RhoGEF2 encodes a member of the Dbl family of oncogenes and controls cell shape changes during gastrulation in Drosophila. Genes Dev 12:274–284

Haigo SL, Hildebrand JD, Harland RM et al (2003) Shroom induces apical constriction and is required for hingepoint formation during neural tube closure. Curr Biol 13:2125–2137

Harold D, Abraham R, Hollingworth P et al (2009) Genome-wide association study identifies variants at CLU and PICALM associated with Alzheimer's disease. Nat Genet 41:1088–1093

Harrison OJ, Vendome J, Brasch J et al (2012) Nectin ectodomain structures reveal a canonical adhesive interface. Nat Struct Mol Biol 19:906–915

Häussinger D, Ahrens T, Aberle T et al (2004) Proteolytic E-cadherin activation followed by solution NMR and X-ray crystallography. EMBO J 23:1699–1708

Hejtmancik JF (2008) Congenital cataracts and their molecular genetics. Sem Cell Dev Biol 19:134–149

Hildebrand JD (2005) Shroom regulates epithelial cell shape via the apical positioning of an actomyosin network. J Cell Sci 118:5191–5203

Hirota T, Irie K, Okamoto R et al (2005) Transcriptional activation of the mouse Necl-5/Tage4/PVR/CD155 gene by fibroblast growth factor or oncogenic Ras through the Raf-MEK-ERK-AP-1 pathway. Oncogene 24:2229–2235

Honda T, Sakisaka T, Yamada T et al (2006) Involvement of nectins in the formation of puncta adherentia junctions and the mossy fiber trajectory in the mouse hippocampus. Mol Cell Neurosci 31:315–325

Hoshino T, Sakisaka T, Baba T et al (2005) Regulation of E-cadherin endocytosis by nectin through afadin, Rap1, and p120ctn. J Biol Chem 280:24095–24103

Huang YY, Yu Z, Lin SF et al (2007) Nectin-1 is a marker of thyroid cancer sensitivity to herpes oncolytic therapy. J Clin Endocrinol Metab 92:1965–1970

Ikeda W, Nakanishi H, Miyoshi J et al (1999) Afadin: A key molecule essential for structural organization of cell-cell junctions of polarized epithelia during embryogenesis. J Cell Biol 146:1117–1132

Ikeda W, Kakunaga S, Itoh S et al (2003) Tage4/Nectin-like molecule-5 heterophilically trans-interacts with cell adhesion molecule Nectin-3 and enhances cell migration. J Biol Chem 278:28167–28172

Ikeda W, Kakunaga S, Takekuni K et al (2004) Nectin-like molecule-5/Tage4 enhances cell migration in an integrin-dependent, Nectin-3-independent manner. J Biol Chem 279:18015–18025

Inagaki M, Irie K, Ishizaki H et al (2005) Roles of cell-adhesion molecules nectin 1 and nectin 3 in ciliary body development. Development 132:1525–1537

Inagaki M, Irie K, Ishizaki H et al (2006) Role of cell adhesion molecule nectin-3 in spermatid development. Genes Cells 11:1125–1132

Indra I, Hong S, Troyanovsky R et al (2013) The adherens junction: a mosaic of cadherin and nectin clusters bundled by actin filaments. J Invest Dermatol 133:2546–2554

Indra I, Troyanovsky R, Troyanovsky S (2014) Afadin controls cadherin cluster stability using clathrin-independent mechanism. Tissue Barriers 2:e28687

Inoue T, Fujiwara T, Rikitake Y et al (2015) Nectin-1 spots as a novel adhesion apparatus that tethers mitral cell lateral dendrites in a dendritic meshwork structure of the developing mouse olfactory bulb. J Comp Neurol 523:1824–1839

Irie K, Shimizu K, Sakisaka T et al (2004) Roles and modes of action of nectins in cell-cell adhesion. Semin Cell Dev Biol 15:643–656

Ishiuchi T, Takeichi M (2012) Nectins localize Willin to cell-cell junctions. Genes Cells 17:387–397

Iwasawa N, Negishi M, Oinuma I (2012) R-Ras controls axon branching through afadin in cortical neurons. Mol Biol Cell 23:2793–2804

Izumi H, Hirabayashi K, Nakamura N et al (2015) Nectin expression in pancreatic adenocarcinoma: nectin-3 is associated with a poor prognosis. Surg Today 45:487–494

Jelani M, Chishti MS, Ahmad W (2011) Mutation in PVRL4 gene encoding nectin-4 underlies ectodermal-dysplasia-syndactyly syndrome (EDSS1). J Hum Genet 56:352–357

Jensen S, Goldschmidt E (1971) Genetic counselling in sporadic cases of congenital cataract. Acta Ophthalmol (Copenh) 49:572–576

Jossin Y, Cooper JA (2011) Reelin, Rap1 and N-cadherin orient the migration of multipolar neurons in the developing neocortex. Nat Neurosci 14:697–703

Kajita M, Ikeda W, Tamaru Y et al (2007) Regulation of platelet-derived growth factor-induced Ras signaling by poliovirus receptor Necl-5 and negative growth regulator Sprouty2. Genes Cell 12:345–357

Kakunaga S, Ikeda W, Shingai T et al (2004) Enhancement of serum- and platelet-derived growth factor-induced cell proliferation by Necl-5/Tage4/poliovirus receptor/CD155 through the Ras-Raf-MEK-ERK signaling. J Biol Chem 279:36419–36425

Kälin M, Cima I, Schiess R et al (2011) Novel prognostic markers in the serum of patients with castration-resistant prostate cancer derived from quantitative analysis of the pten conditional knockout mouse proteome. Eur Urol 60:1235–1243

Kawakatsu T, Shimizu K, Honda T et al (2002) Trans-interactions of nectins induce formation of filopodia and Lamellipodia through the respective activation of Cdc42 and Rac small G proteins. J Biol Chem 277:50749–50755

Kawakatsu T, Ogita H, Fukuhara T et al (2005) Vav2 as a Rac-GDP/GTP exchange factor responsible for the nectin-induced, c-Src- and Cdc42-mediated activation of Rac. J Biol Chem 280:4940–4947

Kawauchi T, Sekine K, Shikanai M et al (2010) Rab GTPases-dependent endocytic pathways regulate neuronal migration and maturation through N-cadherin trafficking. Neuron 67:588–602

Kelley MW (2006) Regulation of cell fate in the sensory epithelia of the inner ear. Nat Rev Neurosci 7:837–849

Kim HJ, Bar-Sagi D (2004) Modulation of signalling by Sprouty: a developing story. Nat Rev Mol Cell Biol 5:441–450

Kinugasa M, Amano H, Satomi-Kobayashi S et al (2012) Necl-5/poliovirus receptor interacts with VEGFR2 and regulates VEGF-induced angiogenesis. Circ Res 110:716–726

Kobayashi R, Kurita S, Miyata M et al (2014) s-Afadin binds more preferentially to the cell adhesion molecules nectins than l-afadin. Genes Cells 19:853–863

Koch AW, Pokutta S, Lustig A et al (1997) Calcium binding and homoassociation of E-cadherin domains. Biochemistry 36:7697–7705

Koike S, Horie H, Ise I et al (1990) The poliovirus receptor protein is produced both as membrane-bound and secreted forms. EMBO J 9:3217–3224

Komura H, Ogita H, Ikeda W et al (2008) Establishment of cell polarity by afadin during the formation of embryoid bodies. Genes Cells 13:79–90

Kurokawa Y, Honma K, Takemasa I et al (2006) Central genetic alterations common to all HCV-positive, HBV-positive and non-B, non-C hepatocellular carcinoma: a new approach to identify novel tumor markers. Int J Oncol 28:383–391

Lachke SA, Higgins AW, Inagaki M et al (2012) The cell adhesion gene PVRL3 is associated with congenital ocular defects. Hum Genet 131:235–250

Lecuit T, Lenne PF (2007) Cell surface mechanics and the control of cell shape, tissue patterns and morphogenesis. Nat Rev Mol Cell Biol 8:633–644

Lee C, Scherr HM, Wallingford JB (2007) Shroom family proteins regulate gamma-tubulin distribution and microtubule architecture during epithelial cell shape change. Development 134:1431–1441

Letessier A, Garrido-Urbani S, Ginestier C et al (2007) Correlated break at PARK2/FRA6E and loss of AF-6/Afadin protein expression are associated with poor outcome in breast cancer. Oncogene 26:298–307

Levi G, Gumbiner B, Thiery JP (1991) The distribution of E-cadherin during Xenopus laevis development. Development 111:159–169

Logue MW, Schu M, Vardarajan BN et al (2011) A comprehensive genetic association study of Alzheimer disease in African Americans. Arch Neurol 68:1569–1579

Lopez M, Eberle F, Mattei MG et al (1995) Complementary DNA characterization and chromosomal localization of a human gene related to the poliovirus receptor-encoding gene. Gene 155:261–265

Lorger M, Moelling K (2006) Regulation of epithelial wound closure and intercellular adhesion by interaction of AF6 with actin cytoskeleton. J Cell Sci 119:3385–3398

Maddala R, Chauhan BK, Walker C et al (2011) Rac1 GTPase-deficient mouse lens exhibits defects in shape, suture formation, fiber cell migration and survival. Dev Biol 360:30–43

Mandai K, Nakanishi H, Satoh A et al (1997) Afadin: A novel actin filament-binding protein with one PDZ domain localized at cadherin-based cell-to-cell adherens junction. J Cell Biol 139:517–528

Mandai K, Nakanishi H, Satoh A et al (1999) Ponsin/SH3P12: an l-afadin- and vinculin-binding protein localized at cell-cell and cell-matrix adherens junctions. J Cell Biol 144:1001–1017

Mandai K, Rikitake Y, Mori M et al (2015) Nectins and nectin-like molecules in development and disease. Curr Top Dev Biol 112:197–231

Maniwa Y, Nishio W, Okita Y et al (2012) Expression of nectin 3: Novel prognostic marker of lung adenocarcinoma. Thoracic Can 3:175–181

Martin TA, Lane J, Harrison GM et al (2013) The expression of the Nectin complex in human breast cancer and the role of Nectin-3 in the control of tight junctions during metastasis. PLoS One 8:e82696

Martinez-Rico C, Pincet F, Perez E et al (2005) Separation force measurements reveal different types of modulation of E-cadherin-based adhesion by nectin-1 and -3. J Biol Chem 280:4753–4760

Martinez-Sanz E, Del Rio A, Barrio C et al (2008) Alteration of medial-edge epithelium cell adhesion in two Tgf-beta3 null mouse strains. Differentiation 76:417–430

Matsushima H, Utani A, Endo H et al (2003) The expression of nectin-1alpha in normal human skin and various skin tumours. Br J Dermatol 148:755–762

Maurin H, Seymour CM, Lechat B et al (2013) Tauopathy differentially affects cell adhesion molecules in mouse brain: early down-regulation of nectin-3 in stratum lacunosum moleculare. PLoS One 8:e63589

Mayor R, Carmona-Fontaine C (2010) Keeping in touch with contact inhibition of locomotion. Trends Cell Biol 20:319–328

McAvoy JW (1980) Induction of the eye lens. Differentiation 17:137–149

McClatchey AI, Yap AS (2012) Contact inhibition (of proliferation) redux. Curr Opin Cell Biol 24:685–694

McEwen BS (2003) Mood disorders and allostatic load. Biol Psychiatry 54:200–207

McKenzie E, Krupin A, Kelley MW (2004) Cellular growth and rearrangement during the development of the mammalian organ of Corti. Dev Dyn 229:802–812

Mendelsohn CL, Wimmer E, Racaniello VR (1989) Cellular receptor for poliovirus: molecular cloning, nucleotide sequence, and expression of a new member of the immunoglobulin superfamily. Cell 56:855–865

Miao X, Yang ZL, Xiong L et al (2013) Nectin-2 and DDX3 are biomarkers for metastasis and poor prognosis of squamous cell/adenosquamous carcinomas and adenocarcinoma of gallbladder. Int J Clin Exp Pathol 6:179–190

Minami Y, Ikeda W, Kajita M et al (2007) Necl-5/poliovirus receptor interacts in cis with integrin alphaVbeta3 and regulates its clustering and focal complex formation. J Biol Chem 282:18481–18496

Minami A, Mizutani K, Waseda M et al (2010) Necl-5/PVR enhances PDGF-induced attraction of growing microtubules to the plasma membrane of the leading edge of moving NIH3T3 cells. Genes Cells 15:1123–1135

Miyahara M, Nakanishi H, Takahashi K et al (2000) Interaction of nectin with afadin is necessary for its clustering at cell-cell contact sites but not for its cis dimerization or trans interaction. J Biol Chem 275:613–618

Miyata M, Rikitake Y, Takahashi M et al (2009) Regulation by afadin of cyclical activation and inactivation of Rap1, Rac1, and RhoA small G proteins at leading edges of moving NIH3T3 cells. J Biol Chem 284:24595–24609

Miyoshi J, Takai Y (2005) Molecular perspective on tight-junction assembly and epithelial polarity. Adv Drug Deliv Rev 57:815–855

Miyoshi J, Takai Y (2007) Nectin and nectin-like molecules: biology and pathology. Am J Nephrol 27:590–604

Mizoguchi A, Nakanishi H, Kimura K et al (2002) Nectin: an adhesion molecule involved in formation of synapses. J Cell Biol 156:555–565

Montcouquiol M, Kelley MW (2003) Planar and vertical signals control cellular differentiation and patterning in the mammalian cochlea. J Neurosci 23:9469–9478

Montcouquiol M, Rachel RA, Lanford PJ et al (2003) Identification of Vangl2 and Scrb1 as planar polarity genes in mammals. Nature 423:173–177

Mori M, Rikitake Y, Mandai K et al (2014) Roles of nectins and nectin-like molecules in the nervous system. Adv Neurobiol 8:91–116

Morita H, Nandadasa S, Yamamoto TS et al (2010) Nectin-2 and N-cadherin interact through extracellular domains and induce apical accumulation of F-actin in apical constriction of Xenopus neural tube morphogenesis. Development 137:1315–1325

Motanis H, Maroun M (2010) Exposure to a novel context following contextual fear conditioning enhances the induction of hippocampal long-term potentiation. Eur J Neurosci 32:840–846

Mueller S, Rosenquist TA, Takai Y et al (2003) Loss of nectin-2 at Sertoli-spermatid junctions leads to male infertility and correlates with severe spermatozoan head and midpiece malformation, impaired binding to the zona pellucida, and oocyte penetration. Biol Reprod 69:1330–1340

Muhlebach MD, Mateo M, Sinn PL et al (2011) Adherens junction protein nectin-4 is the epithelial receptor for measles virus. Nature 480:530–533

Nadarajah B, Brunstrom JE, Grutzendler J et al (2001) Two modes of radial migration in early development of the cerebral cortex. Nat Neurosci 4:143–150

Nakanishi H, Takai Y (2004) Roles of nectins in cell adhesion, migration and polarization. Biol Chem 385:885–892

Nandadasa S, Tao Q, Menon NR et al (2009) N- and E-cadherins in Xenopus are specifically required in the neural and non-neural ectoderm, respectively, for F-actin assembly and morphogenetic movements. Development 136:1327–1338

Narita H, Yamamoto Y, Suzuki M et al (2011) Crystal Structure of the cis-Dimer of Nectin-1: implications for the architecture of cell-cell junctions. J Biol Chem 286:12659–12669

Nejsum LN, Nelson WJ (2009) Epithelial cell surface polarity: the early steps. Front Biosci (Landmark Ed) 14:1088–1098

Nelson WJ, Nusse R (2004) Convergence of Wnt, beta-catenin, and cadherin pathways. Science 303:1483–1487

Niessen CM, Leckband D, Yap AS (2011) Tissue organization by cadherin adhesion molecules: dynamic molecular and cellular mechanisms of morphogenetic regulation. Physiol Rev 91:691–731

Nishimura T, Takeichi M (2008) Shroom3-mediated recruitment of Rho kinases to the apical cell junctions regulates epithelial and neuroepithelial planar remodeling. Development 135:1493–1502

Nishiwada S, Sho M, Yasuda S et al (2015) Nectin-4 expression contributes to tumor proliferation, angiogenesis and patient prognosis in human pancreatic cancer. J Exp Clin Cancer Res 34:30

Noren NK, Liu BP, Burridge K et al (2000) p120 catenin regulates the actin cytoskeleton via Rho family GTPases. J Cell Biol 150:567–580

Noyce RS, Bondre DG, Ha MN et al (2011) Tumor cell marker PVRL4 (nectin 4) is an epithelial cell receptor for measles virus. PLoS Pathog 7:e1002240

Noyce RS, Delpeut S, Richardson CD (2013) Dog nectin-4 is an epithelial cell receptor for canine distemper virus that facilitates virus entry and syncytia formation. Virology 436:210–220

Ogita H, Takai Y (2006) Nectins and nectin-like molecules: roles in cell adhesion, polarization, movement, and proliferation. IUBMB Life 58:334–343

Ogita H, Takai Y (2008) Cross-talk among integrin, cadherin, and growth factor receptor: roles of nectin and nectin-like molecule. Int Rev Cytol 265:1–54

Ogita H, Ikeda W, Takai Y (2008) Roles of cell adhesion molecules nectin and nectin-like molecule-5 in the regulation of cell movement and proliferation. J Microsc 231:455–465

Ogita H, Rikitake Y, Miyoshi J et al (2010) Cell adhesion molecules nectins and associating proteins: Implications for physiology and pathology. Proc Jpn Acad Ser B Phys Biol Sci 86:621–629

Ohka S, Nihei C, Yamazaki M et al (2012) Poliovirus trafficking toward central nervous system via human poliovirus receptor-dependent and -independent pathway. Front Microbiol 3:147

Okabe N, Ozaki-Kuroda K, Nakanishi H et al (2004a) Expression patterns of nectins and afadin during epithelial remodeling in the mouse embryo. Dev Dyn 230:174–186

Okabe N, Shimizu K, Ozaki-Kuroda K et al (2004b) Contacts between the commissural axons and the floor plate cells are mediated by nectins. Dev Biol 273:244–256

Ooshio T, Irie K, Morimoto K et al (2004) Involvement of LMO7 in the association of two cell-cell adhesion molecules, nectin and E-cadherin, through afadin and alpha-actinin in epithelial cells. J Biol Chem 279:31365–31373

Ooshio T, Fujita N, Yamada A et al (2007) Cooperative roles of Par-3 and afadin in the formation of adherens and tight junctions. J Cell Sci 120:2352–2365

Ooshio T, Kobayashi R, Ikeda W et al (2010) Involvement of the interaction of afadin with ZO-1 in the formation of tight junctions in Madin-Darby canine kidney cells. J Biol Chem 285:5003–5012

Oshima T, Sato S, Kato J et al (2013) Nectin-2 is a potential target for antibody therapy of breast and ovarian cancers. Mol Cancer 12:60

Ozaki-Kuroda K, Nakanishi H, Ohta H et al (2002) Nectin couples cell-cell adhesion and the actin scaffold at heterotypic testicular junctions. Curr Biol 12:1145–1150

Pavlova NN, Pallasch C, Elia AE et al (2013) A role for PVRL4-driven cell-cell interactions in tumorigenesis. Elife 2:e00358

Pende D, Bottino C, Castriconi R et al (2005) PVR (CD155) and Nectin-2 (CD112) as ligands of the human DNAM-1 (CD226) activating receptor: involvement in tumor cell lysis. Mol Immunol 42:463–469

Perez-Moreno M, Fuchs E (2006) Catenins: keeping cells from getting their signals crossed. Dev Cell 11:601–612

Perez-Moreno M, Jamora C, Fuchs E (2003) Sticky business: orchestrating cellular signals at adherens junctions. Cell 112:535–548

Petrova YI, Spano MM, Gumbiner BM (2012) Conformational epitopes at cadherin calcium-binding sites and p120-catenin phosphorylation regulate cell adhesion. Mol Biol Cell 23:2092–2108

Pokutta S, Weis WI (2007) Structure and mechanism of cadherins and catenins in cell-cell contacts. Annu Rev Cell Dev Biol 23:237–261

Pokutta S, Drees F, Takai Y et al (2002) Biochemical and structural definition of the l-afadin- and actin-binding sites of alpha-catenin. J Biol Chem 277:18868–18874
Pratakpiriya W, Seki F, Otsuki N et al (2012) Nectin4 is an epithelial cell receptor for canine distemper virus and involved in neurovirulence. J Virol 86:10207–10210
Rakic P (1972) Mode of cell migration to the superficial layers of fetal monkey neocortex. J Comp Neurol 145:61–83
Restivo L, Vetere G, Bontempi B et al (2009) The formation of recent and remote memory is associated with time-dependent formation of dendritic spines in the hippocampus and anterior cingulated cortex. J Neurosci 29:8209–8214
Reymond N, Fabre S, Lecocq E et al (2001) Nectin4/PRR4, a new afadin-associated member of the nectin family that trans-interacts with nectin1/PRR1 through V domain interaction. J Biol Chem 276:43205–43215
Reymond N, Garrido-Urbani S, Borg JP et al (2005) PICK-1: a scaffold protein that interacts with Nectins and JAMs at cell junctions. FEBS Lett 579:2243–2249
Rikitake Y, Takai Y (2008) Interactions of the cell adhesion molecule nectin with transmembrane and peripheral membrane proteins for pleiotropic functions. Cell Mol Life Sci 65:253–263
Rikitake Y, Mandai K, Takai Y (2012) The role of nectins in different types of cell-cell adhesion. J Cell Sci 125:3713–3722
Sakamoto Y, Ogita H, Komura H et al (2008) Involvement of nectin in inactivation of integrin alpha(v)beta(3) after the establishment of cell-cell adhesion. J Biol Chem 283:496–505
Sakisaka T, Takai Y (2004) Biology and pathology of nectins and nectin-like molecules. Curr Opin Cell Biol 16:513–521
Sakisaka T, Taniguchi T, Nakanishi H et al (2001) Requirement of interaction of nectin-1alpha/HveC with afadin for efficient cell-cell spread of herpes simplex virus type 1. J Virol 75:4734–4743
Sakisaka T, Ikeda W, Ogita H et al (2007) The roles of nectins in cell adhesions: cooperation with other cell adhesion molecules and growth factor receptors. Curr Opin Cell Biol 19:593–602
Samanta D, Ramagopal UA, Rubinstein R et al (2012) Structure of Nectin-2 reveals determinants of homophilic and heterophilic interactions that control cell-cell adhesion. Proc Natl Acad Sci U S A 109:14836–14840
Sandi C, Richter-Levin G (2009) From high anxiety trait to depression: a neurocognitive hypothesis. Trends Neurosci 32:312–320
Sato T, Fujita N, Yamada A et al (2006) Regulation of the assembly and adhesion activity of E-cadherin by nectin and afadin for the formation of adherens junctions in Madin-Darby canine kidney cells. J Biol Chem 281:5288–5299
Satoh-Horikawa K, Nakanishi H, Takahashi K et al (2000) Nectin-3, a new member of immunoglobulin-like cell adhesion molecules that shows homophilic and heterophilic cell-cell adhesion activities. J Biol Chem 275:10291–10299
Sawyer JK, Harris NJ, Slep KC et al (2009) The Drosophila afadin homologue Canoe regulates linkage of the actin cytoskeleton to adherens junctions during apical constriction. J Cell Biol 186:57–73
Sekine K, Honda T, Kawauchi T (2011) The outermost region of the developing cortical plate is crucial for both the switch of the radial migration mode and the Dab1-dependent "inside-out" lamination in the neocortex. J Neurosci 31:9426–9439
Shimizu K, Takai Y (2003) Roles of the intercellular adhesion molecule nectin in intracellular signaling. J Biochem 134:631–636
Shimono Y, Rikitake Y, Mandai K et al (2012) Immunoglobulin superfamily receptors and adherens junctions. Subcell Biochem 60:137–170
Sipe CW, Lu X (2011) Kif3a regulates planar polarization of auditory hair cells through both ciliary and non-ciliary mechanisms. Development 138:3441–3449
Sipe CW, Liu L, Lee J et al (2013) Lis1 mediates planar polarity of auditory hair cells through regulation of microtubule organization. Development 140:1785–1795

Soriano E, Del Rio JA (2005) The cells of cajal-retzius: still a mystery one century after. Neuron 46:389–394

Sozen MA, Suzuki K, Tolarova MM et al (2001) Mutation of PVRL1 is associated with sporadic, non-syndromic cleft lip/palate in northern Venezuela. Nat Genet 29:141–142

Spacek J, Lieberman AR (1974) Ultrastructure and three-dimensional organization of synaptic glomeruli in rat somatosensory thalamus. J Anat 117:487–516

Spear PG, Eisenberg RJ, Cohen GH (2000) Three classes of cell surface receptors for alphaherpesvirus entry. Virology 275:1–8

Stanietsky N, Simic H, Arapovic J et al (2009) The interaction of TIGIT with PVR and PVRL2 inhibits human NK cell cytotoxicity. Proc Natl Acad Sci U S A 106:17858–17863

Sun TT, Wang Y, Cheng H et al (2014) Disrupted interaction between CFTR and AF-6/afadin aggravates malignant phenotypes of colon cancer. Biochim Biophys Acta 1843:618–628

Suzuki K, Bustos T, Spritz RA (1998) Linkage disequilibrium mapping of the gene for Margarita Island ectodermal dysplasia (ED4) to 11q23. Am J Hum Genet 63:1102–1107

Suzuki K, Hu D, Bustos T et al (2000) Mutations of PVRL1, encoding a cell-cell adhesion molecule/herpesvirus receptor, in cleft lip/palate-ectodermal dysplasia. Nat Genet 25:427–430

Tabata H, Nakajima K (2003) Multipolar migration: the third mode of radial neuronal migration in the developing cerebral cortex. J Neurosci 23:9996–10001

Tabata H, Kanatani S, Nakajima K (2009) Differences of migratory behavior between direct progeny of apical progenitors and basal progenitors in the developing cerebral cortex. Cereb Cortex 19:2092–2105

Tachibana K, Nakanishi H, Mandai K et al (2000) Two cell adhesion molecules, nectin and cadherin, interact through their cytoplasmic domain-associated proteins. J Cell Biol 150:1161–1176

Takahashi K, Nakanishi H, Miyahara M et al (1999) Nectin/PRR: an immunoglobulin-like cell adhesion molecule recruited to cadherin-based adherens junctions through interaction with Afadin, a PDZ domain-containing protein. J Cell Biol 145:539–549

Takai Y, Nakanishi H (2003) Nectin and afadin: novel organizers of intercellular junctions. J Cell Sci 116:17–27

Takai Y, Irie K, Shimizu K et al (2003a) Nectins and nectin-like molecules: roles in cell adhesion, migration, and polarization. Cancer Sci 94:655–667

Takai Y, Shimizu K, Ohtsuka T (2003b) The roles of cadherins and nectins in interneuronal synapse formation. Curr Opin Neurobiol 13:520–526

Takai Y, Ikeda W, Ogita H et al (2008a) The immunoglobulin-like cell adhesion molecule nectin and its associated protein afadin. Annu Rev Cell Dev Biol 24:309–342

Takai Y, Miyoshi J, Ikeda W et al (2008b) Nectins and nectin-like molecules: roles in contact inhibition of cell movement and proliferation. Nat Rev Mol Cell Biol 9:603–615

Takano A, Ishikawa N, Nishino R et al (2009) Identification of nectin-4 oncoprotein as a diagnostic and therapeutic target for lung cancer. Cancer Res 69:6694–6703

Takei N, Miyashita A, Tsukie T et al (2009) Genetic association study on in and around the APOE in late-onset Alzheimer disease in Japanese. Genomics 93:441–448

Takeichi M (1991) Cadherin cell adhesion receptors as a morphogenetic regulator. Science 251:1451–1455

Takeichi M (2014) Dynamic contacts: rearranging adherens junctions to drive epithelial remodelling. Nat Rev Mol Cell Biol 15:397–410

Takekuni K, Ikeda W, Fujito T et al (2003) Direct binding of cell polarity protein PAR-3 to cell-cell adhesion molecule nectin at neuroepithelial cells of developing mouse. J Biol Chem 278:5497–5500

Taya Y, O'Kane S, Ferguson MW (1999) Pathogenesis of cleft palate in TGF-beta3 knockout mice. Development 126:3869–3879

Taylor JM, Lin E, Susmarski N et al (2007) Alternative entry receptors for herpes simplex virus and their roles in disease. Cell Host Microbe 2:19–28

Togashi H, Miyoshi J, Honda T et al (2006) Interneurite affinity is regulated by heterophilic nectin interactions in concert with the cadherin machinery. J Cell Biol 174:141–151
Togashi H, Kominami K, Waseda M et al (2011) Nectins establish a checkerboard-like cellular pattern in the auditory epithelium. Science 333:1144–1147
Toyoshima D, Mandai K, Maruo T et al (2014) Afadin regulates puncta adherentia junction formation and presynaptic differentiation in hippocampal neurons. PLoS One 9:e89763
Tsukasaki Y, Kitamura K, Shimizu K et al (2007) Role of multiple bonds between the single cell adhesion molecules, nectin and cadherin, revealed by high sensitive force measurements. J Mol Biol 367:996–1006
Tsukita S, Yamazaki Y, Katsuno T et al (2008) Tight junction-based epithelial microenvironment and cell proliferation. Oncogene 27:6930–6938
Uchida N, Honjo Y, Johnson KR et al (1996) The catenin/cadherin adhesion system is localized in synaptic junctions bordering transmitter release zones. J Cell Biol 135:767–779
Umeda K, Iwasawa N, Negishi M et al (2015) A short splicing isoform of afadin suppresses the cortical axon branching in a dominant-negative manner. Mol Biol Cell 26:1957–1970
van der Kooij MA, Fantin M, Rejmak E et al (2014) Role of MMP-9 in stress-induced downregulation of nectin-3 in hippocampal CA1 and associated behavioural alterations. Nat Commun 5:4995
Wagner C, Batiz LF, Rodríguez S et al (2003) Cellular mechanisms involved in the stenosis and obliteration of the cerebral aqueduct of hyh mutant mice developing congenital hydrocephalus. J Neuropathol Exp Neurol 62:1019–1040
Wang XD, Chen Y, Wolf M et al (2011) Forebrain CRHR1 deficiency attenuates chronic stress-induced cognitive deficits and dendritic remodeling. Neurobiol Dis 42:300–310
Wang XD, Su YA, Wagner KV et al (2013) Nectin-3 links CRHR1 signaling to stress-induced memory deficits and spine loss. Nat Neurosci 16:706–713
Wei L, Roberts W, Wang L et al (2001) Rho kinases play an obligatory role in vertebrate embryonic organogenesis. Development 128:2953–2962
Wildenberg GA, Dohn MR, Carnahan RH et al (2006) p120-catenin and p190RhoGAP regulate cell-cell adhesion by coordinating antagonism between Rac and Rho. Cell 127:1027–1039
Xie Z, Huganir RL, Penzes P (2005) Activity-dependent dendritic spine structural plasticity is regulated by small GTPase Rap1 and its target AF-6. Neuron 48:605–618
Yagi T, Takeichi M (2000) Cadherin superfamily genes: functions, genomic organization, and neurologic diversity. Genes Dev 14:1169–1180
Yamada T, Kuramitsu K, Rikitsu E et al (2013) Nectin and junctional adhesion molecule are critical cell adhesion molecules for the apico-basal alignment of adherens and tight junctions in epithelial cells. Genes Cells 18:985–998
Yamamoto H, Maruo T, Majima T et al (2013) Genetic deletion of afadin causes hydrocephalus by destruction of adherens junctions in radial glial and ependymal cells in the midbrain. PLoS One 8:e80356
Yamamoto H, Mandai K, Konno D et al (2015) Impairment of radial glial scaffold-dependent neuronal migration and formation of double cortex by genetic ablation of afadin. Brain Res 1620:139–152
Yamana S, Tokiyama A, Mizutani K et al (2015) The cell adhesion molecule Necl-4/CADM4 serves as a novel regulator for contact inhibition of cell movement and proliferation. PLoS One 10:e0124259
Yonemura S, Itoh M, Nagafuchi A et al (1995) Cell-to-cell adherens junction formation and actin filament organization: similarities and differences between non-polarized fibroblasts and polarized epithelial cells. J Cell Sci 108:127–142
Yoshida N, Liberman MC (1999) Stereociliary anomaly in the guinea pig: effects of hair bundle rotation on cochlear sensitivity. Hear Res 131:29–38
Yoshida T, Miyoshi J, Takai Y et al (2010) Cooperation of nectin-1 and nectin-3 is required for normal ameloblast function and crown shape development in mouse teeth. Dev Dyn 239:2558–2569

Yoshida M, Shimono Y, Togashi H et al (2012) Periderm cells covering palatal shelves have tight junctions and their desquamation reduces the polarity of palatal shelf epithelial cells in palatogenesis. Genes Cells 17:455–472
Yu Z, Chan MK, O-charoenrat P et al (2005) Enhanced nectin-1 expression and herpes oncolytic sensitivity in highly migratory and invasive carcinoma. Clin Cancer Res 11:4889–4897
Yu Z, Adusumilli PS, Eisenberg DP et al (2007) Nectin-1 expression by squamous cell carcinoma is a predictor of herpes oncolytic sensitivity. Mol Ther 15:103–113
Yuste R, Bonhoeffer T (2001) Morphological changes in dendritic spines associated with long-term synaptic plasticity. Annu Rev Neurosci 24:1071–1089
Zhang Y, Sivasankar S, Nelson WJ et al (2009) Resolving cadherin interactions and binding cooperativity at the single-molecule level. Proc Natl Acad Sci U S A 106:109–114

Part III
Atypical Cadherins

Chapter 7
Desmosomal Cadherins

Martyn Chidgey and David Garrod

Abstract The desmosomal cadherins (DCs) are adhesion molecules of desmosomes, intercellular junctions of epithelia and cardiac muscle. DCs have the unique ability to adopt a hyperadhesive state that is characterised by enhanced stability and adhesive strength. Hyperadhesion is important in embryonic development and wound healing, and DC expression is tightly regulated at the transcriptional and posttranslational levels. Desmosomes act as signalling centres and DCs have been linked to an array of intracellular signal transduction pathways that control cell proliferation and differentiation. DCs are targets of autoantibodies, bacterial toxins and mutations, resulting in skin blistering disease, cardiomyopathy and sometimes both. Here we consider the structure and function of the desmosomal cadherins, and their role in normal tissue biology and human disease.

Keywords Desmosomal cadherin • Desmoglein • Desmocollin • Plakoglobin • Plakophilin • Desmoplakin • Hyperadhesion • Pemphigus • Cardiomyopathy

7.1 Introduction

The desmosomal cadherins (DCs), desmocollin and desmoglein (Dsc and Dsg), are adhesion molecules of desmosomes, widely distributed intercellular junctions of epithelia and other tissues of vertebrates. Desmosomes provide strong cell–cell adhesion as well as membrane anchors for the intermediate filament (IF) cytoskeleton, thus forming scaffolding that supports and maintains tissue integrity. In addition to this vitally important but somewhat prosaic role, desmosomes appear to contribute to tissue development and differentiation by playing a part in cell positioning and, directly or indirectly, participating in the regulation of cell proliferation and gene expression. Our understanding of the molecular basis of adhesion by Dsc and Dsg is rudimentary but some recent developments are

M. Chidgey
School of Cancer Sciences, University of Birmingham, Birmingham, UK

D. Garrod (✉)
Faculty of Life Sciences, University of Manchester, Oxford Road, Manchester M13 9PT, UK
e-mail: d.garrod@manchester.ac.uk

S.T. Suzuki, S. Hirano (eds.), *The Cadherin Superfamily*,
DOI 10.1007/978-4-431-56033-3_7

providing the first exciting clues. Equally poorly understood are the signalling processes in which desmosomes participate, and the signals that regulate the formation and breakdown of these strongly adhesive, but clearly highly dynamic, junctions. In both areas stimulating recent discoveries have been made. We refer readers to other recent related reviews (Harmon and Green 2013; Nekrasova and Green 2013; Kowalczyk and Green 2013; Berika and Garrod 2014; Garrod and Tabernero 2014; Johnson et al. 2014).

7.2 Desmosome Composition and Structure

Desmosomes are composed principally of a small number of well-defined molecular components (Fig. 7.1). These are the DCs, Dsg and Dsc, the plakin protein desmoplakin (DP) that links the complex to intermediate filaments (IFs), and the armadillo proteins plakoglobin (PG) and plakophilin (PKP) that link the DCs to DP and appear to regulate desmosomal assembly and size. In humans there are four isoforms of Dsg, and three each of Dsc and PKP. Alternatively spliced versions of the Dsc, PKP, and DP proteins have been described. Other proteins that are associated with desmosomes in a tissue-specific manner include Perp, corneodesmosin, envoplakin, periplakin, kasrin, and the newly identified regulator of desmosomes, inhibitor of apoptosis-stimulating protein of p53 (iASPP; Notari et al. 2015).

It is generally accepted that five proteins (i.e., a Dsg, a Dsc, PG, a PKP, and DP) are both necessary and sufficient for desmosome structure and adhesive function. Transfection of nonadhesive L929 cells with Dsg1, Dsc1a and b, and PG caused cell aggregation with desmosomal components clustered at the membrane, but no intracellular plaques were detectable by electron microscopy (EM; Tselepis et al. 1998). Keratinocytes lacking DP form rudimentary desmosomes with thin plaques but complete desmosomes with IF attachment are only formed upon transfection of DP into the cells (Vasioukhin et al. 2001). It may be that desmosome assembly is not entirely dependent on the presence of both a Dsg and a Dsc as transfection of PG, PKP2, and DP into Dsg2 expressing cells results in the formation of desmosome-like structures by EM (Koeser et al. 2003), and targeting either Dsg2 or Dsc2 transport to the membrane has minimal effect on the distribution of plaque proteins although intercellular adhesion is weakened (Nekrasova et al. 2011). Desmosomes in keratinocytes bearing C-terminal DP mutations that block IF attachment appear to have normal structure, and desmosomes form in keratinocytes lacking keratin (Jonkman et al. 2005; Kroger et al. 2013), so IF attachment is not required for desmosome formation.

Desmosomes have an extremely regular structure. In the transverse or '*z*' direction the structure is layered and symmetrical (Odland 1958). The space between the plasma membranes is sometimes called the 'desmosomal core' or 'desmoglea'. Halfway between the membranes lies a density, the midline, where N-termini of the DCs are located and where adhesive binding occurs (Shimizu

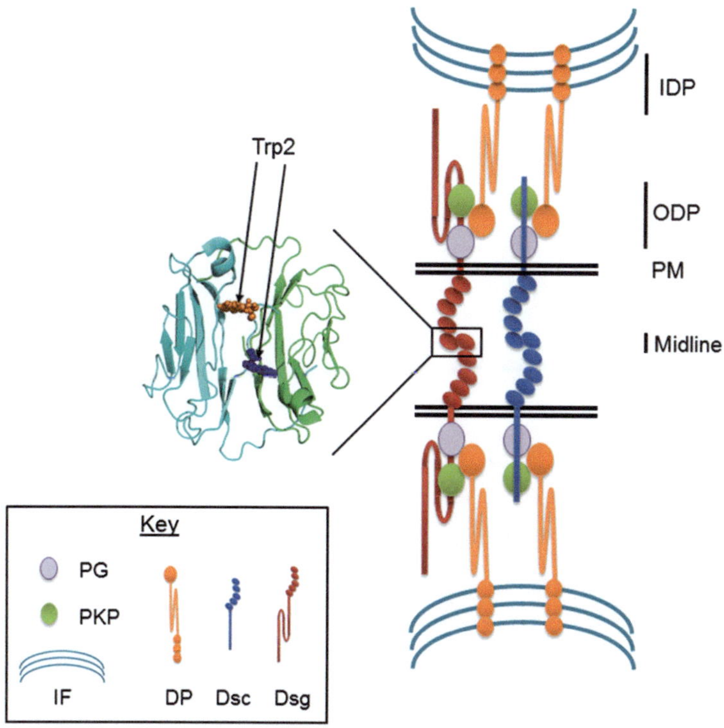

Fig. 7.1 Schematic model of a desmosome showing the relative positions of the major components. For simplicity the Dsc 'b' and alternatively spliced versions of PKP and DP are not shown. The diagram shows homophilic Dsg–Dsg and Dsc–Dsc interactions in the intercellular space although heterophilic Dsg–Dsc interactions may be present. A model of a potential homophilic interaction between Dsg EC1 domains is shown, based on the Dsg2 EC1 domain structure (Protein Data Bank 2YQG). Each EC1 domain forms a seven stranded β-sandwich structure. The two interacting EC1 domains are shown in cyan blue and green with their respective Trp2 residues in orange and dark blue. IDP, inner dense plaque; ODP, outer dense plaque; PM, plasma membrane; PG, plakoglobin; PKP, plakophilin; IF, intermediate filament; DP, desmoplakin; Dsc, desmocollin; Dsg, desmoglein

et al. 2005; Al-Amoudi et al. 2007). Electron tomography (ET) of vitreous sections and lanthanum infiltration have demonstrated that the DC extracellular (EC) domains are arranged in the *en face* or '*x*–*y*' plane in a quadratic array with a repeat of ~70 Å, which is very close to that found in the crystal structure of a Type1 cadherin, C-cadherin (Al-Amoudi et al. 2007; Garrod et al. 2005; Rayns et al. 1969; Boggon et al. 2002). The intermembrane distance is ~35 nm, which is in reasonable agreement with the 38.5 nm membrane-to-membrane distance calculated from the C-cadherin crystal structure (Al-Amoudi et al. 2007).

Small-angle X-ray scattering studies of mouse Dsg2 show that EC domains of DCs are shorter and more flexible than those of Type 1 cadherins (Tariq et al. 2015). These properties provide a better fit with ET data (Al-Amoudi et al. 2007) than the

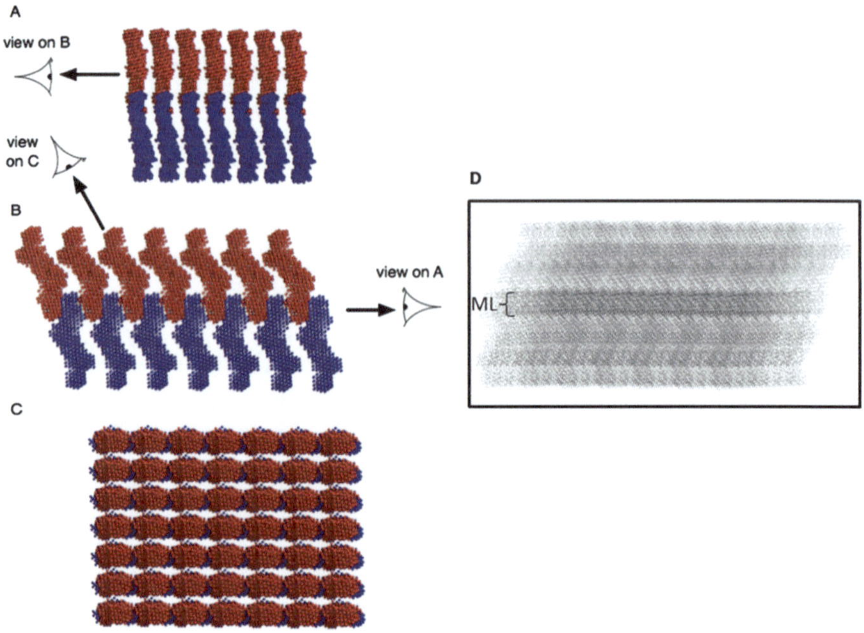

Fig. 7.2 Idealised array of EC domains of DCs. *Front* (**a**), *side* (**b**) and *top* (**c**) views of the array generated with the DAMMIN models from SAXS data obtained by fitting to ET maps of the desmosomal intercellular space. The EC domains originating from opposite cell surfaces are in red and blue, respectively. Note that there are no cis interfaces because the molecules are too far apart. (**d**) Side-view projection clearly showing midline (ML) (Images kindly provided by Dr Jordi Bella) (For further details see Tariq et al. 2015)

C-cadherin structure and suggest a model of the desmosomal interspace that is consistent with our previous findings that DCs adhere homophilically and isoform specifically but lack cis interactions (Nie et al. 2011; Fig. 7.2).

The desmosomal plaques may be resolved into an outer dense plaque (OPD), with its inner face ~20 nm from the membrane, and an inner dense plaque (IDP), which joins the IFs ~50 nm from the membrane (Al-Amoudi et al. 2011; North et al. 1999). Immuno-gold labelling showed the ODP as a region of multiple protein–protein interactions (North et al. 1999) as follows: (i) PKP lies close to the plasma membrane; (ii) PG and the N-terminus of DP are farther from the membrane overlapping with the C-terminus of Dsc 'a' and the entire cytoplasmic domain of Dsg3, which lies in the ODP; (iii) the C-terminus of Dsc 'b' is closer to the membrane and spatially separated from PG and DP; (iv) the C-terminus of DP lies in the IDP, ~40 nm from the membrane. This is consistent with the predicted length of the shorter spliced form, DPII, and suggests that DPI is coiled or folded. The locations of these molecules appear consistent with their interactions determined in vitro.

ET of the ODP produced a molecular map showing a 2D interconnected quasi-periodic lattice with a similar organisation to the EC side (Al-Amoudi et al. 2011).

The transverse organisation was resolved into an outer 4 nm-thick PKP layer and an inner, denser, 8 nm-thick PG layer, which also contains the N-termini of DP molecules. It is not clear how the cytoplasmic domains of the DCs fit into the ODP. They are likely to be extremely flexible, acting as scaffolds during plaque assembly (Kami et al. 2009), wrapping around PG and PKP as the E-cadherin cytoplasmic domain wraps around β-catenin and PG (Huber and Weis 2001; Choi et al. 2009).

7.3 DCs and Their Cytoplasmic Partners

The DCs are single-pass transmembrane proteins (Fig. 7.3). There are significant differences between DC cytoplasmic domains and those of the Type 1 cadherins. For example, in mammals Dscs exhibit alternative mRNA splicing generating a longer ‘a’ form and a shorter ‘b’ form. The size difference depends upon the presence of a mini-exon containing a stop codon which when spliced out extends the open reading frame to encode the longer ‘a’ form C-terminus. Both ‘a’ and ‘b’ form cytoplasmic domains differ in size from those of Type 1 cadherins. Thus human Dsc2a and Dsc2b cytoplasmic domains are 194 and 140 residues, respectively, whereas that of human E-cadherin is 151 residues. The unique C-terminus of the ‘b’ form consists of 11 amino acids in Dsc1 and Dsc2, and 8 in Dsc3. The organisation of the Dsc ‘a’ form resembles that of Type 1 cadherins comprising two subdomains, an intracellular anchor (IA) and an intracellular cadherin-like sequence (ICS) (which is truncated in the ‘b’ form). Dsg cytoplasmic domains are considerably longer than those of Type 1 cadherins and Dscs (483 amino acids in human Dsg2). As well as IA and ICS subdomains they possess a C-terminal Dsg-specific cytoplasmic region (DSCR) consisting of an intracellular proline-

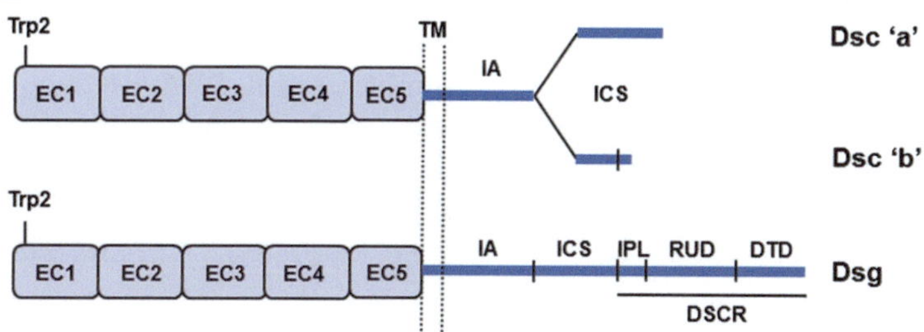

Fig. 7.3 Structure of the DCs. The location of the conserved Trp2 residue in the first extracellular subdomain of both Dscs and Dsgs is indicated. Dsc and Dsg cytoplasmic domains are thought to be intrinsically disordered and are depicted as thick lines. In Dscs alternative splicing leads to a truncated ‘b’ form that contains a number of unique residues at the extreme C-terminus. The desmoglein-specific cytoplasmic region (DSCR) is unique to Dsgs and consists of a number of subdomains including the repeat unit domain (RUD). The number of repeats within the RUD varies between the various Dsg proteins. *IA* intracellular anchor, *ICS* intracellular cadherin-like sequence, *IPL* intracellular proline-rich linker, *DTD* desmoglein terminal domain, TM transmembrane

rich linker (IPL) domain, a repeat unit domain (RUD), and a desmoglein terminal domain (DTD). The RUD repeats are 29 amino acids long and their number varies with 5 in human Dsg1, 6 in Dsg2, 2 in Dsg3, and 3 in Dsg4. Dsg3 lacks a DTD.

7.3.1 Interactions of DC Cytoplasmic Domains with Other Desmosomal Proteins

Intrinsic disorder is characteristic of many protein interaction hubs. The E-cadherin cytoplasmic domain is intrinsically disordered (Huber et al. 2001) and sequence analysis suggests the same is true of Dsc and the corresponding region of Dsg (i.e., the IA and ICS subdomains). In AJs E-cadherin can interact with either β-catenin or PG. The E-cadherin cytoplasmic domain binds to both virtually identically (Huber and Weis 2001; Choi et al. 2009). DC cytoplasmic domains almost certainly interact with PG in a similar manner. The core PG binding region resides within the Dsc and Dsg ICS domains (Troyanovsky et al. 1994a, b). DC cytoplasmic domains also bind to all three PKPs (Hatzfeld et al. 2000; Chen et al. 2002; Bonne et al. 2003). The PKP1 binding site is located within the Dsg1 ICS domain and to a lesser extent the DSCR (Hatzfeld et al. 2000) and thus may differ from that of PG. The DSCR is also intrinsically disordered (Kami et al. 2009) and as well as acting as a binding site for PKP1 it facilitates tail–tail interactions and inhibits internalization (Chen et al. 2012). The truncated ICS domain of Dsc 'b' does not bind PG but instead binds PKP3 (Bonne et al. 2003).

PG links the DCs with DP in desmosomes, and Type 1 cadherins and α-catenin in AJs. All these proteins bind to the central *arm* repeat domain of PG. Desmosomal and Type 1 cadherins are likely to interact with PG at multiple overlapping points (Huber and Weis 2001). The N-terminal head domain of DP also interacts with the *arm* repeat domain of PG (Kowalczyk et al. 1997; Bornslaeger et al. 2001); its C-terminal tail domain engages with IFs, thus completing the link to the cytoskeleton (Stappenbeck and Green 1992; Bornslaeger et al. 1996). DC cytoplasmic domains also bind to PKPs, but via the unstructured N-terminus rather than the central *arm* repeats (Hatzfeld et al. 2000). This unstructured region also interacts with DP (Kowalczyk et al. 1999).

7.4 Adhesion and 'Hyperadhesion'

It has long been established that desmosomes are involved in maintaining adhesion between cells. Recently, the concept of 'hyperadhesion' has emerged. Hyperadhesive desmosomes are characterised by enhanced stability and adhesive strength, a property that distinguishes them from other types of intercellular junction.

7.4.1 Mechanism of Adhesion

Much more is known about adhesion by Type 1 cadherins than by DCs. Type 1 cadherins, the principal adhesion molecules of AJs, have EC domains consisting of five subdomains (EC1–5) each containing just over 100 amino acids and having a Greek key type structure. Adhesive binding involves strand exchange between the EC1 domains of apposed molecules, mediated by insertion of the hydrophobic side chain of conserved tryptophan residues (Trp2) into hydrophobic pockets on the opposing molecules. In addition there are cis interactions between the EC1 domains and the EC2–EC3 linker regions of adjacent molecules on the same cell. Homology modelling suggests that DCs may adopt a similar EC domain structure to Type 1 cadherins and experimental evidence supports a similar mechanism of adhesive binding involving Trp2 (Garrod et al. 2005; Nie et al. 2011; Fig. 7.1). However, cis binding was not detected and cis interactions are not possible in our new model of desmosome structure because the intermolecular distances are too great (Nie et al. 2011; Tariq et al. 2015; Fig. 7.2). The lack of cis interactions may arise because the cis interface conserved in Type 1 cadherins is only partially present in the EC1 of DCs and absent in their EC2 domains (Tariq et al. 2015).

7.4.2 Calcium Dependence and 'Hyperadhesion'

The interdomain regions of Type 1 cadherins are stabilised by conserved Ca^{2+} binding sites, three Ca^{2+} ions being co-ordinated at each interface, which maintain the entire EC domain in an extended rigid configuration, enabling adhesive binding (Pokutta et al. 1994; Nagar et al. 1996). Because of this, cadherin adhesion is referred to as 'Ca^{2+} dependent'. It seems important to be clear about the likely significance of this. Cadherin adhesion is Ca^{2+} dependent in the sense that the cadherin EC domain adopts a globular, noninteractive configuration in the absence of Ca^{2+}. Also, cells do not form AJs when cultured at EC Ca^{2+} concentrations <0.1 mM and AJs are dissociated by Ca^{2+} chelation. However, it seems unlikely that Ca^{2+} regulates cell adhesion in vivo because the Ca^{2+} concentration in animal sera and tissue fluids is maintained well above that necessary for cadherin adhesion.

A striking difference between Type 1 cadherin adhesion and DC adhesion is revealed by *experimental* chelation of EC Ca^{2+}. AJs invariably lose adhesion and are dissociated by Ca^{2+} removal. By contrast, desmosomes are generally resistant to Ca^{2+} chelation, maintaining both their adhesion and structure (Wallis et al. 2000; Kimura et al. 2012). Thus desmosomal adhesion is functionally Ca^{2+} independent and this is a general characteristic of desmosomes in tissues and those in mature confluent epithelial monolayers in culture. Desmosomes exhibit a default condition in which their adhesion is Ca^{2+} dependent. Thus, Ca^{2+} dependence is characteristic of desmosomes in subconfluent or newly confluent cell cultures, in early embryonic tissues and in epidermal keratinocytes at acute wound edges (Garrod et al. 2005; Kimura et al. 2012).

We refer to Ca^{2+}-independent desmosomal adhesion as 'hyperadhesion' because it appears to be stronger than the Ca^{2+}-dependent form (Kimura et al. 2007). In cultured keratinocytes maturation of desmosomes from Ca^{2+} dependence to hyperadhesion occurred without any quantitative or qualitative change in the major desmosomal components, except a slight increase in the amount of Dsc2. Recent evidence suggests that Dsc2 may play a role in regulating hyperadhesion in the epidermis (Kurinna et al. 2014), and overexpression of PKP1 enhances desmosomal hyperadhesiveness (Tucker et al. 2014). Moreover, loss of PKP1 or of PG promotes Ca^{2+} dependence (Caldelari et al. 2001; South et al. 2003).

Hyperadhesive desmosomes switch rapidly to Ca^{2+} dependence. This occurs on wounding a cultured monolayer or the epidermis, or by experimental activation of protein kinase C (PKC) (Wallis et al. 2000; Kimura et al. 2012). By contrast, inhibition of PKC switches Ca^{2+}-dependent desmosomes to hyperadhesion. PKC is a family of serine/threonine kinases, but the key isoform in desmosomal adhesion switching appears to be PKCα because: (1) it localises to plaques of Ca^{2+}-dependent desmosomes; (2) depletion or loss of PKCα promotes hyperadhesion; and (3) expression of constitutively active PKCα promotes Ca^{2+} dependence (Wallis et al. 2000; Garrod et al. 2005; Kimura et al. 2012). Another conventional PKC isoform, PKCβ, may also contribute to the regulation of desmosome adhesive function in embryonic development and tyrosine kinases may play a role in what is clearly a complex process (Garrod et al. 2008; Kimura et al. 2012).

The EC domains of DCs are more flexible than those of Type 1 cadherins at physiological Ca^{2+} concentrations (Tariq et al. 2015). This increased flexibility may facilitate (i) more ordered packing of the EC domains of DCs and thus the ability to adopt hyperadhesion and (ii) rapid switching between adhesive states. We speculate that increased flexibility may have three possible causes: the DCs are less glycosylated than Type 1 cadherins; the interdomain Ca^{2+} binding sites of DCs, particularly Dsgs, are less well conserved than those of Type 1 cadherins; and differences in primary sequence of the EC domains.

Hyperadhesion may have considerable functional importance. Evidence from human disease and gene disruption in mice shows that the integrity of epithelial tissues is dependent on the strength of the desmosome–IF complex (Garrod and Chidgey 2008; Thomason et al. 2010). This complex has several key elements, all of which must be strong and functioning normally: (1) the IFs themselves; (2) the linkage of IFs to desmosomes; (3) the desmosomal plaque; and (4) the intercellular adhesive bonds of the desmosome. Hyperadhesion provides the latter, ensuring continuity of the complex throughout the epithelium.

7.4.3 Specificity

Like Type 1 cadherins, DCs can exhibit both homophilic and heterophilic adhesion in recombinant form or when expressed out of context (Niessen and Gumbiner 2002; Duguay et al. 2003; Katsamba et al. 2009; Chitaev et al. 1998; Chitaev and Troyanovsky 1997; Spindler et al. 2009; Syed et al. 2002; Lowndes et al. 2014). Also desmosomes can form between cells of different tissues and diverse species including mammals, birds, and amphibians (Mattey and Garrod 1985; Overton 1977). Although heterotypic AJs can also form under some circumstances (Volk et al. 1987), Type 1 cadherin adhesion is generally specific or homophilic at the cellular level (Pla et al. 2001; Takeichi 1995; Vleminckx and Kemler 1999). The binding affinities in Type 1 cadherin dimers are weak and the homophilic binding free energies for N- and E-cadherin differ by only ~1 kcal/mol (Haussinger et al. 2002; Katsamba et al. 2009). However, such differences may generate homophilic cell–cell adhesion because they are amplified by the high density of molecules on the cell surface, producing substantial differences in adhesive strength (Chen et al. 2012; Katsamba et al. 2009).

Different isoforms of Dsc and Dsg occur in individual epidermal desmosomes and Dsc2, Dsc3, Dsg2, and Dsg3 colocalise in desmosomes of the human Keratinocyte line HaCaT (North et al. 1996; Nuber et al. 1996; Shimizu et al. 1995). Such proximity affords great potential for heterophilic interaction. However, EC covalent cross-linking of HaCaT desmosomes revealed only homophilic, isoform-specific trans binding (Nie et al. 2011). It may be, therefore, that adhesive binding by DCs is more specific in vivo than that of Type 1 cadherins. Specificity may arise because of differences in the EC1 A-strand sequences of Dscs and Dsgs (Tariq et al. 2015). Alternatively, some regular arrangement, such as alternate rows of Dscs and Dsgs may occur in desmosomes (Nie et al. 2011; Tariq et al. 2015). The EC domains of the DCs form a regular array (Rayns et al. 1969; Garrod et al. 2005; Al-Amoudi et al. 2007) and such an arrangement could be maintained from within the plaque, which is also highly ordered (Miller et al. 1987; North et al. 1999; Al-Amoudi et al. 2011). Paradoxically, adhesive specificity may therefore be determined by the cytoplasmic domains of Dscs and Dsgs, which have their own unique features. If this were so, and if DCs such as Type 1 cadherins exhibit only slight differences in their homophilic and heterophilic binding affinities, it might explain why recombinant DC EC domains and DC expressed out of context can sometimes participate in heterophilic binding. Their homophilic, isoform-specific binding in desmosome-forming cells may depend upon the DCs being located in desmosomes.

7.5 Desmosome Dynamics: Assembly, Maturation, and Downregulation

Formation of desmosomes requires (1) cell–cell contact, (2) adequate EC Ca^{2+} concentration (1–2 mM), and (3) cells possessing the necessary molecular constituents. Descriptive studies suggest desmosome assembly is gradual and symmetrical both in culture and in vivo, as follows: (1) apposition of cell membranes; (2) appearance of symmetrical small membrane densities and a slight increase in density of the intercellular space; and (3) attachment of intermediate filaments, widening of the intercellular space, and maturation of structure (Lentz and Trinkaus 1971; Overton 1962; Mattey and Garrod 1986). Another possibility is that half desmosomes bind to half desmosomes on opposing cells or act as templates for assembly of the other half (Hennings and Holbrook 1983; Overton 1973; Duden and Franke 1988; Demlehner et al. 1995). However, half desmosomes generated by Ca^{2+} depletion are internalised and degraded, not recycled to the cell surface (McHarg et al. 2014).

The Ca^{2+}-switch model has been extensively used to study desmosome assembly (Hennings and Holbrook 1983). Cells are grown in low Ca^{2+} medium (LCM: $[Ca^{2+}]$ <0.1 mM) and junction assembly is induced by raising $[Ca^{2+}]$ to a physiological level (1.8 mM). Cells synthesise desmosomal components in LCM at rates comparable to those in normal Ca^{2+} medium (NCM; Mattey et al. 1990; Penn et al. 1987; Pasdar and Nelson 1988a, 1989; Duden and Franke 1988). However, the proteins are unstable with half-lives between 1.7 h for Dsc (originally called DG II/III) and 22 h for DP. Dsg (originally called DG I) and Dsc are transported to the cell surface after entering the Triton X-100–insoluble phase. Dsg enters the insoluble pool in a compartment located between the medial stack of the Golgi and the vesicles that transport it to the membrane. Soluble DP is diffuse and the insoluble pool forms dots localised to keratin IFs (Jones and Goldman 1985; Mattey and Garrod 1986; Pasdar and Nelson 1988b; Duden and Franke 1988; Trevor and Steben 1992).

Switching cells to NCM initiates rapid desmosome assembly (Mattey and Garrod 1986; Hennings and Holbrook 1983; Bologna et al. 1986; Watt et al. 1984). Desmosomal proteins are stabilised, accumulating in the insoluble pool at cell–cell contacts (Penn et al. 1987; Pasdar et al. 1991; Pasdar and Nelson 1989; Pasdar and Nelson 1988b; Pasdar and Nelson 1988a; Mattey et al. 1990). The protein half-lives increase to >72 h for DP, >24 h for Dsg and 20 h for Dsc (Penn et al. 1987; Pasdar and Nelson 1988a, 1989). DP dots are cleared from the cytoplasm and the soluble pool of DP decreases (Watt et al. 1984; Pasdar and Nelson 1988b; Mattey and Garrod 1986; Jones and Goldman 1985; Duden and Franke 1988). Live imaging suggests that DP rapidly (within 5 min) accumulates at cell contacts, then, 10–15 min later, DP-containing particles, appearing at the cell cortex, move to fuse with the initial DP deposits. This process is microtubule, actin, and Rho dependent (Godsel et al. 2005; Godsel et al. 2010). In cultured keratinocytes PKP2 forms a complex with DP and PKCα, which is then recruited to desmosomes and forms IF attachments (Bass-Zubek et al. 2008). PKCα is not required for desmosome assembly or IF attachment in mice, suggesting either that there is compensation from other PKC isoforms or that the mechanism of desmosome assembly differs in vivo and in culture (Thomason et al. 2012). Fluorescence recovery after photobleaching (FRAP) showed that DP exhibits low turnover both in culture and epidermis, whereas considerable turnover of YFP-tagged Dsc2a occurs in MDCK cells, 30–60 % recovery occurring within 30 min (Windoffer et al. 2002; Foote et al. 2013).

Live imaging of fluorescently tagged Dsc2 and Dsg2 showed that they localise to distinct vesicles that move independently to the membrane, Dsc2 slightly preceding Dsg2 (Nekrasova et al. 2011). Transport is microtubule dependent using distinct motors, kinesin-1 and -2 for Dsg and Dsc, respectively. Localisation of DCs to cholesterol-rich rafts may also be required for junction assembly and targeting of DCs to extant puncta may require Sec-3–containing exocyst complexes (Andersen and Yeaman 2010; Resnik et al. 2011). Dsg is then stabilised at the membrane by dimerisation mediated through its cytoplasmic domain (Chen et al. 2012). The ODP of the desmosome is a major zone of protein clustering (North et al. 1999). PG probably contributes to clustering whereas PKP1 recruits DP to the plaque by mediating interaction between the DC cytoplasmic domains and DP (Kowalczyk et al. 1999; Hatzfeld et al. 2000). The N-terminus of DP also clusters PG and DCs into discrete complexes (Kowalczyk et al. 1997) and the DP rod domain may self-aggregate (Stappenbeck and Green 1992; Bornslaeger et al. 1996; Meng et al. 1997).

Desmosomes are punctate membrane domains. What maintains their discreteness and their stability? PG is involved in maintaining desmosomal integrity; loss of PG permits intermixing of desmosomal and adherens junction components (Ruiz et al. 1996). Intermixing may involve the EC domains of DCs and E-cadherin, inasmuch as Dsg3 lacking most of its cytoplasmic domain is localised to adherens junctions, whereas the PG binding domain was sufficient to target Dsg3 to desmosomes (Andl and Stanley 2001). Desmosomes in MDKC cells are extremely persistent structures that occasionally fuse whereas A431 desmosomes exhibited independent lateral mobility and fusion (Windoffer et al. 2002; Gloushankova et al. 2003). Desmosomes generally persist during mitosis exhibiting only minor destabilisation and fusion into larger structures (Baker and Garrod 1993; Windoffer et al. 2002). Keratinocyte desmosomes change their DC composition as they ascend the epidermis (North et al. 1996). This could arise by turnover of whole desmosomes or by turnover of the cadherins alone.

Once desmosome assembly has been initiated by Ca^{2+} switching and the component proteins have become stabilised, cells accumulate desmosomal material (Penn et al. 1987; Pasdar and Nelson 1988a; Mattey et al. 1990). In MDCK cells the amount of desmosomal material reaches a plateau about 36 h after the switch, through an increase in desmosome number rather than desmosome size (Mattey et al. 1990). It is not known how the number of desmosomes per cell is regulated. Following initial assembly, desmosome mature to become hyperadhesive, both in culture and in vivo (see Section 7.4.2).

There are two alternative views of desmosome downregulation. The first, commonly referred to as 'disassembly', appears to mean the opposite of 'assembly' or the dissolution of desmosomes into their component molecules. Disassembly commonly occurs in cultured cells with calcium-dependent desmosomes, but there is as yet no evidence that it occurs in vivo, as far as we are aware, though of course it may. The second is the internalisation of intact whole desmosomes or desmosomal halves. This occurs in vivo, for example, at the wound edge (Garrod et al. 2005), but is difficult to reproduce in culture. In an attempt to begin studying the latter, we investigated internalisation of desmosomal halves formed by Ca^{2+} chelation from Ca^{2+}-dependent desmosomes in culture (McHarg et al. 2014). The half desmosomes were internalised in a PKC-, actin-, and microtubule-dependent manner. After internalisation, they were not recycled to the cell surface but the remained intact until degraded by a combination of lysosomal and proteasomal activity. Heightened PKC activity also promotes the internalisation of whole desmosomes in keratin-free epidermis (Kroger et al. 2013).

7.5.1 Posttranslational Regulation of DCs in Desmosome Dynamics

Assembly and downregulation of desmosomes must be tightly controlled following the synthesis of desmosomal proteins. Protein phosphorylation can both drive desmosome assembly and trigger downregulation. Most of the work in this area has been concentrated on phosphorylation of PG and DP (Yin and Green 2004) but some work on the DCs has been carried out. For instance, in squamous cell carcinoma keratinocyte Dsc3 interacts with PG, is serine phosphorylated, and then interacts with Dsg3 to promote desmosome assembly (Aoyama et al. 2009). Binding of PV antibodies to the same cells causes phosphorylation of Dsg3 and its dissociation from PG (Aoyama et al. 1999), suggesting that differential phosphorylation of DCs can drive both assembly and remodelling. The epidermal growth factor receptor (EGFR) modulates desmosome assembly and downregulation. Blockage of EGFR kinase activity with inhibitors prevents tyrosine phosphorylation of Dsg2 and PG and promotes desmosome assembly and adhesive strength (Lorch et al. 2004).

DCs are targeted by proteases, cleaved, and presumably degraded during apoptosis. Dsg1 is a target of proteases during keratinocyte apoptosis. Thus the EC domain is cleaved by metalloproteinase and the cytoplasmic domain is cleaved by caspase-3 (Dusek et al. 2006). Similarly, Dsg2 is cleaved during apoptosis of intestinal epithelial cells (Nava et al. 2007). Cleavage of Dsgs may promote apoptosis as downregulation of either Dsg1 or Dsg2 protects cells from apoptosis (Dusek et al. 2006; Nava et al. 2007). EGFR may play a role in desmosome turnover, acting upstream of matrix metalloproteinases to promote cleavage and internalisation of Dsg2 (Klessner et al. 2009). Cleavage of Dsg1, Dsc1, and corneodesmosin, a glycoprotein that is incorporated into desmosomes in the later stages of epidermal differentiation, is necessary for desquamation (Ishida-Yamamoto and Igawa 2015). However, the process must be tightly controlled. Mutations in *SPINK5*, which encodes the serine protease inhibitor lympho-epithelial kazal type related inhibitor type 5 (LEKTI-1), cause Netherton syndrome, a rare autosomal recessive skin disease that is characterised by severe skin inflammation and scaling, hair abnormalities, and allergic manifestations (Hovnanian 2013). LEKTI deficiency causes premature Dsg1 degradation due to overactivity of stratum corneum proteases, leading to desmosome dissolution and stratum corneum defects (Descargues et al. 2005).

7.6 Transcriptional Regulation of DC Gene Expression

All seven DCs are expressed in the epidermis. Dscs 1 and 3, and Dsgs 1 and 3, show unusual reciprocally graded distributions in the skin (Shimizu et al. 1995; North et al. 1996). No other adhesive epidermal proteins exhibit similar patterns of expression and dissecting the mechanisms by which the remarkable patterns of expression of the DCs in epidermis are achieved is an important research goal.

DC genes are clustered and it may be that their expression is coordinated. In humans and mice the DC gene cluster is located at chromosome 18q12 with the Dscs and Dsgs arranged in two tandem arrays. The human gene order is 'centromere-Dsc3-Dsc2-Dsc1-Dsg1-Dsg4-Dsg3-Dsg2-telomere' and in mice the order is 'centromere-Dsc3-Dsc2-Dsc1-Dsg6-Dsg1-Dsg5-Dsg4-Dsg3-Dsg2-telomere' (Whittock 2003; Hunt et al. 1999; Kljuic et al. 2003). (Dsgs 1, 5 and 6 are sometimes referred to as Dsg1α, Dsg1β, and Dsg1γ, respectively.) In both humans and mice transcription occurs outward from the centre of the arrays. The close proximity of the Dsc and Dsg genes may result from amplification and mutation of an ancestral DC gene, itself evolved from a common ancestor with Type 1 cadherins. Subsequently the Dsgs have evolved further from the Type 1 cadherins than the Dscs (Greenwood et al. 1997).

Long-range genetic elements may be important in transcriptional control of DC expression. Many eucaryotic genes are organised into multigene loci, where functionally related genes are co-ordinately expressed or differentially regulated by one genetic element such as the locus control region (LCR) that controls globin gene expression (Kim and Dean 2012). Some evidence suggests that DC gene expression could be controlled by such an element. Statistical analysis of the amounts of Dsc1 and Dsc3 in various layers of bovine nasal epidermis suggests that their distributions are consistent with a process of linked gene expression (North et al. 1996), and there is some correlation between the spatial order in which DC mRNAs are expressed during morphogenesis of stratified epithelia and the gene order on chromosome 18 (King et al. 1997).

In the skin numerous signalling pathways are known to control keratinocytes differentiation (Lopez-Pajares et al. 2013), but links between these and either a putative LCR or specific transcription factors that regulate DC gene expression are generally not clear. A direct link between the Notch pathway, which regulates epidermal differentiation, and the transcriptional regulation of desmosomal genes is provided by p63. There is cross-talk between Notch and p63 (Okuyama et al. 2008), and p63 regulates expression of Dsc3, Dsg1, and DP (Ferone et al. 2013). Mutation of the p63 gene and loss of expression of these desmosomal proteins is responsible for severe skin fragility associated with ankyloblepharon, ectodermal defects, cleft lip/palate (AEC) syndrome, a rare autosomal dominant disorder (Ferone et al. 2013). The Grainy head family of transcription factors is important for epidermal development and repair (Pare et al. 2012) and grainy-head–like 1 (Grhl1) regulates Dsg1 expression (Wilanowski et al. 2008). Grhl1-null mice exhibit hair loss and palmoplantar keratoderma due to reduction of Dsg1 expression and concomitant reduction in cells (Wilanowski et al. 2008). Other transcription factors implicated in Dsg1 expression include serum response factor (Dubash et al. 2013) and the Kruppel-like factor Klf5 (Kenchegowda et al. 2012). How disparate transcription factors cooperate to regulate Dsg1 expression remains unknown. Transcription factor Smad 4 activates Dsg4 expression (Owens et al. 2008) whereas HOXC13, LEF1, and FOXN1 repress its expression in keratinocytes (Bazzi et al. 2009).

Transcription factors implicated in Dsc gene regulation in the epidermis include two CCAAT/enhancer binding proteins (C/EBPs). C/EBPβ activates Dsc3, but not Dsc1, transcription in keratinocytes, whereas C/EBPα does the opposite (Smith et al. 2004). These observations may partly explain why Dsc3 is most strongly expressed in lower cell layers, where C/EBPβ predominates and Dsc1 expression is strongest in upper layers where C/EBPα predominates (Maytin and Habener 1998; Maytin et al. 1999). However, the situation is undoubtedly more complicated. Lymphoid enhancer-binding factor 1 (Lef-1) acts as a switch activating Dsc2 and repressing Dsc3 in the presence of PG in keratinocytes (Tokonzaba et al. 2013). Homeobox transcription factors Cdx1 and Cdx2 regulate Dsc2 (Funakoshi et al. 2008) and Dsc3 is a p53 target gene (Oshiro et al. 2003; Cui et al. 2011).

Several attempts have been made to identify regulatory regions that direct tissue-specific expression of DC genes in vivo. The amount of 4.2 kb of the human Dsg1 promoter was sufficient to direct expression of a β-galactosidase reporter to the epidermis of transgenic mice, but not other stratified epithelia (Adams et al. 1998). Similarly, 4.0 kb of mouse Dsc3 5′ flanking DNA contained some elements responsible for tissue-specific expression of Dsc3, but not all (Merritt et al. 2007). It may be that additional regulatory elements, such a LCR, are required in reporter vectors to recapitulate DC gene expression patterns fully in transgenic mice.

7.7 DCs in Differentiation and Signalling

Why do the Dsc and Dsg proteins exhibit their distinctive tissue- and differentiation-specific patterns of expression? Why are all of these proteins expressed in some tissues such as the epidermis? Are they simply adhesion molecules or do they have some regulatory function in epithelial biology? It seems increasingly certain that the latter is true, that desmosomes serve as regulators of differentiation and morphogenesis.

7.7.1 Knockouts, Dominant Negatives, and Misexpression: Evidence for a Role in Regulating Differentiation

That desmosomal adhesion does more than simply bind cells together is suggested by some surprising phenotypic features of DC knockout mice (Table 7.1). Disruption of genes encoding DCs that are expressed in upper epidermis (i.e., Dsc1 and Dsg4) results in defective desmosomal adhesion, but also causes altered keratinocyte differentiation and increased proliferation (Chidgey et al. 2001; Kljuic et al. 2003). It is possible that damage to the skin's barrier function could be partially responsible for these changes (Vidémont et al. 2012). However, altered differentiation and increased proliferation are seen in some transgenic mice that show no obvious loss of barrier function. These include transgenic keratin 1 (K1)-Dsc3, involucrin (Inv)-Dsg2, and K1-Dsg3 mice that misexpress Dsc3, Dsg2, and Dsg3 in the upper layers (Merritt et al. 2002; Hardman et al. 2005; Brennan et al. 2007). In contrast to K1-Dsg3 mice, Inv Dsg3 mice exhibit normal differentiation and proliferation in the nucleated epidermis, despite severe barrier defects resulting from a thin stratum corneum with gross scaling and an abnormal histology (Elias et al. 2001).

Disruption of genes normally expressed in lower epidermis, such as Dsc3 (Chen et al. 2008) and Dsg3 (Koch et al. 1997), does not cause changes in differentiation/proliferation, nor are they seen upon misexpression of Dsc1 and Dsg1in the basal layer (Henkler et al. 2001; Hanakawa et al. 2002). Expression of an N-terminally truncated Dsg3 protein in the basal epidermal layer has a dramatic effect on differentiation and proliferation but this difficult to interpret as the expressed protein may have a dominant negative effect (Allen et al. 1996). Thus perturbations in DC ratios in the upper epidermis generally seem to affect keratinocyte behaviour whereas perturbations in the lower epidermis do not.

Table 7.1 Effect of loss of expression of DCs on mouse phenotype. Null mice were genetically engineered with targeted disruptions in the Dsg2, Dsg3, Dsc1, and Dsc3 genes. In the case of Dsg4 the spontaneously arising *lanceolate hair* ($Dsg4^{lahJ}/Dsg4^{lahJ}$) mouse serves as a null mouse model as a result of a homozygous single base deletion that creates a premature stop codon and results in nonsense-mediated mRNA decay. Note that Dsg1, Dsg5, Dsg6, and Dsc2 knockout experiments have yet to be performed. The asterisk indicates a conditional knockout, with loss of gene expression restricted to the epidermis

Gene	Phenotype	References
Dsg2	Embryonic lethal, around implantation	Eshkind et al. (2002)
	Changes in embryonic stem cell proliferation	
Dsg3	Suprabasal blistering in lower layers of oral mucosa	Koch et al. (1997)
	Suprabasal blistering of traumatised skin	
	Hair loss at weaning	
Dsg4	Abnormalities in hair growth	Kljuic et al. (2003)
	Alterations in differentiation, proliferation	
Dsc1	Skin blistering in upper epidermis	Chidgey et al. (2001)
	Development of chronic dermatitis and hair loss	
	Alterations in differentiation, proliferation	
Dsc3	Embryonic lethal, prior to preimplantation	Den et al. (2006)
*Dsc3**	Severe suprabasal skin blistering	Chen et al. (2008)
	Hair loss at weaning	

7.7.2 Role of DCs in Intracellular Signal Transduction Pathways

How might altered patterns of DC gene expression in the epidermis regulate differentiation and proliferation? One mechanism could be via Wnt/β-catenin signalling. The downstream effector of this pathway, β-catenin, is pro-proliferative, and elevated levels of cytoplasmic and nuclear β-catenin were observed in skin of K1-Dsc3 mice (Hardman et al. 2005). β-catenin signalling activity in K1-Dsc3 keratinocytes was double that of wild-type cells (Hardman et al. 2005). Similar changes have been found in Dsc1−/− and K1-Dsg3 keratinocytes (unpublished data). The mechanism generating elevated β-catenin signalling is unclear. PG, released from desmosomes because of changes in DC ratios, may either displace β-catenin from AJs or prevent its degradation. Alternatively, PG may have a role in Wnt/β-catenin signalling, although whether as a positive or negative regulator remains uncertain (Swope et al. 2013). PKP2 potentiates β-catenin signalling (Chen et al. 2002) and it is conceivable that release of PKPs from desmosomes may be important.

At least one DC, Dsg1, promotes epidermal differentiation when overexpressed in a raft model of human epidermis (Getsios et al. 2009). Activation of the mitogen-activated protein kinase (MAPK) pathway via the EGFR and ERK causes cell proliferation. Overexpression of Dsg1 represses the pathway and allows progression of differentiation. This does not require the EC domain of Dsg1, nor is it dependent on interactions with PG or Dsc1 (Getsios et al. 2009). It may depend on its ability to bind Erbin, a LAP protein family member and a known ERK regulator (Harmon et al. 2013). It appears that Dsg1 and Erbin cooperate to repress MAPK signalling and promote differentiation. If expression of Dsg1 in the upper epidermis facilitates keratinocyte terminal differentiation, the question is what controls Dsg1 expression.

7.8 DCs in Development

Desmosomes first assemble at early stages in the development of vertebrate embryos. In the killifish, *Fundulus heteroclitus,* 'nascent' desmosomes are found between the outer blastomeres of the blastula, and fully-formed desmosomes between enveloping layer cells of the mid gastrula, that is, during epiboly (Lentz and Trinkaus 1971). In the chick, desmosomes first appear in the area opaca at Hamburger and Hamilton stage 3, when the cell sheet commences spreading over the yolk, but are absent from the area pellucida (Garrod and Fleming 1990). In the mouse early embryo desmosomes appear between the trophectoderm cells of the morula or early blastocyst (Ducibella et al. 1975; Jackson et al. 1980; Jackson et al. 1981; Fleming et al. 1991). The timing of expression of various desmosomal components in the mouse embryo is shown in Fig. 7.4.

Desmosomes are essential for embryonic development. Deletion of mouse PG, PKP2, or DP causes embryonic death principally because of failure of intercellular adhesion (Gallicano et al. 1998; Ruiz et al. 1996; Bierkamp et al. 1996; Grossmann et al. 2004). However, the early embryonic lethality caused by deletion of Dsg2 or Dsc3 are probably due to signalling defects. Dsg2 deletion causes death at implantation because of embryonic stem cell proliferation defects whereas Dsc3 deletion causes death before E2.5, preceding desmosome assembly (Den et al. 2006; Eshkind et al. 2002).

In zebrafish both zfDsc and zfDsgα are present as maternal and zygotic transcripts whereas zfDsgβ is first expressed from 8 h postfertilisation (hpf; Goonesinghe et al. 2012). All three are present throughout subsequent stages. Knockdown of zfDsc or zfDsgα produced similar defects in epiboly, axis elongation, and somite formation, associated with abnormal desmosomes or reduced desmosome numbers. These results demonstrate an important role for DCs and desmosomes in the early morphogenesis of the zebrafish. Knockdown of PG in zebrafish gave rise to heart defects at 48 hpf (Martin et al. 2009). In *Xenopus* knockdown of PKP3 caused skin fragility, disruption in peripheral neural structures, altered establishment and migration of neural crest, and defects in ectodermal multiciliated cells (Munoz et al. 2012).

Because desmosomes appear early in mouse tissues and inasmuch as developing tissues must remain malleable to participate in morphogenetic movements we

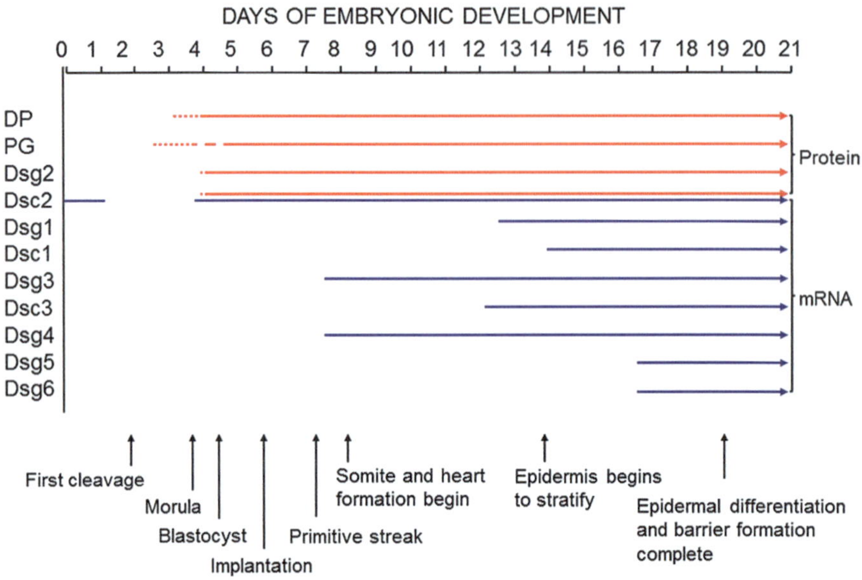

Fig. 7.4 Expression of desmosomal constituents during murine early development (Data for the PKPs are not yet available)

hypothesised that initial weak adhesion would be superseded by hyperadhesion (Kimura et al. 2012). Epidermal desmosomes were Ca^{2+} dependent until E12 and became hyperadhesive by E14. Similarly, blastocyst trophectodermal desmosomes were Ca^{2+} dependent on E3 but became hyperadhesive by E4.5. By contrast, AJs remained Ca^{2+} dependent throughout development but tight junctions became Ca^{2+} independent as desmosomes matured. Conventional PKC isoforms are involved in progression to hyperadhesiveness (Sect. 7.4.2) and regulation of desmosomal adhesion by PKC may be important in trophoblast migration during implantation. It appears that tissue stabilisation is one of several roles played by desmosomes in animal development.

7.9 DCs in Wound Healing

It seems likely that migration and remodelling of the epidermis during wound healing involves modulation of cell–cell adhesion and there is some EM evidence for desmosome downregulation at the wound edge (Allen and Potten 1975; Croft and Tarin 1970; Garrod et al. 2005). On the other hand lateral cell–cell adhesion appears essential for cell sheet migration (Danjo and Gipson 1998). Desmosomal hyperadhesion seems incompatible with cell migration and, indeed, desmosomal adhesion changes from hyperadhesive to Ca^{2+} dependent in wound edge epithelium, indicating a weakening of cell–cell adhesion (Wallis et al. 2000; Garrod et al. 2005; Kimura et al. 2007; Thomason et al. 2012). This change is regulated by PKCα and PKCα −/− mice show delayed re-epithelialisation, whereas mice overexpressing constitutively active PKCα at the wound edge show accelerated re-epithelialisation (Thomason et al. 2012). Association of PKCα with the plaques of Ca^{2+}-dependent desmosomes may presage desmosome downregulation through internalisation (Garrod et al. 2005; Kimura et al. 2007). PKC activity is required for internalisation of half desmosomes and activation of PKC through loss of keratin promotes internalisation of epidermal desmosomes (Kroger et al. 2013; McHarg et al. 2014). These results suggest that manipulation of PKC signalling could provide a novel therapeutic approach for human chronic wounds. Mice lacking the tetraspan desmosomal protein Perp from epidermis showed structurally defective desmosomes and delayed epidermal wound healing (Beaudry et al. 2010).

Live imaging of GFP-tagged Dsc2 has shown that desmosomes assemble between the lateral edges of epithelial cells migrating into scratch wounds (Roberts et al. 2011). The nascent desmosomes are then transported retrogradely in an actin- and PKC-dependent fashion, becoming stabilised and keratin associated. Such assembly is presumably essential for maintaining cell sheet integrity.

7.10 DCs in Disease: Skin Blistering and Cardiomyopathy

Desmosomes are targeted in a number of diseases that affect skin, heart, and sometimes both. In some cases the symptoms can be ascribed to loss of cellular adhesion and tissue integrity. However, often the situation is not straightforward; symptoms may result from alterations in signalling pathways that involve desmosomes. This is true of pemphigus, a skin blistering disease, and arrhythmogenic right ventricular dysplasia (ARVD; also known as arrhythmogenic right ventricular cardiomyopathy), a heart muscle disorder characterised by life-threatening arrhythmias and sudden heart failure.

7.10.1 Autoimmune Skin Blistering Diseases

Pemphigus is a rare autoimmune blistering disease of epidermis and mucous membranes. It usually develops in mid-life and patients are prone to remissions and relapses. It is caused by pathogenic antibodies against Dsgs (Table 7.2). A combination of genetic and environmental factors is probably responsible for triggering production of these autoantibodies. There are two major forms, pemphigus vulgaris (PV) and pemphigus foliaceus (PF). PV is the more severe although the mortality rate has been much reduced by systemic corticosteroids together with corticosteroid-sparing immunosuppressive drugs. Additional new therapeutic options include high-dose immunoglobulins, immunoadsorption, and treatment

Table 7.2 Acquired and genetic diseases of DCs

Gene	Protein	Type	Condition
DSG1	Desmoglein 1	Autoimmune	Pemphigus foliaceus
		Autoimmune	Pemphigus vulgaris (mucocutaneous)
		Infectious	Bullous impetigo/Staphylococcal scalded skin syndrome (*S.Aureus*)
		Genetic	Palmoplantar keratoderma 1 (148700)
		Genetic	Severe dermatitis, multiple allergies, and metabolic wasting syndrome (615508)
DSG2	Desmoglein 2	Genetic	Arrhythmogenic right ventricular dysplasia 10 (610193)
		Infectious	Respiratory and urinary tract infection (Adenovirus)
DSG3	Desmoglein 3	Autoimmune	Pemphigus vulgaris (mucosal-dominant)
		Autoimmune	Pemphigus vulgaris (mucocutaneous)
DSG4	Desmoglein 4	Genetic	Hypotrichosis 6 (607903)
DSC1	Desmocollin 1	Autoimmune	IgA pemphigus (subcorneal pustular dermatitis-type)
DSC2	Desmocollin 2	Genetic	Arrhythmogenic right ventricular dysplasia 11 (610476)
DSC3	Desmocollin 3	Genetic	Hypotrichosis and recurrent skin vesicles (613102)

For genetic diseases the online Mendelian inheritance in humans (http://www.ncbi.nlm.nih.gov/omim) reference is given in parentheses

with the anti-CD20 antibody rituximab, which targets mature B cells (Kasperkiewicz et al. 2012).

Both PV and PF antibodies cause epidermal blistering due to acantholysis or splitting between the epidermal layers. PF is caused by autoantibodies against Dsg1 and blistering, exclusively epidermal, occurs between the superficial layers, where Dsg1 expression is highest. Many PV patients develop only oral lesions (mucosal-dominant PV) whereas as others have both oral and epidermal lesions (mucocutaneous PV). Mucosal-dominant PV is caused by autoantibodies against Dsg3 and acantholysis occurs between the basal layer and first suprabasal layer where Dsg3 expression is highest. In mucocutaneous PV epidermal lesions are caused by acquisition of additional antibodies to Dsg1. The Dsg compensation theory offers an explanation of this complex clinical manifestation (Stanley and Amagai 2006). It states that Dsg1 and Dsg3 have compensatory adhesive functions when coexpressed. Hence sera containing Dsg1 antibodies alone (as in PF) cause blisters only in superficial epidermis where Dsg1, but not Dsg3, is expressed. In the unaffected lower epidermis, Dsg3 compensates for loss of Dsg1. Sera that contain Dsg3 antibodies alone (as in mucosal-dominant PV) do not cause skin blisters because Dsg1 is expressed in all epidermal layers. However, they do cause oral blisters as the low levels of Dsg1 in mucous membranes cannot compensate for loss of Dsg3. Sera containing both Dsg1 and Dsg3 antibodies (as in mucocutaneous PV) cause blistering of both the skin and mucous membranes. The theory does not fully explain why acantholysis characteristically occurs immediately above the basal layer in PV, rather than more extensively throughout the epidermis, and it may be necessary to postulate a plane of weak cell–cell adhesion at this level (Dmochowski et al. 1995).

A major question is whether pemphigus blistering is caused by direct inhibition of Dsg adhesive binding, endocytosis of Dsgs, modification of intracellular signal transduction pathways, or some combination of these. Blockage of trans interaction appears important in PV as PV IgG directly inhibits Dsg3 trans interaction in atomic force microscopy (AFM; Heupel et al. 2008), pemphigus antibodies target EC1 and EC2 subdomains (Di Zenzo et al. 2012), and PV IgG-induced acantholysis is reduced by a peptide that prevents IgG blocking of Dsg3 trans interaction (Heupel et al. 2009). However, PF antibodies failed to block Dsg1 trans interaction by AFM (Waschke et al. 2005) so the blistering mechanisms in PV and PF may differ. Moreover, EM of PV-affected skin appears to show that direct disruption of desmosomes is not the primary event (Diercks et al. 2009). Rather, extensive loss of cell–cell adhesion in interdesmosomal regions and intracellular cleavage behind the desmosomal plaque might indicate weakening of the cytoskeleton, perhaps through signalling involving PG (Diercks et al. 2009; Muller et al. 2008). By contrast abundant split desmosomes with inserted keratin filaments were found in a mouse model of pemphigus (Shimizu et al. 2004).

Endocytosis of Dsgs may be important in pemphigus pathogenesis. Exposure of keratinocyte cultures to PV IgG causes rapid internalisation of Dsg3-PG complexes and results in keratin filament retraction, desmosome disassembly and a loss of adhesive strength (Calkins et al. 2006). Inhibition of Dsg3 endocytosis with genistein, a tyrosine kinase inhibitor that inhibits clathrin-independent endocytosis, prevents disruption of desmosomes and loss of adhesion in the presence of PV IgG (Delva et al. 2008).

Pemphigus IgG triggers several intracellular signalling pathways, including the p38MAPK, PKC, RhoA, c-myc, and tyrosine kinase pathways (Waschke and Spindler 2014). A convincing case has been established for p38MAPK signalling in pemphigus because p38MAPK inhibition prevents blistering when either PV or PF IgG is injected into neonatal mice (Berkowitz et al. 2008). Silencing of PG causes activation of p38MAPK signalling, keratin filament collapse, and loss of cell adhesion (Spindler et al. 2014) so PG may have a regulatory role in the pathway. Activation of p38MAPK is probably secondary to loss of adhesion but whether it is required for Dsg3 endocytosis remains unclear (Jolly et al. 2010; Mao et al. 2011).

Most work with PV antibodies has presumably been carried out with cultured keratinocytes possessing Ca^{2+}-dependent desmosomes. Comparison of keratinocytes with Ca^{2+}-dependent and hyperadhesive desmosomes showed that hyperadhesion inhibited PV autoantibody-induced acantholysis and internalisation of Dsg3 and E-cadherin (Cirillo et al. 2010). Furthermore, overexpression of PKP1 in keratinocytes promoted desmosomal hyperadhesiveness and blocked dissociation by PV antibodies. However, keratinocytes in confluent culture became hyperadhesive without increased PKP1 expression (Kimura et al. 2007). Desmosome disruption by PV IgG can also be prevented by expressing exogenous Dsg3 but the mechanism may involve replacing internalised Dsg3 at the membrane with newly synthesized protein and desmosome assembly, rather than by any effect on hyperadhesiveness (Jennings et al. 2011).

DCs are targeted in other autoimmune blistering diseases. IgA pemphigus is rare and characterised by neutrophil infiltration and IgA antibody deposition at epidermal cell surfaces (Tsuruta et al. 2011). There are two types, subcorneal pustular dermatosis (SPD) and intraepidermal neutrophilic (IEN). SPD is manifested by subcorneal pustules whereas IEN is characterised by pustules throughout the epidermis. It is thought that Dsc1 is the target of autoantibodies in SPD (Yasuda et al. 2000; Ishii et al. 2004), although definitive evidence for their pathogenicity is lacking and antibodies against other adhesion molecules, including DCs, may be present. The target of autoantibodies in the IEN type of IgA pemphigus is not known.

7.10.2 Infectious Diseases

DCs are targeted by bacterial toxins in staphylococcal scalded skin syndrome (SSSS) and bullous impetigo. These are blistering diseases caused by infection with *Staphylococcus aureus* (Stanley and Amagai 2006). The bacterium secretes serine protease toxins that specifically cleave residues 1-381 from the EC domain of Dsg1 (Amagai et al. 2000; Hanakawa et al. 2004). In SSSS the pathology is widespread whereas in bullous impetigo it is localised to the infection site. Histology is often indistinguishable between SSSS and PF with blistering in superficial epidermis where Dsg1 is strongly expressed. Mechanistically, the pathogenic potential of cleaved Dsg1 may depend on its interaction with PG, in addition to the loss of adhesion caused by its lack of interaction with neighbouring cells. Thus, expression of truncated Dsg1 in cells disrupts desmosomes, mimicking the toxin-cleaved protein. However, a mutant truncated Dsg1 protein that cannot bind PG does not impair adhesion (Simpson et al. 2010).

Dsg2 acts as a high-affinity receptor for adenovirus serotypes 3, 7, 11, and 14 which cause respiratory and urinary tract infections (Wang et al. 2011). In cultured cells adenovirus interaction with Dsg2 causes dissociation of intercellular junctions and activation of intracellular signalling pathways reminiscent of those activated by pemphigus autoantibodies (Wang et al. 2011).

7.10.3 Genetic Diseases

Skin diseases are also caused by DC mutations. For example, Dsg1 EC domain mutations cause autosomal dominant striate palmoplantar keratoderma (SPPK), characterised by development of focal hyperkeratosis of the palms and soles during the first or second decade (Rickman et al. 1999; Hunt et al. 2001). Several family-specific Dsg1 mutations lead to premature truncation within the Dsg1 EC domain.

EM of affected skin shows fewer smaller desmosomes with abnormal attachment of keratin filaments (Wan et al. 2004). Dsg4 mutation causes localised autosomal recessive hypotrichosis, a rare form of alopecia, characterised by fragile hairs, leading to widespread hair loss (Kljuic et al. 2003). Histology reveals abnormal hair follicles and shafts, often failing to penetrate the epidermis (Kljuic et al. 2003). In some patients monilethrix hairs, which appear beaded as a result of constrictions along their length, and scalp erosions have been observed (Schaffer et al. 2006). Defective intercellular adhesion is probably at least partly responsible for the phenotypes observed in both SPPK and hypotrichosis. Skin thickening, as in PPK, is a common response to defects in the skin's permeability barrier, which in turn is often caused by defective cell adhesion. Similarly, loss of adhesion within the hair follicle may cause hair loss in hypotrichosis. However, DCs also regulate differentiation and signalling, changes in which may account for some aspects of patient phenotypes.

A new syndrome, characterised by severe dermatitis, multiple allergies, and metabolic wasting (SAM syndrome) and observed in two families, is caused by homozygous mutations in Dsg1, involving loss of expression and uneven desmosome distribution in the upper epidermis (Samuelov et al. 2013). Compromised barrier formation, allowing penetration of allergens may cause development of severe allergic dermatitis. Hereditary hypotrichosis with recurrent skin vesicles, described in one family is caused by autosomal recessive mutations in Dsg3 and characterised by sparse, fragile scalp hair, absence of eyebrows and eyelashes, and skin vesicles containing watery fluid (Ayub et al. 2009). Dsc3 is expressed in hair follicles and loss of cell adhesion in the follicle, perhaps accompanied by altered follicle differentiation, may be responsible for hair loss in this syndrome.

7.10.4 Cardiomyopathy

ARVD is one of the most prevalent cardiomyopathies and a common form of ventricular arrhythmias, cardiac failure, and sudden death in young adults, particularly athletes. Its inheritance is usually autosomal dominant with variable penetrance (Awad et al. 2008) and it can be caused by mutations in genes encoding any of the five desmosomal proteins expressed by cardiomyocytes (i.e., *DSG2, DSC2, JUP, PKP2,* and *DSP*). The ARVD database (http://www.arvcdatabase.info/) currently lists a large number of DC variants, of which many are thought to have pathogenic effects. Large deletions or premature stop codons are likely to abrogate surface expression. Less dramatic mutations could affect precursor processing, domain structure, ligand interactions, trafficking to the membrane, and/or protein stability (Al-Jassar et al. 2013). One Dsg2 mutation, N266S, affects a residue critical for binding a Ca^{2+} ion required for EC domain function (Pilichou et al. 2006). Overexpression of the equivalent mutation in mice (i.e., N271S) recapitulates some features of ARVD including myocardial damage, ventricular arrhythmias, and fibrofatty replacement (Pilichou et al. 2009; Rizzo et al. 2012).

How do mutations in Dsg2, Dsc2, and other desmosomal genes cause the characteristic replacement of cardiomyocytes by fibrofatty tissue in ARVD? Impaired desmosomal adhesion could lead to cell detachment and cardiomyocyte death, followed by inflammation and fibrofatty replacement. That mutations in all heart desmosomal protein genes cause a similar disease may indicate that a general failure of desmosomal adhesion ultimately results in all other aspects of ARVD. If so, loss of desmosomal adhesion might be an early event; indeed overexpression of N271S-Dsg2 caused intercellular space widening at intercalated discs, preceding necrosis, which in turn preceded inflammation and fibrofatty replacement (Pilichou et al. 2009). Furthermore, some electrophysiological abnormalities and increased susceptibility to arrhythmias precede the onset of necrosis and fibrosis, and could be accounted for by interaction between Dsg2 and the Na^+ channel protein $Na_V1.5$ (Rizzo et al. 2012).

That alterations in intracellular signalling pathways could also account for the characteristic appearance of adipocytes in the hearts of ARVD patients cannot be discounted. Both Dsg2 and Dsc2 interact with PG and signalling changes may occur through release of PG from defective desmosomes. Loss of PG from the intercalated disc is frequently observed in ARVD patients (Asimaki et al. 2009). Some evidence suggests that PG suppresses Wnt/β-catenin signalling (Li et al. 2011) which normally enhances myogenesis (Nakamura et al. 2003) and inhibits adipogenesis (Christodoulides et al. 2009), and it may be that suppression of Wnt/β-catenin signalling by PG allows activation of adipogenic genes. Indeed, suppression of DP expression in atrial myocyte cell lines causes redistribution of PG to the nucleus, reduced Wnt/β-catenin signalling, and increased expression of adipogenic and fibrogenic genes (Garcia-Gras et al. 2006). Cardiac-specific DP suppression in mice causes development of dysfunctional enlarged hearts and ventricular arrhythmias (Garcia-Gras et al. 2006). A similar phenotype is seen in PG overexpressing mice (Lombardi et al. 2011). Nuclear localisation of PG and reduced β-catenin signalling has also been reported in cardiomyocytes derived from induced pluripotent stem cell (iPSC) lines generated from fibroblasts from ARVD patients with *PKP2* mutations (Kim et al. 2013).

7.11 Conclusion

Substantial progress in our understanding of the structure and function of desmosomes has clearly been made since they were first isolated (Skerrow and Matoltsy 1974). However, because of their great insolubility and extreme complexity, working out the details of how their adhesiveness is regulated, and how they participate in signal transduction and disease still represents a major technical challenge but provides an exciting basis for future research.

References

Adams MJ, Reichel MB, King IA, Marsden MD, Greenwood MD, Thirlwell H, Arnemann J, Buxton RS, Ali RR (1998) Characterization of the regulatory regions in the human desmoglein genes encoding the pemphigus foliaceous and pemphigus vulgaris antigens. Biochem J 329(Pt 1):165–174

Al-Amoudi A, Diez DC, Betts MJ, Frangakis AS (2007) The molecular architecture of cadherins in native epidermal desmosomes. Nature 450:832–837

Al-Amoudi A, Castano-Diez D, Devos DP, Russell RB, Johnson GT, Frangakis AS (2011) The three-dimensional molecular structure of the desmosomal plaque. Proc Natl Acad Sci U S A 108:6480–6485

Al-Jassar C, Bikker H, Overduin M, Chidgey M (2013) Mechanistic Basis of Desmosome-Targeted Diseases. J Mol Biol 425(21):4006–4022

Allen TD, Potten CS (1975) Desmosomal form, fate, and function in mammalian epidermis. J Ultrastruct Res 51:94–105

Allen E, Yu QC, Fuchs E (1996) Mice expressing a mutant desmosomal cadherin exhibit abnormalities in desmosomes, proliferation, and epidermal differentiation. J Cell Biol 133: 1367–1382

Amagai M, Matsuyoshi N, Wang ZH, Andl C, Stanley JR (2000) Toxin in bullous impetigo and staphylococcal scalded-skin syndrome targets desmoglein 1. Nat Med 6:1275–1277

Andersen NJ, Yeaman C (2010) Sec3-containing exocyst complex is required for desmosome assembly in mammalian epithelial cells. Mol Biol Cell 21:152–164

Andl CD, Stanley JR (2001) Central role of the plakoglobin-binding domain for desmoglein 3 incorporation into desmosomes. J Invest Dermatol 117:1068–1074

Aoyama Y, Owada MK, Kitajima Y (1999) A pathogenic autoantibody, pemphigus vulgaris-IgG, induces phosphorylation of desmoglein 3, and its dissociation from plakoglobin in cultured keratinocytes. Eur J Immunol 29:2233–2240

Aoyama Y, Yamamoto Y, Yamaguchi F, Kitajima Y (2009) Low to high Ca^{2+} -switch causes phosphorylation and association of desmocollin 3 with plakoglobin and desmoglein 3 in cultured keratinocytes. Exp Dermatol 18:404–408

Asimaki A, Tandri H, Huang H, Halushka MK, Gautam S, Basso C, Thiene G, Tsatsopoulou A, Protonotarios N, Mckenna WJ, Calkins H, Saffitz JE (2009) A new diagnostic test for arrhythmogenic right ventricular cardiomyopathy. N Engl J Med 360:1075–1084

Awad MM, Calkins H, Judge DP (2008) Mechanisms of disease: molecular genetics of arrhythmogenic right ventricular dysplasia/cardiomyopathy. Nat Clin Pract Cardiovasc Med 5:258–267

Ayub M, Basit S, Jelani M, Ur Rehman F, Iqbal M, Yasinzai M, Ahmad W (2009) A homozygous nonsense mutation in the human desmocollin-3 (DSC3) gene underlies hereditary hypotrichosis and recurrent skin vesicles. Am J Hum Genet 85:515–520

Baker J, Garrod D (1993) Epithelial cells retain junctions during mitosis. J Cell Sci 104(Pt 2): 415–425

Bass-Zubek AE, Hobbs RP, Amargo EV, Garcia NJ, Hsieh SN, Chen X, Wahl JK 3rd, Denning MF, Green KJ (2008) Plakophilin 2: a critical scaffold for PKC alpha that regulates intercellular junction assembly. J Cell Biol 181:605–613

Bazzi H, Demehri S, Potter CS, Barber AG, Awgulewitsch A, Kopan R, Christiano AM (2009) Desmoglein 4 is regulated by transcription factors implicated in hair shaft differentiation. Differentiation 78:292–300

Beaudry VG, Ihrie RA, Jacobs SB, Nguyen B, Pathak N, Park E, Attardi LD (2010) Loss of the desmosomal component perp impairs wound healing in vivo. Dermatol Res Pract 2010:759731

Berika M, Garrod D (2014) Desmosomal adhesion in vivo. Cell Commun Adhes 21:65–75

Berkowitz P, Chua M, Liu Z, Diaz LA, Rubenstein DS (2008) Autoantibodies in the autoimmune disease pemphigus foliaceus induce blistering via p38 mitogen-activated protein kinase-dependent signaling in the skin. Am J Pathol 173:1628–1636

Bierkamp C, Mclaughlin KJ, Schwarz H, Huber O, Kemler R (1996) Embryonic heart and skin defects in mice lacking plakoglobin. Dev Biol 180:780–785

Boggon TJ, Murray J, Chappuis-Flament S, Wong E, Gumbiner BM, Shapiro L (2002) C-cadherin ectodomain structure and implications for cell adhesion mechanisms. Science 296:1308–1313

Bologna M, Allen R, Dulbecco R (1986) Organization of cytokeratin bundles by desmosomes in rat mammary cells. J Cell Biol 102:560–567

Bonne S, Gilbert B, Hatzfeld M, Chen X, Green KJ, Van Roy F (2003) Defining desmosomal plakophilin-3 interactions. J Cell Biol 161:403–416

Bornslaeger EA, Corcoran CM, Stappenbeck TS, Green KJ (1996) Breaking the connection: displacement of the desmosomal plaque protein desmoplakin from cell-cell interfaces disrupts anchorage of intermediate filament bundles and alters intercellular junction assembly. J Cell Biol 134:985–1001

Bornslaeger EA, Godsel LM, Corcoran CM, Park JK, Hatzfeld M, Kowalczyk AP, Green KJ (2001) Plakophilin 1 interferes with plakoglobin binding to desmoplakin, yet together with plakoglobin promotes clustering of desmosomal plaque complexes at cell-cell borders. J Cell Sci 114:727–738

Brennan D, Hu Y, Joubeh S, Choi YW, Whitaker-Menezes D, O'brien T, Uitto J, Rodeck U, Mahoney MG (2007) Suprabasal Dsg2 expression in transgenic mouse skin confers a hyperproliferative and apoptosis-resistant phenotype to keratinocytes. J Cell Sci 120:758–771

Caldelari R, De Bruin A, Baumann D, Suter MM, Bierkamp C, Balmer V, Muller E (2001) A central role for the armadillo protein plakoglobin in the autoimmune disease pemphigus vulgaris. J Cell Biol 153:823–834

Calkins CC, Setzer SV, Jennings JM, Summers S, Tsunoda K, Amagai M, Kowalczyk AP (2006) Desmoglein endocytosis and desmosome disassembly are coordinated responses to pemphigus autoantibodies. J Biol Chem 281:7623–7634

Chen X, Bonne S, Hatzfeld M, Van Roy F, Green KJ (2002) Protein binding and functional characterization of plakophilin 2. Evidence for its diverse roles in desmosomes and beta-catenin signaling. J Biol Chem 277:10512–10522

Chen J, Den Z, Koch PJ (2008) Loss of desmocollin 3 in mice leads to epidermal blistering. J Cell Sci 121:2844–2849

Chen J, Nekrasova OE, Patel DM, Klessner JL, Godsel LM, Koetsier JL, Amargo EV, Desai BV, Green KJ (2012) The C-terminal unique region of desmoglein 2 inhibits its internalization via tail-tail interactions. J Cell Biol 199:699–711

Chidgey M, Brakebusch C, Gustafsson E, Cruchley A, Hail C, Kirk S, Merritt A, North A, Tselepis C, Hewitt J, Byrne C, Fassler R, Garrod D (2001) Mice lacking desmocollin 1 show epidermal fragility accompanied by barrier defects and abnormal differentiation. J Cell Biol 155:821–832

Chitaev NA, Troyanovsky SM (1997) Direct Ca^{2+}–dependent heterophilic interaction between desmosomal cadherins, desmoglein and desmocollin, contributes to cell-cell adhesion. J Cell Biol 138:193–201

Chitaev NA, Averbakh AZ, Troyanovsky RB, Troyanovsky SM (1998) Molecular organization of the desmoglein-plakoglobin complex. J Cell Sci 111(Pt 14):1941–1949

Choi HJ, Gross JC, Pokutta S, Weis WI (2009) Interactions of plakoglobin and beta-catenin with desmosomal cadherins: basis of selective exclusion of alpha- and beta-catenin from desmosomes. J Biol Chem 284:31776–31788

Christodoulides C, Lagathu C, Sethi JK, Vidal-Puig A (2009) Adipogenesis and WNT signalling. Trends Endocrinol Metab 20:16–24

Cirillo N, Lanza A, Prime SS (2010) Induction of hyper-adhesion attenuates autoimmune-induced keratinocyte cell-cell detachment and processing of adhesion molecules via mechanisms that involve PKC. Exp Cell Res 316:580–592

Croft CB, Tarin D (1970) Ultrastructural studies of wound healing in mouse skin. I. Epithelial behaviour. J Anat 106:63–77

Cui T, Chen Y, Yang L, Knosel T, Zoller K, Huber O, Petersen I (2011) DSC3 expression is regulated by p53, and methylation of DSC3 DNA is a prognostic marker in human colorectal cancer. Br J Cancer 104:1013–1019

Danjo Y, Gipson IK (1998) Actin 'purse string' filaments are anchored by E-cadherin-mediated adherens junctions at the leading edge of the epithelial wound, providing coordinated cell movement. J Cell Sci 111(Pt 22):3323–3332

Delva E, Jennings JM, Calkins CC, Kottke MD, Faundez V, Kowalczyk AP (2008) Pemphigus vulgaris IgG-induced desmoglein-3 endocytosis and desmosomal disassembly are mediated by a clathrin- and dynamin-independent mechanism. J Biol Chem 283:18303–18313

Demlehner MP, Schafer S, Grund C, Franke WW (1995) Continual assembly of half-desmosomal structures in the absence of cell contacts and their frustrated endocytosis: a coordinated Sisyphus cycle. J Cell Biol 131:745–760

Den Z, Cheng X, Merched-Sauvage M, Koch PJ (2006) Desmocollin 3 is required for pre-implantation development of the mouse embryo. J Cell Sci 119:482–489

Descargues P, Deraison C, Bonnart C, Kreft M, Kishibe M, Ishida-Yamamoto A, Elias P, Barrandon Y, Zambruno G, Sonnenberg A, Hovnanian A (2005) Spink5-deficient mice mimic Netherton syndrome through degradation of desmoglein 1 by epidermal protease hyperactivity. Nat Genet 37:56–65

Di Zenzo G, Di Lullo G, Corti D, Calabresi V, Sinistro A, Vanzetta F, Didona B, Cianchini G, Hertl M, Eming R, Amagai M, Ohyama B, Hashimoto T, Sloostra J, Sallusto F, Zambruno G, Lanzavecchia A (2012) Pemphigus autoantibodies generated through somatic mutations target the desmoglein-3 cis-interface. J Clin Invest 122:3781–3790

Diercks GF, Pas HH, Jonkman MF (2009) The ultrastructure of acantholysis in pemphigus vulgaris. Br J Dermatol 160:460–461

Dmochowski M, Hashimoto T, Chidgey MA, Yue KK, Wilkinson RW, Nishikawa T, Garrod DR (1995) Demonstration of antibodies to bovine desmocollin isoforms in certain pemphigus sera. Br J Dermatol 133:519–525

Dubash AD, Koetsier JL, Amargo EV, Najor NA, Harmon RM, Green KJ (2013) The GEF Bcr activates RhoA/MAL signaling to promote keratinocyte differentiation via desmoglein-1. J Cell Biol 202:653–666

Ducibella T, Albertini DF, Anderson E, Biggers JD (1975) The preimplantation mammalian embryo: characterization of intercellular junctions and their appearance during development. Dev Biol 45:231–250

Duden R, Franke WW (1988) Organization of desmosomal plaque proteins in cells growing at low calcium concentrations. J Cell Biol 107:1049–1063

Duguay D, Foty RA, Steinberg MS (2003) Cadherin-mediated cell adhesion and tissue segregation: qualitative and quantitative determinants. Dev Biol 253:309–323

Dusek RL, Getsios S, Chen F, Park JK, Amargo EV, Cryns VL, Green KJ (2006) The differentiation-dependent desmosomal cadherin desmoglein 1 is a novel caspase-3 target that regulates apoptosis in keratinocytes. J Biol Chem 281:3614–3624

Elias PM, Matsuyoshi N, Wu H, Lin C, Wang ZH, Brown BE, Stanley JR (2001) Desmoglein isoform distribution affects stratum corneum structure and function. J Cell Biol 153:243–249

Eshkind L, Tian Q, Schmidt A, Franke WW, Windoffer R, Leube RE (2002) Loss of desmoglein 2 suggests essential functions for early embryonic development and proliferation of embryonal stem cells. Eur J Cell Biol 81:592–598

Ferone G, Mollo MR, Thomason HA, Antonini D, Zhou H, Ambrosio R, De Rosa L, Salvatore D, Getsios S, Van Bokhoven H, Dixon J, Missero C (2013) p63 control of desmosome gene expression and adhesion is compromised in AEC syndrome. Hum Mol Genet 22:531–543

Fleming TP, Garrod DR, Elsmore AJ (1991) Desmosome biogenesis in the mouse preimplantation embryo. Development 112:527–539

Foote HP, Sumigray KD, Lechler T (2013) FRAP analysis reveals stabilization of adhesion structures in the epidermis compared to cultured keratinocytes. PLoS One 8:e71491

Funakoshi S, Ezaki T, Kong J, Guo r j, Lynch JP (2008) Repression of the desmocollin 2 gene expression in human colon cancer cells is relieved by the homeodomain transcription factors Cdx1 and Cdx2. Mol Cancer Res 6:1478–1490

Gallicano GI, Kouklis P, Bauer C, Yin M, Vasioukhin V, Degenstein L, Fuchs E (1998) Desmoplakin is required early in development for assembly of desmosomes and cytoskeletal linkage. J Cell Biol 143:2009–2022

Garcia-Gras E, Lombardi R, Giocondo MJ, Willerson JT, Schneider MD, Khoury DS, Marian AJ (2006) Suppression of canonical Wnt/beta-catenin signaling by nuclear plakoglobin recapitulates phenotype of arrhythmogenic right ventricular cardiomyopathy. J Clin Invest 116: 2012–2021

Garrod D, Chidgey M (2008) Desmosome structure, composition and function. Biochim Biophys Acta 1778:572–587

Garrod DR, Fleming S (1990) Early expression of desmosomal components during kidney tubule morphogenesis in human and murine embryos. Development 108:313–321

Garrod D, Tabernero L (2014) Hyper-adhesion: a unique property of desmosomes. Cell Commun Adhes 21:249–256

Garrod DR, Berika MY, Bardsley WF, Holmes D, Tabernero L (2005) Hyper-adhesion in desmosomes: its regulation in wound healing and possible relationship to cadherin crystal structure. J Cell Sci 118:5743–5754

Garrod DR, Fisher C, Smith A, Nie Z (2008) Pervanadate stabilizes desmosomes. Cell Adh Migr 2:161–166

Getsios S, Simpson CL, Kojima S, Harmon R, Sheu LJ, Dusek RL, Cornwell M, Green KJ (2009) Desmoglein 1-dependent suppression of EGFR signaling promotes epidermal differentiation and morphogenesis. J Cell Biol 185:1243–1258

Gloushankova NA, Wakatsuki T, Troyanovsky RB, Elson E, Troyanovsky SM (2003) Continual assembly of desmosomes within stable intercellular contacts of epithelial A-431 cells. Cell Tissue Res 314:399–410

Godsel LM, Hsieh SN, Amargo EV, Bass AE, Pascoe-Mcgillicuddy LT, Huen AC, Thorne ME, Gaudry CA, Park JK, Myung K, Goldman RD, Chew TL, Green KJ (2005) Desmoplakin assembly dynamics in four dimensions: multiple phases differentially regulated by intermediate filaments and actin. J Cell Biol 171:1045–1059

Godsel LM, Dubash AD, Bass-Zubek AE, Amargo EV, Klessner JL, Hobbs RP, Chen X, Green KJ (2010) Plakophilin 2 couples actomyosin remodeling to desmosomal plaque assembly via RhoA. Mol Biol Cell 21:2844–2859

Goonesinghe A, Luan XM, Hurlstone A, Garrod D (2012) Desmosomal cadherins in zebrafish epiboly and gastrulation. BMC Dev Biol 12:1

Greenwood MD, Marsden MD, Cowley CM, Sahota VK, Buxton RS (1997) Exon-intron organization of the human type 2 desmocollin gene (DSC2): desmocollin gene structure is closer to "classical" cadherins than to desmogleins. Genomics 44:330–335

Grossmann KS, Grund C, Huelsken J, Behrend M, Erdmann B, Franke WW, Birchmeier W (2004) Requirement of plakophilin 2 for heart morphogenesis and cardiac junction formation. J Cell Biol 167:149–160

Hanakawa Y, Matsuyoshi N, Stanley JR (2002) Expression of desmoglein 1 compensates for genetic loss of desmoglein 3 in keratinocyte adhesion. J Invest Dermatol 119:27–31

Hanakawa Y, Schechter NM, Lin C, Nishifuji K, Amagai M, Stanley JR (2004) Enzymatic and molecular characteristics of the efficiency and specificity of exfoliative toxin cleavage of desmoglein 1. J Biol Chem 279:5268–5277

Hardman MJ, Liu K, Avilion AA, Merritt A, Brennan K, Garrod DR, Byrne C (2005) Desmosomal cadherin misexpression alters beta-catenin stability and epidermal differentiation. Mol Cell Biol 25:969–978

Harmon RM, Green KJ (2013) Structural and functional diversity of desmosomes. Cell Commun Adhes 20:171–187

Harmon RM, Simpson CL, Johnson JL, Koetsier JL, Dubash AD, Najor NA, Sarig O, Sprecher E, Green KJ (2013) Desmoglein-1/Erbin interaction suppresses ERK activation to support epidermal differentiation. J Clin Invest 123:1556–1570

Hatzfeld M, Haffner C, Schulze K, Vinzens U (2000) The function of plakophilin 1 in desmosome assembly and actin filament organization. J Cell Biol 149:209–222

Haussinger D, Ahrens T, Sass HJ, Pertz O, Engel J, Grzesiek S (2002) Calcium-dependent homoassociation of E-cadherin by NMR spectroscopy: changes in mobility, conformation and mapping of contact regions. J Mol Biol 324:823–839

Henkler F, Strom M, Mathers K, Cordingley H, Sullivan K, King I (2001) Trangenic misexpression of the differentiation-specific desmocollin isoform 1 in basal keratinocytes. J Invest Dermatol 116:144–149

Hennings H, Holbrook KA (1983) Calcium regulation of cell-cell contact and differentiation of epidermal cells in culture. An ultrastructural study. Exp Cell Res 143:127–142

Heupel WM, Zillikens D, Drenckhahn D, Waschke J (2008) Pemphigus vulgaris IgG directly inhibit desmoglein 3-mediated transinteraction. J Immunol 181:1825–1834

Heupel WM, Muller T, Efthymiadis A, Schmidt E, Drenckhahn D, Waschke J (2009) Peptides Targeting the Desmoglein 3 Adhesive Interface Prevent Autoantibody-induced Acantholysis in Pemphigus. J Biol Chem 284:8589–8595

Hovnanian A (2013) Netherton syndrome: skin inflammation and allergy by loss of protease inhibition. Cell Tissue Res 351:289–300

Huber AH, Weis WI (2001) The structure of the beta-catenin/E-cadherin complex and the molecular basis of diverse ligand recognition by beta-catenin. Cell 105:391–402

Huber AH, Stewart DB, Laurents DV, Nelson WJ, Weis WI (2001) The cadherin cytoplasmic domain is unstructured in the absence of beta-catenin. A possible mechanism for regulating cadherin turnover. J Biol Chem 276:12301–12309

Hunt DM, Sahota VK, Taylor K, Simrak D, Hornigold N, Arnemann J, Wolfe J, Buxton RS (1999) Clustered cadherin genes: a sequence-ready contig for the desmosomal cadherin locus on human chromosome 18. Genomics 62:445–455

Hunt DM, Rickman L, Whittock NV, Eady RA, Simrak D, Dopping-Hepenstal PJ, Stevens HP, Armstrong DK, Hennies HC, Kuster W, Hughes AE, Arnemann J, Leigh IM, Mcgrath JA, Kelsell DP, Buxton RS (2001) Spectrum of dominant mutations in the desmosomal cadherin desmoglein 1, causing the skin disease striate palmoplantar keratoderma. Eur J Hum Genet 9:197–203

Ishida-Yamamoto A, Igawa S (2015) The biology and regulation of corneodesmosomes. Cell Tissue Res 360:477–482

Ishii N, Ishida-Yamamoto A, Hashimoto T (2004) Immunolocalization of target autoantigens in IgA pemphigus. Clin Exp Dermatol 29:62–66

Jackson BW, Grund C, Schmid E, Burki K, Franke WW, Illmensee K (1980) Formation of cytoskeletal elements during mouse embryogenesis. Intermediate filaments of the cytokeratin type and desmosomes in preimplantation embryos. Differentiation 17:161–179

Jackson BW, Grund C, Winter S, franke WW, illmensee K (1981) Formation of cytoskeletal elements during mouse embryogenesis. II. Epithelial differentiation and intermediate-sized filaments in early postimplantation embryos. Differentiation 20:203–216

Jennings JM, Tucker DK, Kottke MD, Saito M, Delva E, Hanakawa Y, Amagai M, Kowalczyk AP (2011) Desmosome disassembly in response to pemphigus vulgaris IgG occurs in distinct phases and can be reversed by expression of exogenous Dsg3. J Invest Dermatol 131:706–718

Johnson JL, Najor NA, Green KJ (2014) Desmosomes: regulators of cellular signalling and adhesion in epidermal health and disease. Cold Spring Harb Perspect Med 4:a015297

Jolly PS, Berkowitz P, Bektas M, Lee HE, Chua M, Diaz LA, Rubenstein DS (2010) p38MAPK signaling and desmoglein-3 internalization are linked events in pemphigus acantholysis. J Biol Chem 285:8936–8941

Jones JC, Goldman RD (1985) Intermediate filaments and the initiation of desmosome assembly. J Cell Biol 101:506–517

Jonkman MF, Pasmooij AM, Pasmans SG, Van Den Berg MP, Ter Horst HJ, Timmer A, Pas HH (2005) Loss of desmoplakin tail causes lethal acantholytic epidermolysis bullosa. Am J Hum Genet 77:653–660

Kami K, Chidgey M, Dafforn T, Overduin M (2009) The desmoglein-specific cytoplasmic region is intrinsically disordered in solution and interacts with multiple desmosomal protein partners. J Mol Biol 386:531–543

Kasperkiewicz M, Schmidt E, Zillikens D (2012) Current therapy of the pemphigus group. Clin Dermatol 30:84–94

Katsamba P, Carroll K, Ahlsen G, Bahna F, Vendome J, Posy S, Rajebhosale M, Price S, Jessell TM, Ben-Shaul A, Shapiro L, Honig BH (2009) Linking molecular affinity and cellular specificity in cadherin-mediated adhesion. Proc Natl Acad Sci U S A 106:11594–11599

Kenchegowda D, Harvey SA, Swamynathan S, Lathrop KL, Swamynathan SK (2012) Critical role of Klf5 in regulating gene expression during post-eyelid opening maturation of mouse corneas. PLoS One 7:e44771

Kim A, Dean A (2012) Chromatin loop formation in the beta-globin locus and its role in globin gene transcription. Mol Cells 34:1–5

Kim C, Wong J, Wen J, Wang S, Wang C, Spiering S, Kan NG, Forcales S, Puri PL, Leone TC, Marine JE, Calkins H, Kelly DP, Judge DP, Chen HS (2013) Studying arrhythmogenic right ventricular dysplasia with patient-specific iPSCs. Nature 494:105–110

Kimura TE, Merritt AJ, Garrod DR (2007) Calcium-independent desmosomes of keratinocytes are hyper-adhesive. J Invest Dermatol 127:775–781

Kimura TE, Merritt AJ, Lock FR, Eckert JJ, Fleming TP, Garrod DR (2012) Desmosomal adhesiveness is developmentally regulated in the mouse embryo and modulated during trophectoderm migration. Dev Biol 369:286–297

King IA, Angst BD, Hunt DM, Kruger M, Arnemann J, Buxton RS (1997) Hierarchical expression of desmosomal cadherins during stratified epithelial morphogenesis in the mouse. Differentiation 62:83–96

Klessner JL, Desai BV, Amargo EV, Getsios S, Green KJ (2009) EGFR and ADAMs cooperate to regulate shedding and endocytic trafficking of the desmosomal cadherin desmoglein 2. Mol Biol Cell 20:328–337

Kljuic A, Bazzi H, Sundberg JP, Martinez-Mir A, O'shaughnessy R, Mahoney MG, Levy M, Montagutelli X, Ahmad W, Aita VM, Gordon D, Uitto J, Whiting D, Ott J, Fischer S, Gilliam TC, Jahoda CA, Morris RJ, Panteleyev AA, Nguyen VT, Christiano AM (2003) Desmoglein 4 in hair follicle differentiation and epidermal adhesion: evidence from inherited hypotrichosis and acquired pemphigus vulgaris. Cell 113:249–260

Koch PJ, Mahoney MG, Ishikawa H, Pulkkinen L, Uitto J, Shultz L, Murphy GF, Whitaker-Menezes D, Stanley JR (1997) Targeted disruption of the pemphigus vulgaris antigen (desmoglein 3) gene in mice causes loss of keratinocyte cell adhesion with a phenotype similar to pemphigus vulgaris. J Cell Biol 137:1091–1102

Koeser J, Troyanovsky SM, Grund C, Franke WW (2003) De novo formation of desmosomes in cultured cells upon transfection of genes encoding specific desmosomal components. Exp Cell Res 285:114–130

Kowalczyk AP, Green KJ (2013) Structure, function, and regulation of desmosomes. Prog Mol Biol Transl Sci 116:95–118

Kowalczyk AP, Bornslaeger EA, Borgwardt JE, Palka HL, Dhaliwal AS, Corcoran CM, Denning MF, Green KJ (1997) The amino-terminal domain of desmoplakin binds to plakoglobin and clusters desmosomal cadherin-plakoglobin complexes. J Cell Biol 139:773–784

Kowalczyk AP, Hatzfeld M, Bornslaeger EA, Kopp DS, Borgwardt JE, Corcoran CM, Settler A, Green KJ (1999) The head domain of plakophilin-1 binds to desmoplakin and enhances its recruitment to desmosomes. Implications for cutaneous disease. J Biol Chem 274: 18145–18148

Kroger C, Loschke F, Schwarz N, Windoffer R, Leube RE, Magin TM (2013) Keratins control intercellular adhesion involving PKC-alpha-mediated desmoplakin phosphorylation. J Cell Biol 201:681–692

Kurinna S, Schafer M, Ostano P, Karouzakis E, Chiorino G, Bloch W, Bachmann A, Gay S, Garrod D, Lefort K, Dotto GP, Beer HD, Werner S (2014) A novel Nrf2-miR-29-desmocollin-2 axis regulates desmosome function in keratinocytes. Nat Commun 5:5099

Lentz TL, Trinkaus JP (1971) Differentiation of the junctional complex of surface cells in the developing Fundulus blastoderm. J Cell Biol 48:455–472

Li J, Swope D, Raess N, Cheng L, Muller EJ, Radice GL (2011) Cardiac tissue-restricted deletion of plakoglobin results in progressive cardiomyopathy and activation of {beta}-catenin signaling. Mol Cell Biol 31:1134–1144

Lombardi R, Da Graca Cabreira-Hansen M, Bell A, Fromm RR, Willerson JT, Marian AJ (2011) Nuclear plakoglobin is essential for differentiation of cardiac progenitor cells to adipocytes in arrhythmogenic right ventricular cardiomyopathy. Circ Res 109:1342–1353

Lopez-Pajares V, Yan K, Zarnegar BJ, Jameson KL, Khavari PA (2013) Genetic pathways in disorders of epidermal differentiation. Trends Genet 29:31–40

Lorch JH, Klessner J, Park JK, Getsios S, Wu YL, Stack MS, Green KJ (2004) Epidermal growth factor receptor inhibition promotes desmosome assembly and strengthens intercellular adhesion in squamous cell carcinoma cells. J Biol Chem 279:37191–37200

Lowndes M, Rakshit S, Shafraz O, Borghi N, Harmon RM, Green KJ, Sivisankar S, Nelson WJ (2014) Different roles of cadherins in the assembly and structural integrity of the desmosome complex. J Cell Sci 127:2339–2350

Mao X, Sano Y, Park JM, Payne AS (2011) p38 MAPK activation is downstream of the loss of intercellular adhesion in pemphigus vulgaris. J Biol Chem 286:1283–1291

Martin ED, Moriarty MA, Byrnes L, Grealy M (2009) Plakoglobin has both structural and signalling roles in zebrafish development. Dev Biol 327:83–96

Mattey DL, Garrod DR (1985) Mutual desmosome formation between all binary combinations of human, bovine, canine, avian and amphibian cells: desmosome formation is not tissue- or species-specific. J Cell Sci 75:377–399

Mattey DL, Garrod DR (1986) Calcium-induced desmosome formation in cultured kidney epithelial cells. J Cell Sci 85:95–111

Mattey DL, Burdge G, Garrod DR (1990) Development of desmosomal adhesion between MDCK cells following calcium switching. J Cell Sci 97(Pt 4):689–704

Maytin EV, Habener JF (1998) Transcription factors C/EBP alpha, C/EBP beta, and CHOP (Gadd153) expressed during the differentiation program of keratinocytes in vitro and in vivo. J Invest Dermatol 110:238–246

Maytin EV, Lin JC, Krishnamurthy R, Batchvarova N, Ron D, Mitchell PJ, Habener JF (1999) Keratin 10 gene expression during differentiation of mouse epidermis requires transcription factors C/EBP and AP-2. Dev Biol 216:164–181

Mcharg S, Hopkins G, Lim L, Garrod D (2014) Down-regulation of desmosomes in cultured cells: the roles of PKC, microtubules and lysosomal/proteasomal degradation. PLoS One 9:e108570

Meng JJ, Bornslaeger EA, Green KJ, Steinert PM, Ip W (1997) Two-hybrid analysis reveals fundamental differences in direct interactions between desmoplakin and cell type-specific intermediate filaments. J Biol Chem 272:21495–21503

Merritt AJ, Berika MY, Zhai W, Kirk SE, Ji B, Hardman MJ, Garrod DR (2002) Suprabasal desmoglein 3 expression in the epidermis of transgenic mice results in hyperproliferation and abnormal differentiation. Mol Cell Biol 22:5846–5858

Merritt AJ, Zhu K, Garrod DR, Chidgey MA (2007) A 4 kb fragment of the desmocollin 3 promoter directs reporter gene expression to parakeratotic epidermis and primary hair follicles. J Invest Dermatol 127:245–247

Miller K, Mattey D, Measures H, Hopkins C, Garrod D (1987) Localisation of the protein and glycoprotein components of bovine nasal epithelial desmosomes by immunoelectron microscopy. EMBO J 6:885–889

Muller EJ, Williamson L, Kolly C, Suter MM (2008) Outside-in signaling through integrins and cadherins: a central mechanism to control epidermal growth and differentiation? J Invest Dermatol 128:501–516

Munoz WA, Kloc M, Cho K, Lee M, Hofmann I, Sater A, Vleminckx K, Mccrea PD (2012) Plakophilin-3 is required for late embryonic amphibian development, exhibiting roles in ectodermal and neural tissues. PLoS One 7:e34342

Nagar B, Overduin M, Ikura M, Rini JM (1996) Structural basis of calcium-induced E-cadherin rigidification and dimerization. Nature 380:360–364

Nakamura T, Sano M, Songyang Z, Schneider MD (2003) A Wnt- and beta -catenin-dependent pathway for mammalian cardiac myogenesis. Proc Natl Acad Sci U S A 100:5834–5839

Nava P, Laukoetter MG, Hopkins AM, Laur O, Gerner-Smidt K, Green KJ, Parkos CA, Nusrat A (2007) Desmoglein-2: a novel regulator of apoptosis in the intestinal epithelium. Mol Biol Cell 18:4565–4578

Nekrasova O, Green KJ (2013) Desmosome assembly and dynamics. Trends Cell Biol 23:537–546

Nekrasova OE, Amargo EV, Smith WO, Chen J, Kreitzer GE, Green KJ (2011) Desmosomal cadherins utilize distinct kinesins for assembly into desmosomes. J Cell Biol 195:1185–1203

Nie Z, Merritt A, Rouhi-Parkouhi M, Tabernero L, Garrod D (2011) Membrane-impermeable cross-linking provides evidence for homophilic, isoform-specific binding of desmosomal cadherins in epithelial cells. J Biol Chem 286:2143–2154

Niessen CM, Gumbiner BM (2002) Cadherin-mediated cell sorting not determined by binding or adhesion specificity. J Cell Biol 156:389–399

North AJ, Chidgey MA, Clarke JP, Bardsley WG, Garrod DR (1996) Distinct desmocollin isoforms occur in the same desmosomes and show reciprocally graded distributions in bovine nasal epidermis. Proc Natl Acad Sci U S A 93:7701–7705

North AJ, Bardsley WG, Hyam J, Bornslaeger EA, Cordingley HC, Trinnaman B, Hatzfeld M, Green KJ, Magee AI, Garrod DR (1999) Molecular map of the desmosomal plaque. J Cell Sci 112(Pt 23):4325–4336

Notari M, Hu Y, Sutendra G, Dedeic Z, Lu M, Dupays L, Yavari A, Carr CA, Zhong S, Opel A, Tinker A, Clarke K, Watkins H, Ferguson DJ, Kelsell DP, De Noronha S, Sheppard MN, Hollinshead M, Mohun TJ, Lu X (2015) HiASPP, a previously unidentified regulator of desmosomes, prevents arrhythmogenic right ventricular cardiomyopathy (ARVC)-induced sudden death. Proc Natl Acad Sci U S A 112:E973–E981

Nuber UA, Schafer S, Stehr S, Rackwitz HR, Franke WW (1996) Patterns of desmocollin synthesis in human epithelia: immunolocalization of desmocollins 1 and 3 in special epithelia and in cultured cells. Eur J Cell Biol 71:1–13

Odland GF (1958) The fine structure of the interrelationship of cells in the human epidermis. J Biophys Biochem Cytol 4:529–538

Okuyama R, Tagami H, Aiba S (2008) Notch signaling: its role in epidermal homeostasis and in the pathogenesis of skin diseases. J Dermatol Sci 49:187–194

Oshiro MM, Watts GS, Wozniak RJ, Junk DJ, Munoz-Rodriguez JL, Domann FE, Futscher BW (2003) Mutant p53 and aberrant cytosine methylation cooperate to silence gene expression. Oncogene 22:3624–3634

Overton J (1962) Desmosome development in normal and reassociating cells in the early chick blastoderm. Dev Biol 4:532–548

Overton J (1973) Experimental manipulation of desmosome formation. J Cell Biol 56:636–646

Overton J (1977) Formation of junctions and cell sorting in aggregates of chick and mouse cells. Dev Biol 55:103–116

Owens P, Bazzi H, Engelking E, Han G, Christiano AM, Wang XJ (2008) Smad4-dependent desmoglein-4 expression contributes to hair follicle integrity. Dev Biol 322:156–166

Pare A, Kim M, Juarez MT, Brody S, Mcginnis W (2012) The functions of grainy head-like proteins in animals and fungi and the evolution of apical extracellular barriers. PLoS One 7: e36254

Pasdar M, Nelson WJ (1988a) Kinetics of desmosome assembly in Madin-Darby canine kidney epithelial cells: temporal and spatial regulation of desmoplakin organization and stabilization upon cell-cell contact. I. Biochemical analysis. J Cell Biol 106:677–685

Pasdar M, Nelson WJ (1988b) Kinetics of desmosome assembly in Madin-Darby canine kidney epithelial cells: temporal and spatial regulation of desmoplakin organization and stabilization upon cell-cell contact. II. Morphological analysis. J Cell Biol 106:687–695

Pasdar M, Nelson WJ (1989) Regulation of desmosome assembly in epithelial cells: kinetics of synthesis, transport, and stabilization of desmoglein I, a major protein of the membrane core domain. J Cell Biol 109:163–177

Pasdar M, Krzeminski KA, Nelson WJ (1991) Regulation of desmosome assembly in MDCK epithelial cells: coordination of membrane core and cytoplasmic plaque domain assembly at the plasma membrane. J Cell Biol 113:645–655

Penn EJ, Burdett ID, Hobson C, Magee AI, Rees DA (1987) Structure and assembly of desmosome junctions: biosynthesis and turnover of the major desmosome components of Madin-Darby canine kidney cells in low calcium medium. J Cell Biol 105:2327–2334

Pilichou K, Nava A, Basso C, Beffagna G, Bauce B, Lorenzon A, Frigo G, Vettori A, Valente M, Towbin J, Thiene G, Danieli GA, Rampazzo A (2006) Mutations in desmoglein-2 gene are associated with arrhythmogenic right ventricular cardiomyopathy. Circulation 113:1171–1179

Pilichou K, Remme CA, Basso C, Campian ME, Rizzo S, Barnett P, Scicluna BP, Bauce B, Van Den Hoff MJ, De Bakker JM, Tan HL, Valente M, Nava A, Wilde AA, Moorman AF, Thiene G, Bezzina CR (2009) Myocyte necrosis underlies progressive myocardial dystrophy in mouse dsg2-related arrhythmogenic right ventricular cardiomyopathy. J Exp Med 206: 1787–1802

Pla P, Moore R, Morali OG, Grille S, Martinozzi S, Delmas V, Larue L (2001) Cadherins in neural crest cell development and transformation. J Cell Physiol 189:121–132

Pokutta S, Herrenknecht K, Kemler R, Engel J (1994) Conformational changes of the recombinant extracellular domain of E-cadherin upon calcium binding. Eur J Biochem 223:1019–1026

Rayns DG, simpson FO, Ledingham JM (1969) Ultrastructure of desmosomes in mammalian intercalated disc; appearances after lanthanum treatment. J Cell Biol 42:322–326

Resnik N, Sepcic K, Plemenitas A, Windoffer R, Leube R, Veranic P (2011) Desmosome assembly and cell-cell adhesion are membrane raft-dependent processes. J Biol Chem 286: 1499–1507

Rickman L, Simrak D, Stevens HP, Hunt DM, King IA, Bryant SP, Eady RA, Leigh IM, Arnemann J, Magee AI, Kelsell DP, Buxton RS (1999) N-terminal deletion in a desmosomal cadherin causes the autosomal dominant skin disease striate palmoplantar keratoderma. Hum Mol Genet 8:971–976

Rizzo S, Lodder EM, Verkerk AO, Wolswinkel R, Beekman L, Pilichou K, Basso C, Remme CA, Thiene G, Bezzina CR (2012) Intercalated disc abnormalities, reduced Na(+) current density, and conduction slowing in desmoglein-2 mutant mice prior to cardiomyopathic changes. Cardiovasc Res 95:409–418

Roberts BJ, Pashaj A, Johnson KR, Wahl JK 3rd (2011) Desmosome dynamics in migrating epithelial cells requires the actin cytoskeleton. Exp Cell Res 317:2814–2822

Ruiz P, Brinkmann V, Ledermann B, Behrend M, Grund C, Thalhammer C, Vogel F, Birchmeier C, Gunthert U, Franke WW, Birchmeier W (1996) Targeted mutation of plakoglobin in mice reveals essential functions of desmosomes in the embryonic heart. J Cell Biol 135:215–225

Samuelov L, Sarig O, Harmon RM, Rapaport D, Ishida-Yamamoto A, Isakov O, Koetsier JL, Gat A, Goldberg I, Bergman R, Spiegel R, Eytan O, Geller S, Peleg S, Shomron N, Goh CS, Wilson NJ, Smith FJ, Pohler E, Simpson MA, Mclean WH, Irvine AD, Horowitz M, Mcgrath JA, Green KJ, Sprecher E (2013) Desmoglein 1 deficiency results in severe dermatitis, multiple allergies and metabolic wasting. Nat Genet 45:1244–1248

Schaffer JV, Bazzi H, Vitebsky A, Witkiewicz A, Kovich OI, Kamino H, Shapiro LS, Amin SP, Orlow SJ, Christiano AM (2006) Mutations in the desmoglein 4 gene underlie localized

autosomal recessive hypotrichosis with monilethrix hairs and congenital scalp erosions. J Invest Dermatol 126:1286–1291
Shimizu H, Masunaga T, Ishiko A, Kikuchi A, Hashimoto T, Nishikawa T (1995) Pemphigus vulgaris and pemphigus foliaceus sera show an inversely graded binding pattern to extracellular regions of desmosomes in different layers of human epidermis. J Invest Dermatol 105:153–159
Shimizu A, Ishiko A, Ota T, Tsunoda K, Amagai M, Nishikawa T (2004) IgG binds to desmoglein 3 in desmosomes and causes a desmosomal split without keratin retraction in a pemphigus mouse model. J Invest Dermatol 122:1145–1153
Shimizu A, Ishiko A, Ota T, Saito H, Oka H, Tsunoda K, Amagai M, Nishikawa T (2005) In vivo ultrastructural localization of the desmoglein 3 adhesive interface to the desmosome mid-line. J Invest Dermatol 124:984–989
Simpson CL, Kojima S, Cooper-Whitehair V, Getsios S, Green KJ (2010) Plakoglobin rescues adhesive defects induced by ectodomain truncation of the desmosomal cadherin desmoglein 1: implications for exfoliative toxin-mediated skin blistering. Am J Pathol 177:2921–2937
Skerrow CJ, Matoltsy AG (1974) Isolation of epidermal desmosomes. J Cell Biol 63:515–523
Smith C, Zhu K, Merritt A, Picton R, Youngs D, Garrod D, Chidgey M (2004) Regulation of desmocollin gene expression in the epidermis: CCAAT/enhancer-binding proteins modulate early and late events in keratinocyte differentiation. Biochem J 380:757–765
South AP, Wan H, Stone MG, Dopping-Hepenstal PJ, Purkis PE, Marshall JF, Leigh IM, Eady RA, Hart IR, Mcgrath JA (2003) Lack of plakophilin 1 increases keratinocyte migration and reduces desmosome stability. J Cell Sci 116:3303–3314
Spindler V, Heupel WM, Efthymiadis A, Schmidt E, Eming R, Rankl C, Hinterdorfer P, Muller T, Drenckhahn D, Waschke J (2009) Desmocollin 3-mediated binding is crucial for keratinocyte cohesion and is impaired in pemphigus. J Biol Chem 284:30556–30564
Spindler V, Dehner C, Hubner S, Waschke J (2014) Plakoglobin but not desmoplakin regulates keratinocyte cohesion via modulation of p38MAPK signaling. J Invest Dermatol 134: 1655–1664
Stanley JR, Amagai M (2006) Pemphigus, bullous impetigo, and the staphylococcal scalded-skin syndrome. N Engl J Med 355:1800–1810
Stappenbeck TS, Green KJ (1992) The desmoplakin carboxyl terminus coaligns with and specifically disrupts intermediate filament networks when expressed in cultured cells. J Cell Biol 116:1197–1209
Swope D, Li J, Radice GL (2013) Beyond cell adhesion: the role of armadillo proteins in the heart. Cell Signal 25:93–100
Syed SE, Trinnaman B, Martin S, Major S, Hutchinson J, Magee AI (2002) Molecular interactions between desmosomal cadherins. Biochem J 362:317–327
Takeichi M (1995) Morphogenetic roles of classic cadherins. Curr Opin Cell Biol 7:619–627
Tariq H, Bella J, Jowitt TA, Holmes D, Rouhi M, Nie Z, Baldock C, Garrod D, Tabernero L (2015) Cadherin flexibility provides a key difference between desmosomes and adherens junctions. Proc Natl Acad Sci U S A 112(17):5395–5400
Thomason HA, Scothern A, Mcharg S, Garrod DR (2010) Desmosomes: adhesive strength and signalling in health and disease. Biochem J 429:419–433
Thomason HA, Cooper NH, Ansell DM, Chiu M, Merrit AJ, Hardman MJ, Garrod DR (2012) Direct evidence that PKCalpha positively regulates wound re-epithelialization: correlation with changes in desmosomal adhesiveness. J Pathol 227:346–356
Tokonzaba E, Chen J, Cheng X, Den Z, Ganeshan R, Muller EJ, Koch PJ (2013) Plakoglobin as a regulator of desmocollin gene expression. J Invest Dermatol 133:2732–2740
Trevor KT, Steben LS (1992) Distribution of desmosomal proteins in F9 embryonal carcinoma cells and epithelial cell derivatives. J Cell Sci 103(Pt 1):69–80
Troyanovsky SM, Troyanovsky RB, Eshkind LG, krutovskikh VA, Leube RE, Franke WW (1994a) Identification of the plakoglobin-binding domain in desmoglein and its role in plaque assembly and intermediate filament anchorage. J Cell Biol 127:151–160

Troyanovsky SM, Troyanovsky RB, Eshkind LG, Leube RE, Franke WW (1994b) Identification of amino acid sequence motifs in desmocollin, a desmosomal glycoprotein, that are required for plakoglobin binding and plaque formation. Proc Natl Acad Sci U S A 91:10790–10794

Tselepis C, Chidgey M, North A, Garrod D (1998) Desmosomal adhesion inhibits invasive behavior. Proc Natl Acad Sci U S A 95:8064–8069

Tsuruta D, Ishii N, Hamada T, Ohyama B, Fukuda S, Koga H, Imamura K, Kobayashi H, Karashima T, Nakama T, Dainichi T, Hashimoto T (2011) IgA pemphigus. Clin Dermatol 29:437–442

Tucker DK, Stahley SN, Kowalczyk AP (2014) Plakophilin-1 protects keratinocytes from pemphigus vulgaris IgG by forming calcium-independent desmosomes. J Invest Dermatol 134: 1033–1043

Vasioukhin V, Bowers E, Bauer C, Degenstein L, Fuchs E (2001) Desmoplakin is essential in epidermal sheet formation. Nat Cell Biol 3:1076–1085

Vidémont E, Mariani C, Vidal S, Pin D (2012) Characterization of the canine skin barrier restoration following acute disruption by tape stripping. Vet Dermatol 23:103–109

Vleminckx K, Kemler R (1999) Cadherins and tissue formation: integrating adhesion and signaling. Bioessays 21:211–220

Volk T, Cohen O, Geiger B (1987) Formation of heterotypic adherens-type junctions between L-CAM-containing liver cells and A-CAM-containing lens cells. Cell 50:987–994

Wallis S, Lloyd S, Wise I, Ireland G, Fleming TP, Garrod D (2000) The alpha isoform of protein kinase C is involved in signaling the response of desmosomes to wounding in cultured epithelial cells. Mol Biol Cell 11:1077–1092

Wan H, Dopping-Hepenstal PJ, Gratian MJ, Stone MG, Zhu G, Purkis PE, South AP, Keane F, Armstrong DK, Buxton RS, Mcgrath JA, Eady RA (2004) Striate palmoplantar keratoderma arising from desmoplakin and desmoglein 1 mutations is associated with contrasting perturbations of desmosomes and the keratin filament network. Br J Dermatol 150:878–891

Wang H, Li ZY, Liu Y, Persson J, Beyer I, Moller T, Koyuncu D, Drescher MR, Strauss R, Zhang XB, Wahl JK, Urban N, Drescher C, Hemminki A, Fender P, Lieber A (2011) Desmoglein 2 is a receptor for adenovirus serotypes 3, 7, 11 and 14. Nat Med 17:96–104

Waschke J, Spindler V (2014) Desmosomes and extradesmosomal adhesive signaling contacts in Pemphigus. Med Res Rev 34:1127–1145

Waschke J, Bruggeman P, Baumgartner W, Zillikens D, Drenckhahn D (2005) Pemphigus foliaceus IgG causes dissociation of desmoglein 1-containing junctions without blocking desmoglein 1 transinteraction. J Clin Invest 115:3157–3165

Watt FM, Mattey DL, Garrod DR (1984) Calcium-induced reorganization of desmosomal components in cultured human keratinocytes. J Cell Biol 99:2211–2215

Whittock NV (2003) Genomic sequence analysis of the mouse desmoglein cluster reveals evidence for six distinct genes: characterization of mouse DSG4, DSG5, and DSG6. J Invest Dermatol 120:970–980

Wilanowski T, Caddy J, Ting SB, Hislop NR, Cerruti L, Auden A, Zhao LL, Asquith S, Ellis S, Sinclair R, Cunningham JM, Jane SM (2008) Perturbed desmosomal cadherin expression in grainy head-like 1-null mice. EMBO J 27:886–897

Windoffer R, Borchert-Stuhltrager M, Leube RE (2002) Desmosomes: interconnected calcium-dependent structures of remarkable stability with significant integral membrane protein turnover. J Cell Sci 115:1717–1732

Yasuda H, Kobayashi H, Hashimoto T, Itoh K, Yamane M, Nakamura J (2000) Subcorneal pustular dermatosis type of IgA pemphigus: demonstration of autoantibodies to desmocollin-1 and clinical review. Br J Dermatol 143:144–148

Yin T, Green KJ (2004) Regulation of desmosome assembly and adhesion. Semin Cell Dev Biol 15:665–677

Chapter 8
Clustered Protocadherins

Kar Men Mah and Joshua A. Weiner

Abstract Nearly 60 cadherin superfamily adhesion molecules are encoded by the *Pcdha, Pcdhb,* and *Pcdhg* gene clusters. These so-called clustered protocadherins (Pcdhs) are broadly expressed throughout the nervous system, with lower levels found in a few nonneuronal tissues. Each neuron expresses a limited repertoire of clustered Pcdh genes, in a complicated process controlled by differential methylation and promoter choice. The clustered Pcdh proteins interact homophilically in *trans* as *cis*-multimers, which has the potential to generate a combinatorially explosive number of distinct adhesive interfaces that may give neurons unique molecular identities important for circuit formation. Functional studies of animals in which clustered Pcdhs have been deleted or disrupted demonstrate that these proteins play critical roles in neuronal survival, axon and dendrite arborization, and synaptogenesis. Additionally, they have been implicated in the progression of several cancers, suggesting that basic studies of their function and signaling pathways will have important future clinical applications. This chapter reviews the extant literature on this fascinating and important group of cell adhesion molecules, the most diverse within the larger cadherin superfamily.

Keywords Pcdh • Cell adhesion • Synaptogenesis • Dendrite arborization • Dendritic spine • Synaptic specificity • Molecular diversity • Homophilic • Apoptosis

K.M. Mah, M.S. (✉)
Integrated Biology Graduate Program, Department of Biology, The University of Iowa, 355 Biology Building, Iowa City, IA 52242, USA
e-mail: karmen-mah@uiowa.edu

J.A. Weiner, Ph.D. (✉)
Department of Biology, The University of Iowa, 348 Biology Building, Iowa City, IA 52242, USA
e-mail: joshua-weiner@uiowa.edu

S.T. Suzuki, S. Hirano (eds.), *The Cadherin Superfamily*,
DOI 10.1007/978-4-431-56033-3_8

8.1 Introduction

The organization of functional neural circuits requires the coordinated control of cell–cell interactions at many stages of development, including neuronal differentiation and migration, axon outgrowth, dendrite arborization, and synaptogenesis. This coordination involves the concerted action of many cell surface receptors, whose regulation enables neurons to adopt their proper functions within developing neural circuits. Each neuron is estimated to make thousands of synapses with partner neurons, and it is generally believed that a large and diverse array of cell surface recognition molecules is required for this to occur with the proper specificity in complex nervous systems. Because the synapse is, in many ways, a specialized type of adhesive junction, it is not surprising that many of the molecular cues thus far shown to influence the formation of neural circuits, and to be disrupted in a variety of human neurodevelopmental disorders, are cell adhesion molecules. One of the most diverse and functionally important of the many cell adhesion molecule families is that encoded by the clustered protocadherin (Pcdh) gene loci. Clustered Pcdh family members are widely, but differentially, expressed by neurons and astrocytes within the developing and mature nervous system, and have been implicated in critical processes such as neuronal survival, dendritic and axonal arborization, and synaptogenesis. Additionally, although they are primarily found in neural tissues, it has recently become clear that they also play potentially clinically relevant roles in cancer of several types. In this chapter, we review the growing literature on the clustered Pcdhs, and highlight their important roles in both neuronal and nonneuronal biology. We also discuss the fascinating structure and regulation of the Pcdh gene clusters, and the molecular mechanisms through which their encoded proteins act.

8.2 Protein Structure and Adhesive Interactions

The clustered Pcdhs represent the largest group within the cadherin superfamily, and are expressed primarily in the nervous system, though they are detectable at low levels in other organs such as the lungs and kidneys (Frank et al. 2005). The cadherin superfamily is defined by extracellular cadherin (EC) motifs that are approximately 100 amino acids long. Classical cadherins are type I transmembrane proteins with 5 EC repeats and a conserved cytoplasmic domain that engages in well-defined interactions with catenins (Gumbiner 2005; Takeichi 2007; Nelson 2008; Niessen et al. 2011). Shintaro Suzuki and colleagues used degenerate PCR to search for additional cadherin molecules, and discovered and named the first protocadherins in the early 1990s (Sano et al. 1993). Protocadherins, also being type I transmembrane proteins, are structurally similar to classical cadherins, but contain six (clustered Pcdhs, δ2-Pcdhs) or seven (δ1-Pcdhs) EC domains, and distinct cytoplasmic domains that lack catenin-binding sites (Sano et al. 1993;

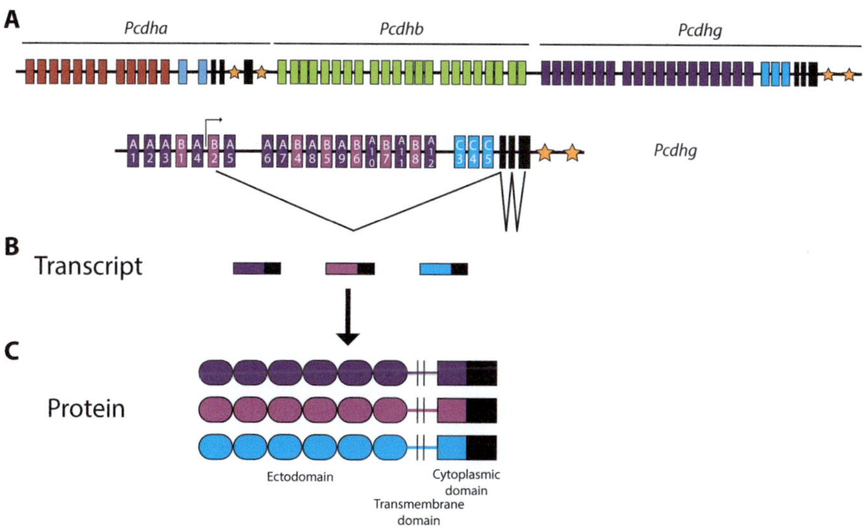

Fig. 8.1 The clustered Pcdhs. (**a**) Schematic of the murine *Pcdha, Pcdhb,* and *Pcdhg* gene clusters on chromosome 18. A very similar structure is observed for the human clusters at chromosome 5q31. The exon structure of the *Pcdhg* cluster is shown below, with an example of the transcription initiation and splicing pattern (for B2). (**b**) Schematic of the *Pcdhg* spliced transcripts generated by the cluster; each mature transcript consists of one large variable exon and the three small constant exons. (**c**) Protein structure of the γ-Pcdhs (α-Pcdhs are identical in structure; β-Pcdhs lack any constant domain). Six EC ectodomains, a transmembrane domain, and a variable cytoplasmic domain are encoded by each variable exon; the constant exons encode a 125 amino acid C-terminal domain. *Colored boxes* are variable exons; *black boxes* are constant exons. *Stars* indicate the sites of the HS5-1, HS7 (*Pcdha* cluster), HS16/17, and HS18-20 (*Pcdhg* cluster) cluster control regions

Wu and Maniatis 1999; Nollet et al. 2000; Vanhalst et al. 2001). We discuss only the clustered Pcdhs, which are encoded by the *Pcdha*, *Pcdhb*, and *Pcdhg* gene clusters in mammals; the nonclustered δ-Pcdhs are discussed in Chapter 9 and in several recent reviews (Morishita and Yagi 2007; Hulpiau and van Roy 2009; Hulpiau and van Roy 2011; Hayashi and Takeichi 2015; Keeler et al. 2015).

The clustered Pcdh genes are arranged in three tandem arrays encompassing about 1 MB at human chromosome 5q31 and mouse chromosome 18 (Wu and Maniatis 1999; Sugino et al. 2000; Wu et al. 2001) (Fig. 8.1a). Multiple large variable exons (14 in the mouse *Pcdha* cluster, 22 in the mouse *Pcdhb* cluster, and 22 in the mouse *Pcdhg* cluster) encode six EC domains, a transmembrane domain, and a variable cytoplasmic domain of approximately 90 amino acids. Each variable exon is expressed from its own promoter and, in the *Pcdha* and *Pcdhg* clusters, spliced to three small constant exons that encode a shared ~125 amino acid C-terminal domain (the *Pcdhb* cluster contains no such constant exons and thus expresses single-exon transmembrane molecules) (Tasic et al. 2002; Wang et al. 2002a; Fig. 8.1b, c).

The *trans*-interaction of classical cadherin EC1 domains mediates calcium-dependent, primarily homophilic adhesion. The EC1 domain of one α-Pcdh for

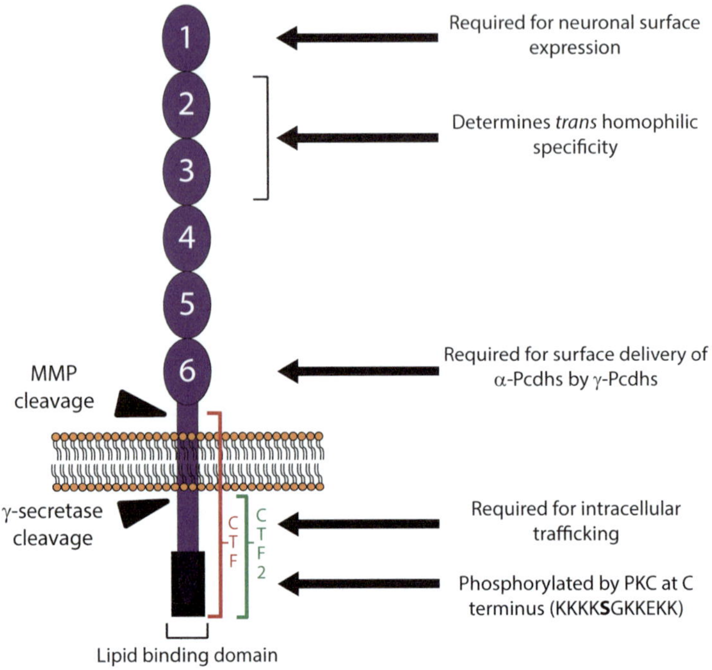

Fig. 8.2 Functions of clustered Pcdh protein domains. Schematic of the protein domains found in α- and γ-Pcdhs, with known roles of several indicated at right. EC1 is required for efficient surface expression (and therefore homophilic interaction) of both α- and γ-Pcdhs. EC2 and EC3 determine the specificity of homophilic binding. EC6 of the γ-Pcdhs is required for interaction with α-Pcdhs and their efficient delivery to the cell surface. The variable cytoplasmic domain is involved in γ-Pcdh intracellular trafficking, and a serine at the lipid-binding C-terminus of the constant domain has recently been identified as a phosphorylation target of PKC. Ectodomain shedding of the γ-Pcdhs can occur due to matrix metalloproteinase (MMP) cleavage, which is followed by γ-secretase/presenillin cleavage at the intracellular side of the membrane. The former generates the CTF, and the latter cleaves this fragment to generate CTF2 (See text for details)

which partial structure is available, however, does not have the tryptophan residue or the hydrophobic pocket known to be required for classical cadherin adhesion (Boggon et al. 2002; Morishita et al. 2006). The α-Pcdh4 EC1 domain also possesses variations in the loop regions, which includes the protocadherin-specific disulfide bonded Cys–$(X)_5$–Cys motif. The presence of the Cys –$(X)_5$–Cys motif in place of the hydrophobic pocket was suggested to indicate an alternative adhesion interface for protocadherins (Morishita et al. 2006). However, Schreiner and Weiner (2010) demonstrated that although γ-Pcdhs do require EC1 for trans interactions (possibly for efficient cell-surface expression), the specificity of such interactions is mediated, rather, through EC2 and EC3 (Fig. 8.2). Consistent with this, the EC2/3 domains are the most divergent among mouse and human α-Pcdh and γ-Pcdh family members (Schreiner and Weiner 2010). *Trans*-interactions of γ-Pcdhs were completely homophilic: both qualitative and quantitative cell adhesion assays found no evidence of any adhesive cross-talk between the seven γ-Pcdh proteins tested in all combinations. Furthermore, chimeric molecules containing EC2 from one isoform and EC3 from another were able to interact homophilically, despite no longer binding to either "parent" isoform (Schreiner and Weiner 2010).

In contrast to the high degree of specificity in γ-Pcdh trans interactions, they engage in promiscuous cis interactions to form multimers in the membrane that, based on molecular weight, initially appeared to be tetramers (Schreiner and Weiner 2010). Isoform matching experiments in the K562 cell line suggest that cell adhesion between two cells requires a high degree of overlap in their γ-Pcdh isoform repertoire, which would allow for enough matches to form (Schreiner and Weiner 2010). Interestingly, the Cys–(X)–Cys motif found in the EC1 domains of γ-Pcdhs may be important for *cis*-multimerization, as mutation of these residues led to greater proportions of bands at monomer and dimer molecular weights; however, even γ-Pcdhs lacking EC1 were able to multimerize, suggesting a multidomain *cis*-interaction (Schreiner and Weiner 2010). Subsequent protein structure and biochemical data showed that *cis*-interactions between clustered Pcdhs require EC6; in the model proposed by Rubinstein et al. (2015), the Pcdhs form *cis*-dimers that then interact in *trans*. Greater Pcdh repertoire matching between two membranes would lead to an enlarged, zipper-like interaction structure that may be sufficient to trigger intracellular signaling (Rubinstein et al. 2015).

A paper by the Maniatis group elaborated on Schreiner and Weiner's 2010 report on γ-Pcdhs' ability to form homophilic interactions, by studying all 58 clustered Pcdh isoforms. Also using a cell-aggregation assay in K562 cells, Thu et al. (2014) found that all but one of the 58 clustered Pcdhs mediate highly specific homophilic trans interactions. Initially, a failure of α-Pcdhs to localize to the plasma membrane, as previously reported (Murata et al. 2004; Bonn et al. 2007), led to the lack of aggregation in the assay; however, when γ-Pcdhs were cotransfected to act as a carrier protein with these α-Pcdhs, homophilic interactions through α-Pcdhs were observed. Thu et al. (2014) confirmed for α- and β-Pcdhs, that, as for γ-Pcdhs (Schreiner and Weiner 2010), homophilic interaction requires EC1, but specificity is mediated by EC2/3. Importantly, a series of EC-domain deletions of γ-Pcdh B6 and chimeric domain shuffling of various α- and γ-Pcdh isoforms revealed that the EC6 domain regulates cell surface delivery (Fig. 8.2).

8.3 Protein Cleavage Events and Trafficking

As type I transmembrane proteins, γ-Pcdhs are susceptible to cleavage by matrix metalloproteinases, much like E-cadherin and N-cadherin (George and Dwivedi 2004). This cleavage generates a soluble extracellular fragment and a carboxyl-terminal fragment (CTF) including the transmembrane domain (Haas et al. 2005; Hambsch et al. 2005). Subsequently, this CTF is a substrate for presenilins (γ-secretase), which produce an even smaller fragment termed CTF2 that is rapidly processed by proteasomes (Haas et al. 2005; Fig. 8.2). Similar to the intracellular domain of Notch, which is translocated to the nucleus to regulate the transcriptional activity of target genes, the CTF2 fragment produced by γ-secretase cleavage has been found to translocate to the nucleus in heterologous cells in vitro (Haas et al. 2005). Hambsch et al. (2005) presented evidence that this CTF2 fragment could increase *Pcdhg* promoter activity in exogenously introduced luciferase constructs, though there is, as yet, no further evidence to suggest that this could happen at the endogenous locus in vivo. Work in the Maniatis laboratory expanded on this to show that the cleavage of the extracellular domain of α-Pcdhs requires endocytosis, as inhibition of this process through dynasore or chloroquine decreased CTF production (Buchanan et al. 2010). Pcdh–Pcdh interaction may stabilize these proteins in the membrane, because increased interaction lowers levels of cleavage products in differentiated CAD cells (Buchanan et al. 2010). In addition, Buchanan et al. (2010) found that full-length α-Pcdh4 and its CTF interacts with ESCRT-0, a regulator of vesicular sorting, which suggests a mechanism for Pcdh trafficking.

Immuno-EM experiments have demonstrated both synaptic and extrasynaptic dendritic and axonal localization of α-Pcdhs (Kohmura et al. 1998) and γ-Pcdhs (Phillips et al. 2003). Studies of transfected neurons show that exogenous, tagged γ-Pcdhs accumulate at axonal and dendritic cell contacts in neurons during development, consistent with their demonstrated role in homophilic interaction (Fernandez-Monreal et al. 2009). It is clear, however, that much γ-Pcdh protein is found intracellularly in neurons. The trafficking of γ-Pcdhs was regulated by their cytoplasmic domains (Fig. 8.2), because deletion of these domains led to a reduction in the intracellular pool and increased targeting to cell–cell interfaces (Fernandez-Monreal et al. 2009; Schreiner and Weiner 2010). The intracellular pool of γ-Pcdhs found in vesicles was associated with COPII and ERGIC proteins, which are known to target proteins from ER to Golgi and to modulate receptor surface expression in dendrites (Fernandez-Monreal et al. 2010). There is also a possibility that these vesicles traffic γ-Pcdhs to surface membranes. A further study revealed that extensive networks of juxta-nuclear membrane tubules were generated in heterologous cells upon expression of exogenous constructs encoding γ-Pcdhs A3 and B2 (but not those encoding N-cadherin), which recruit the autophagy marker LC3 but are not associated with autophagic vesicles (Hanson et al. 2010). Deletion of the cytoplasmic domains of γ-Pcdh A3 abolished formation of the tubules and caused a shift of γ-Pcdh A3 to a conventional secretory pathway (O'Leary et al. 2011).

8.4 Protein-Binding Partners

Pcdhs have been found to interact with a number of proteins, although many interactions were observed in in vitro experiments and await confirmation of an in vivo function. Links between α-Pcdhs and the cytoskeleton have been proposed based on their interaction with neurofilament M and fascin in a yeast two-hybrid screen (Triana-Baltzer and Blank 2006). Additionally, γ-Pcdh B1 was found to interact with a microtubule-destabilizing protein, SCG10, in another yeast two-hybrid screen. However, these interactors have not been successfully validated in vivo using brain lysates. Several α-Pcdhs (then termed "CNRs") were initially discovered in a yeast two-hybrid screen for interactors of Fyn kinase; this interaction was validated by coimmunoprecipitation in mice brain lysates (Kohmura et al. 1998).

The α- and γ-Pcdhs have been found, both in vitro and in vivo, to bind two tyrosine kinases, focal adhesion kinase (FAK), and the related Pyk2 (also referred to as FAK2) through their respective constant regions (Chen et al. 2009). This interaction suppresses the autoactivation of these kinases by phosphorylation, the FAK and Pyk2 that are immunoprecipitated with α- or γ-Pcdhs are inactivated (unphosphorylated; Chen et al. 2009). Overexpression of Pyk2 in the embryonic chicken spinal cord led to neuronal death; thus, if α-Pcdh and γ-Pcdh interact with and inhibit the kinase activity of Pyk2, this could provide an explanation for the widespread neuronal apoptosis that is observed in *Pcdhg* mutant animals (Wang et al. 2002b; Weiner et al. 2005; Prasad et al. 2008; see below). In addition, studies by Garrett et al. (2012) showed a hyperactivation of a signaling pathway involving FAK, protein kinase C (PKC), and the PKC target MARCKS in *Pcdhg* mutant cerebral cortex (Fig. 8.3). Multiple PKC isoforms were hyperactive in mutant cortex, suggesting the involvement of PLC, which was also found to be hyperphosphorylated/activated (Garrett et al. 2012). Suo et al. (2012) complement this study, showing that a targeted deletion of the *Pcdha* gene cluster led to the hyperactivation of Pyk2 and FAK, in addition to the inactivation of Rac1 and RhoA in vivo (Suo et al. 2012). Furthermore, they were able to rescue dendrite and dendritic spine defects observed in *Pcdha* mutant and *Pcdhg* knockdown hippocampal neurons by overexpressing constitutively active Rac1, implicating Pcdhs in the regulation of dendritic development through Rho GTPases (Suo et al. 2012).

A proteomics survey performed by Han et al. (2010) revealed several proteins that were found in macromolecular complexes with γ-Pcdhs. Mass spectrometry and further validation by in vivo co-IP experiments identified these other proteins to include 14-3-3 proteins, R-cadherin, CamKII-β, CamKII-γ, α-catenin, β-catenin, PSD-95, P140CAP/SNIP, and β-tubulin (Han et al. 2010). This report also confirmed previous findings that α-Pcdhs and γ-Pcdhs interact with each other (Murata et al. 2004; Han et al. 2010). Finally, Lin et al. (2010) identified the interaction of the γ-Pcdhs with PDCD10 (programmed cell death 10; also known as CCM [cerebral cavernous malformation]) in vitro and in vivo (see below, Section 8.7.1, Neuronal Survival).

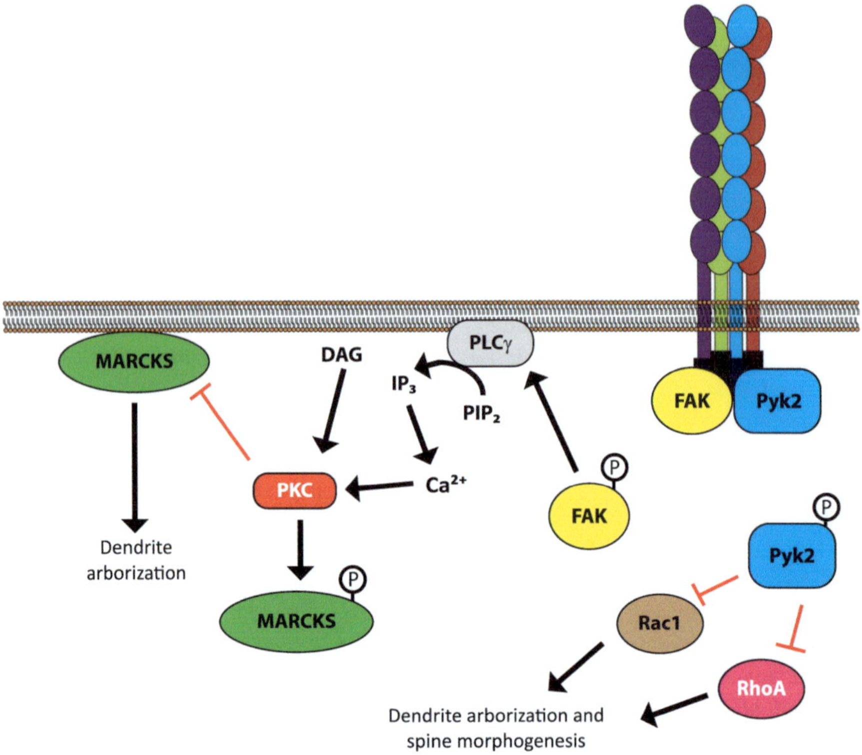

Fig. 8.3 Schematic of signaling pathways through which the γ-Pcdhs (and α-Pcdhs) regulate dendrite arborization in forebrain neurons. The α- and γ-Pcdh constant domain binds FAK and Pyk2 and prevents their activation by autophosphorylation. In the absence of these Pcdhs, FAK and Pyk2 are hyperactive, which leads to disruption of signaling pathways involving the Rho GTPases and PKC, which results in reduced dendrite arborization. It remains unclear whether homophilic *trans*-interaction is a required step in this intracellular signaling

8.5 Molecular Evolution

Clustered protocadherins (Pcdhs) have been described in many vertebrate species, including coelacanth, pufferfish, zebrafish, chicken, mouse, rat, chimpanzee, and human (Wu and Maniatis 1999; Sugino et al. 2000, 2004; Noonan et al. 2004a; Noonan et al. 2004b; Tada et al. 2004; Yanase et al. 2004; Yu et al. 2007). Invertebrates possess several cadherin-related genes, such as Fat, Dachsous, and 7-transmembrane Flamingo cadherins, but not any clustered Pcdh genes (Hill et al. 2001). Clustered Pcdhs are predominantly expressed in the brain; hence, it has been postulated that the Pcdh gene clusters originated in the genome of a vertebrate ancestor with the formation of the neural tube, an elaborate central nervous system, and diversified from this point (Tada et al. 2004). Teleost fishes have two unlinked clustered Pcdh loci, likely as a result of the whole genome duplication in this lineage. Fish cluster composition varies across species: whereas zebrafish has a set of *Pcdha* and *Pcdhg* in each locus, pufferfish has one set of *Pcdha* and *Pcdhg* in one locus, and the other locus only has a *Pcdha* cluster (Noonan et al. 2004b; Tada et al. 2004; Yu et al. 2007).

Comparisons of the constant exons in *Pcdha* and *Pcdhg* gene clusters between human, rat, and mouse reveal a high percentage of identities, and amino acid sequences are 99 % identical for *Pcdha* and *Pcdhg* constant regions, which suggests that intracellular signal transduction pathways should be well conserved in mammals (Wu 2001; Sugino et al. 2004). The gene order of the *Pcdha* gene cluster is essentially conserved across these three species as well, albeit with one fewer variable exon in the mouse cluster, and phylogenetic analysis reveals that the majority of individual *Pcdha* genes are orthologous between human and mouse (Wu et al. 2001; Yanase et al. 2004). On the other hand, α-Pcdh C1 and α-Pcdh C2 are distinct from the other α-Pcdhs, being homologous rather to γ-Pcdh C3, C4, and C5, but are still highly conserved between human and mouse. This suggests that they have specific functions different from those of the other Pcdhα genes, which is consistent too with their ubiquitous expression (see below). Interestingly, sequence relationships of EC1-3 of *Pcdh* genes in humans reveal that the *Pcdhg* C3 is distinctly different from *Pcdhg* C4 and C5, both of which are instead, much more similar to *Pcdha* C2 (Sotomayor et al. 2014).

The various mammalian *Pcdhb* genes display both orthologous and paralogous relationships, which is evidence for an expansion of the *Pcdhb* gene cluster in mouse after divergence of mouse and human (Wu et al. 2001). The *Pcdhg* gene cluster, however, is tightly conserved between mouse and human, save for *Pcdhg*-B3, which has degenerated into a pseudogene in mice, and *Pcdhg*-B8, which is a pseudogene in humans (Wu et al. 2001). Comparisons between the human and chimpanzee Pcdh cluster showed the expected high level of conservation between the species; however, there are a few Pcdh genes (*Pcdhb17, Pcdhb18* in humans, and *Pcdhg*-B3 in chimpanzee) with one- or two-nucleotide insertions leading to frameshifts (Wu 2005). Across the clusters, the orthologous relationships described above are found among mammals, but not among chicken, zebrafish, pufferfish, and coelacanth (Sugino et al. 2000, 2004; Noonan et al. 2004a; Noonan et al. 2004b; Tada et al. 2004; Wu 2005; Yu et al. 2007). In addition to each species having a varying number of Pcdh genes, certain species such as zebrafish have alternative splice sites within variable and constant exons; this has been proposed to generate a higher level of diversity, although this has not yet been shown (Wu 2005). *Pcdha, Pcdhb,* and *Pcdhg* exons have individual promoters (see below, Section 6, Regulation of Gene Expression), found upstream of the translation start site, which are also well conserved across species (Tada et al. 2004).

Alignment of EC1-3 sequences of the classical C-cadherins with corresponding Pcdh sequences of human, chimpanzee, mouse, and rat revealed that the majority of positively selected sites were located on the surface of EC2 and EC3 (Wu 2005), consistent with the analysis of clustered Pcdh homophilic binding discussed above (Schreiner and Weiner 2010; Thu et al. 2014). A higher nonsynonymous substitution rate is suggestive of a diversifying selection that actively creates differences among Pcdh paralogues in mammalian species, which is consistent with the observation that EC2 and EC3 of zebrafish and mammalian Pcdhs seldom undergo a sequence homogenization process (Noonan et al. 2004b; Wu 2005).

8.6 Regulation of Gene Expression

The mechanism for the transcription of *Pcdha* and *Pcdhg* clusters was reported concurrently by Tasic et al. (2002) and Wang et al. (2002a). Each variable exon has its own promoter region that includes a ~20 base pair conserved sequence element (CSE) required for expression. A single variable exon promoter initiates transcription, and the entire remaining downstream *Pcdha* or *Pcdhg* cluster is transcribed. Variable exons that are located in between are removed when the 5' variable exon is *cis*-spliced to the three downstream constant exons (Tasic et al. 2002; Wang et al. 2002a; Fig. 8.1a, b). Although the *Pcdhb* gene cluster does not have constant exons, a 5' splice site exists near the end of the gene cluster, which suggests that *Pcdhb* exons can be spliced to constant exons of other clusters (Wu and Maniatis 1999). *Trans*-splicing between distinct pre-mRNA intermediates, producing intercluster spliced transcripts, has been observed, albeit infrequently (Tasic et al. 2002; Wang et al. 2002a). Studies found that the majority of *Pcdha*, *Pcdhb,* and *Pcdhg* variable exons are monoallelically expressed (Esumi et al. 2005; Kaneko et al. 2006). Both cluster alleles are transcriptionally active in a given cell, but any particular variable exon promoter is "chosen" from only one of the two alleles (Esumi et al. 2005; Kaneko et al. 2006). However, the C isoforms of α-Pcdh (α-PcdhC1 and α-Pcdh-C2) and the related γ-PcdhC3, C4, and C5, all of which are nearly ubiquitously expressed, are biallelically expressed (Esumi et al. 2005; Kaneko et al. 2006). Analysis of cerebellar Purkinje cell neurons using single-cell RT-PCR revealed that each neuron expresses approximately 4 α-Pcdh isoforms, 2 β-Pcdh isoforms, and 7 γ-Pcdh isoforms (Hirano et al. 2012; Yagi 2012).

Although the mechanisms of promoter choice are still not entirely clear, long-range regulatory elements have been found by the Yagi and Maniatis laboratories to play a role in the control of clustered Pcdh expression. Ribich et al. (2006) discovered two long-range regulatory elements located near the 3' end of the *Pcdha* cluster through sequence conservation and hypersensitivity to DNase I degradation (Fig. 8.1a). These HS5-1 and HS7 sites were found to promote gene expression independently in regions of the CNS known to express *Pcdha* genes (Kehayova et al. 2011). Genetic deletion of the HS5-1 site led to significantly reduced expression of most *Pcdha* genes; more 3' genes were greatly reduced, whereas genes lying more 5' were moderately reduced (Kehayova et al. 2011; Yokota et al. 2011). Additional hypersensitive sites were found downstream of the *Pcdhg* cluster and termed HS16-20 by Yokota et al. (2011; Fig. 8.1a). Deletion of these sites in mice resulted in a nearly complete loss of expression of the *Pcdhb* cluster, and moderately affected the *Pcdhg* cluster, without affecting the *Pcdha* cluster. Results from mice in which *Pcdha* variable exons were duplicated or deleted indicate that

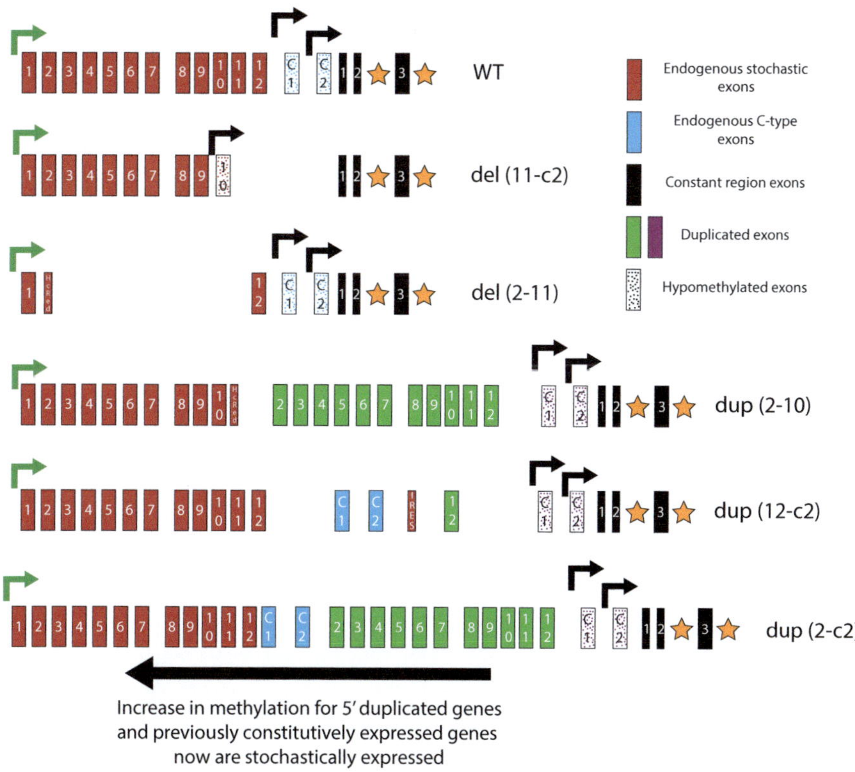

Fig. 8.4 Gene regulation in the wild-type, deleted, and duplicated *Pcdha* locus. In the wild-type (WT) *Pcdha* locus, most variable exons (*red*) are stochastically (and monoallelically) expressed (*green arrows*). This is controlled by the methylation state of their promoters. The aC1 and aC2 exons (*blue*), in contrast, are hypomethylated (stippling) and constitutively (and biallelically) expressed (*black arrow*). Across five lines of mice harboring deletions (del) and duplications (dup) within the cluster, this basic pattern is maintained: the 3′-most variable exon(s) is hypomethylated and constitutively expressed, whereas any exon that is farther 5′ (even C-type exons) becomes stochastically expressed. (Adapted, redrawn, and reannotated from Noguchi et al. (2009) and Kaneko et al. (2014); work from the laboratory of Takeshi Yagi)

proximity to the HS sites downstream of the *Pcdha* gene cluster is responsible for the regulation of expression levels of individual exons and by extension, the varying expression levels across the *Pcdha* cluster (Noguchi et al. 2009). Across five lines of mice (Fig. 8.4), total levels of *Pcdha* transcripts remained constant despite a varying number of variable exons. The key observation is that, in all cases, the 3'-most variable exons become ubiquitously expressed (similar to *Pcdha*-C1 and -C2 exons in the wild-type mouse), whereas duplicate *Pcdha*-C1 and -C2 exons become stochastically expressed when they lie farther away from the HS sites (Noguchi et al. 2009; Kaneko et al. 2014; Fig. 8.4).

Several recent papers have found that the clustered *Pcdh* genes are subject to epigenetic modifications, specifically silencing by methylation (Tasic et al. 2002;

Kawaguchi et al. 2008; Toyoda et al. 2014). Kawaguchi et al. (2008) demonstrated that degree of methylation of *Pcdha* promoters, along with the 5' regions of each variable exon, had a negative correlation with that exon's expression level; *Pcdha*-C1 and *Pcdha*-C2, both of which are ubiquitously expressed by neurons, have hypomethylated promoters (Fig. 8.4). Inducing demethylation using 5-azacytidine was sufficient to increase transcription of *Pcdha* genes, and experimental hypermethylation of a promoter repressed its transcriptional activity (Kawaguchi et al. 2008). In the many lines of *Pcdha* deletion and duplication mice described above, too, methylation patterns changed as expected based on the observed alteration in variable exon expression. Variable exons situated at the 3' end of the cluster become hypomethylated and ubiquitously expressed, whereas the normally ubiquitous *Pcdha*-C1 and *Pcdha*-C2 exons become hypermethylated and stochastically expressed when situated farther 5' (Noguchi et al. 2009; Kaneko et al. 2014; Fig. 8.4). A recent study from the Yagi laboratory identified Dnmt3b as the DNA methyltransferase responsible for regulating methylation patterns of stochastically expressed Pcdh isoforms in neural cells at early embryonic stages (Toyoda et al. 2014). As Dnmt3b-KO mice die at around E15.5, these researchers generated a chimeric mouse through induced pluripotent stem cells (iPSCs) generated from Dnmt3b-KO embryos that expressed EGFP (Toyoda et al. 2014). These iPSCs were injected into the blastocysts of wild-type mice to create a mosaic pattern of cells that were Dnmt3b-KO. In Dnmt3b-KO cerebellar Purkinje cells, each cell's *Pcdha* transcript repertoire was greatly increased, with some cells expressing every variable exon isoform, confirming that this methylase is critical for stochastic expression within Pcdh clusters (Toyoda et al. 2014).

As noted above, a key motif within each clustered Pcdh promoter is the CSE. Chromatin organizers CCCTC-binding factor (CTCF) and cohesin subunit Rad21 are responsible for binding to the CSE to regulate normal expression of the *Pcdha* gene cluster by mediating promoter-enhancer interactions. A series of studies (Golan-Mashiach et al. 2011; Kehayova et al. 2011) has shown that: (1) siRNA knockdown of CTCF or Rad 21 reduces the expression of several stochastically expressed *Pcdha* genes; (2) CTCF and Rad21 are bound to transcriptionally active Pcdhα promoters; and (3) deletion of the HS5-1 site, which also binds CTCF and Rad21, leads to poor binding of CTCF at *Pcdha* promoters. Consistent with the aforementioned role of DNA methylation on Pcdh gene cluster expression, it was found that CTCF binding to Pcdh promoter regions (in a human neuroblastoma cell line, SK-N-SH) is sensitive to CpG methylation (Guo et al. 2012). Another regulator of clustered Pcdh gene expression is the neuron-restrictive silencer factor/RE-1 silencing transcription factor (NRSF/REST), which binds neuron-restrictive silencer elements (NRSEs) to repress neuronal gene expression in nonneuronal cells (Chong et al. 1995; Schoenherr and Anderson 1995). NRSEs are conserved across species as well; they are found within pufferfish, mouse, and human Pcdh clusters (Tan et al. 2010). Exogenously introduced Pcdh constructs lacking NRSEs were found to be expressed ubiquitously, whereas their wild-type counterparts exhibited a neural-specific expression (Tan et al. 2010).

8.7 Roles in the Nervous System

The clustered Pcdhs are most prominently expressed in the developing and adult nervous system, and the vast majority of functional studies of these molecules has focused on roles in neuronal development. Initial studies found that α-Pcdhs were localized to dendrites and synapses (Kohmura et al. 1998), and that γ-Pcdhs were present at some pre-and postsynaptic membranes and concentrated in synaptosome and postsynaptic density (PSD) fractions biochemically (Phillips et al. 2001; Wang et al. 2002b; Phillips et al. 2003). Despite this, γ-Pcdhs are detectable at only a fraction of synapses, with most being found in dendrites, along axons, and perisynaptically at sites of neuron–astrocyte contact (Phillips et al. 2001, 2003; Wang et al. 2002b; Frank et al. 2005; Lefebvre et al. 2008; Fernandez-Monreal et al. 2009; Garrett and Weiner 2009; Li et al. 2010). Although the β-Pcdhs have not been studied extensively, β-Pcdh 16 and β-Pcdh 22 have been shown to be preferentially localized to dendritic and PSD compartments in retinal and cerebellar neurons (Junghans et al. 2008; Puller and Haverkamp 2010). These observations are consistent with the functions revealed by analyses of α-Pcdh and γ-Pcdh mutant animals: regulation of neuronal apoptosis, synaptogenesis, dendrite arborization, and proper targeting of axons (Wang et al. 2002b; Weiner et al. 2005; Fukuda et al. 2008; Hasegawa et al. 2008; Prasad et al. 2008; Garrett and Weiner 2009; Lin et al. 2010; Prasad and Weiner 2011; Chen et al. 2012; Hasegawa et al. 2012; Lefebvre et al. 2012; Suo et al. 2012; Ledderose et al. 2013).

8.7.1 Neuronal Survival

Early analyses of mice lacking the entire *Pcdhg* gene cluster (constitutive null mutant mice) found that spinal interneurons, which make up over 95% of the spinal cord, were especially sensitive to the loss of γ-Pcdhs, and underwent massive apoptosis during late embryogenesis (Wang et al. 2002b). Newborn null mice lacked coordinated movements and reflexes, and pups died within hours after birth. Cultured mutant spinal interneurons also underwent extensive apoptosis followed by degeneration of their processes and synapses after 7 days in vitro, suggesting this phenotype is intrinsic to the CNS and not due to any other physiological abnormalities in mutant pups (Wang et al. 2002b). Prasad et al. (2008) showed that the loss of molecularly defined spinal interneuron populations in *Pcdhg* null mice was variable: only 20% of some neuronal types died, whereas over 80% of others did. This represents an exacerbation of a previously undescribed pattern of normal developmental spinal interneuron apoptosis, as revealed by analysis of *Bax* mutant mice, in which apoptosis is genetically blocked (Prasad et al. 2008). Further analyses using a conditional mutant *Pcdhg* allele revealed that the survival of interneurons is non-cell–autonomous: mutant neurons can survive if surrounded by wild-type neurons, and wild-type neurons can undergo apoptosis if surrounded by mutant neurons (Prasad et al. 2008). This suggests the likelihood that γ-Pcdhs regulate neuronal survival via homophilic *trans*-interactions.

A requirement for the γ-Pcdhs in regulating neuronal survival was subsequently found in other regions of the CNS. Studies using a conditional *Pcdhg* mutant allele in the retina found that γ-Pcdhs are required for the survival of many, but not all, retinal cell types (Lefebvre et al. 2008). Interestingly, in contrast to Prasad et al. (2008), Lefebvre et al. (2008) presented evidence that the requirement of γ-Pcdh for retinal neuron survival was cell autonomous. Su et al. (2010) found that hypothalamic neurons lacking γ-Pcdhs exhibited increased apoptosis, and that this contributed to an obesity phenotype in mice. In zebrafish, neuronal apoptosis throughout the developing brain and spinal cord was observed when splice-blocking antisense morpholinos were used to prevent the addition of the constant domain to the entire *Pcdha* gene cluster (Emond and Jontes 2008).

How might the clustered Pcdhs regulate neuronal survival? A study by Xiaozhong Wang and colleagues found that, as noted above, the γ-Pcdhs interact with PDCD10 (programmed cell death protein 10) via the constant domain in vitro and in vivo (Lin et al. 2010). This is consistent with PDCD10's role as an intracellular adaptor protein involved in regulating apoptosis, found throughout the soma and processes of neuronal cells (Wang et al. 1999). Using electroporation in the embryonic chicken spinal cord, Lin et al. (2010) showed that knockdown of PDCD10 using shRNA attenuated the increase in apoptosis induced by knockdown of γ-Pcdhs, and ectopic expression of wild-type PDCD10, but not a membrane-bound PDCD10, triggered neuronal apoptosis (Lin et al. 2010). Further studies also revealed that PDCD10 and PYK2 can cooperate to induce neuronal apoptosis, leading the authors to propose a model in which γ-Pcdhs (possibly with other clustered Pcdhs) form protein complexes that sequester PDCD10 and lead to the inhibition of PYK2 kinase at the plasma membrane. Loss of γ-Pcdhs would mean more free PDCD10 and abnormal activation of PYK2, both of which would contribute to the excessive neuronal death observed in γ-Pcdh mutant animals (Lin et al. 2010).

An interesting point to note is that not all neurons respond to loss of the γ-Pcdhs with increased apoptosis. In the initial functional analyses in the spinal cord, it was clear that, in contrast to interneurons, neither motor neurons nor dorsal root ganglion sensory neurons died in greater numbers in *Pcdhg* mutants (Wang et al. 2002b; Prasad et al. 2008). Similarly, Garrett et al. (2012) found no increase in apoptosis when *Pcdhg* mutation was restricted to the cerebral cortex (primary neurons and astrocytes) using Emx1-Cre, and Lefebvre et al. (2012) did not report any increase in the apoptosis of cerebellar Purkinje cells lacking the γ-Pcdhs. It appears that interneurons, which by definition make short-range connections within a particular region of the CNS, may be more sensitive than long-range projection neurons such as cortical pyramidal neurons or spinal motor neurons. Whether this has to do with the given neuron's response to defects in circuit formation, or to differential expression of particular γ-Pcdh signal transduction partners in different neuronal subtypes, remains to be examined.

8.7.2 Synaptic Development

Early studies showed that both the α- and γ-Pcdhs were present in synaptosomes, enriched in postsynaptic density protein fractions, and found immunohistochemically at some synapses (Kohmura et al. 1998; Phillips et al. 2001, 2003; Wang et al. 2002b), suggesting roles in synaptic development. Though immunostaining for synaptic markers was severely decreased in the constitutive *Pcdhg* knockout mice we studied initially (Wang et al. 2002b), extensive apoptosis of spinal interneurons made the interpretation of this unclear. Subsequently, we (Weiner et al. (2005)) crossed constitutive knockout mice to mice harboring a null allele of the proapoptotic gene *Bax* to block neuronal apoptosis. This revealed that although the apoptosis, neurodegeneration, and spinal cord hypoplasia were rescued, double mutant animals still died within hours of birth and exhibited a significantly reduced number of both excitatory and inhibitory synapses in the developing spinal cord (Weiner et al. 2005). The synaptic defects observed in these animals were recapitulated in a separate, *Pcdhg* hypomorphic line harboring a truncated allele lacking 57 amino acids from the carboxyl terminus of all γ-Pcdh isoforms. Spinal cord neurons cultured from these hypomorphic animals exhibited normal neuronal differentiation and survival in vitro, but developed fewer, and physiologically weaker, synapses (Weiner et al. 2005).

One puzzling aspect about the γ-Pcdhs was that, although they appear concentrated in synaptic protein fractions and clearly have synaptic functions, most of the protein actually localizes extrasynaptically: antibody staining or GFP signal from fused proteins in transgenic mice fills the neuropil with dotty labeling, which is often found to be perisynaptic but not directly overlapping with markers such as PSD-95. Following up on these observations, work in our laboratory found that γ-Pcdhs are highly expressed by astrocytes, which localize these molecules to their perisynaptic processes, accounting for much of the observed CNS labeling (Garrett and Weiner 2009). Experiments using neuron/astrocyte cocultures, in which wild-type or *Pcdhg* mutant cells can be combined, showed that astrocytic γ-Pcdhs are important for synapse formation or stabilization in developing cultures; when astrocytes are mutant but neurons are wild-type, synaptogenesis is delayed, although it eventually reaches wild-type levels (Garrett and Weiner 2009). In contrast, when neurons are mutant, synaptogenesis never occurs normally, as reported previously (Weiner et al. 2005). This control of synaptogenesis by astrocytes is contact-dependent, as the use of paraformaldehyde-fixed astrocytes in a coculture experiment, which removes the influence of signaling or secreted factors, still supported synaptic development, but only if astrocytes were wild-type. These findings were confirmed in vivo as well, through the use of a GFAP promoter controlling the expression of Cre recombinase to cause the restricted loss of γ-Pcdhs in astrocytes. Both excitatory and inhibitory synaptogenesis were significantly delayed in embryos with *Pcdhg* mutant astrocytes in vivo (Garrett and Weiner 2009), confirming astrocytic Pcdhs in the developing CNS.

8.7.3 Mistargeting of Axons

Several of the clustered Pcdh proteins have been localized to axons (Phillips et al. 2003; Blank et al. 2004; Morishita et al. 2004a, b). Studies from the Yagi lab have discovered a role for the α-Pcdhs in the sorting of olfactory sensory axons (Hasegawa et al. 2008, 2012). Olfactory sensory neurons normally project their axons to specific glomeruli in the main olfactory bulb. However, in *Pcdha* mutant mice, their axons were found to project to multiple, small extraneous glomeruli instead of coalescing into a single glomerulus. In addition, these smaller glomeruli did not disappear with time, which is the case in wild-type animals when glomeruli are innervated by a small number of olfactory sensory neurons of the same odorant receptor type; this implicates the α-Pcdhs in the elimination of axon branches projecting to ectopic glomeruli. The authors also found that although the early stages of olfactory sensory neuron axon guidance are not dependent on α-Pcdhs, it appears that the final stages of axonal coalescence in glomeruli require the presence of α-Pcdhs (Hasegawa et al. 2008). More recently, the same group determined that the α-Pcdh cytoplasmic region was essential for this process, through the use of a transgenic mouse line that produces truncated α-Pcdhs isoforms lacking their cytoplasmic regions. They also showed, interestingly, that the constitutive expression of even a single α-Pcdh isoform could rescue the axonal coalescence of olfactory sensory neurons (Hasegawa et al. 2012).

It is important to note that other work from the Yagi laboratory (2009) showed that the deletion of *Pcdha* constant region exons led to mistargeting of serotonergic neurons in several regions of the brain. Serotonergic axons in *Pcdha* mutant animals approached their targets but did not form extensive axonal arbors within them, as are observed in wild-type animals (Katori et al. 2009). Because the serotonergic system is involved in a wide variety of mood and anxiety disorders, these observations suggest the potential of a clinically relevant role for the α-Pcdhs. Intriguingly, epigenetic studies have shown that the clustered Pcdh loci are hypermethylated in rat pups that experienced poor maternal care, as well as in tissue from the brains of human suicide completers (McGowan et al. 2011; Suderman et al. 2012), although it remains to be seen if there is a link between the serotonergic *Pcdha* mutant phenotype in mice and these studies.

More recently, Meguro et al. (2015) reported that retinogeniculate terminals in the dorsal lateral geniculate nucleus of α-Pcdh mutant animals formed large aggregates. These terminals project correctly in early development, appearing to form aggregates only between P10 and P14, which is just before the eye opens and retinogeniculate projections begin to be refined. Mutant α-Pcdh mice had reduced visual acuity, although the orientation and direction selectivity of neurons in the primary visual cortex appeared to be normal (Yoshitake et al. 2013; Meguro et al. 2015). Additionally, Yamashita et al. (2012) found that the corticocortical pathways that connect the primary somatosensory cortices of both hemispheres were also disrupted in α-Pcdh mutant mice. Together, these studies demonstrate an important role for the α-Pcdhs particularly in the proper projection and refinement of long axonal projections in the CNS.

The γ-Pcdhs have been implicated primarily in dendrite, rather than axon, development, however, Prasad and Weiner (2011) found that the formation of spinal Ia afferent terminal arbors in the ventral horn was disrupted in *Pcdhg* mutant mice. Mutant Ia terminals are aberrantly expanded and aggregated around their primary motor neuron targets, a phenotype not dissimilar to that reported in *Pcdha* mutant retinogeniculate projections (Meguro et al. 2015). The ancillary projection of Ia afferent axon branches to ventral interneurons was correspondingly reduced in mutant animals, even when apoptosis was not an issue (Prasad and Weiner 2011). Using various Cre transgenic lines to restrict the loss of γ-Pcdhs to (1) dorsal root ganglion neurons, (2) Pax2+ spinal interneurons, and (3) motor neurons, Prasad and Weiner (2011) provided evidence that γ-Pcdhs provide a homophilic cue between Ia afferents and their target ventral interneurons, but not motor neurons directly, to regulate formation of the total afferent field. The aggregation of terminals might also represent a self-avoidance phenotype analogous to that demonstrated by Lefebvre et al. (2012) in the dendrites of particular *Pcdhg* mutant neurons (see Section 7.4, Dendrite Arborization). Experiments seeking to rescue this Ia afferent phenotype by re-expressing a single γ-Pcdh isoform are underway to address this possibility (P. Bosch and J.A. Weiner, unpublished data).

8.7.4 Dendrite Arborization

Using Emx1-Cre mice to restrict mutation of a conditional *Pcdhg* allele to the cerebral cortex, we found that γ-Pcdhs are essential for the proper formation of dendritic arbors in cortical pyramidal neurons (Garrett et al. 2012). Most aspects of neurodevelopment such as neurogenesis, neuronal migration, and establishment of major axonal tracts appear to be normal in the *Pcdhg* mutant cortex, but the cortex is thinner due to loss of the cell-sparse, apical dendritic tuft-rich layer I (Garrett et al. 2012). Detailed Sholl analysis of mutant cortical layer V and layer II/III pyramidal neurons revealed a significant loss of dendrite arbor complexity in the absence of the γ-Pcdhs (Fig. 8.5). Garrett et al. (2012) determined that the γ-Pcdhs are critical for the formation of cortical dendrite arbors, but apparently not for their maintenance: when the *Pcdhg* locus was mutated only after 2 months of age (using a tamoxifen-inducible Cre-ER line), arbors that had already formed were stable. Analysis of cell signaling both in vivo and in vitro showed that this phenotype was due in part to aberrant activation of FAK in the absence of γ-Pcdhs, as predicted by the earlier studies of Chen et al. (2009). Downstream of FAK, PKC was hyperactive and its target MARCKS hyperphosphorylated, both of which are known to regulate dendrite arborization negatively (Fig. 8.3). Pharmacological inhibition of FAK or PKC or overexpression of nonphosphorylatable MARCKS in cultured neurons rescued this knockout phenotype (Garrett et al. 2012). Subsequent work from our laboratory has identified a serine residue within the γ-Pcdh constant domain that is

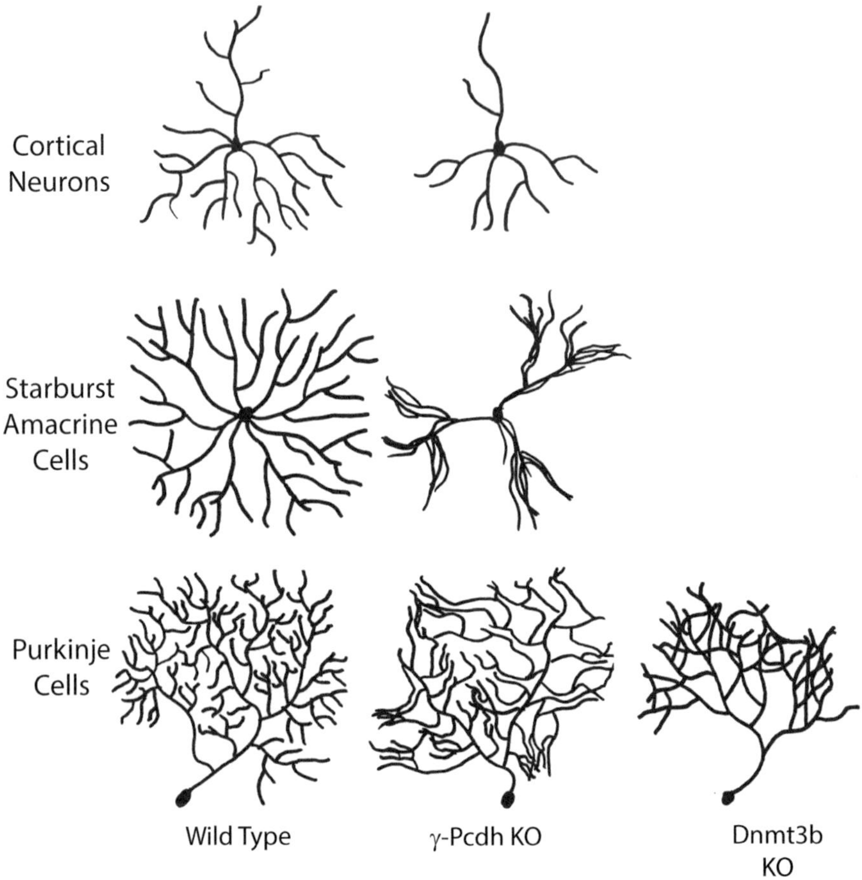

Fig. 8.5 Summary of γ-Pcdh dendrite arborization phenotypes. In cortical and hippocampal neurons, loss of the γ-Pcdhs results in less complex dendritic arbors. In two types of neurons with planar dendrite arbors, starburst amacrine cells of the retina and cerebellar Purkinje cells, γ-Pcdh loss leads to aberrant dendrite self-crossing and/or fasciculation. A combination of these two phenotypes is observed in Purkinje cells when clustered Pcdh repertoire is increased due to loss of the methyltransferase Dnmt3b

phosphorylated by PKC itself; this phosphorylation disrupts the ability of γ-Pcdhs to inhibit FAK, providing a potential feedback mechanism (Keeler, Schreiner, and Weiner, in revision, 2015).

Concurrent work from Qiang Wu's laboratory offers complementary observations on the role of α-Pcdhs in dendrite arborization (Suo et al. 2012). This work showed that *Pcdha* mutant hippocampal neurons both in vivo and in vitro had simplified dendritic arbors as well as a reduction in dendritic spine density (Suo et al. 2012). The authors also performed a knockdown of the *Pcdhg* gene cluster in vitro; the effects on dendrites were in line with those of Garrett et al. (2012) and more severe than that in the *Pcdha* mutants. Suo et al. (2012) found that PYK2 and

FAK were hyperactivated, whereas Rho GTPases were inhibited, in *Pcdha* mutant neurons (Fig. 8.3). Thus, together Garrett et al. (2012) and Suo et al. (2012) demonstrate the neuronal importance of the earlier results of Chen et al. (2009) identifying FAK and PYK2 as protein interactors of the clustered Pcdhs. Additionally, a report by Ledderose et al. (2013) suggested a reduction in dendrite arborization and dendritic spine density in olfactory bulb neurons derived from progenitor neurons of the subventricular zone following lentiviral-Cre-mediated *Pcdhg* mutation. Taken together, all of these data provide strong support for the role of clustered Pcdhs in the elaboration of complex dendritic arbors (Chen et al. 2009; Garrett et al. 2012; Suo et al. 2012; Ledderose et al. 2013).

Interestingly, Lefebvre et al. (2012) demonstrated a distinct dendritic phenotype in *Pcdhg*-mutant retinal starburst amacrine cells (SACs) and cerebellar Purkinje cells: defective self-avoidance between a mutant neuron's own dendrites, which led to aberrant self-crossing and thus failure to cover the appropriate territory (Fig. 8.5). Proper self-avoidance was restored even when only a single γ-Pcdh isoform was re-expressed by using a Cre-inducible transgene. When all SACs expressed this single γ-Pcdh isoform they no longer properly intermingled with their neighbors, suggesting that in these cells, the γ-Pcdhs can mediate repulsive interactions (Lefebvre et al. 2012), though this was not shown directly. Gibson et al. (2014) subsequently showed that the role of γ-Pcdhs in mediating Purkinje dendrite self-avoidance is genetically independent of a similar self-avoidance role mediated by Slit/Robo signaling; the signaling pathways through which the γ-Pcdhs might promote repulsive self-avoidance thus remain unknown. An interesting point is that the two neurons in which a self-avoidance role for the γ-Pcdhs has been established have entirely planar dendritic arbors. We, and others, have not found evidence of self-avoidance defects in nonplanar, bushy dendritic arbors such as those of cortical neurons in the absence of γ-Pcdhs (Garrett et al. 2012, Suo et al. 2012). In multiple heterologous cell lines such as HEK293, COS7, H1299, and K562, transfection of constructs encoding γ-Pcdhs results in robust cell junctions at which these proteins are localized (Frank et al. 2005; Schreiner and Weiner 2010; Thu et al. 2014; M. Molumby and J.A. Weiner, unpublished observations). Similarly, neurons expressing tagged γ-Pcdhs clearly localize these proteins to sites of dendrite and axon contact that are apparently stable (Fernandez-Monreal et al. 2009). Although these observations do not preclude a repulsive function for the γ-Pcdhs, they do suggest that any such function would require distinct signaling partners in planar neurons that exhibit self-avoidance, such as SACs and Purkinje cells.

Studies on CTCF, a major regulator of the three Pcdh gene clusters, have also indirectly implicated Pcdhs in dendrite development (Hirayama et al. 2012). A forebrain-restricted conditional CTCF knockout mouse in which the clustered Pcdhs, among many other genes, are severely downregulated, was found to have reduced cortical and hippocampal neuron dendrite arborization, similar to Garrett et al. (2012) and Suo et al. (2012), along with disrupted barrel formation in the somatosensory cortex (Hirayama et al. 2012). Analogously, as noted above, a study on the methylation of Pcdh genes by DNA methyltransferase Dnmt3b showed that loss of Dnmt3b led to the constitutive expression of many Pcdh genes that are normally stochastically expressed, resulting in an aberrantly large clustered Pcdh gene repertoire in neurons (Toyoda et al. 2014). Dnmt3b mutant Purkinje cells displayed increased self-crossing in Purkinje cells (similar to Lefebvre et al. 2012), but also reduced dendrite arborization and total length, which was not observed previously (Toyoda et al. 2014; Fig. 8.5). These results are intriguing, but it is important to remember that loss of CTCF or Dnmt3b leads to dysregulation of other hundreds of non-Pcdh genes that might also contribute to the phenotypes observed.

8.8 Pcdhs in Cancer

Considering that clustered Pcdhs are thought of primarily as neuronal cell adhesion molecules, and neurons are intrinsically postmitotic and terminally differentiated, it is perhaps surprising that several studies have also implicated them in various types of cancer. A microarray-based methylation study of astrocytomas (World Health Organization grade II) found a CpG island with frequent methylation within the first exon of the *Pcdhg* A11 gene (Waha et al. 2005). As previously discussed, hypermethylation is strongly correlated with decreased transcription in the Pcdh gene locus (Tasic et al. 2002; Kawaguchi et al. 2008). Consistent with this, Waha et al. (2005) found decreased transcription of γ-Pcdh A11 when *Pcdhg* A11 is hypermethylated in these astrocytomas, and treatment with a demethylating agent restored transcript levels of *Pcdhg* A11.

In addition to the brain, Pcdhs have been found to be expressed at low levels in several other organs. A genome-wide analysis of promoter methylation in Wilm's tumor (WiT), a human pediatric kidney cancer, identified a region spanning 800 kilobases at chromosome 5q31 that was hypermethylated; this region contains more than 50 genes from the Pcdh gene cluster (Dallosso et al. 2009). Wilm's tumor precursor cells exhibited no Pcdh gene methylation, with hypermethylation occurring only during malignant progression of the tumor. Knockdown of *Pcdhg* genes using siRNA in a WiT cell line led to an increase in β-catenin/TCF reporter gene activity and corresponding increased expression of target genes in the Wnt signaling pathway, which is implicated in cancer and is constitutively active in these cell lines (Dallosso et al. 2009). Conversely, overexpression of individual γ-Pcdhs in HEK293 and WiT cell lines led to a decrease in Wnt signaling activity, and inhibition of colony formation and tumor cell growth in vitro (Dallosso et al. 2009). The authors subsequently expanded on this study to show that γ-Pcdh C3 can also inhibit Wnt signaling activity in colon cancer tumor cells in vitro (Dallosso et al. 2012). This may involve γ-Pcdh C3 inhibition of the mTOR signaling pathway, although the molecular mechanisms by which the γ-Pcdhs affect the Wnt and mTOR pathways remain to be explored (Dallosso et al. 2012).

Additionally, the *Pcdhb* gene cluster, which remains lamentably understudied functionally, has been implicated in neuroblastoma. Banelli et al. (2011) proposed the *Pcdhb* gene cluster as a candidate of CpG island methylator phenotype, which identifies clinically distinct groups of patients via differential methylation signatures. The *Pcdhb* gene cluster is especially informative as its methylation was found to be highly predictive of two groups of neuroblastoma patients at opposite ends of the International Neuroblastoma Risk Group classification system. Severson et al. (2012) employed an in vitro model of toxicant-induced malignant transformation and found several aberrant DNA methylation events that arise during malignant transformation, including ones that occur in an agglomerative fashion. One such locus that was targeted for agglomerative DNA methylation was the entire Pcdh gene cluster region. The *Pcdhg* gene cluster, in particular, was found to harbor 18 differentially methylated regions, which have been observed to accumulate during the immortalization of human mammary epithelial cells (Severson et al. 2012). Clustered Pcdhs have also been found to be differentially methylated in prostate cancers, and hypermethylated in breast cancers (Novak et al. 2009; Kobayashi et al. 2011). As more is learned about the molecular mechanisms through which the clustered Pcdhs affect tumor progression, the exciting possibility exists that these cell adhesion molecules might provide targets for future therapeutic approaches.

References

Banelli B, Brigati C, Di Vinci A, Casciano I, Forlani A, Borzì L, Allemanni G, Romani M (2011) A pyrosequencing assay for the quantitative methylation analysis of the PCDHB gene cluster, the major factor in neuroblastoma methylator phenotype. Lab Investig 92(3):458–465. doi:10.1038/labinvest.2011.169

Blank M, Triana-Baltzer GB, Richards CS, Berg DK (2004) Alpha-protocadherins are presynaptic and axonal in nicotinic pathways. Mol Cell Neurosci 26(4):530–543

Boggon TJ, Murray J, Chappuis-Flament S, Wong E, Gumbiner BM, Shapiro L (2002) C-cadherin ectodomain structure and implications for cell adhesion mechanisms. Science (New York, NY) 296 (5571):1308. doi:10.1126/science.1071559

Bonn S, Seeburg P, Schwarz M (2007) Combinatorial Expression of {alpha}- and {gamma}-Protocadherins Alters Their Presenilin-Dependent Processing. Mol Cell Biol 27(11):4121. doi:10.1128/MCB.01708-06

Buchanan SM, Schalm SS, Maniatis T (2010) Proteolytic processing of protocadherin proteins requires endocytosis. Proc Natl Acad Sci USA 107(41):17774. doi:10.1073/pnas.1013105107

Chen J, Lu Y, Meng S, Han MH, Lin C, Wang X (2009) alpha- and gamma-Protocadherins negatively regulate PYK2. J Biol Chem 284(5):2880–2890. doi:10.1074/jbc.M807417200

Chen WV, Alvarez FJ, Lefebvre JL, Friedman B, Nwakeze C, Geiman E, Smith C, Thu CA, Tapia JC, Tasic B, Sanes JR, Maniatis T (2012) Functional significance of isoform diversification in the protocadherin gamma gene cluster. Neuron 75(3):402–409. doi:10.1016/j.neuron.2012.06.039

Chong JA, Tapia-Ramirez J, Kim S, Toledo-Aral JJ, Zheng Y, Boutros MC, Altshuller YM, Frohman MA, Kraner SD, Mandel G (1995) REST: a mammalian silencer protein that restricts sodium channel gene expression to neurons. Cell 80(6):949–957

Dallosso AR, Hancock AL, Szemes M, Moorwood K, Chilukamarri L, Tsai HH, Sarkar A, Barasch J, Vuononvirta R, Jones C, Pritchard-Jones K, Royer-Pokora B, Lee SB, Owen C, Malik S, Feng Y, Frank M, Ward A, Brown KW, Malik K (2009) Frequent long-range

epigenetic silencing of protocadherin gene clusters on chromosome 5q31 in Wilms' tumor. PLoS Genet 5(11):e1000745. doi:10.1371/journal.pgen.1000745
Dallosso AR, Oster B, Greenhough A, Thorsen K, Curry TJ, Owen C, Hancock AL, Szemes M, Paraskeva C, Frank M, Andersen CL, Malik K (2012) Long-range epigenetic silencing of chromosome 5q31 protocadherins is involved in early and late stages of colorectal tumorigenesis through modulation of oncogenic pathways. Oncogene 31(40):4409–4419. doi:10.1038/onc.2011.609
Emond MR, Jontes JD (2008) Inhibition of protocadherin-alpha function results in neuronal death in the developing zebrafish. Dev Biol 321(1):175
Esumi S, Kakazu N, Taguchi Y, Hirayama T, Sasaki A, Hirabayashi T, Koide T, Kitsukawa T, Hamada S, Yagi T (2005) Monoallelic yet combinatorial expression of variable exons of the protocadherin-α gene cluster in single neurons. Nat Genet 37(2):171–176. doi:10.1038/ng1500
Fernandez-Monreal M, Kang S, Phillips GR (2009) Gamma-protocadherin homophilic interaction and intracellular trafficking is controlled by the cytoplasmic domain in neurons. Mol Cell Neurosci 40(3):344–353. doi:10.1016/j.mcn.2008.12.002
Fernandez-Monreal M, Oung T, Hanson HH, O'Leary R, Janssen WG, Dolios G, Wang R, Phillips GR (2010) Gamma-protocadherins are enriched and transported in specialized vesicles associated with the secretory pathway in neurons. Eur J Neurosci 32(6):921. doi:10.1111/j.1460-9568.2010.07386.x
Frank M, Ebert M, Shan W, Phillips GR, Arndt K, Colman DR, Kemler R (2005) Differential expression of individual gamma-protocadherins during mouse brain development. Mol Cell Neurosci 29(4):603–616
Fukuda E, Hamada S, Hasegawa S, Katori S, Sanbo M, Miyakawa T, Yamamoto T, Yamamoto H, Hirabayashi T, Yagi T (2008) Down-regulation of protocadherin-alpha A isoforms in mice changes contextual fear conditioning and spatial working memory. Eur J Neurosci 28(7):1362
Garrett AM, Weiner JA (2009) Control of CNS synapse development by {gamma}-protocadherin-mediated astrocyte-neuron contact. J Neurosci 29(38):11723–11731. doi:10.1523/JNEUROSCI.2818-09.2009
Garrett AM, Schreiner D, Lobas MA, Weiner JA (2012) gamma-protocadherins control cortical dendrite arborization by regulating the activity of a FAK/PKC/MARCKS signaling pathway. Neuron 74(2):269–276. doi:10.1016/j.neuron.2012.01.028
George SJ, Dwivedi A (2004) MMPs, cadherins, and cell proliferation. Trends Cardiovasc Med 14 (3):100–105. doi:http://dx.doi.org/10.1016/j.tcm.2003.12.008
Gibson DA, Tymanskyj S, Yuan RC, Leung HC, Lefebvre JL, Sanes JR, Chedotal A, Ma L (2014) Dendrite self-avoidance requires cell-autonomous slit/robo signaling in cerebellar purkinje cells. Neuron 81(5):1040–1056. doi:10.1016/j.neuron.2014.01.009
Golan-Mashiach M, Grunspan M, Emmanuel R, Gibbs-Bar L, Dikstein R, Shapiro E (2011) Identification of CTCF as a master regulator of the clustered protocadherin genes. Nucleic Acids Res. doi:10.1093/nar/gkr1260
Gumbiner BM (2005) Regulation of cadherin-mediated adhesion in morphogenesis. Nat Rev Mol Cell Biol 6(8):622. doi:10.1038/nrm1699
Guo Y, Monahan K, Wu H, Gertz J, Varley KE, Li W, Myers RM, Maniatis T, Wu Q (2012) CTCF/cohesin-mediated DNA looping is required for protocadherin alpha promoter choice. Proc Natl Acad Sci U S A 109(51):21081–21086. doi:10.1073/pnas.1219280110
Haas IG, Frank M, Veron N, Kemler R (2005) Presenilin-dependent processing and nuclear function of gamma-protocadherins. J Biol Chem 280(10):9313–9319
Hambsch B, Grinevich V, Seeburg PH, Schwarz MK (2005) gamma -protocadherins: Presenilin-mediated release of C-terminal fragment promotes locus expression. J Biol Chem 280:15888–15897
Han MH, Lin C, Meng S, Wang X (2010) Proteomics analysis reveals overlapping functions of clustered protocadherins. Mol Cell Proteomics 9(1):71. doi:10.1074/mcp.M900343-MCP200
Hanson HH, Kang S, Fernandez-Monreal M, Oung T, Yildirim M, Lee R, Suyama K, Hazan RB, Phillips GR (2010) LC3-dependent intracellular membrane tubules induced by gamma-

protocadherins A3 and B2: a role for intraluminal interactions. J Biol Chem 285 (27):20982–20992. doi:10.1074/jbc.M109.092031

Hasegawa S, Hamada S, Kumode Y, Esumi S, Katori S, Fukuda E, Uchiyama Y, Hirabayashi T, Mombaerts P, Yagi T (2008) The protocadherin-alpha family is involved in axonal coalescence of olfactory sensory neurons into glomeruli of the olfactory bulb in mouse. Mol Cell Neurosci 38:66–79

Hasegawa S, Hirabayashi T, Kondo T, Inoue K, Esumi S, Okayama A, Hamada S, Yagi T (2012) Constitutively expressed Protocadherin-alpha regulates the coalescence and elimination of homotypic olfactory axons through its cytoplasmic region. Front Mol Neurosci 5:97. doi:10.3389/fnmol.2012.00097

Hayashi S, Takeichi M (2015) Emerging roles of protocadherins: from self-avoidance to enhancement of motility. J Cell Sci 128:1455–1464. doi:10.1242/jcs.166306

Hill E, Broadbent ID, Chothia C, Pettitt J (2001) Cadherin superfamily proteins in Caenorhabditis elegans and Drosophila melanogaster. J Mol Biol 305(5):1011. doi:10.1006/jmbi.2000.4361

Hirano K, Kaneko R, Izawa T, Kawaguchi M, Kitsukawa T, Yagi T (2012) Single-neuron diversity generated by Protocadherin-β cluster in mouse central and peripheral nervous systems. Front Mol Neurosci 5:90. doi:10.3389/fnmol.2012.00090

Hirayama T, Tarusawa E, Yoshimura Y, Galjart N, Yagi T (2012) CTCF is required for neural development and stochastic expression of clustered Pcdh genes in neurons. Cell Rep 2(2):345. doi:10.1016/j.celrep.2012.06.014

Hulpiau P, van Roy F (2009) Molecular evolution of the cadherin superfamily. Int J Biochem Cell Biol 41(2):349–369. doi:10.1016/j.biocel.2008.09.027

Hulpiau P, van Roy F (2011) New insights into the evolution of metazoan cadherins. Mol Biol Evol 28(1):647–657. doi:10.1093/molbev/msq233

Junghans D, Heidenreich M, Hack I, Taylor V, Frotscher M, Kemler R (2008) Postsynaptic and differential localization to neuronal subtypes of protocadherin beta16 in the mammalian central nervous system. Eur J Neurosci 27(3):559. doi:10.1111/j.1460-9568.2008.06052.x

Kaneko R, Kato H, Kawamura Y, Esumi S, Hirayama T, Hirabayashi T, Yagi T (2006) Allelic gene regulation of Pcdh- and Pcdh- clusters involving both monoallelic and biallelic expression in single Purkinje cells. J Biol Chem 281(41):30551–30560. doi:10.1074/jbc.M605677200

Kaneko R, Abe M, Hirabayashi T, Uchimura A, Sakimura K, Yanagawa Y, Yagi T (2014) Expansion of stochastic expression repertoire by tandem duplication in mouse Protocadherin-alpha cluster. Sci Rep 4:6263. doi:10.1038/srep06263

Katori S, Hamada S, Noguchi Y, Fukuda E, Yamamoto T, Yamamoto H, Hasegawa S, Yagi T (2009) Protocadherin-alpha family is required for serotonergic projections to appropriately innervate target brain areas. J Neurosci 29(29):9137–9147. doi:10.1523/JNEUROSCI.5478-08.2009

Kawaguchi M, Toyama T, Kaneko R, Hirayama T, Kawamura Y, Yagi T (2008) Relationship between DNA methylation states and transcription of individual isoforms encoded by the protocadherin-alpha gene cluster. J Biol Chem 283:12064–12075

Keeler AB, Molumby MJ, Weiner JA (2015) Protocadherins branch out: Multiple roles in dendrite development. Cell Adhes Migr 9(3):214–226. doi:10.1080/19336918.2014.1000069

Kehayova P, Monahan K, Chen W, Maniatis T (2011) Regulatory elements required for the activation and repression of the protocadherin-{alpha} gene cluster. Proc Natl Acad Sci 108:17195–17200. doi:10.1073/pnas.1114357108

Kobayashi Y, Absher DM, Gulzar ZG, Young SR, McKenney JK, Peehl DM, Brooks JD, Myers RM, Sherlock G (2011) DNA methylation profiling reveals novel biomarkers and important roles for DNA methyltransferases in prostate cancer. Genome Res 21(7):1017–1027. doi:10.1101/gr.119487.110

Kohmura N, Senzaki K, Hamada S, Kai N, Yasuda R, Watanabe M, Ishii H, Yasuda M, Mishina M, Yagi T (1998) Diversity revealed by a novel family of cadherins expressed in neurons at a synaptic complex. Neuron 20(6):1137–1151

Ledderose J, Dieter S, Schwarz MK (2013) Maturation of postnatally generated olfactory bulb granule cells depends on functional gamma-protocadherin expression. Sci Rep 3:1514. doi:10.1038/srep01514

Lefebvre J, Zhang Y, Meister M, Wang X, Sanes J (2008) {gamma}-Protocadherins regulate neuronal survival but are dispensable for circuit formation in retina. Development 135 (24):4141. doi:10.1242/dev.027912

Lefebvre JL, Kostadinov D, Chen WV, Maniatis T, Sanes JR (2012) Protocadherins mediate dendritic self-avoidance in the mammalian nervous system. Nature 488(7412):517–521. doi:10.1038/nature11305

Li Y, Serwanski DR, Miralles CP, Fiondella CG, Loturco JJ, Rubio ME, De Blas AL (2010) Synaptic and nonsynaptic localization of protocadherin-gammaC5 in the rat brain. J Comp Neurol 518(17):3439–3463. doi:10.1002/cne.22390

Lin C, Meng S, Zhu T, Wang X (2010) PDCD10/CCM3 acts downstream of {gamma}-protocadherins to regulate neuronal survival. J Biol Chem 285(53):41675–41685. doi:10.1074/jbc.M110.179895

McGowan PO, Suderman M, Sasaki A, Huang TCT, Hallett M, Meaney MJ, Szyf M (2011) PLOS ONE: broad epigenetic signature of maternal care in the brain of adult rats. PLoS One 6(2): e14739. doi:10.1371/journal.pone.0014739

Meguro R, Hishida R, Tsukano H, Yoshitake K, Imamura R, Tohmi M, Kitsukawa T, Hirabayashi T, Yagi T, Takebayashi H, Shibuki K (2015) Impaired clustered protocadherin-alpha leads to aggregated retinogeniculate terminals and impaired visual acuity in mice. J Neurochem 33:66–72. doi:10.1111/jnc.13053

Morishita H, Yagi T (2007) Protocadherin family: diversity, structure, and function. Curr Opin Cell Biol 19(5):584

Morishita H, Kawaguchi M, Murata Y, Seiwa C, Hamada S, Asou H, Yagi T (2004a) Myelination triggers local loss of axonal CNR/protocadherin alpha family protein expression. Eur J Neurosci 20(11):2843–2847

Morishita H, Murata Y, Esumi S, Hamada S, Yagi T (2004b) CNR/Pcdhalpha family in subplate neurons, and developing cortical connectivity. Neuroreport 15(17):2595–2599

Morishita H, Umitsu M, Murata Y, Shibata N, Udaka K, Higuchi Y, Akutsu H, Yamaguchi T, Yagi T, Ikegami T (2006) Structure of the cadherin-related neuronal receptor/protocadherin-alpha first extracellular cadherin domain reveals diversity across cadherin families. J Biol Chem 281(44):33650

Murata Y, Hamada S, Morishita H, Mutoh T, Yagi T (2004) Interaction with protocadherin-gamma regulates the cell surface expression of protocadherin-alpha. J Biol Chem 279 (47):49508

Nelson WJ (2008) Regulation of cell–cell adhesion by the cadherin–catenin complex. Biochem Soc Trans 36(Pt 2):149–155. doi:10.1042/BST0360149

Niessen CM, Leckband D, Yap AS (2011) Tissue organization by cadherin adhesion molecules: dynamic molecular and cellular mechanisms of morphogenetic regulation. Physiol Rev 91 (2):691–731. doi:10.1152/physrev.00004.2010

Noguchi Y, Hirabayashi T, Katori S, Kawamura Y, Sanbo M, Hirabayashi M, Kiyonari H, Nakao K, Uchimura A, Yagi T (2009) Total expression and dual gene-regulatory mechanisms maintained in deletions and duplications of the Pcdha cluster. J Biol Chem 284 (46):32002–32014. doi:10.1074/jbc.M109.046938

Nollet F, Kools P, van Roy F (2000) Phylogenetic analysis of the cadherin superfamily allows identification of six major subfamilies besides several solitary members. J Mol Biol 299 (3):551–572

Noonan J, Grimwood J, Danke J, Schmutz J, Dickson M, Amemiya C, Myers RM (2004a) Coelacanth genome sequence reveals the evolutionary history of vertebrate genes. Genome Res 14(1):2397–2405

Noonan JP, Grimwood J, Schmutz J, Dickson M, Myers RM (2004b) Gene conversion and the evolution of protocadherin gene cluster diversity. Genome Res 14(3):354–366

Novak P, Jensen TJ, Garbe JC, Stampfer MR, Futscher BW (2009) Stepwise DNA methylation changes are linked to escape from defined proliferation barriers and mammary epithelial cell immortalization. Cancer Res 69(12):5251–5258. doi:10.1158/0008-5472.can-08-4977

O'Leary R, Reilly JE, Hanson HH, Kang S, Lou N, Phillips GR (2011) A variable cytoplasmic domain segment is necessary for gamma-protocadherin trafficking and tubulation in the endosome/lysosome pathway. Mol Biol Cell 22(22):4362–4372. doi:10.1091/mbc.E11-04-0283

Phillips GR, Huang JK, Wang Y, Tanaka H, Shapiro L, Zhang W, Shan WS, Arndt K, Frank M, Gordon RE, Gawinowicz MA, Zhao Y, Colman DR (2001) The presynaptic particle web: ultrastructure, composition, dissolution, and reconstitution. Neuron 32(1):63–77

Phillips GR, Tanaka H, Frank M, Elste A, Fidler L, Benson DL, Colman DR (2003) Gamma-protocadherins are targeted to subsets of synapses and intracellular organelles in neurons. J Neurosci 23(12):5096–5104

Prasad T, Weiner JA (2011) Direct and indirect regulation of spinal cord Ia afferent terminal formation by the gamma-protocadherins. Front Mol Neurosci 4:54. doi:10.3389/fnmol.2011.00054

Prasad T, Wang X, Gray PA, Weiner JA (2008) A differential developmental pattern of spinal interneuron apoptosis during synaptogenesis: insights from genetic analyses of the protocadherin-gamma gene cluster. Development 135(24):4153–4164. doi:10.1242/dev.026807

Puller C, Haverkamp S (2010) Cell-type-specific localization of protocadherin β16 at AMPA and AMPA/kainate receptor-containing synapses in the primate retina. J Comp Neurol 519(3):467–479. doi:10.1002/cne.22528

Ribich S, Tasic B, Maniatis T (2006) Identification of long-range regulatory elements in the protocadherin-alpha gene cluster. Proc Natl Acad Sci USA 103(52):19719. doi:10.1073/pnas.0609445104

Rubinstein R, Thu CA, Goodman KM, Wolcott HN, Bahna F, Mannepalli S, Ahlsen G, Chevee M, Halim A, Clausen H, Maniatis T, Shapiro L, Honig B (2015) Molecular logic of neuronal self-recognition through protocadherin domain interactions. Cell 163(3):629–642. doi:10.1016/j.cell.2015.09.026

Sano K, Tanihara H, Heimark RL, Obata S, Davidson M, St John T, Taketani S, Suzuki S (1993) Protocadherins: a large family of cadherin-related molecules in central nervous system. Embo J 12(6):2249–2256

Schoenherr CJ, Anderson DJ (1995) The neuron-restrictive silencer factor (NRSF): a coordinate repressor of multiple neuron-specific genes. Science 267(5202):1360–1363

Schreiner D, Weiner JA (2010) Combinatorial homophilic interaction between gamma-protocadherin multimers greatly expands the molecular diversity of cell adhesion. Proc Natl Acad Sci U S A 107(33):14893–14898. doi:10.1073/pnas.1004526107

Severson PL, Tokar EJ, Vrba L, Waalkes MP, Futscher BW (2012) Agglomerates of aberrant DNA methylation are associated with toxicant-induced malignant transformation. Epigenetics 7(11):1238–1248. doi:10.4161/epi.22163

Sotomayor M, Gaudet R, Corey DP (2014) Sorting out a promiscuous superfamily: towards cadherin connectomics. Trends Cell Biol 24 (9):524–536. doi:http://dx.doi.org/10.1016/j.tcb.2014.03.007

Su H, Marcheva B, Meng S, Liang FA, Kohsaka A, Kobayashi Y, Xu AW, Bass J, Wang X (2010) Gamma-protocadherins regulate the functional integrity of hypothalamic feeding circuitry in mice. Dev Biol 339(1):38. doi:10.1016/j.ydbio.2009.12.010

Suderman M, McGowan PO, Sasaki A, Huang TCT, Hallett MT, Meaney MJ, Turecki G, Szyf M (2012) Conserved epigenetic sensitivity to early life experience in the rat and human hippocampus. Proc Natl Acad Sci U S A 109:17266–17272

Sugino H, Hamada S, Yasuda R, Tuji A, Matsuda Y, Fujita M, Yagi T (2000) Genomic organization of the family of CNR cadherin genes in mice and humans. Genomics 63(1):75–87. doi:10.1006/geno.1999.6066

Sugino H, Yanase H, Hamada S, Kurokawa K, Asakawa S, Shimizu N, Yagi T (2004) Distinct genomic sequence of the CNR/Pcdhalpha genes in chicken. Biochem Biophys Res Commun 316(2):437–445. doi:10.1016/j.bbrc.2004.02.067

Suo L, Lu H, Ying G, Capecchi MR, Wu Q (2012) Protocadherin clusters and cell adhesion kinase regulate dendrite complexity through Rho GTPase. J Mol Cell Biol 4(6):362–376. doi:10.1093/jmcb/mjs034

Tada MN, Senzaki K, Tai Y, Morishita H, Tanaka YZ, Murata Y, Ishii Y, Asakawa S, Shimizu N, Sugino H, Yagi T (2004) Genomic organization and transcripts of the zebrafish Protocadherin genes. Gene 340(2):197–211. doi:10.1016/j.gene.2004.07.014

Takeichi M (2007) The cadherin superfamily in neuronal connections and interactions. Nat Rev Neurosci 8(1):11. doi:10.1038/nrn2043

Tan F, Ghosh S, Mbeunkui F, Thomas R, Weiner JA, Ofori-Acquah SF (2010) Essential role for ALCAM gene silencing in megakaryocytic differentiation of K562 cells. BMC Mol Biol 11:91. doi:10.1186/1471-2199-11-91

Tasic B, Nabholz CE, Baldwin KK, Kim Y, Rueckert EH, Ribich SA, Cramer P, Wu Q, Axel R, Maniatis T (2002) Promoter choice determines splice site selection in protocadherin alpha and gamma pre-mRNA splicing. Mol Cell 10(1):21–33

Thu CA, Chen WV, Rubinstein R, Chevee M, Wolcott HN, Felsovalyi KO, Tapia JC, Shapiro L, Honig B, Maniatis T (2014) Single-cell identity generated by combinatorial homophilic interactions between alpha, beta, and gamma protocadherins. Cell 158(5):1045–1059. doi:10.1016/j.cell.2014.07.012

Toyoda S, Kawaguchi M, Kobayashi T, Tarusawa E, Toyama T, Okano M, Oda M, Nakauchi H, Yoshimura Y, Sanbo M, Hirabayashi M, Hirayama T, Hirabayashi T, Yagi T (2014) Developmental epigenetic modification regulates stochastic expression of clustered protocadherin genes, generating single neuron diversity. Neuron 82(1):94–108. doi:10.1016/j.neuron.2014.02.005

Triana-Baltzer GB, Blank M (2006) Cytoplasmic domain of protocadherin-alpha enhances homophilic interactions and recognizes cytoskeletal elements. J Neurobiol 66(4):393. doi:10.1002/neu.20228

Vanhalst K, Kools P, Vanden Eynde E, van Roy F (2001) The human and murine protocadherin-beta one-exon gene families show high evolutionary conservation, despite the difference in gene number. FEBS Lett 495(1-2):120–125

Waha A, Guntner S, Huang TH, Yan PS, Arslan B, Pietsch T, Wiestler OD, Waha A (2005) Epigenetic silencing of the protocadherin family member PCDH-gamma-A11 in astrocytomas. Neoplasia 7(3):193–199. doi:10.1593/neo.04490

Wang Y, Liu H, Zhang Y, Ma D (1999) cDNA cloning and expression of an apoptosis-related gene, humanTFAR15 gene. Sci China Ser C-Life Sci 42(3):323–329. doi:10.1007/BF03183610

Wang X, Su H, Bradley A (2002a) Molecular mechanisms governing Pcdh-gamma gene expression: evidence for a multiple promoter and cis-alternative splicing model. Genes Dev 16 (15):1890–1905

Wang X, Weiner JA, Levi S, Craig AM, Bradley A, Sanes JR (2002b) Gamma protocadherins are required for survival of spinal interneurons. Neuron 36(5):843–854

Weiner JA, Wang X, Tapia JC, Sanes JR (2005) Gamma protocadherins are required for synaptic development in the spinal cord. Proc Natl Acad Sci U S A 102(1):8–14. doi:10.1073/pnas.0407931101

Wu Q (2001) Comparative DNA sequence analysis of mouse and human protocadherin gene clusters. Genome Res 11(3):389–404. doi:10.1101/gr.167301

Wu Q (2005) Comparative Genomics and Diversifying Selection of the Clustered Vertebrate Protocadherin Genes. Genetics 169:2179–2188

Wu Q, Maniatis T (1999) A striking organization of a large family of human neural cadherin-like cell adhesion genes. Cell 97:779–790

Wu Q, Zhang T, Cheng JF, Kim Y, Grimwood J, Schmutz J, Dickson M, Noonan JP, Zhang MQ, Myers RM, Maniatis T (2001) Comparative DNA sequence analysis of mouse and human protocadherin gene clusters. Genome Res 11(3):389–404

Yagi T (2012) Molecular codes for neuronal individuality and cell assembly in the brain. Front Mol Neurosci 5:45. doi:10.3389/fnmol.2012.00045

Yamashita H, Chen S, Komagata S, Hishida R, Iwasato T, Itohara S, Yagi T, Endo N, Shibata M, Shibuki K (2012) Restoration of Contralateral Representation in the Mouse Somatosensory Cortex after Crossing Nerve Transfer. PLoS One 7(4):e35676. doi:10.1371/journal.pone.0035676

Yanase H, Sugino H, Yagi T (2004) Genomic sequence and organization of the family of CNR/Pcdhalpha genes in rat. Genomics 83(4):717–726. doi:10.1016/j.ygeno.2003.09.022

Yokota S, Hirayama T, Hirano K, Kaneko R, Toyoda S, Kawamura Y, Hirabayashi M, Hirabayashi T, Yagi T (2011) Identification of the cluster control region for the protocadherin-genes located beyond the protocadherin- cluster. J Biol Chem 286(36):31885–31895. doi:10.1074/jbc.M111.245605

Yoshitake K, Tsukano H, Tohmi M, Komagata S, Hishida R, Yagi T, Shibuki K (2013) Visual map shifts based on whisker-guided cues in the young mouse visual cortex. Cell Rep 5(5):1365–1374. doi:10.1016/j.celrep.2013.11.006

Yu WP, Yew K, Rajasegaran V, Venkatesh B (2007) Sequencing and comparative analysis of fugu protocadherin clusters reveal diversity of protocadherin genes among teleosts. BMC Evol Biol 7:49. doi:10.1186/1471-2148-7-49

Chapter 9
The Nonclustered Protocadherins

James D. Jontes

Abstract The protocadherins comprise the largest family of proteins within the cadherin superfamily, and are themselves further subdivided into the clustered and nonclustered protocadherins. The nonclustered protocadherins are evolutionarily conserved and vary in their regional patterns of expression within the nervous system, leading to the hypothesis that they participate in neural circuit assembly. Although the nonclustered protocadherins are strongly linked to both neurodevelopmental disorders and multiple forms of cancer, their physiological roles are poorly understood. Recent work is providing new insights into the nonclustered protocadherins, resulting in emerging themes. Here, data revealing roles for these molecules at multiple stages of development and that suggest an involvement in regulating proliferation and cell differentiation is discussed.

Keywords Nonclustered protocadherins • Cancer • Neural development • Adhesion • Cell motility • Neurodevelopmental disorders

9.1 Overview of Nonclustered Protocadherins

Protocadherins were discovered by Suzuki and colleagues in a degenerate PCR screen for additional cadherin family members (Sano et al. 1993). Subsequently, the cadherin superfamily has expanded immensely (Hulpiau and van Roy 2009; Nollet et al. 2000). Within this extended constellation of cadherin-related genes, the protocadherins (pcdhs) are the largest group, consisting of both clustered (C) and nonclustered (NC) families. The NC-Pcdhs comprise the δ-Pcdhs (δ0: *pcdh20*; δ1: *pcdh1*, *pcdh7*, *pcdh9*, and *pcdh11*; δ2: *pcdh8*, *papc*, *pcdh10*, *pcdh17*, *pcdh18*, *pcdh19*), and *pcdh12*. Most ideas regarding the function of the NC-Pcdhs have derived from analogies to classical cadherins. As members of the cadherin superfamily with a domain organization similar to classical cadherins, it was widely presumed that NC-Pcdhs would function as homophilic adhesion molecules.

J.D. Jontes (✉)
Department of Neuroscience, Ohio State University, 1060 Carmack Road 119 Rightmire Hall, Columbus, OH 43210, USA
e-mail: jontes.1@osu.edu

S.T. Suzuki, S. Hirano (eds.), *The Cadherin Superfamily*,
DOI 10.1007/978-4-431-56033-3_9

Similarly, the strong and differential expression patterns of the NC-Pcdhs in the nervous system, as well as their presumed homophilic adhesion, suggested that they could participate in establishing connectivity between brain regions expressing common protocadherins, similar to what has been proposed for classical cadherins. Although some evidence indicates that NC-Pcdhs can interact homophilically, their roles in adhesion remain unclear, as they may be contingent, vary among family members, and in some cases be antiadhesive. Moreover, data supporting a role for NC-Pcdhs in either synaptogenesis or connectivity patterns is sparse. The most compelling argument for the involvement of these molecules in neural circuit formation comes from human genetics, which has strongly linked a number of NC-Pcdhs to neurodevelopmental disorders.

Thus, the NC-Pcdhs remain an intriguing family of molecules, although initial expectations regarding this family have proven to be too simplistic. Clear principles defining the function of these molecules have not emerged. This is due to the relatively subtle phenotypes observed in rodent and zebrafish knockouts, the expanding array of extracellular cofactors and coreceptors, and the relatively small number of identified intracellular binding partners. Much remains to be done in order to understand this conserved family of neuronal receptors.

9.2 Structure and Biochemical Properties

The NC-Pcdhs have a similar domain organization to classical cadherins with an amino-terminal signal peptide, multiple extracellular cadherin (EC) repeats, a single-pass transmembrane segment, and an intracellular domain (ICD; Fig. 9.1). Although the δ1-Pcdhs have 7EC repeats, the δ0-, δ2-Pcdhs, and Pcdh12 each contain 6 EC repeats (Nollet et al. 2000). As is the case for the C-Pcdhs, the entire ectodomain, the transmembrane domain, and a small portion of the ICD are encoded by a large first exon, with subsequent exons encoding the remainder of the ICD. This arrangement differs slightly for *pcdh1* and *pcdh11,* as the ectodomain-encoding exon is split into two: exon 1 codes for an amino-terminal portion of the ectodomain and exon 2 encodes the rest of the ectodomain (Vanhalst et al. 2005). Although the functional significance is unknown, the NC-Pcdhs exhibit some alternative splicing of the downstream exons, resulting in isoforms with variation in the ICD. In contrast to the classical cadherins, there is a dearth of information regarding the structure of protocadherins. To date, available structural data are limited to NMR solution structures for EC3 of Pcdh7 and Pcdh9, making the NC-Pcdhs a rich target for structural analysis.

The NC-Pcdhs exhibit a number of important sequence features that are likely to be of functional significance. As do classical cadherins, the ~110 amino acid EC repeats of NC-Pcdhs contain conserved sequence motifs constituting the calcium-binding sites that form the interfaces between adjacent EC repeats. Chen and Gumbiner (2008) identified cysteines present in EC1, EC3, and EC4 of PAPC that are conserved among the protocadherins. Two conserved cysteines within

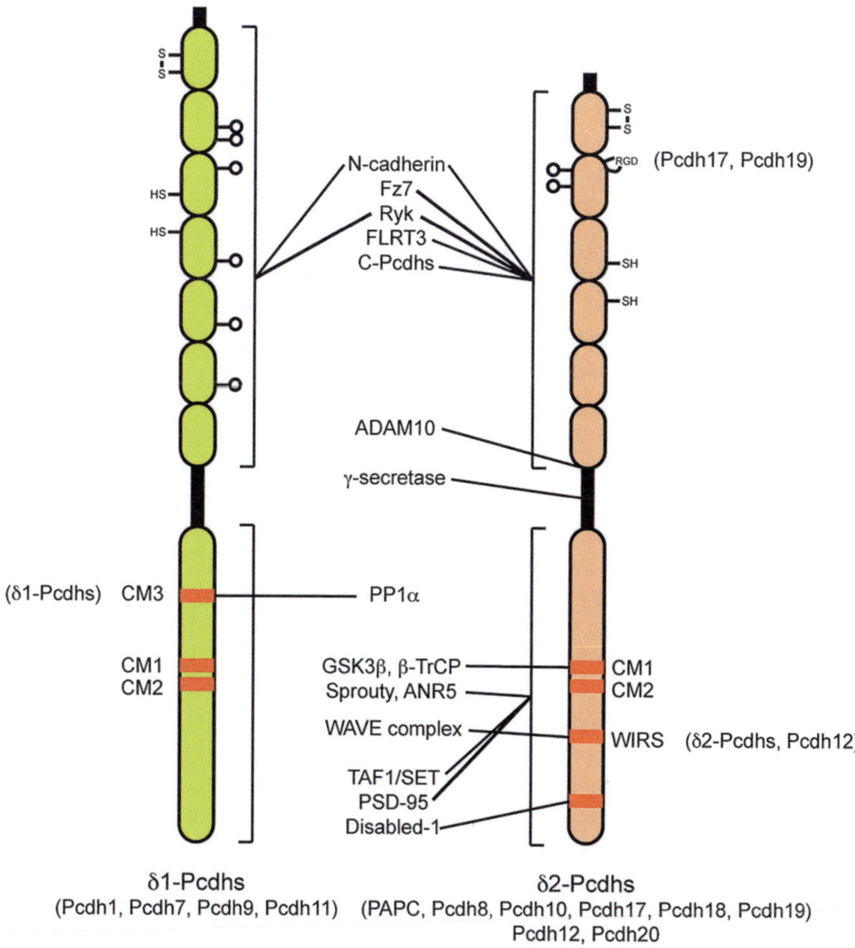

Fig. 9.1 *Structure and organization of the nonclustered protocadherins.* The δ1-Pcdhs, left, are distinct from the remaining NC-Pcdhs in that they have 7 EC repeats, rather than the 6 present in the δ2-Pcdhs, Pcdh12 and Pcdh20. The extracellular domain included conserved cysteines in EC1, EC3, and EC4. The cysteines in EC1 form an intramolecular disulfide, which appears to be important for trafficking. The intermolecular disulfides formed by the cysteines in EC3 and EC4 are believed to stabilize *cis*-oligomers. Both the δ1-Pcdhs and δ2-Pcdhs are subject to O-mannose glycosylation, although there are differences in the patterns between the different subfamilies. Both Pcdh17 and Pcdh19 have a conserved RGD motif present in EC2, suggesting that they may interact with integrins. Pcdh12 undergoes sequential processing by ADAM10 and γ-secretase; similar processing has not yet been demonstrated for other NC-Pcdhs. The ectodomains of NC-Pcdhs mediate a variety of protein–protein interactions. Thus far, δ2-Pcdhs have been shown to form *cis*-homo-oligomers, and to interact with N-cadherin, Frizzled-7, Ryk, FLRT3, clustered protocadherins, and other nonclustered protocadherins (unpublished observations). This list is likely to expand as more systematic proteomic approaches are used to discover interaction partners. Although the intracellular domains of the NC-Pcdhs share some conserved sequence motifs (CM1 and CM2), it does not, thus far, appear that they share downstream cytoplasmic interactions. The CM3 motif is specific to the δ1-Pcdhs and has been shown to bind PP1α. The WIRS site, which is absent from the δ1-Pcdhs (apart from Pcdh9), interacts with the WAVE complex. A number of proteins have been shown to interact with PAPC, including Sprouty and

EC1 form an intramolecular disulfide, which appears to be essential for protein stability and efficient trafficking. The cysteines in EC3 and EC4 form intermolecular disulfides that stabilize *cis*-homo-oligomers, but are dispensable, as one or the other may be missing in some protocadherins. A recent study has also identified O-mannose glycosylation sites in several of the NC-Pcdhs (Vester-Christensen et al. 2013). These were present at several sites throughout the ectodomains of δ1-Pcdhs, EC2, EC3, EC4, EC5, and EC6. A more limited number of sites were found in the δ2-Pcdhs, limited primarily to EC2. As NC-Pcdhs exhibit a multitude of *cis*-interactions as well as their presumed *trans*-adhesive interactions, glycosylation could play an important role in modulating their function (Langer et al. 2012). In addition, both Pcdh17 and Pcdh19 have a conserved RGD motif present on a predicted surface loop in EC2, which suggests that these proteins could associate with integrins. Although the classical cadherins, C-Pcdhs and calsyntenins have been shown to undergo proteolytic processing and ectodomain shedding, much less is known about the processing of NC-Pcdhs. However, Pcdh12 was recently shown to undergo sequential cleavage by ADAM10 and γ-secretase (Bouillot et al. 2011). This is likely to be physiologically relevant, as shed Pcdh12 ectodomain was found in human serum and urine. It will be important to determine whether other NC-Pcdhs are processed similarly, the distinct roles for intact Pcdhs and the shed ectodomains, and what signals regulate cleavage. Much remains to be learned about the biochemistry and structural biology of the NC-Pcdhs, which is still in its infancy.

As members of the cadherin superfamily, it is generally assumed that NC-Pcdhs participate in homophilic cell adhesion. Despite this presumption, the evidence for a role in adhesion is ambiguous. Data from cell sorting and cell aggregation assays have been used as evidence that NC-Pcdhs mediate weak homophilic adhesion. Pcdh1, Pcdh8, and Pcdh18a have been shown to affect cell adhesion using cell sorting assays (Aamar and Dawid 2008; Kuroda et al. 2002; Yamagata et al. 1999). Similarly, Pcdh7, Pcdh10, Pcdh19, and Pcdh17 have been shown to mediate adhesion in cell aggregation assays (Hirano et al. 1999; Hoshina et al. 2013; Yoshida 2003). However, adhesion generally appears to be much weaker than that exhibited by the classical cadherins. When investigating in more detail, Chen and Gumbiner (2006) found no evidence for homophilic adhesion by PAPC. Instead, the authors demonstrated that the cell sorting behavior was mediated indirectly through antagonism of a classical cadherin (Chen and Gumbiner 2006). Cells that expressed Pcdh8/PAPC downregulated C-cadherin, and the differential adhesive activity of these cells caused them to segregate from the cells with normal

---

Fig. 9.1 (continued) ANR5. A conserved region within CM1 is phosphorylated by GSK3β, which promotes polyubiquitination by the E3 ubiquitin ligase, β-TrCP. As this site is present in other NC-Pcdhs, it is possible that other protocadherins could be subject to similar regulation. In addition, binding partners have been found that appear to be specific for individual family members. Pcdh7 binds to TAF1/Set, Pcdh18 binds Disabled-1, and Pcdh10 appears to associate with ubiquitinated PSD-95

C-cadherin activity. A similar antagonism of classical cadherins has been found for Pcdh8, the mammalian paralogue of PAPC (Yasuda et al. 2007) and Pcdh10 (Nakao et al. 2008). Thus, the protocadherins may not act as cell adhesion molecules in a manner analogous to the classical cadherins. In addition, it was shown that zebrafish Pcdh19 is not adhesive in bead aggregation assays. However, zebrafish Pcdh19 can associate with N-cadherin to form a *cis*-complex (Biswas et al. 2010; Emond et al. 2011) and this *cis*-complex exhibits homophilic adhesion that is mediated by Pcdh19, not N-cadherin (Emond et al. 2011). Similar behavior has been observed for other NC-Pcdhs, which can interact with N-cadherin or other classical cadherins to mediate adhesion in bead aggregation assays (C.Blevins, M. Emond and J.Jontes, unpublished observations). Thus, the roles of NC-Pcdhs in cell–cell adhesion are likely to be complex. Bead aggregation assays assess the intrinsic adhesive properties of ectodomains (Emond and Jontes 2014), as has been done for classical cadherins (Chappuis-Flament et al. 2001; Zhang et al. 2009) and other proteins, such as Dscam (Wojtowicz et al. 2004). However, these assays are nonphysiological, as the proteins have been stripped of their cellular and tissue environments, which may include important coreceptors or intracellular cofactors. Additionally, these assays do not rule out the existence of heterophilic adhesive interactions. Although closer to an in vivo environment, cell-based studies cannot rule out indirect effects attributable to modulation of other adhesive systems by any of a number of mechanisms. Moreover, such assays are performed in heterologous cell types, whose specific complement of other cell surface proteins may influence the outcome. Thus, data from bead-based or cell-based experiments need to be interpreted carefully in the context of data from in vivo functional studies. In those cases where homophilic interactions can be verified, these may turn out to be repulsive or antiadhesive, as described below. Moreover, given the proclivity of the NC-Pcdhs to associate in *cis*-complexes with a variety of partners, their participation in cell–cell interactions may be context dependent.

9.2.1 Interactions of the Extracellular Domain

The prevailing working model for classical cadherins is that individual cadherin molecules form homophilic dimers in trans with a cadherin molecule on the adjacent cell. Then, through both association with the action cytoskeleton through their intracellular domains and homo-oligomerization through an extracellular cis interface, cadherins cluster to strengthen cell–cell junctions. In this scenario, cadherins function as autonomous adhesive units driven by the intrinsic adhesive character of the ectodomain, which are modulated through intracellular interactions or by regulation of surface levels. In contrast, the function of NC-Pcdhs may largely be driven by their association with a shifting array of coreceptors and the formation of multiprotein complexes. As discussed above, the ectodomains of NC-Pcdhs do not exhibit strong adhesive capacity. Most of the evidence for NC-Pcdh adhesion has come from cell-based assays (cell sorting or cell aggregation), which cannot exclude the participation of other cofactors with NC-Pcdhs.

NC-Pcdhs associate with a variety of other cell surface proteins. As discussed in more detail below, PAPC interacts with the Wnt receptor Frizzled-7 (Fz7) to regulate convergent extension cell movements and tissue separation (Kraft et al. 2012). Moreover, *Xenopus* PAPC was shown to associate with the leucine-rich repeat protein, FLRT3, to influence adhesion and cell sorting (Chen et al. 2009b). Evidence has also begun to accumulate which indicates that the NC-Pcdhs associate with classical cadherins. Yasuda et al. (2007) showed that Pcdh8 forms a *cis*-complex with N-cadherin, which inhibits N-cadherin adhesion by promoting its endocytosis. Zebrafish Pcdh19 interacts in cis with N-cadherin, as revealed both by coimmunoprecipitation and bimolecular fluorescence complementation (Biswas et al. 2010). NC-Pcdhs have been shown to associate with C-Pcdhs (Chen et al. 2009a), and NC-Pcdhs can form *cis*-hetero-oligomers (unpublished observations). A recent proteomic study using tandem affinity purification of the Wnt receptor, Ryk, revealed interactions with both N-cadherin and several NC-Pcdhs (Berndt et al. 2011). It is also worth noting that, although an interaction with integrins has not yet been demonstrated, both Pcdh17 and Pcdh19 have conserved RGD sequences present in EC2 of their ectodomains. Thus, NC-Pcdhs exhibit a growing range of protein–protein interactions mediated through their ectodomains, and these interactions likely influence both the function of the NC-Pcdhs and their various coreceptors.

9.2.2 Intracellular Domain

To date, a surprisingly small number of binding partners have been identified for the intracellular domains. This has limited our insight into pathways downstream of the NC-Pcdh function and into the potential biological roles of these molecules. Like the classical cadherins and C-Pcdhs, the intracellular domains of the NC-Pcdhs appear to be intrinsically disordered proteins (IDPs): they have a high charge-to-hydrophobicity ratio, little predicted secondary structure, and no predicted folding motifs (Babu et al. 2011; Uversky 2011). The intracellular domains of classical cadherins are disordered, but adopt an extended ordered structure upon associating with their armadillo family binding partners, $p120^{ctn}$ and β-catenin (Huber et al. 2001; Huber and Weis 2001; Ishiyama et al. 2010). It is possible that the intracellular domains of NC-Pcdhs become similarly ordered upon complex formation, although few binding partners have been identified. Initially, two conserved sequence motifs were identified, CM1 and CM2, that helped define the δ-Pcdh subfamily (Wolverton and Lalande 2001). Recently, Kai et al. (2015) identified a conserved region in PAPC, the DSR domain, as it is rich in aspartate (D) and serine (S) residues. This region, which is within the conserved CM1 motif, is phosphorylated by GSK3β, which then promotes polyubiquitination by the E3 ubiquitin ligase, β-TrCP (Kai et al. 2015). It will be interesting to investigate whether other

NC-Pcdhs are similarly regulated. A CM3 motif has also been identified, which is specific to the δ1 subfamily. The CM3 motif of human Pcdh7 interacts with protein phosphatase 1-α (PP1α), a molecule that is implicated in synaptic plasticity (Yoshida et al. 1999). PP1α activity is inhibited upon binding to the Pcdh7 cytoplasmic domain. The site of interaction has been mapped to an RRVTF sequence element within the CM3 motif. The other δ1-Pcdhs, Pcdh1, Pcdh9, and Pcdh11, each contain the RRVTF sequence, and each has been shown to interact with PP1α (Vanhalst et al. 2005). Recently, an additional conserved motif has been defined (WIRS; WAVE interacting receptor sequence) that binds to the WAVE complex (Chen et al. 2014). Comprising five proteins, WAVE1 (or 2,3), Nap1, Sra1/Cyfip1 (or 2), Abi2 (or 1,3), and HSPC300, the WAVE complex promotes Arp2/3-dependent actin assembly upon activation by Rac1. The WIRS peptide interacts with a composite binding site made up of Sra1/Cyfip1 and Abi2. This site is present in all of the δ2-Pcdhs, Pcdh12, and Pcdh9, as well as α-Pcdhs and other cell surface receptors. The presence of the WIRS site in these Pcdhs indicates that regulation of actin dynamics is a common downstream pathway for these family members. In *Xenopus*, Pcdh7 was first identified as NF-Protocadherin (Bradley et al. 1998). Pcdh7/NF-Pcdh was shown with the histone binding protein TAF1/Set (Heggem and Bradley 2003). It is not known whether TAF1/Set interacts with other NC-Pcdh family members, nor what role TAF1/Set plays in Pcdh7 function. Pcdh18 was identified as a binding partner for Disabled-1 (Dab1), which is a component of the Reelin signaling pathway that mediates neuronal migration during cortical lamination (Homayouni et al. 2001). Dab1 binds to a conserved NPTS motif in Pcdh18, similar to the NPxY motif found in other Dab1 interacting proteins. This motif appears to be unique to Pcdh18 among the δ-pcdhs. A pair of binding partners has been identified for PAPC, although the PAPC intracellular domain doesn't share the conserved motifs found in the other NC-Pcdhs. Wang et al. (2008) showed that PAPC antagonizes the activity of Sprouty during convergent extension movements in *Xenopus*. The interaction of PAPC with Sprouty appears to require phosphorylation of serines 741 and 955 (Wang et al. 2008). In addition, PAPC also interacts with the ankyrin repeat protein, ANR5, a scaffolding protein that regulates cell movements during gastrulation (Chung et al. 2007). Collectively, the downstream effectors of the NC-Pcdhs that have been identified thus far indicate important involvement of the NC-Pcdhs in cell movements and motility.

9.3 Expression Patterns

Although present in other tissues, the expression patterns of NC-Pcdhs have been most intensively investigated in the nervous system (Blevins et al. 2011; Kim et al. 2007; Vanhalst et al. 2005). In general, the NC-Pcdhs are present in most major subdivisions of the CNS: retina, forebrain, midbrain, hindbrain, and spinal cord. Broad patterns of expression have been investigated in mouse, rat (Kim et al. 2007), chicken (Lin et al. 2012), ferret (Krishna et al. 2009), and zebrafish (Blevins et al. 2011). The expression of NC-Pcdhs is broadly consistent with a role in circuit connectivity, as they show regionalized expression in the basal ganglia (Hertel et al. 2008), cerebellum (Redies et al. 2011), cortex (Krishna and Redies 2009; Krishna et al. 2011), spinal cord (Lin et al. 2012), and hippocampus (Kim et al. 2010). The differential expression of NC-Pcdhs, along with classical cadherins, could constitute a molecular code to define subdivisions within the developing and mature brain. In particular, *Pcdh7* and *Pcdh10*, as well as *Cdh8*, are expressed in parasagittal stripes in the chicken cerebellum (Redies et al. 2011). These stripes represented restricted Purkinje cell domains. Tracing of axon projections revealed that *Pcdh10* labeled domains in the cerebellar cortex, and selected deep cerebellar nuclei and the tracts that connected them. These data are consistent with the idea that NC-Pcdhs partition the developing nervous system and may play a role in the development of interregional connectivity (Redies et al. 2011). In addition to their expression within neuronal populations, both classical cadherins and NC-Pcdhs are expressed in neural vasculature (Krishna and Redies 2009), suggesting that they may also play a role in angiogenesis in the developing nervous system.

9.4 Biological Roles of the Nonclustered Protocadherins

9.4.1 Paraxial Protocadherin (PAPC)

Perhaps the most extensively investigated and best understood protocadherin is paraxial protocadherin (PAPC). As mentioned, PAPC does not have an orthologue in mammals, alhough it is most similar to Pcdh8, which both zebrafish and frog lack. Moreover, the intracellular domain of PAPC differs substantially from Pcdh8 and other δ-protocadherins, lacking the conserved motifs, CM2 and WIRS. As PAPC both lacks a clear mammalian orthologue and is unlikely to act through intracellular pathways shared with other δ-protocadherins, it has been unclear what PAPC can reveal about protocadherin function generally. However, accumulating evidence suggests that the data on PAPC provide important clues to the mechanisms underlying δ-protocadherin function; the indirect influence on adhesion, the involvement in cell motility and the functional interaction with Wnt signaling are all mirrored in other NC-Pcdhs.

PAPC was initially isolated in a screen for molecules expressed in the dorsal lip of the *Xenopus* blastopore and was subsequently also identified in zebrafish (Kim et al. 1998; Yamamoto et al. 1998). In both fish and frog, *papc* has been implicated in both cell movements during convergent-extension and in the formation of tissue boundaries. In the initial characterization of PAPC in both zebrafish and Xenopus, expression of DN-PAPC caused defects in somitogenesis (Kim et al. 2000; Kim et al. 1998; Yamamoto et al. 1998). Initially, *papc* is expressed uniformly in paraxial mesoderm. As somitogenesis proceeds, *papc* expression appears banded, becoming restricted to the anterior edge of the newly forming somites and disappearing from formed somites. In *Xenopus*, injection of DN-PAPC or M-PAPC impairs somite formation, causing disorganization of somitomeric cells (Kim et al. 1998). In mouse embryos, expression of DN-Pcdh8 was shown to disrupt the epithelia that form at the borders between somites (Rhee et al. 2003).

During gastrulation, cells involute and migrate to form the germ layers and shape the embryo. The boundary that forms between the ectoderm and the migrating mesendodermal cells is called Brachet's cleft. Knockdown of *papc* in *Xenopus* using antisense morpholino oligonucleotides abolishes cleft formation (Medina et al. 2004). Loss of Frizzled-7 (Fz7) has a similar effect on cleft formation and, importantly, PAPC and Fz7 act synergistically in this process, suggesting a functional interaction. PAPC and Fz7 interact physically, as shown by coimmunoprecipitation and by bimolecular fluorescence complementation (BiFC; Kraft et al. 2012; Medina et al. 2004). As signaling through Frizzled often requires a coreceptor, it is possible that PAPC forms a complex with Fz7 to mediate Wnt signaling. In support of this, it was recently shown that PAPC forms a ternary complex with Wnt11/Fz7, which stabilizes PAPC at the plasma membrane (Kraft et al. 2012). PAPC is ordinarily present both on the plasma membrane and in internal membrane compartments. In the absence of either Wnt11 or Fz7, PAPC is lost from the membrane; in the presence of added Wnt11 and Fz7, PAPC becomes enriched at the cell surface, which appears to be due to Wnt11/Fz blocking receptor-mediated endocytosis of PAPC. It remains to be determined whether Wnt11/Fz7 affects the phosphorylation of PAPC by GSK3β and polyubiquitination by β-TrCP, causing internalization of PAPC and targeting for degradation (Kai et al. 2015).

PAPC also influences gastrulation through interaction with C-cadherin (Chen and Gumbiner 2006). As discussed above, PAPC antagonizes C-cadherin adhesion, although they have not been shown to interact physically (Chen and Gumbiner 2006). In addition to forming a Wnt11/Fz7/PAPC complex, Wnt11 also forms a separate Wnt11/Fz7/C-cadherin complex (Kraft et al. 2012). It was proposed that the weakening of adhesion during gastrulation was due to sequestration of C-cadherin into a Wnt11/Fz7/C-cadherin complex, which inhibited lateral clustering of C-cadherin at cell contacts and prevented strong adhesion. Although PAPC and C-cadherin were not found in the same complexes, both PAPC and Wnt11/Fz7 were required to block C-cadherin clustering. Thus, not only does PAPC influence cell adhesion indirectly through modulating a classical cadherin, it does so in concert with a Wnt signaling pathway.

A number of intracellular pathways have been identified downstream of PAPC. Consistent with an intimate involvement in Wnt signaling, PAPC acts upstream of both RhoA and c-jun N-terminal kinase (JNK). PAPC promoted phosphorylation and activation of JNK, and activated RhoA (Medina et al. 2004). Moreover, the effects of PAPC depletion could be mimicked by expression of dominant-negative RhoA, and partially rescued by expression of constitutively active RhoA or an active form of Rho kinase. In addition, PAPC has been shown to antagonize

Sprouty (Wang et al. 2008), which is an inhibitor of the PCP pathway. Thus, PAPC appears to be a node in a molecular network, having multiple connections to noncanonical Wnt signaling to coordinate cell adhesion and cell movements during gastrulation. As mentioned, the ankyrin repeat protein ANR5 also associates with the intracellular domain of PAPC and is required for PAPC function during the formation of Brachet's cleft (Chung et al. 2007).

9.4.2 Involvement of Other NC-Protocadherins in Morphogenesis

In addition to PAPC, other NC-Pcdhs are involved in cell movements during embryonic morphogenesis. Pcdh7/NF-Pcdh plays a role in ectodermal development in *Xenopus*, as expression of a dominant-interfering mutant disrupted the deep layer of the developing ectoderm (Bradley et al. 1998). Similarly, Pcdh1 is expressed in the bronchial epithelium and is important for its differentiation (Koning et al. 2012). Upregulation of Pcdh1 appears to play a role in bronchial hyperresponsiveness, which contributes to asthma (Koppelman et al. 2009; Mortensen et al. 2014). Moreover, Pcdh7 also participates in neural tube closure in Xenopus (Rashid et al. 2006). As with PAPC, depletion of Pcdh10 in zebrafish and expression of dominant-interfering mutants interferes with cell movements in the paraxial mesodersm, perturbing somitogenesis (Murakami et al. 2006). Similarly, Pcdh18a contributes to convergence cell movements during zebrafish gastrulation (Aamar and Dawid 2008). Over- and misexpression of Pcdh18a resulted in a duplicated neural axis, which is likely due to impaired convergent extension (Tawk et al. 2007). Further analysis suggested that cells overexpressing Pcdh18a were more adhesive, less motile, and exhibited fewer cell protrusions. By contrast, depletion of Pcdh18a, using antisense morpholinos, interfered with gastrulation and appeared to reduce cell adhesion. Depletion of zebrafish Pcdh19 using morpholinos also affects convergence movements (Emond et al. 2009). However, in contrast to the other NC-Pcdhs discussed, these movements are at later stages of anterior neurulation, when the lateral neural plate converges toward the dorsal midline. Knockdown of *pcdh19* causes a delay in convergence, resulting in defects in forebrain and midbrain morphogenesis. This phenotype is remarkably similar to that of N-cadherin mutants and morphants (Lele et al. 2002). Strikingly, Pcdh19 interacts with N-cadherin both physically and functionally, and these proteins appear to collaborate during morphogenesis (Biswas et al. 2010). NC-Pcdhs have also been implicated in eye development, as they are strongly expressed in the developing retina (Blevins et al. 2011). In addition, antisense morpholino studies have shown the involvement of *pcdh17* (Chen et al. 2013) and *pcdh9* (Izuta et al. 2015) in retinal morphogenesis.

9.4.3 NC-Protocadherins in Actin Dynamics and Axon Guidance

Pcdh10/OL-Pcdh was first identified as a molecule expressed in the olfactory and limbic systems of the mouse brain (Hirano et al. 1999), and has more recently been identified as a potential susceptibility gene in autism spectrum disorders (Morrow et al. 2008). Using GST-pulldowns, the cytoplasmic domain of Pcdh10/OL-Pcdh was shown to interact with Nck-associated protein 1 (Nap1) and Cyfip2 (Nakao et al. 2008), which are core components of the WAVE complex. When expressed in heterologous cells, Pcdh10/OL-Pcdh recruits Nap-1 and WAVE to the plasma membrane and mediates contact-dependent increases in cell motility. Subsequently, both Pcdh19 (Tai et al. 2010) and Pcdh18 (Biswas et al. 2014) were also shown to interact with Nap1, suggesting that the WAVE complex and actin assembly is a downstream pathway common to all the δ2-Pcdhs. This was confirmed by the discovery of a conserved binding site, the WIRS motif F(S/A/C)TFGK, in the intracellular domains of all the δ2-Pcdhs, Pcdh9, as well as the α-Pcdhs (Chen et al. 2014). However, when used in in vitro actin assembly assays, the intracellular domains of δ2-Pcdhs differed in their activity. The intracellular domains of Pcdh10 and Pcdh19 acted synergistically with Rac1 to promote actin assembly, however, those of Pcdh17 and Pcdh18 were slightly inhibitory. This suggests a complex relationship between δ2-Pcdhs and actin dynamics, as some may promote Rac1/WAVE-dependent assembly and others may repress assembly. Additionally, the intracellular domains may also be dependent on other associated proteins and appropriate phosphorylation states to function appropriately, none of which would be present in these assays.

The association of the δ-Pcdhs with the WAVE complex through the conserved WIRS motif suggests that these molecules play important roles in motility. As mentioned, Nakao et al. (2008) showed that Pcdh10 promotes contact-dependent motility. In line with these observations, recent studies with Pcdh17 suggest that this contact-dependent motility could provide the basis of selective axon fasciculation (Hayashi et al. 2014). Pcdh17 is present on the axons of amygdala neurons. Loss of *pcdh17* in mouse knockouts revealed misoriented axons, suggesting a role for Pcdh17 in axon outgrowth or guidance. Relying on *lhx6:GFP* as a marker of *pcdh17*+ amygdala neurons (~90% of *lhx6*+ neurons also expressed *pcdh17*), the authors showed that growth of *lhx6:GFP* growth cones along other axons was impaired. Ectopic expression of truncated Pcdh17, lacking either the entire intracellular domain or the WIRS site, resulted in misrouted amygdala axons. In heterologous U251 cells, contact between Pcdh17-expressing cells promoted the recruitment of the WAVE complex, as well as Ena/VASP and lamellipodin. The activity of Rac1 was required for Pcdh17-dependent recruitment of the WAVE complex and Ena/VASP. Based on these observations, Hayashi et al. (2014) proposed a model in which homophilic contact mediated by Pcdh17 promotes coextension of axons. In addition, recent evidence in zebrafish suggests that Pcdh18b acts with the WAVE complex to promote motor axonal arbor growth and branching (Biswas et al. 2014). Depletion of either Pcdh18b or Nap1 reduces the density of filopodia in developing motor axons, which reduces both the size and complexity of arbors at later times. These results suggest that NC-Pcdhs may be important for regulating actin dynamics and cell motility in different contexts and at different stages of development.

Pcdh7/NF-Pcdh was first found to play a role in early *Xenopus* development (Bradley et al. 1998; Rashid et al. 2006), but more recently has been implicated in retinal ganglion cell axon outgrowth and guidance (Piper et al. 2008). Pcdh7 is expressed in the deep layer of the bilayered ectoderm in Xenopus embryos (Bradley et al. 1998). Overexpression of a dominant-negative form of Pcdh7 disrupts the deep ectodermal layer, resulting in blistering of the ectoderm. Pcdh7 is also expressed in a restricted region of the neural folds in Xenopus, and inhibition of Pcdh7 function causes a failure of neural tube closure (Rashid et al. 2006). There are three splice variants of Pcdh7 (Pcdh7a, b, and c). Pcdh7a and Pcdh7b have been shown to mediate weak cell adhesion when transfected into mouse L-cells (Yoshida 2003). More recently, Pcdh7 was shown to play a role in axon initiation and elongation in retinal ganglion cells (Piper et al. 2008). Interestingly, some evidence suggests a potential interaction between Pcdh7 and semaphorin signaling during RGC axon guidance (Leung et al. 2013). Depletion of Pcdh7 results in pathfinding defects at an intermediate choice point in the retinotectal projection (the caudal turn of the mid optic tract). Signaling by the secreted guidance molecule semaphorin 3A promotes increased local translation of Pcdh7 in RGC growth cones. Disruption of SEMA3A/neuropilin-1 signaling results in decreased Pcdh7 levels and defects in guidance. Pcdh10 may also contribute to axon guidance. As mentioned previously, Pcdh10 is present in distinct domains within the cerebellar cortex, as well as in deep cerebellar nuclei and the axon tracts that connect them. The mouse olfactory system provides further evidence for a role in axon guidance, as misexpression causes defects in axon convergence on glomeruli (Williams et al. 2011). As well as affecting axon growth, NC-Pcdhs can affect dendrite development, as Pcdh11x negatively regulates dendrite growth through activation of PI3K and AKT (Wu et al. 2015). Thus, it is very likely that NC-Pcdhs play multiple roles in neurite growth and guidance.

9.4.4 Synaptic Roles for NC-Protocadherins

As putative adhesion molecules with differential and distinctive expression patterns in the developing nervous system, it has been widely presumed that the NC-Pcdhs would play a role at synapses, possibly specifying synaptic connections during neural circuit formation. Although compelling evidence for such a role is lacking and NC-Pcdhs appear more broadly distributed in neurons, several studies have shown that these molecules can act at synapses. Pcdh8/Arcadlin was identified in a screen for genes upregulated in the hippocampus in response to seizures (Yamagata et al. 1999). Yasuda et al. (2007) further investigated Pcdh8 by inducing its expression in cultured hippocampal neurons by tetanic stimulation. Upon enhanced expression, Pcdh8 becomes synaptically localized and interacts directly with N-cadherin. This interaction promotes internalization of N-cadherin through a pathway that involves the kinases TAO2β and p38 MAP kinase. Homozygous knockouts lacking *pcdh8* exhibited an increased spine density. Collectively, these data suggest that Pcdh8 could promote synapse elimination through controlling the surface levels of synaptic N-cadherin (Yasuda et al. 2007). A further role for NC-Pcdhs in synapse elimination has emerged from the study of the transcription factor MEF2 (Tsai et al. 2012). MEF2 is activated by neuronal activity, leading to synapse elimination in a process that requires FMRP (Pfeiffer et al. 2010). A search for MEF2 target genes identified *pcdh10*. MEF2-induced synapse elimination was proposed to be due to ubiquitin-dependent degradation of the postsynaptic scaffolding protein, PSD-95. Upon activation of MEF2, PSD-95 becomes ubiquitinated by the E3 ubiquitin ligase Mdm2, and is then targeted to the proteasome for degradation. This was also associated with a decrease in mini-EPSC frequency and a reduction in spine density. Knockdown of Pcdh10, using shRNA, inhibited MEF2-induced synapse elimination and PSD-95 degradation, as did overexpression of the Pcdh10 intracellular domain. The authors proposed that Pcdh10 interacts with ubiquitinated PSD-95 to promote its delivery to the proteasome for degradation. Although several questions about this process remain unresolved, these results parallel those found for Pcdh8: upregulation of a δ2-Pcdh by activity removes an important structural protein from the synapse to initiate synapse disassembly. In the case of Pcdh8, the effects were attributed to loss of N-cadherin, and in the case of Pcdh10, the effects were attributed to the loss of PSD-95. In contrast to the in vitro results with Pcdh8 and Pcdh10, Pcdh17 does not appear to influence synapse number in vivo (Hoshina et al. 2013). Pcdh17 exhibits a perisynaptic localization at synapses in the anterior striatum and lateral globus pallidus. In *pcdh17−/−*mice, there was no difference in the number or size of synaptic contacts in the anterior striatum, although there was an increase in both the total number of synaptic vesicles and in the number of docked vesicles. This was accompanied both by an increase in paired-pulse facilitation and in antidepression-like behaviors. Thus, Pcdh17 does not appear to promote synapse elimination, but does appear to limit the recruitment of synaptic vesicles to presynaptic sites, influencing short-term plasticity. Collectively, these published studies do support a role for the δ2-Pcdhs at mature synapses, though the function(s) appear to be as negative regulators of synapse function and stability.

9.4.5 Role for NC-Protocadherins in Partitioning the Zebrafish Brain

Recent work in zebrafish suggests a role for the δ-Pcdhs in early stages of neural circuit assembly (Cooper et al. 2015). Using a Gal4/UAS transgenic approach to drive Lifeact-GFP in *pcdh19*+ cells, Cooper et al. (2015) found that expression of *pcdh19* in the zebrafish optic tectum was restricted to discrete columns of neurons, which were closely associated with one or more *pcdh19*+ radial glia. At earlier stages of development, expression of *pcdh19* in the neuroepithelium was also striped, suggesting that columns arise by proliferation of *pcdh19*+ neural progenitors (Fig. 9.2). This hypothesis was supported by two lines of evidence. First, cellular mosaics generated, either by BAC injections, or by cell transplantation from transgenic donors, exhibited columns of labeled neurons, even when no other labeled cells were present in the tectum. This strongly argues that the cells within those columns were generated by proliferation of a common precursor. Second, time-lapse imaging in live embryos demonstrated the proliferation of *pcdh19*+ cells to populate columns. The columnar organization and mechanism of formation is likely to extend to other NC-Pcdhs, as horizontal sections through the optic tectum revealed striped patterns of expression for *pcdh19*, *pcdh9*, and *pcdh10b* and injection of recombinant BAC clones for *pcdh18b* and *pcdh1a* also labeled columns of neurons. Thus, both δ1- and δ2-Pcdhs are expressed in tightly organized columns in the optic tectum, and NC-Pcdhs are expressed in radial glia throughout the zebrafish brain. Both the presence of columns and their origins are reminiscent of the columns found in mammalian cortex, which represent both developmental and functional modules (Mountcastle 1997; Rakic 1988). Though unanticipated, these results are consistent with retroviral lineage studies, which revealed radial clones in the optic tectum of chicken (Gray and Sanes 1991) and medaka (Nguyen et al. 1999). Moreover, in mammals, NC-Pcdhs are expressed in the ventricular zone of cortex (Krishna et al. 2009; Zhang et al. 2014), and NC-Pcdhs appear to participate in cell differentiation in other contexts (Koning et al. 2012; Zhang et al. 2014). Thus, differential, striped expression of NC-Pcdhs in the developing zebrafish neuroepithelium partitions the tectum into discrete domains.

To assess the function of *pcdh19*, Cooper et al. (2015) used transcriptional activator-like effector nucleases (TALENs) to introduce germline lesions in zebrafish *pcdh19*. Homozygous mutants completely lack Pcdh19, but are morphologically normal and are both viable and fertile. However, when crossed into the Lifeact-GFP BAC transgenic background, the columnar organization of *pcdh19*+ cells is lost (Fig. 9.2). This is likely due to a failure of *pcdh19*+ cells to remain in close association, as well as increased proliferation. Interestingly, *pcdh11x* is found in the ventricular zone and subventricular zone, and has been shown to influence neural proliferation, both in vitro and in vivo (Zhang et al. 2014). Moreover, tectal neurons exhibited reduced fasciculation, as well as arborization defects. Thus, Pcdh19 appears to perform multiple functions, regulating proliferation/differentiation, maintaining cell contacts, controlling neurite fasciculation and arborization. Quantitative behavioral analysis revealed that visually guided behaviors are impaired in *pcdh19* mutants. These results demonstrate that Pcdh19 does play a role in the assembly of neural circuitry in zebrafish and also highlight the fact that the loss of NC-Pcdhs is likely to be very subtle; mutants appear normal morphologically and behaviorally and require sensitive assays to reveal either cellular or behavioral phenotypes. This is unsurprising, particularly in the case of *pcdh19*,

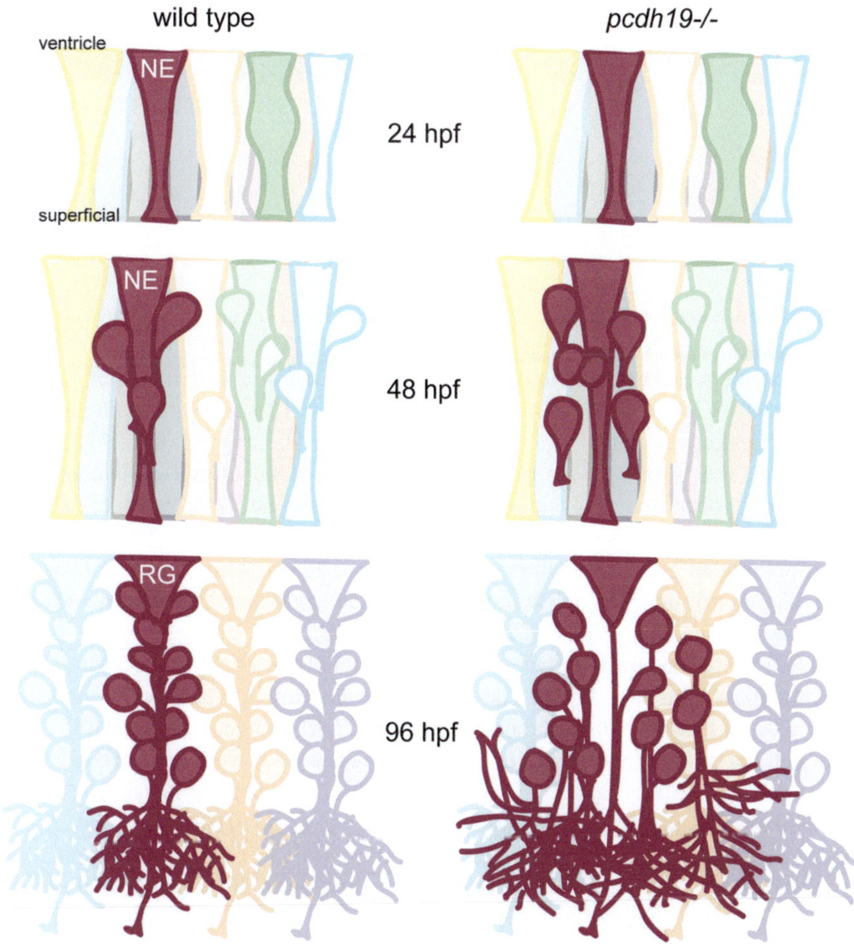

Fig. 9.2 *Model for protocadherin function during modular assembly of the zebrafish optic tectum.* Transgenic zebrafish lines expressing Lifeact-GFP under the control of the *pcdh19* promoter exhibit striped patterns in the midbrain neuroepithelium. Shown at the top right is a small portion of the neuroepithelium in a 24 h postfertilization (hpf) embryo. The burgundy neuroepithelial cell (NE) is expressing *pcdh19*, and the cells in the background are unlabeled and presumably express other NC-Pcdhs. Proliferation of NE cells gives rise to neurons and intermediate precursors, which transform the individual *pcdh19*+ NE cell into a growing column of *pcdh19*+ cells. By 48 hpf, neurites project to the forming synaptic neuropil and begin to arborize, while proliferation continues and the tectum expands. Labeled cells remain closely associated and neuronal processes tend to fasciculate. By 96 hpf, the labeled neurons remain tightly associated, giving the appearance of a radial column. These columns are associated with one or more *pcdh19*+ radial glia (RG) that continue to function as progenitors, as well as a scaffold around which labeled cells associate and along which neuronal processes migrate. Both axons and dendrites arborize immediately adjacent to the column, forming a compact functional unit. As other NC-Pcdhs exhibit similar columnar expression patterns, these molecules partition the developing tectum into discrete developmental modules. Genomic lesioning of *pcdh19* with TALENs disrupts the formation of neuronal columns when crossed into the Lifeact-GFP transgenic background. Mutants appear largely normal at 24 hpf, but exhibit increased rates of proliferation at 48 hpf. By 96 hpf, the columnar organization in the tectum is masked, due to the increased number of *pcdh19*+ neurons and their dispersion,

inasmuch as human males that completely lack *PCDH19* are largely normal (see below). This study indicates that although NC-Pcdhs may contribute to interregional patterns of connectivity, their loss also affects the assembly of local microcircuitry.

9.5 Nonclustered Protocadherins in Cancer

There is now a sizable literature linking the NC-Pcdhs to cancer (Berx and van Roy, 2009; Kahr et al., 2013). In many instances, NC-Pcdh expression is silenced, as their promoters are hypermethylated in primary tumors or cell lines. Of the δ1-Pcdhs, both *PCDH1* and *PCDH7* have been reported to be hypermethylated in cancers (Beukers et al. 2013; Vasilatos et al. 2013). The expression of *PCDH1* is silenced in the breast cancer cell line, MDA-MB-468, but is reactivated upon combined knockdown of lysine-specific demethylase and inhibition of histone deacetylases (Vasilatos et al. 2013). Similarly, the *PCDH7* promoter is hypermethylated in bladder tumors (Beukers et al. 2013). Although *pcdh9* has not been shown to undergo silencing, it has been implicated in glioma (Wang et al. 2012), gastric cancer (Chen et al. 2015b), and ovarian cancer (Asad et al. 2014) and has been shown to inhibit epithelial-mesenchymal transition and cell migration (Zhu et al. 2014). The effect on expression is even more pronounced in the case of the δ2-Pcdhs. The promoter for *PCDH8* is hypermethylated in clear renal cell carcinoma (Morris et al. 2011), nonmuscle invasive bladder cancer (Lin et al. 2014), gastric cancer (Zhang et al. 2012), nasopharyngeal carcinoma (He et al. 2012), mantle cell lymphoma (Leshchenko et al. 2010), and breast cancer (Yu et al. 2008). Similarly, epigenetic inactivation of *PCDH10* has been implicated in gastric cancer (Yu et al. 2009), prostate cancer (Li et al. 2011), colorectal cancer (Yu et al. 2010), pancreatic cancer (Yu et al. 2010) cervical cancer (Narayan et al. 2009), testicular cancer (Cheung et al. 2010), non-small–cell lung cancer (Tang et al. 2012), bladder cancer (Lin et al. 2013), multiple myeloma (Li et al. 2012), lymphoblastic leukemia (Narayan et al. 2011), medulloblastoma (Bertrand et al., 2011), nasopharyngeal carcinoma (Ying et al. 2006), breast cancer (Miyamoto et al. 2005), and hepatocellular cancer (Fang et al. 2013). Studies have also found hypermethylation of *PCDH17* in a number of cancers (Costa et al. 2011; Giefing et al. 2011; Haruki et al. 2010; Hu et al. 2013; Wang et al. 2014). There is also experimental evidence for a role of NC-Pcdhs as tumor supressors. Loss of

---

Fig. 9.2 (continued) presumably to loss of adhesion. In addition, fasciculation is reduced among the primary neuronal projections to the synaptic neuropil and arborization within the neuropil is also aberrant. Thus, Pcdh19 appears to be required for a variety of functions, during the development of neuronal columns, possibly mediating contact-inhibition to control proliferation and differentiation, participating in fasciculation of neuronal projections to the tectal neuropil and regulating arborization of tectal neurons

PCDH9 expression correlates with poor prognosis in both gastric cancer and glioma patients (Chen et al. 2015b; Wang et al. 2012). Similarly, Pcdh10 loss is associated with poor outcomes in colorectal cancer and expression of Pcdh10 in vitro suppresses proliferation, migration, and invasion (Jao et al. 2014). *PCDH20* has also been shown to play a role in cancer, both being silenced by methylation in nasopharygenal carcinoma (Chen et al. 2015a) and by acting as a putative tumor suppressor in hepatocellular carcinoma (HCC) (Lv et al. 2015). Expression of Pcdh20 inhibited growth and migration of an HCC cell line, possibly through downregulation of canonical Wnt signaling by activating GSK3β (Lv et al. 2015). Thus, the NC-Pcdhs are targets of epigentic silencing in a wide range of cancers.

As well as roles as tumor suppressors, there is some evidence that PCDH7 can also function as an oncogene. *PCDH7* expression is upregulated in brain metastases of breast cancer (Bos et al. 2009) and medulloblastoma (Hernan et al. 2003). In addition, Pcdh7 appears to play a role in bone metastasis of breast cancer, as it is upregulated and overexpression promotes proliferation and invasion of breast cancer cells in vitro and bone metastasis in vivo (Li et al. 2013). Moreover, knockdown of *pcdh7* inhibits proliferation, migration, and invasion in breast cancer cell lines. Recently, Pcdh7 was shown to localize to the cell surface of mitotic cancer cells (Ozlu et al. 2015). Surface biotinylation was performed on HeLa cells that were synchronized at different stages of the cell cycle, then the surface proteome was determined for mitotic and interphase cells. The surface levels of two NC-Pcdhs (Pcdh1 and Pcdh7) were strongly upregulated during mitosis. Pcdh1 is similarly enriched in mitotically active, undifferentiated neural crest cells in chicken (Bononi et al. 2008). During interphase, Pcdh7 localizes primarily to the ER, but translocates to the plasma membrane during mitosis. Surface localization is controlled by the intracellular domain, as a Pcdh7 lacking the intracellular domain was expressed on the cell surface during interphase (Ozlu et al. 2015). Knockdown of Pcdh1 and Pcdh7 reduced the rounding pressure in these cells, possibly by downregulating cell–cell adhesion. Thus, the roles of NC-Pcdhs in various cancers may be complicated by the fact that they may fulfill various functions that make them both potential tumor suppressors and potential oncogenes.

9.6 Nonclustered Protocadherins in Neurodevelopmental Disorders

Direct experimental data implicating the NC-Pcdhs in neural circuit formation remains sparse, however, strong support for such a role comes from human genetics (Hirano and Takeichi 2012; Redies et al. 2012). *PCDH8*, *PCDH9*, *PCDH10*, and *PCDH19* have all been implicated in autism spectrum disorders (Butler et al. 2015; Dibbens et al. 2008; Marshall et al. 2008; Morrow et al. 2008). Whole exome sequencing of female autism patients identified a missense mutation in the ectodomain of PCDH8 (Butler et al. 2015). In a search for copy number variants

(CNVs) in autism patients, Marshall et al. (2008) found alterations in changes in *PCDH9* in two patients. A large deletion near *PCDH10* was found in a family with an inherited form of autism. In addition, *PCDH9* and *PCDH12* have been linked to schizophrenia (Gregorio et al. 2009; Pedrosa et al. 2010). A screen for copy number variation in dyslexia revealed linkage to *PCDH11X* (Veerappa et al. 2014). In addition, *PCDH18* has been implicated in intellectual disability, as it is within a large genomic deletion in a boy with severe morphological abnormalities, seizures, microcephaly and developmental delay (Kasnauskiene et al. 2012). Rett syndrome is a disorder with autistic features that is associated with mutations in the methyl-CpG-binding protein 2 (MeCP2). Along with *PCDHβ1*, *PCDH7* is repressed by MeCP2, implicating these genes in Rett syndrome (Miyake et al. 2011). Moreover, microdeletions in *PCDH7* are associated with genetic generalized epilepsy (Lal et al. 2015).

The clearest case for an involvement of NC-Pcdhs in neurodevelopmental disorders is the role of *PCDH19* in epilepsy and mental retardation limited to females (EFMR; Ryan et al. 1997). EFMR is unusual in that it is an X-linked dominant disorder, yet males are spared. Although male carriers may exhibit rigid or obsessive behaviors, females suffer seizures beginning in infancy and lasting through early childhood. Later, affected females have an increased incidence of intellectual disability, bipolar disorder, and autistic features. The distinctive inheritance pattern allowed Dibbens et al. (2008) to sequence the X chromosomes in seven families with EFMR. Mutations in *PCDH19* were identified in five of these families. The identified mutations resulted in both premature stop codons, as well as missense mutations in conserved calcium-binding sites within the ectodomain (Fig. 9.3). Subsequently, well over a hundred different *PCDH19* mutations have been identified (e.g., (Depienne et al. 2009; Depienne and LeGuern 2012; Specchio et al. 2011), making *PCDH19* the second most common gene linked to epilepsy (Depienne and Leguern 2012). Approximately half of the recovered mutations are missense mutations in the ectodomain. Many of the affected residues are involved in calcium binding, although others are likely involved in protein–protein interactions. The role of *PCDH19* in EFMR clearly shows the importance of this gene and other NC-Pcdhs to neural development, and this raises many questions and answers few. It remains unclear why females are affected, while males are spared. Due to X-inactivation, females are mosaic for *PCDH19* mutations, with some cells exclusively expressing the wild-type copy and some cells expressing the mutated gene. The prevailing hypothesis is that this mosaicism creates a dominant-interfering effect. A similar explanation, referred to as "cellular interference", was used to explain a similar pattern of inheritance for craniofacial defects associated with mutations in Ephrin-B1 (Wieland et al. 2004). In support of a cellular interference model, a male suffering from EFMR-like symptoms was found to be mosaic for a *de novo PCDH19* mutation (Depienne et al. 2009). Despite the evidence supporting a role for cellular mosaicism in EFMR, the cellular mechanisms underlying this effect are unknown. Does it result in competition or sorting among cells? Does it affect neurogenesis, neuronal migration, axon guidance, or synaptogenesis? How does/do the cellular defect(s) lead to seizures? Are the seizures responsible for the

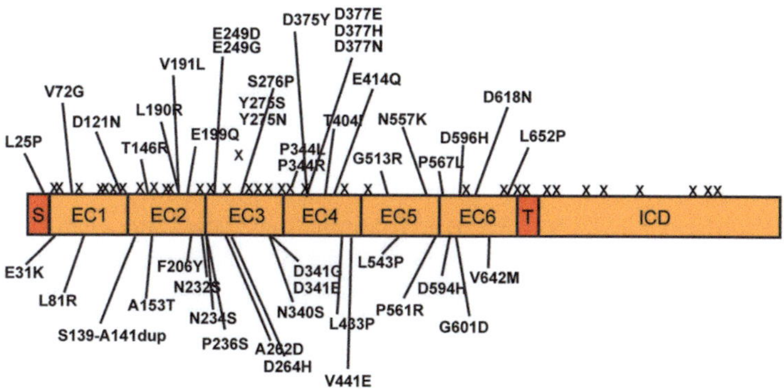

Fig. 9.3 *Mutations in PCDH19 identified in patients with a female-limited form of infantile-onset epilepsy*. To date, over 100 distinct mutations in *PCDH19* have been identified in patients suffering from epilepsy and mental retardation limited to females (EFMR). These mutations range from whole or partial gene deletions to frame-shift and nonsense mutations (x) to missense mutations. The missense mutations have, thus far, been confined to the ectdomain. Although many of these missense mutations occur in the conserved calcium binding sites, a large proportion do not and may affect function through altering important protein–protein interactions. This collection of identified mutations provides an important resource for probing protocadherin function

later neurological problems or other effects on neural assembly? Providing the answers to some of these questions could have a profound influence on our understanding of the importance of protocadherins in orchestrating neural development.

9.7 Conclusion

Although the NC-Pcdhs were first discovered over 20 years ago, they remain poorly understood compared to other members of the cadherin superfamily. This is at least partially due to the fact that their loss-of-function phenotypes are rather subtle. Zebrafish harboring homozygous mutations for *pcdh19* are viable and fertile, and have defects in behavior and neural organization that require sensitive analyses. The same holds true for other zebrafish NC-Pcdhs (unpublished observations). Mice lacking *pcdh17* also have subtle behavioral phenotypes (exhibiting antidepression-like behavior, whereas other behaviors are normal), as well as mild disturbance of neural architecture (disorganization within axon tracts). Another factor potentially complicating analysis of NC-Pcdh function is the fact that they likely function within larger macromolecular complexes. As well as forming *cis*-oligomers, they can form *cis*-complexes with other NC-Pcdhs, with clustered-Pcdhs, and with classical cadherins, as well as other families of cell surface receptors. Thus, their role may partially be to modulate other pathways, and their loss-of-function or gain-of-function phenotypes could be partially due to indirect consequences of disrupting other molecules.

Despite these difficulties, some themes are beginning to emerge. Like their relatives, the classical cadherins, the NC-Pcdhs appear to function at multiple times during development to control a range of fundamental processes: cell movements and morphogenesis, proliferation and differentiation, axon outgrowth and guidance, and synaptic function and elimination. NC-Pcdhs influence cell movements during gastrulation and neurulation, including Pcdh1, Pcdh7/NF-Pcdh, PAPC, Pcdh10, Pcdh18, and Pcdh19. Additionally, Pcdh1 and Pcdh7 exhibit a dynamic localization to the plasma membrane during mitosis and several NC-Pcdhs are expressed in progenitor cells, indicating important roles in proliferation and differentiation. Such a role is also supported by the association of multiple NC-Pcdhs with Ryk/Wnt. A more direct role in brain wiring is supported by results showing that Pcdh7, Pcdh17, Pcdh18, and Pcdh19 participate in axon outgrowth, fasciculation, pathfinding, or arborization. Finally, in the mature nervous system, Pcdh8, Pcdh10, and Pcdh17 can act as negative regulators of synapses, either promoting synapse disassembly or limiting the size of the synaptic vesicle pool. In these different contexts, the cellular activities of the NC-Pcdhs probably vary dramatically, as does the relevant complement of downstream effectors. At the biochemical level, the relationship of NC-Pcdhs to cell adhesion is complex, as there is some evidence that family members can be adhesive, but also that NC-Pcdhs can modulate other adhesive systems and even promote the disassembly of cell–cell contacts. Thus, the diversity of experimental results reflects the diversity of developmental and cellular roles played by these molecules.

References

Aamar E, Dawid IB (2008) Protocadherin-18a has a role in cell adhesion, behavior and migration in zebrafish development. Dev Biol 318:335–346

Asad M, Wong MK, Tan TZ, Choolani M, Low J, Mori S, Virshup D, Thiery JP, Huang RY (2014) FZD7 drives in vitro aggressiveness in Stem-A subtype of ovarian cancer via regulation of non-canonical Wnt/PCP pathway. Cell Death Dis 5:e1346

Babu MM, van der Lee R, de Groot NS, Gsponer J (2011) Intrinsically disordered proteins: regulation and disease. Curr Opin Struct Biol 21:432–440

Berndt JD, Aoyagi A, Yang P, Anastas JN, Tang L, Moon RT (2011) Mindbomb 1, an E3 ubiquitin ligase, forms a complex with RYK to activate Wnt/beta-catenin signaling. J Cell Biol 194:737–750

Beukers W, Hercegovac A, Vermeij M, Kandimalla R, Blok AC, van der Aa MM, Zwarthoff EC, Zuiverloon TC (2013) Hypermethylation of the polycomb group target gene PCDH7 in bladder tumors from patients of all ages. J Urol 190:311–316

Biswas S, Emond MR, Jontes JD (2010) Protocadherin-19 and N-cadherin interact to control cell movements during anterior neurulation. J Cell Biol 191:1029–1041

Biswas S, Emond MR, Duy PQ, Haole T, Beattie CE, Jontes JD (2014) Protocadherin-18b interacts with Nap1 to control motor axon growth and arborization in zebrafish. Mol Biol Cell 25:633–642

Blevins CJ, Emond MR, Biswas S, Jontes JD (2011) Differential expression, alternative splicing, and adhesive properties of the zebrafish delta1-protocadherins. Neuroscience 199:523–534

Bononi J, Cole A, Tewson P, Schumacher A, Bradley R (2008) Chicken protocadherin-1 functions to localize neural crest cells to the dorsal root ganglia during PNS formation. Mech Dev 125:1033–1047

Bos PD, Zhang XH, Nadal C, Shu W, Gomis RR, Nguyen DX, Minn AJ, van de Vijver MJ, Gerald WL, Foekens JA, Massague J (2009) Genes that mediate breast cancer metastasis to the brain. Nature 459:1005–1009

Bouillot S, Tillet E, Carmona G, Prandini MH, Gauchez AS, Hoffmann P, Alfaidy N, Cand F, Huber P (2011) Protocadherin-12 cleavage is a regulated process mediated by ADAM10 protein: evidence of shedding up-regulation in pre-eclampsia. J Biol Chem 286:15195–15204

Bradley RS, Espeseth A, Kintner C (1998) NF-protocadherin, a novel member of the cadherin superfamily, is required for Xenopus ectodermal differentiation. Curr Biol 8:325–334

Butler MG, Rafi SK, Hossain W, Stephan DA, Manzardo AM (2015) Whole exome sequencing in females with autism implicates novel and candidate genes. Int J Mol Sci 16:1312–1335

Chappuis-Flament S, Wong E, Hicks LD, Kay CM, Gumbiner BM (2001) Multiple cadherin extracellular repeats mediate homophilic binding and adhesion. J Cell Biol 154:231–243

Chen X, Gumbiner BM (2006) Paraxial protocadherin mediates cell sorting and tissue morphogenesis by regulating C-cadherin adhesion activity. J Cell Biol 174:301–313

Chen J, Lu Y, Meng S, Han MH, Lin C, Wang X (2009a) alpha- and gamma-Protocadherins negatively regulate PYK2. J Biol Chem 284:2880–2890

Chen X, Koh E, Yoder M, Gumbiner BM (2009b) A protocadherin-cadherin-FLRT3 complex controls cell adhesion and morphogenesis. PLoS One 4:e8411

Chen Y, Londraville R, Brickner S, El-Shaar L, Fankhauser K, Dearth C, Fulton L, Sochacka A, Bhattarai S, Marrs JA, Liu Q (2013) Protocadherin-17 function in Zebrafish retinal development. Dev Neurobiol 73:259–273

Chen B, Brinkmann K, Chen Z, Pak CW, Liao Y, Shi S, Henry L, Grishin NV, Bogdan S, Rosen MK (2014) The WAVE regulatory complex links diverse receptors to the actin cytoskeleton. Cell 156:195–207

Chen T, Long B, Ren G, Xiang T, Li L, Wang Z, He Y, Zeng Q, Hong S, Hu G (2015a) Protocadherin20 acts as a tumor suppressor gene: epigenetic inactivation in nasopharyngeal carcinoma. J Cell Biochem 116:1766–1775

Chen Y, Xiang H, Zhang Y, Wang J, Yu G (2015b) Loss of PCDH9 is associated with the differentiation of tumor cells and metastasis and predicts poor survival in gastric cancer. Clin Exp Metastasis 32:417–428

Cheung HH, Lee TL, Davis AJ, Taft DH, Rennert OM, Chan WY (2010) Genome-wide DNA methylation profiling reveals novel epigenetically regulated genes and non-coding RNAs in human testicular cancer. Br J Cancer 102:419–427

Chung HA, Yamamoto TS, Ueno N (2007) ANR5, an FGF target gene product, regulates gastrulation in Xenopus. Curr Biol 17:932–939

Cooper SR, Emond MR, Duy PQ, Liebau BG, Wolman MA, Jontes JD (2015) Protocadherins control the modular assembly of neuronal columns in the zebrafish optic tectum. J Cell Biol 211:807–814

Costa VL, Henrique R, Danielsen SA, Eknaes M, Patricio P, Morais A, Oliveira J, Lothe RA, Teixeira MR, Lind GE, Jeronimo C (2011) TCF21 and PCDH17 methylation: an innovative panel of biomarkers for a simultaneous detection of urological cancers. Epigenetics 6:1120–1130

Depienne C, LeGuern E (2012) PCDH19-related infantile epileptic encephalopathy: an unusual X-linked inheritance disorder. Hum Mutat 33:627–634

Depienne C, Bouteiller D, Keren B, Cheuret E, Poirier K, Trouillard O, Benyahia B, Quelin C, Carpentier W, Julia S, Afenjar A, Gautier A, Rivier F, Meyer S, Berquin P, Helias M, Py I, Rivera S, Bahi-Buisson N, Gourfinkel-An I, Cazeneuve C, Ruberg M, Brice A, Nabbout R, Leguern E (2009) Sporadic infantile epileptic encephalopathy caused by mutations in PCDH19 resembles Dravet syndrome but mainly affects females. PLoS Genet 5:e1000381

Dibbens LM, Tarpey PS, Hynes K, Bayly MA, Scheffer IE, Smith R, Bomar J, Sutton E, Vandeleur L, Shoubridge C, Edkins S, Turner SJ, Stevens C, O'Meara S, Tofts C, Barthorpe S, Buck G, Cole J, Halliday K, Jones D, Lee R, Madison M, Mironenko T, Varian J, West S, Widaa S, Wray P, Teague J, Dicks E, Butler A, Menzies A, Jenkinson A, Shepherd R, Gusella JF, Afawi Z, Mazarib A, Neufeld MY, Kivity S, Lev D, Lerman-Sagie T, Korczyn AD, Derry CP, Sutherland GR, Friend K, Shaw M, Corbett M, Kim HG, Geschwind DH, Thomas P, Haan E, Ryan S, McKee S, Berkovic SF, Futreal PA, Stratton MR, Mulley JC, Gecz J (2008) X-linked protocadherin 19 mutations cause female-limited epilepsy and cognitive impairment. Nat Genet 40:776–781

Emond MR, Jontes JD (2014) Bead aggregation assays for the characterization of putative cell adhesion molecules. J Vis Exp:e51762

Emond MR, Biswas S, Jontes JD (2009) Protocadherin-19 is essential for early steps in brain morphogenesis. Dev Biol 334:72–83

Emond MR, Biswas S, Blevins CJ, Jontes JD (2011) A complex of Protocadherin-19 and N-cadherin mediates a novel mechanism of cell adhesion. J Cell Biol 195:1115–1121

Fang S, Huang SF, Cao J, Wen YA, Zhang LP, Ren GS (2013) Silencing of PCDH10 in hepatocellular carcinoma via de novo DNA methylation independent of HBV infection or HBX expression. Clin Exp Med 13:127–134

Giefing M, Zemke N, Brauze D, Kostrzewska-Poczekaj M, Luczak M, Szaumkessel M, Pelinska K, Kiwerska K, Tonnies H, Grenman R, Figlerowicz M, Siebert R, Szyfter K, Jarmuz M (2011) High resolution ArrayCGH and expression profiling identifies PTPRD and PCDH17/PCH68 as tumor suppressor gene candidates in laryngeal squamous cell carcinoma. Genes Chromosomes Cancer 50:154–166

Gray GE, Sanes JR (1991) Migratory paths and phenotypic choices of clonally related cells in the avian optic tectum. Neuron 6:211–225

Gregorio SP, Sallet PC, Do KA, Lin E, Gattaz WF, Dias-Neto E (2009) Polymorphisms in genes involved in neurodevelopment may be associated with altered brain morphology in schizophrenia: preliminary evidence. Psychiatry Res 165:1–9

Haruki S, Imoto I, Kozaki K, Matsui T, Kawachi H, Komatsu S, Muramatsu T, Shimada Y, Kawano T, Inazawa J (2010) Frequent silencing of protocadherin 17, a candidate tumour suppressor for esophageal squamous cell carcinoma. Carcinogenesis 31:1027–1036

Hayashi S, Inoue Y, Kiyonari H, Abe T, Misaki K, Moriguchi H, Tanaka Y, Takeichi M (2014) Protocadherin-17 mediates collective axon extension by recruiting actin regulator complexes to interaxonal contacts. Dev Cell 30:673–687

He D, Zeng Q, Ren G, Xiang T, Qian Y, Hu Q, Zhu J, Hong S, Hu G (2012) Protocadherin8 is a functional tumor suppressor frequently inactivated by promoter methylation in nasopharyngeal carcinoma. Eur J Cancer Prev 21:569–575

Heggem MA, Bradley RS (2003) The cytoplasmic domain of Xenopus NF-protocadherin interacts with TAF1/set. Dev Cell 4:419–429

Hernan R, Fasheh R, Calabrese C, Frank AJ, Maclean KH, Allard D, Barraclough R, Gilbertson RJ (2003) ERBB2 up-regulates S100A4 and several other prometastatic genes in medulloblastoma. Cancer Res 63:140–148

Hertel N, Krishna K, Nuernberger M, Redies C (2008) A cadherin-based code for the divisions of the mouse basal ganglia. J Comp Neurol 508:511–528

Hirano S, Takeichi M (2012) Cadherins in brain morphogenesis and wiring. Physiol Rev 92:597–634

Hirano S, Yan Q, Suzuki ST (1999) Expression of a novel protocadherin, OL-protocadherin, in a subset of functional systems of the developing mouse brain. J Neurosci 19:995–1005

Homayouni R, Rice DS, Curran T (2001) Disabled-1 interacts with a novel developmentally regulated protocadherin. Biochem Biophys Res Commun 289:539–547

Hoshina N, Tanimura A, Yamasaki M, Inoue T, Fukabori R, Kuroda T, Yokoyama K, Tezuka T, Sagara H, Hirano S, Kiyonari H, Takada M, Kobayashi K, Watanabe M, Kano M, Nakazawa T,

Yamamoto T (2013) Protocadherin 17 regulates presynaptic assembly in topographic corticobasal Ganglia circuits. Neuron 78:839–854
Hu X, Sui X, Li L, Huang X, Rong R, Su X, Shi Q, Mo L, Shu X, Kuang Y, Tao Q, He C (2013) Protocadherin 17 acts as a tumour suppressor inducing tumour cell apoptosis and autophagy, and is frequently methylated in gastric and colorectal cancers. J Pathol 229:62–73
Huber AH, Weis WI (2001) The structure of the beta-catenin/E-cadherin complex and the molecular basis of diverse ligand recognition by beta-catenin. Cell 105:391–402
Huber AH, Stewart DB, Laurents DV, Nelson WJ, Weis WI (2001) The cadherin cytoplasmic domain is unstructured in the absence of beta-catenin. A possible mechanism for regulating cadherin turnover. J Biol Chem 276:12301–12309
Hulpiau P, van Roy F (2009) Molecular evolution of the cadherin superfamily. Int J Biochem Cell Biol 41:349–369
Ishiyama N, Lee SH, Liu S, Li GY, Smith MJ, Reichardt LF, Ikura M (2010) Dynamic and static interactions between p120 catenin and E-cadherin regulate the stability of cell-cell adhesion. Cell 141:117–128
Izuta Y, Taira T, Asayama A, Machigashira M, Kinoshita T, Fujiwara M, Suzuki ST (2015) Protocadherin-9 involvement in retinal development in Xenopus laevis. J Biochem 157:235–249
Jao TM, Tsai MH, Lio HY, Weng WT, Chen CC, Tzeng ST, Chang CY, Lai YC, Yen SJ, Yu SL, Yang YC (2014) Protocadherin 10 suppresses tumorigenesis and metastasis in colorectal cancer and its genetic loss predicts adverse prognosis. Int J Cancer 135:2593–2603
Kai M, Ueno N, Kinoshita N (2015) Phosphorylation-dependent ubiquitination of paraxial protocadherin (PAPC) controls gastrulation cell movements. PLoS One 10:e0115111
Kasnauskiene J, Ciuladaite Z, Preiksaitiene E, Matuleviciene A, Alexandrou A, Koumbaris G, Sismani C, Pepalyte I, Patsalis PC, Kucinskas V (2012) A single gene deletion on 4q28.3: PCDH18--a new candidate gene for intellectual disability? Eur J Med Genet 55:274–277
Kim SH, Yamamoto A, Bouwmeester T, Agius E, Robertis EM (1998) The role of paraxial protocadherin in selective adhesion and cell movements of the mesoderm during Xenopus gastrulation. Development 125:4681–4690
Kim SH, Jen WC, De Robertis EM, Kintner C (2000) The protocadherin PAPC establishes segmental boundaries during somitogenesis in xenopus embryos. Curr Biol 10:821–830
Kim SY, Chung HS, Sun W, Kim H (2007) Spatiotemporal expression pattern of non-clustered protocadherin family members in the developing rat brain. Neuroscience 147:996–1021
Kim SY, Mo JW, Han S, Choi SY, Han SB, Moon BH, Rhyu IJ, Sun W, Kim H (2010) The expression of non-clustered protocadherins in adult rat hippocampal formation and the connecting brain regions. Neuroscience 170:189–199
Koning H, Sayers I, Stewart CE, de Jong D, Ten Hacken NH, Postma DS, van Oosterhout AJ, Nawijn MC, Koppelman GH (2012) Characterization of protocadherin-1 expression in primary bronchial epithelial cells: association with epithelial cell differentiation. FASEB J 26:439–448
Koppelman GH, Meyers DA, Howard TD, Zheng SL, Hawkins GA, Ampleford EJ, Xu J, Koning H, Bruinenberg M, Nolte IM, van Diemen CC, Boezen HM, Timens W, Whittaker PA, Stine OC, Barton SJ, Holloway JW, Holgate ST, Graves PE, Martinez FD, van Oosterhout AJ, Bleecker ER, Postma DS (2009) Identification of PCDH1 as a novel susceptibility gene for bronchial hyperresponsiveness. Am J Respir Crit Care Med 180:929–935
Kraft B, Berger CD, Wallkamm V, Steinbeisser H, Wedlich D (2012) Wnt-11 and Fz7 reduce cell adhesion in convergent extension by sequestration of PAPC and C-cadherin. J Cell Biol 198:695–709
Krishna K, Redies C (2009) Expression of cadherin superfamily genes in brain vascular development. J Cereb Blood Flow Metab 29:224–229
Krishna K, Nuernberger M, Weth F, Redies C (2009) Layer-specific expression of multiple cadherins in the developing visual cortex (V1) of the ferret. Cereb Cortex 19:388–401
Krishna KK, Hertel N, Redies C (2011) Cadherin expression in the somatosensory cortex: evidence for a combinatorial molecular code at the single-cell level. Neuroscience 175:37–48

Kuroda H, Inui M, Sugimoto K, Hayata T, Asashima M (2002) Axial protocadherin is a mediator of prenotochord cell sorting in Xenopus. Dev Biol 244:267–277

Lal D, Ruppert AK, Trucks H, Schulz H, de Kovel CG, Kasteleijn-Nolst Trenite D, Sonsma AC, Koeleman BP, Lindhout D, Weber YG, Lerche H, Kapser C, Schankin CJ, Kunz WS, Surges R, Elger CE, Gaus V, Schmitz B, Helbig I, Muhle H, Stephani U, Klein KM, Rosenow F, Neubauer BA, Reinthaler EM, Zimprich F, Feucht M, Moller RS, Hjalgrim H, De Jonghe P, Suls A, Lieb W, Franke A, Strauch K, Gieger C, Schurmann C, Schminke U, Nurnberg P, Consortium E, Sander T (2015) Burden analysis of rare microdeletions suggests a strong impact of neurodevelopmental genes in genetic generalised epilepsies. PLoS Genet 11: e1005226

Langer MD, Guo H, Shashikanth N, Pierce JM, Leckband DE (2012) N-glycosylation alters cadherin-mediated intercellular binding kinetics. J Cell Sci 125:2478–2485

Lele Z, Folchert A, Concha M, Rauch GJ, Geisler R, Rosa F, Wilson SW, Hammerschmidt M, Bally-Cuif L (2002) parachute/n-cadherin is required for morphogenesis and maintained integrity of the zebrafish neural tube. Development 129:3281–3294

Leshchenko VV, Kuo PY, Shaknovich R, Yang DT, Gellen T, Petrich A, Yu Y, Remache Y, Weniger MA, Rafiq S, Suh KS, Goy A, Wilson W, Verma A, Braunschweig I, Muthusamy N, Kahl BS, Byrd JC, Wiestner A, Melnick A, Parekh S (2010) Genomewide DNA methylation analysis reveals novel targets for drug development in mantle cell lymphoma. Blood 116:1025–1034

Leung LC, Urbancic V, Baudet ML, Dwivedy A, Bayley TG, Lee AC, Harris WA, Holt CE (2013) Coupling of NF-protocadherin signaling to axon guidance by cue-induced translation. Nat Neurosci 16:166–173

Li Z, Li W, Xie J, Wang Y, Tang A, Li X, Ye J, Gui Y, Cai Z (2011) Epigenetic inactivation of PCDH10 in human prostate cancer cell lines. Cell Biol Int 35:671–676

Li Z, Chim JC, Yang M, Ye J, Wong BC, Qiao L (2012) Role of PCDH10 and its hypermethylation in human gastric cancer. Biochim Biophys Acta 1823:298–305

Li AM, Tian AX, Zhang RX, Ge J, Sun X, Cao XC (2013) Protocadherin-7 induces bone metastasis of breast cancer. Biochem Biophys Res Commun 436:486–490

Lin J, Wang C, Redies C (2012) Expression of delta-protocadherins in the spinal cord of the chicken embryo. J Comp Neurol 520:1509–1531

Lin YL, Li ZG, Guan TY (2013) The clinical significance of PCDH10 promoter methylation in patients with bladder transitional cell carcinoma. Urol Int 90:219–224

Lin YL, Wang YL, Ma JG, Li WP (2014) Clinical significance of protocadherin 8 (PCDH8) promoter methylation in non-muscle invasive bladder cancer. J Exp Clin Cancer Res 33:68

Lv J, Zhu P, Yang Z, Li M, Zhang X, Cheng J, Chen X, Lu F (2015) PCDH20 functions as a tumour-suppressor gene through antagonizing the Wnt/beta-catenin signalling pathway in hepatocellular carcinoma. J Viral Hepat 22:201–211

Marshall CR, Noor A, Vincent JB, Lionel AC, Feuk L, Skaug J, Shago M, Moessner R, Pinto D, Ren Y, Thiruvahindrapduram B, Fiebig A, Schreiber S, Friedman J, Ketelaars CE, Vos YJ, Ficicioglu C, Kirkpatrick S, Nicolson R, Sloman L, Summers A, Gibbons CA, Teebi A, Chitayat D, Weksberg R, Thompson A, Vardy C, Crosbie V, Luscombe S, Baatjes R, Zwaigenbaum L, Roberts W, Fernandez B, Szatmari P, Scherer SW (2008) Structural variation of chromosomes in autism spectrum disorder. Am J Hum Genet 82:477–488

Medina A, Swain RK, Kuerner KM, Steinbeisser H (2004) Xenopus paraxial protocadherin has signaling functions and is involved in tissue separation. Embo J 23:3249–3258

Miyake K, Hirasawa T, Soutome M, Itoh M, Goto Y, Endoh K, Takahashi K, Kudo S, Nakagawa T, Yokoi S, Taira T, Inazawa J, Kubota T (2011) The protocadherins, PCDHB1 and PCDH7, are regulated by MeCP2 in neuronal cells and brain tissues: implication for pathogenesis of Rett syndrome. BMC Neurosci 12:81

Miyamoto K, Fukutomi T, Akashi-Tanaka S, Hasegawa T, Asahara T, Sugimura T, Ushijima T (2005) Identification of 20 genes aberrantly methylated in human breast cancers. Int J Cancer 116:407–414

Morris MR, Ricketts CJ, Gentle D, McRonald F, Carli N, Khalili H, Brown M, Kishida T, Yao M, Banks RE, Clarke N, Latif F, Maher ER (2011) Genome-wide methylation analysis identifies epigenetically inactivated candidate tumour suppressor genes in renal cell carcinoma. Oncogene 30:1390–1401

Morrow EM, Yoo SY, Flavell SW, Kim TK, Lin Y, Hill RS, Mukaddes NM, Balkhy S, Gascon G, Hashmi A, Al-Saad S, Ware J, Joseph RM, Greenblatt R, Gleason D, Ertelt JA, Apse KA, Bodell A, Partlow JN, Barry B, Yao H, Markianos K, Ferland RJ, Greenberg ME, Walsh CA (2008) Identifying autism loci and genes by tracing recent shared ancestry. Science 321:218–223

Mortensen LJ, Kreiner-Moller E, Hakonarson H, Bonnelykke K, Bisgaard H (2014) The PCDH1 gene and asthma in early childhood. Eur Respir J 43:792–800

Mountcastle VB (1997) The columnar organization of the neocortex. Brain 120(Pt 4):701–722

Murakami T, Hijikata T, Matsukawa M, Ishikawa H, Yorifuji H (2006) Zebrafish protocadherin 10 is involved in paraxial mesoderm development and somitogenesis. Dev Dyn 235:506–514

Nakao S, Platek A, Hirano S, Takeichi M (2008) Contact-dependent promotion of cell migration by the OL-protocadherin-Nap1 interaction. J Cell Biol 182:395–410

Narayan G, Scotto L, Neelakantan V, Kottoor SH, Wong AH, Loke SL, Mansukhani M, Pothuri B, Wright JD, Kaufmann AM, Schneider A, Arias-Pulido H, Tao Q, Murty VV (2009) Protocadherin PCDH10, involved in tumor progression, is a frequent and early target of promoter hypermethylation in cervical cancer. Genes Chromosomes Cancer 48:983–992

Narayan G, Freddy AJ, Xie D, Liyanage H, Clark L, Kisselev S, Un Kang J, Nandula SV, McGuinn C, Subramaniyam S, Alobeid B, Satwani P, Savage D, Bhagat G, Murty VV (2011) Promoter methylation-mediated inactivation of PCDH10 in acute lymphoblastic leukemia contributes to chemotherapy resistance. Genes Chromosomes Cancer 50:1043–1053

Nguyen V, Deschet K, Henrich T, Godet E, Joly JS, Wittbrodt J, Chourrout D, Bourrat F (1999) Morphogenesis of the optic tectum in the medaka (Oryzias latipes): a morphological and molecular study, with special emphasis on cell proliferation. J Comp Neurol 413:385–404

Nollet F, Kools P, van Roy F (2000) Phylogenetic analysis of the cadherin superfamily allows identification of six major subfamilies besides several solitary members. J Mol Biol 299:551–572

Ozlu N, Qureshi MH, Toyoda Y, Renard BY, Mollaoglu G, Ozkan NE, Bulbul S, Poser I, Timm W, Hyman AA, Mitchison TJ, Steen JA (2015) Quantitative comparison of a human cancer cell surface proteome between interphase and mitosis. EMBO J 34:251–265

Pedrosa E, Shah A, Tenore C, Capogna M, Villa C, Guo X, Zheng D, Lachman HM (2010) beta-catenin promoter ChIP-chip reveals potential schizophrenia and bipolar disorder gene network. J Neurogenet 24:182–193

Pfeiffer BE, Zang T, Wilkerson JR, Taniguchi M, Maksimova MA, Smith LN, Cowan CW, Huber KM (2010) Fragile X mental retardation protein is required for synapse elimination by the activity-dependent transcription factor MEF2. Neuron 66:191–197

Piper M, Dwivedy A, Leung L, Bradley RS, Holt CE (2008) NF-protocadherin and TAF1 regulate retinal axon initiation and elongation in vivo. J Neurosci 28:100–105

Rakic P (1988) Specification of cerebral cortical areas. Science 241:170–176

Rashid D, Newell K, Shama L, Bradley R (2006) A requirement for NF-protocadherin and TAF1/Set in cell adhesion and neural tube formation. Dev Biol 291:170–181

Redies C, Neudert F, Lin J (2011) Cadherins in cerebellar development: translation of embryonic patterning into mature functional compartmentalization. Cerebellum 10:393–408

Redies C, Hertel N, Hubner CA (2012) Cadherins and neuropsychiatric disorders. Brain Res 1470:130–144

Rhee J, Takahashi Y, Saga Y, Wilson-Rawls J, Rawls A (2003) The protocadherin papc is involved in the organization of the epithelium along the segmental border during mouse somitogenesis. Dev Biol 254:248–261

Ryan SG, Chance PF, Zou CH, Spinner NB, Golden JA, Smietana S (1997) Epilepsy and mental retardation limited to females: an X-linked dominant disorder with male sparing. Nat Genet 17:92–95

Sano K, Tanihara H, Heimark RL, Obata S, Davidson M, John TS, Taketani S, Suzuki S (1993) Protocadherins: a large family of cadherin-related molecules in central nervous system. Embo J 12:2249–2256

Specchio N, Marini C, Terracciano A, Mei D, Trivisano M, Sicca F, Fusco L, Cusmai R, Darra F, Bernardina BD, Bertini E, Guerrini R, Vigevano F (2011) Spectrum of phenotypes in female patients with epilepsy due to protocadherin 19 mutations. Epilepsia 52:1251–1257

Tai K, Kubota M, Shiono K, Tokutsu H, Suzuki ST (2010) Adhesion properties and retinofugal expression of chicken protocadherin-19. Brain Res 1344:13–24

Tang X, Yin X, Xiang T, Li H, Li F, Chen L, Ren G (2012) Protocadherin 10 is frequently downregulated by promoter methylation and functions as a tumor suppressor gene in non-small cell lung cancer. Cancer Biomark 12:11–19

Tawk M, Araya C, Lyons DA, Reugels AM, Girdler GC, Bayley PR, Hyde DR, Tada M, Clarke JD (2007) A mirror-symmetric cell division that orchestrates neuroepithelial morphogenesis. Nature 446:797–800

Tsai NP, Wilkerson JR, Guo W, Maksimova MA, DeMartino GN, Cowan CW, Huber KM (2012) Multiple autism-linked genes mediate synapse elimination via proteasomal degradation of a synaptic scaffold PSD-95. Cell 151:1581–1594

Uversky VN (2011) Intrinsically disordered proteins from A to Z. Int J Biochem Cell Biol 43:1090–1103

Vanhalst K, Kools P, Staes K, van Roy F, Redies C (2005) delta-Protocadherins: a gene family expressed differentially in the mouse brain. Cell Mol Life Sci 62:1247–1259

Vasilatos SN, Katz TA, Oesterreich S, Wan Y, Davidson NE, Huang Y (2013) Crosstalk between lysine-specific demethylase 1 (LSD1) and histone deacetylases mediates antineoplastic efficacy of HDAC inhibitors in human breast cancer cells. Carcinogenesis 34:1196–1207

Veerappa AM, Saldanha M, Padakannaya P, Ramachandra NB (2014) Family based genome-wide copy number scan identifies complex rearrangements at 17q21.31 in dyslexics. Am J Med Genet B Neuropsychiatr Genet 165B:572–580

Vester-Christensen MB, Halim A, Joshi HJ, Steentoft C, Bennett EP, Levery SB, Vakhrushev SY, Clausen H (2013) Mining the O-mannose glycoproteome reveals cadherins as major O-mannosylated glycoproteins. Proc Natl Acad Sci U S A 110:21018–21023

Wang Y, Janicki P, Koster I, Berger CD, Wenzl C, Grosshans J, Steinbeisser H (2008) Xenopus Paraxial Protocadherin regulates morphogenesis by antagonizing Sprouty. Genes Dev 22:878–883

Wang C, Yu G, Liu J, Wang J, Zhang Y, Zhang X, Zhou Z, Huang Z (2012) Downregulation of PCDH9 predicts prognosis for patients with glioma. J Clin Neurosci 19:541–545

Wang XB, Lin YL, Li ZG, Ma JH, Li J, Ma JG (2014) Protocadherin 17 promoter methylation in tumour tissue from patients with bladder transitional cell carcinoma. J Int Med Res 42:292–299

Wieland I, Jakubiczka S, Muschke P, Cohen M, Thiele H, Gerlach KL, Adams RH, Wieacker P (2004) Mutations of the ephrin-B1 gene cause craniofrontonasal syndrome. Am J Hum Genet 74:1209–1215

Williams EO, Sickles HM, Dooley AL, Palumbos S, Bisogni AJ, Lin DM (2011) Delta protocadherin 10 is regulated by activity in the mouse main olfactory system. Front Neural Circuits 5:9

Wojtowicz WM, Flanagan JJ, Millard SS, Zipursky SL, Clemens JC (2004) Alternative splicing of Drosophila Dscam generates axon guidance receptors that exhibit isoform-specific homophilic binding. Cell 118:619–633

Wolverton T, Lalande M (2001) Identification and characterization of three members of a novel subclass of protocadherins. Genomics 76:66–72

Wu C, Niu L, Yan Z, Wang C, Liu N, Dai Y, Zhang P, Xu R (2015) Pcdh11x negatively regulates dendritic branching. J Mol Neurosci 56:822–828

Yamagata K, Andreasson KI, Sugiura H, Maru E, Dominique M, Irie Y, Miki N, Hayashi Y, Yoshioka M, Kaneko K, Kato H, Worley PF (1999) Arcadlin is a neural activity-regulated cadherin involved in long term potentiation. J Biol Chem 274:19473–11979

Yamamoto A, Amacher SL, Kim SH, Geissert D, Kimmel CB, De Robertis EM (1998) Zebrafish paraxial protocadherin is a downstream target of spadetail involved in morphogenesis of gastrula mesoderm. Development 125:3389–3397

Yasuda S, Tanaka H, Sugiura H, Okamura K, Sakaguchi T, Tran U, Takemiya T, Mizoguchi A, Yagita Y, Sakurai T, De Robertis EM, Yamagata K (2007) Activity-induced protocadherin arcadlin regulates dendritic spine number by triggering N-cadherin endocytosis via TAO2beta and p38 MAP kinases. Neuron 56:456–471

Ying J, Li H, Seng TJ, Langford C, Srivastava G, Tsao SW, Putti T, Murray P, Chan AT, Tao Q (2006) Functional epigenetics identifies a protocadherin PCDH10 as a candidate tumor suppressor for nasopharyngeal, esophageal and multiple other carcinomas with frequent methylation. Oncogene 25:1070–1080

Yoshida K (2003) Fibroblast cell shape and adhesion in vitro is altered by overexpression of the 7a and 7b isoforms of protocadherin 7, but not the 7c isoform. Cell Mol Biol Lett 8:735–741

Yoshida K, Watanabe M, Kato H, Dutta A, Sugano S (1999) BH-protocadherin-c, a member of the cadherin superfamily, interacts with protein phosphatase 1 alpha through its intracellular domain. FEBS Lett 460:93–98

Yu JS, Koujak S, Nagase S, Li CM, Su T, Wang X, Keniry M, Memeo L, Rojtman A, Mansukhani M, Hibshoosh H, Tycko B, Parsons R (2008) PCDH8, the human homolog of PAPC, is a candidate tumor suppressor of breast cancer. Oncogene 27:4657–4665

Yu J, Cheng YY, Tao Q, Cheung KF, Lam CN, Geng H, Tian LW, Wong YP, Tong JH, Ying JM, Jin H, To KF, Chan FK, Sung JJ (2009) Methylation of protocadherin 10, a novel tumor suppressor, is associated with poor prognosis in patients with gastric cancer. Gastroenterology 136(640–651):e641

Yu B, Yang H, Zhang C, Wu Q, Shao Y, Zhang J, Guan M, Wan J, Zhang W (2010) High-resolution melting analysis of PCDH10 methylation levels in gastric, colorectal and pancreatic cancers. Neoplasma 57:247–252

Zhang Y, Sivasankar S, Nelson WJ, Chu S (2009) Resolving cadherin interactions and binding cooperativity at the single-molecule level. Proc Natl Acad Sci U S A 106:109–114

Zhang D, Zhao W, Liao X, Bi T, Li H, Che X (2012) Frequent silencing of protocadherin 8 by promoter methylation, a candidate tumor suppressor for human gastric cancer. Oncol Rep 28:1785–1791

Zhang P, Wu C, Liu N, Niu L, Yan Z, Feng Y, Xu R (2014) Protocadherin 11 x regulates differentiation and proliferation of neural stem cell in vitro and in vivo. J Mol Neurosci 54:199–210

Zhu P, Lv J, Yang Z, Guo L, Zhang L, Li M, Han W, Chen X, Zhuang H, Lu F (2014) Protocadherin 9 inhibits epithelial-mesenchymal transition and cell migration through activating GSK-3beta in hepatocellular carcinoma. Biochem Biophys Res Commun 452:567–574

Chapter 10
Seven-Pass Transmembrane Cadherin CELSRs, and Fat4 and Dchs1 Cadherins: From Planar Cell Polarity to Three-Dimensional Organ Architecture

Dongbo Shi, Masaki Arata, Tadao Usui, Toshihiko Fujimori, and Tadashi Uemura

Abstract In this chapter, two subfamilies of atypical cadherins are described: the subfamily of seven-pass transmembrane cadherins (7-TM cadherins) and Fat and Dachsous cadherins. Pioneering genetic studies in Drosophila have defined both subfamilies and dissected their roles in animal development. It is now clear that the founding members in Drosophila and their respective vertebrate homologues make critical and essential contributions to a variety of dynamic behaviors of cell populations, and that malfunctions of those atypical cadherins cause anomalies in embryonic development, resulting in postnatal organ malformation or embryonic demise. Here we discuss how the atypical cadherins control cell behaviors with the emphasis on one particular orchestration of cells along the axes of tissues, organs, or bodies, inclusively designated as planar cell polarity (PCP). Nowadays the purview of PCP ranges from the unidirectional orientation of subcellular structures, such as wing hairs of Drosophila and vertebrate motile cilia, to three-dimensional dynamics of multicellular units, such as tilting hair follicles, neural tube closure, epithelial folding in the oviduct, and collective cell migration. The PCP field is at an extraordinarily exciting juncture, bursting with questions about functions of 7-TM cadherins and Fat and Dachsous cadherins at the cellular and molecular level.

D. Shi
Laboratory of Cell Recognition and Pattern Formation, Graduate School of Biostudies, Kyoto University, Kyoto 606-8501, Japan

Division of Embryology, National Institute for Basic Biology, 5-1 Higashiyama, Myodaiji, Okazaki, Aichi 444-8787, Japan

M. Arata • T. Usui • T. Uemura (✉)
Laboratory of Cell Recognition and Pattern Formation, Graduate School of Biostudies, Kyoto University, Kyoto 606-8501, Japan
e-mail: tauemura@lif.kyoto-u.ac.jp

T. Fujimori
Division of Embryology, National Institute for Basic Biology, 5-1 Higashiyama, Myodaiji, Okazaki, Aichi 444-8787, Japan

S.T. Suzuki, S. Hirano (eds.), *The Cadherin Superfamily*,
DOI 10.1007/978-4-431-56033-3_10

Keywords Seven-pass transmembrane cadherins (7-TM cadherins) • CELSR • Fat and Dachsous cadherins • Planar cell polarity (PCP) • Cilia • Three-dimensional organogenesis • Hair follicle • Neural tube closure • Epithelial folding • Oviduct • Collective cell migration

10.1 Introduction: 7-TM Cadherins, Fat and Dachsous Cadherins, and Planar Cell Polarity

As we know, things inevitably fall apart. But in the meantime, what holds them together? In the hierarchy of the animal body, how do cells connect with each other to realize tissue integrity, and how are the myriad different cell types sorted out? Assembled cells are far from static during development; they sometimes reorient themselves along global axes of tissues, organs, or bodies, dramatically converting simple local geometries into three-dimensional organs or whole bodies. How are these dynamic processes regulated, and what types of problems do we face if the underlying machineries lose control? Behind almost all of these fundamental questions, there exists a large superfamily of cell-to-cell adhesion proteins, the cadherin superfamily.

The cadherin superfamily comprises "classic" cadherins and atypical cadherins (see the chapter by Suzuki and Hirano). It is assumed that atypical cadherins are responsible for cellular and molecular functions distinct from classic members. In this chapter, two subfamilies of atypical cadherins are described (Fig. 10.1). The first is the subfamily of seven-pass transmembrane cadherins (7-TM cadherins), where the founding member is Flamingo (Fmi)/Starry night (Stan) in Drosophila (Chae et al. 1999; Usui et al. 1999) and its vertebrate homologues are CELSRs (cadherin epidermal growth factor laminin G seven-pass G-type receptors; Hadjantonakis et al. 1997; Nakayama et al. 1998). The second subfamily comprises single-pass transmembrane proteins in Drosophila, named Fat (Ft) and Dachsous (Ds; Mahoney et al. 1991; Clark et al. 1995), and their respective vertebrate homologues (Nakajima et al. 2001; Hoeng et al. 2004; Rock et al. 2005), all of which are collectively designated as Fat and Dachsous cadherins (Thomas and Strutt 2012; Matis and Axelrod 2013; Sharma and McNeill 2013; Sadeqzadeh et al. 2014). Ft and Ds (and their vertebrate orthologues Fat4 and Dcsh1, respectively) bind in a heterophilic manner at the cell surface (Strutt and Strutt 2002; Ma et al. 2003; Matakatsu and Blair 2004; Ishiuchi et al. 2009; Tsukasaki et al. 2014).

Understanding the in vivo functions of the two subfamilies has been tightly coupled with deciphering the mechanisms of one particular orchestration of cells, namely planar cell polarity (PCP). In many organs, epithelial cells are polarized not only along the apicobasal axis (the outside–inside axis) of the two-dimensional sheet, but also along a perpendicular, second axis within the plane. The latter polarity is the original definition of planar cell polarity, and is crucial for specialized cellular functions (Fig. 10.2; Adler 2012; Lawrence and Casal 2013). Nowadays the purview of PCP has expanded to include various directional behaviors of

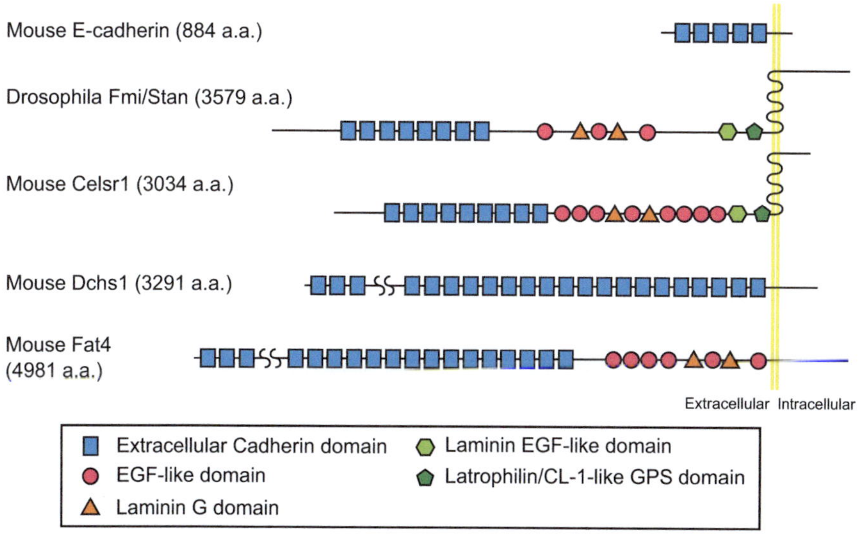

Fig. 10.1 Subfamilies of 7-TM cadherins and Fat and Dachsous cadherins. Schematic drawings of a classic cadherin and representative members of two atypical cadherin subfamilies. Mouse E-cadherin, Drosophila Flamingo (Fmi)/Starry night (Stan), and mouse Celsr1 in the 7-TM cadherin subfamily, and mouse Fat4 and Dchs1 (homologues of Drosophila Fat and Dachsous, respectively). All of them share extracellular cadherin domains (ECs) within their ectodomains in the N-termini. Other motifs in their ectodomains are also illustrated

rearranging cell populations of both epithelial and nonepithelial tissues (Vladar et al. 2009; Goodrich and Strutt 2011; Gray et al. 2011; Vichas and Zallen 2011; Yang 2012; Wallingford 2012). Below we discuss Drosophila and vertebrate model systems that have distinguished the two subfamilies in PCP, the remarkable subcellular localization of 7-TM cadherins, roles of the two subfamilies in generating polarity at multiple levels from cells to organs in various vertebrate developmental contexts, and a number of outstanding questions that are being addressed.

10.2 Landmarks of Planar Cell Polarity: Drosophila Wing Hairs and Vertebrate Cilia

Roles of Fmi/Stan, Ft, and Ds in PCP have been best studied in the Drosophila wing, where individual epidermal cells localize an assembly of actin filaments at the distal cell vertices on the apical surface, and produce single wing hairs pointing distally (Figs. 10.2a, b, and 10.3a; Adler 2012; Carvajal-Gonzalez and Mlodzik 2014). If any of these genes malfunction, the mutant cells no longer respect the proximal–distal axis of the wing and fail to select the correct sites on the apical surface for hair formation (Adler 2012). For example, *fmi/stan* mutant cells form wing hairs at the center of the apical surface (Fig. 10.3b), supporting the hypothesis

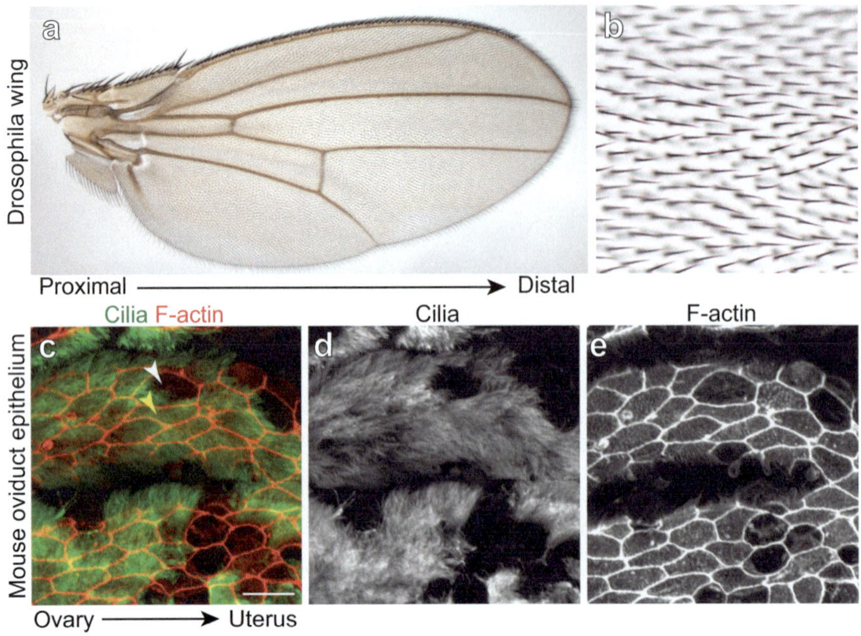

Fig. 10.2 Landmarks of planar cell polarity. (**a** and **b**) The Drosophila wing. (**a**) The whole image where the proximal–distal (P–D) axis is running from left to right. (**b**) The high-power image shows hairs pointing in a distal direction along the tissue axis. (**c**–**e**) A maximum intensity projection of confocal images of the oviduct epithelium, where the ovary–uterus axis is running from left to right. The oviduct is opened longitudinally and the epithelium is stained with antiacetylated α-tubulin antibodies (**d**, and *green* in **c**; cilia) and phalloidin (**e**, and *red* in **c**; F-actin). (**c**) A secretory cell and a multiciliated cell are marked by *white* and *yellow arrowheads*, respectively. The epithelium of the infundibular region (close to the ovary end of the oviduct) is mostly covered by multiciliated cells. *Scale bar*: 10 μm

that Fmi/Stan functions to limit the region where the cytoskeleton is activated to form a hair.

In vertebrate epithelia, directional alignments of apical appendages are seen in the inner ear, where actin filament-based stereocilia protrude from the apical surface of hair cells (May-Simera and Kelley 2012; see the chapter by El-Amraoui and Petit), and in many other epithelia that develop microtubule-based motile cilia (Brooks and Wallingford 2014). Cilia-forming epithelial tissues or organs include the node that gives rise to axial mesoderm (Sulik et al. 1994), the oviduct (Fig. 10.2c–e; Agduhr 1927; Boisvieuxulrich et al. 1985), the airway (Lucas and Douglas 1934), the lateral ventricle in the brain (Del Bigio 1995), and tadpole epidermis (Assheton 1896). The planar axis along which cells are polarized is referred to as the tissue, organ, or body axis, or more simply the global axis, depending on the context. In the oviduct and the airway, cilia cover the entire apical surface of each epithelial cell (Figs. 10.2c–e and 10.3c); by contrast, in the ependymal cells in the ventricle they are clustered and off-centered at a fixed corner

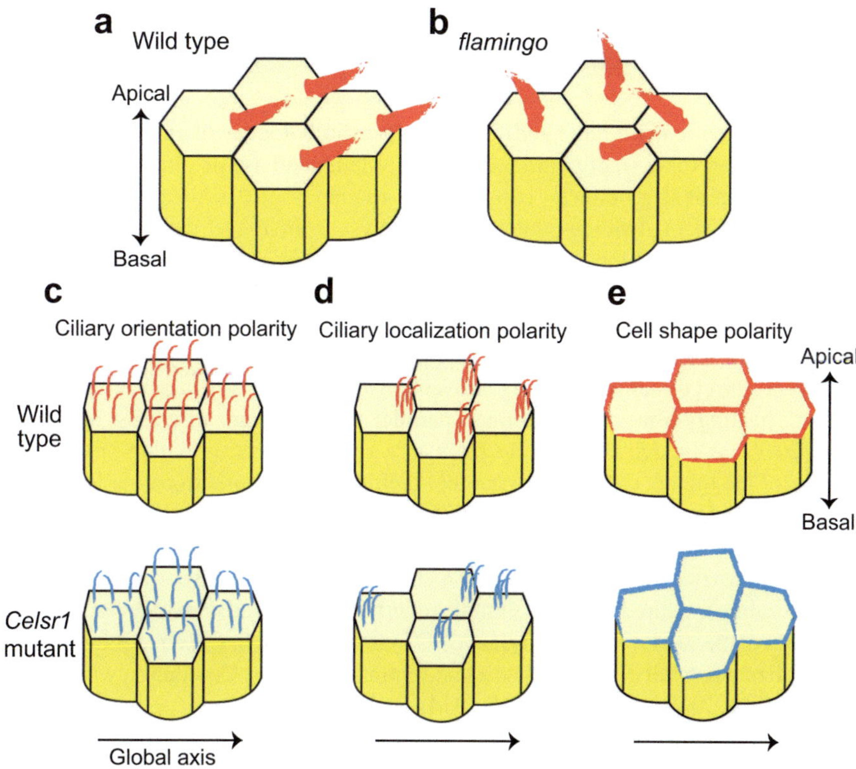

Fig. 10.3 Cellular phenotypes caused by loss of function of 7-TM cadherins. (**a** and **b**) (**a**) Diagrams of epidermal cells of the Drosophila wing. In the wild-type wing, individual epidermal cells localize an assembly of actin filaments at the distal cell vertexes on the apical surface, and produce single wing hairs pointing distally. (**b**) In contrast, *fmi/stan* mutant cells mislocalize actin filaments at the center of the apical surface and the generated hairs are not necessarily aligned along the proximal–distal axis. (**c–e**) Diagrams of multiciliated cells in the oviduct epithelium (**c** and **e**) and ependymal cells in the ventricle (**d**), comparing wild-type (*top*) and a *Celsr1* mutant (*bottom*). *Arrows* indicate the ovary–uterus axis of the oviduct (**c** and **e**) or the direction of cerebrospinal fluid (CSF) circulation in the ventricle (**d**). Phenotypic landmarks of the wild-type (*red*) and *Celsr1* mutant (*blue*) are oriented cilia (**c** and **d**) and the elongated shape of the apical cell surface (**e**). Cilia cover the entire apical surface of each cell in the oviduct (**c**), whereas in ependymal cells they are clustered and off-centered at a fixed vertex (**d**). (**e**) To highlight the elongated shape of the apical cell surface along the axis in the wild-type, cilia are not included in the drawing; similarly, in "c" the phenotype of abnormal cell shape in the *Celsr1* mutant is not drawn in order to emphasize the misoriented cilia (See descriptions of the mutant phenotypes in the text)

along the organ axis (Fig. 10.3d), much like the wing hairs in Drosophila that are generated at the distal cell vertex (Guirao et al. 2010; Hirota et al. 2010; Mirzadeh et al. 2010). Each multiciliated cell forms tens to hundreds of cilia, all of which beat in a defined direction (Fig. 10.2c, d).

10.3 Phenotypes of Stereocilia and Motile Cilia Caused by Loss of Function of Atypical Cadherins

Genetic studies in mice have addressed whether homologues of Fmi/Stan, Ft, and Ds are required for correct polarization of stereocilia and motile cilia. As expected, misorientation of stereocilia is seen in *Celsr1*-mutant mice (Curtin et al. 2003). The oviduct in *Celsr1*-deficient mice also shows a PCP-specific phenotype of multiciliated cells, which is reminiscent of the polarity defect found in the Drosophila wing epidermis (Fig. 10.3b, c). Concretely, multiciliated cells of the knockout (KO) mice are differentiated with no apparent differences in the structure, length, density, and motility of cilia when compared to the wild-type oviduct; yet, the overall directionality of ciliary movements and the ultrastructure of the basal feet of beating cilia are no longer parallel to the ovary-to-uterus axis (Shi et al. 2014). Although the cilia do point in a similar direction within small local regions of the mutant cell, they are neither facing the same direction over the whole cell nor between neighboring cells. Thus Celsr1 is dispensable for ciliogenesis itself but is indispensable for the proper orientation of the cilia. As a result, the directional flow from the ovary to the uterus is not established, and the transportation of ova is demonstrably impaired in the cultured, longitudinally opened oviduct of the KO mice (Shi et al. 2014). The ependymal cells in the ventricle in *Celsr1*-deficient mice mislocalize clustered cilia, which is again reminiscent of the Drosophila wing in *fmi* mutants (Fig. 10.3b, d; Boutin et al. 2014).

Among the three Celsr paralogues in mice, their roles in the development of ciliated epithelia are not necessarily restricted to PCP. Development and planar organization of ependymal cilia are compromised in *Celsr2*-deficient mice, leading to defective circulation of the cerebrospinal fluid (CSF) and hydrocephalus. The hydrocephalus phenotype is accelerated in *Celsr2* and *Celsr3* double mutant embryos due to a more markedly impaired ciliogenesis in ependymal cells, showing contributions of Celsr2 and Celsr3 to ciliogenesis itself (Tissir et al. 2010; Tissir and Goffinet, 2013; Boutin et al. 2014).

Mammalian orthologues of Drosophila Ft and Ds are Fat4 and Dchs1, respectively, and the corresponding KO mice have also been investigated with respect to PCP defects. Loss of Fat4 or Dchs1 results in a subtle misoriented phenotype of stereocilia and various morphological phenotypes in other tissues and organs that are discussed in detail later, but abnormality of motile cilia has not been reported.

10.4 7-TM Cadherins Flamingo/Starry Night and Celsr1 Control Planar Polarity at Selective Plasma Membrane Domains

Not only genetic but also cell biological studies on Fmi/Stan have brought about breakthroughs in our understanding of the molecular mechanisms of PCP. A key finding was its subcellular localization in the Drosophila wing. Along the apicobasal axis of the epithelium in general, Fmi/Stan is localized at adherens junctions where classic E-cadherin is also located. Remarkably, when looking down at the epidermal plane, Fmi/Stan is redistributed selectively to both proximal and distal cell boundaries before the onset of hair formation, which is referred to as the polarized localization of Fmi/Stan or the "zigzag" for short (Fig. 10.4a, d; Usui et al. 1999). Similarly, on many polarizing epithelia in vertebrates, Celsr1 zigzags emerge prior to morphologically visible cell polarization such as directional cilia motility (Fig. 10.4f), providing a striking contrast to the uniform, honeycomb-like distribution of classic E-cadherin (Fig. 10.4e, g; Davies et al. 2005; Devenport and Fuchs 2008; Shi et al. 2014).

In addition to Fmi/Stan and Celsr1, several other evolutionarily conserved PCP regulators localize at the same plasma membrane domains, all of which are now categorized into the "core" group of PCP. It should be noted, however, that except for the 7-TM cadherins, the other core-group members are localized to only one end of the cell or the other along the tissue axis (Fig. 10.4b; Vladar et al. 2009; Goodrich and Strutt 2011; Devenport et al. 2011; Wallingford 2012; Vladar et al. 2012). For example, the four-pass transmembrane protein Van Gogh (Vang; also known as Strabismus; Wolff and Rubin 1998; Taylor et al. 1998) is localized at the cellular proximal end (Bastock et al. 2003), whereas the seven-pass transmembrane protein Frizzled (Fz; Wong and Adler 1993) is at the opposite end, distal in the Drosophila wing (Strutt 2001). The core-group members assemble into a complex, where the homophilic interaction of Fmi/Stan or Celsr1 across the boundary is essential for the junctional recruitment of Vang and Fz, and the resultant complex is an asymmetric one that straddles cell boundaries almost perpendicular to the global axis (Fig. 10.4c; Chen et al. 2008; Strutt and Strutt 2008; Devenport et al. 2011; Vladar et al. 2012). This asymmetric complex recruits downstream polarity effectors to restrict reorganization of the cytoskeleton spatially, leading to the array of distally pointing wing hairs in Drosophila (Adler 2012; Wang et al. 2014b; Lu et al. 2015). Less is known about how exactly Celsr1 on cell boundaries reorients cilia along the organ axis, although it has been observed that cytoskeletal reorganization coincides with and is required for the ciliary polarity formation (Vladar et al. 2012; Boutin et al. 2014). Further proteomic approaches and subsequent assays at organ and/or cell-culture levels may complete the picture downstream of Celsr1, including ciliogenesis and orientation. In fact, previous studies have identified relevant binding partners or downstream molecular events in the context of a role of Fmi/Stan in neuronal cell morphogenesis (Matsubara et al. 2011) and a control of

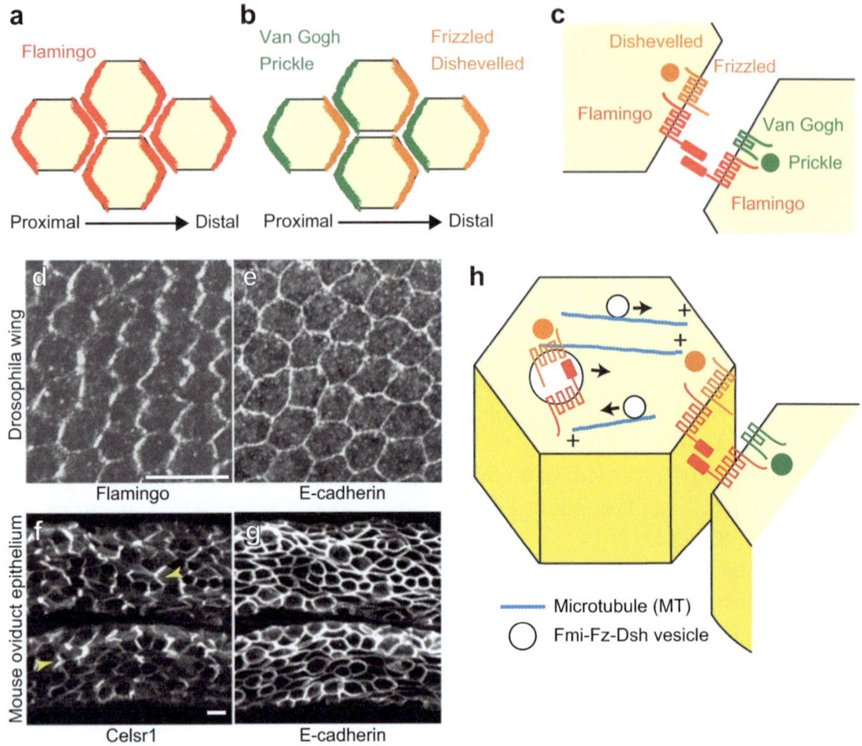

Fig. 10.4 Localization of core-group proteins to selective cell boundaries. All diagrams and confocal images are oriented as in Fig. 10.2. (**a–c**) Diagrams of apical views of wing epidermal cells show the localization of core-group proteins. (**a**) Flamingo (Fmi)/Starry night (Stan). (**b**) Transmembrane proteins, Van Gogh (Vang)/Strabismus (Stbm) and Frizzled (Fz), and undercoat proteins Prickle (Pk) and Dishvelled (Dsh). (**c**) The enlarged diagram shows the disposition of the core group at the cell junction between abutting cells along the axis. (**d–g**) Double staining of the Drosophila wing epidermis for Flamingo (**d**) and E-cadherin (**e**) and the mouse oviduct epithelium for Celsr1 (**f**) and E-cadherin (**g**). The uniform cell-boundary labeling of E-cadherin (**e** and **g**) and the zigzag pattern of Flamingo (**d**) and Celsr1 (*yellow arrowheads* in **f**). (**f** and **g**) Along the ovary–uterus axis of the oviduct, two longitudinal folds can be seen. *Scale bar*: 10 μm. (**e**) Reprinted from "Flamingo, a Seven-Pass Transmembrane Cadherin, Regulates Planar Cell Polarity under the Control of Frizzled" by Usui et al., *Cell*, Volume 98, 585–595, 1999, with permission from Elsevier. (**h**) The model for the polarized transport of Fz–Dsh and Fmi-containing vesicles along the proximal–distal (P–D)-oriented microtubules (MTs). + end-distal MTs are slightly more abundant than + end-proximal MTs, which biases the vesicles towards the distal cell cortex with the help of + end motor proteins (*arrows*). Once Fmi reaches the distal boundary together with Fz and Dsh, Fmi can engage and lock with a counterpart on the proximal membrane of the adjacent cell through its homophilic binding property. Formation of this Fmi–Fmi bridge across the cell boundary could anchor the Fz–Dsh complex at the distal cortex. This input may initiate recruitment of Vang/Stbm–Pk on the opposing proximal cortex by means of mutual exclusion between Dsh and Pk, and by means of ectodomain interactions between Fz and Vang/Stbm

subcellular localization of Celsr1 during mitosis (Shrestha et al. 2015; discussed below).

It is important to note that the polarized localization of Fmi/Stan, Celsr1, and other core-group members is thought to be a functional requirement of PCP signaling and plays an instructive role in polarity establishment (Goodrich and Strutt 2011). Fundamental unanswered questions include how such an asymmetric redistribution of the core group is achieved and why Vang and Fz are localized at opposite cell boundaries in the first place. Across the boundary, it is proposed that a positive feedback plays a critical role in the assembly of the complex; however, it is still unknown how the initial bias is generated. This question and multiple hypotheses are discussed in the end.

10.5 Celsr1 Angles Hair Follicles Along the Body Axis

Because the discoveries that polarization of apical cell appendages is under the control of the core group, the purview of PCP has expanded dramatically to include various directional behaviors of cell populations. One example alluded to earlier is convergent-extension in early embryos, more specifically, the intercalation of cells towards the midline causing elongation of the body along the anterior–posterior (A–P) axis (Gray et al. 2011; Tada and Kai 2012; Wallingford 2012; Tada and Heisenberg 2012). Below, we discuss Celsr1-dependent dynamic behaviors of cells, with the emphasis on large-scale and/or three-dimensional organogenesis in vertebrates.

In addition to unicellular apical appendages, Celsr1 regulates the polarization of multicellular units such as hair follicles, each of which comprises hundreds of proliferative basal epidermal stem cells (Fig. 10.5a; Fuchs 2007). Development of the hair follicle in mice is initiated by budding about 10 cells from the epithelium; and then invaginating nascent hair follicles become anteriorly angled and morphologically polarized through marked changes in cell shape and cytoskeletal reorganization, producing hairs that point to the posterior side (Fig. 10.5a). Individual hair follicles acquire asymmetric patterns of gene expression along the A–P axis, with P-cadherin, ZO-1, and *Shh* mRNA upreglated in anterior cells and NCAM upreglated in posterior cells. All of these events depend on Celsr1 and at least one more member of the core group Vangl2 (Devenport and Fuchs 2008). These manifestations of planar polarization within the basal layer closely resemble what happens in epithelia that give rise to Drosophila ommatidia or sensory bristles (Lu et al. 1999; Gaengel and Mlodzik 2003), which are also polarized multicellular units. Hair follicle initiation coincides with asymmetric redistribution of Celsr1 to anterior and posterior cell boundaries of basal epidermal stem cells and also in the interfollicular epidermis (whereas Vangl2 is restricted to the anterior boundary), suggesting the polarity along the body axis has spread throughout the tissue. The *Celsr1* mutant misaligns hair follicles in the entire body so that it shows a whorled

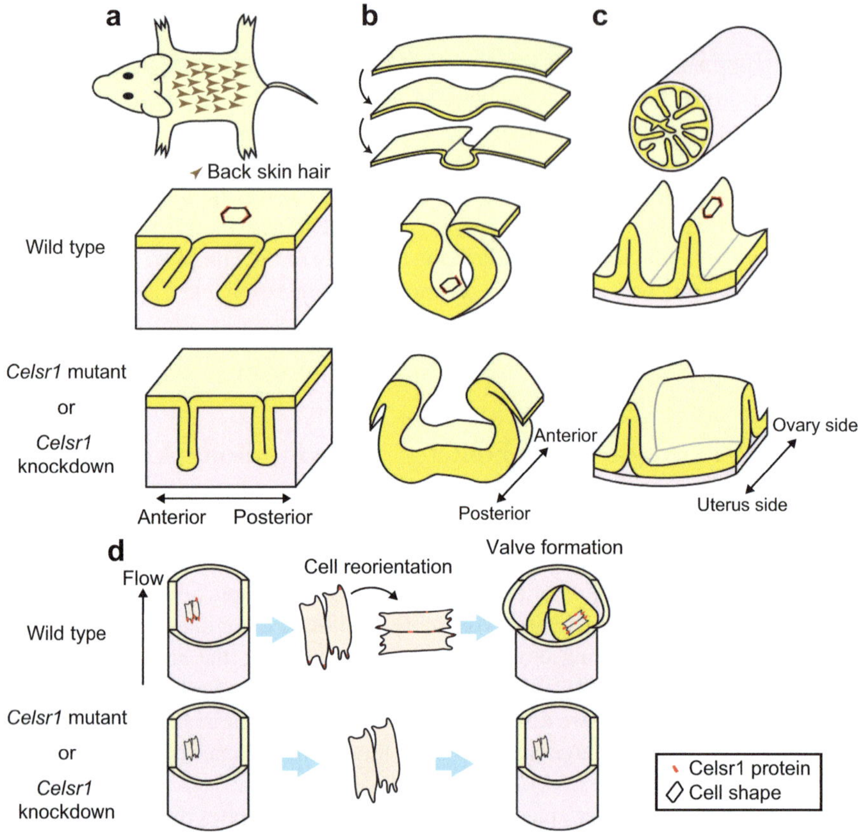

Fig. 10.5 7-TM cadherin Celsr1-dependent three-dimensional organogenesis in vertebrates. Diagrams showing anterior angling of hair follicles in mouse epidermis (**a**), neural tube closure in chick and mice (**b**), longitudinal epithelial folding in the mouse oviduct (**c**), and formation of a lymph valve in mice (**d**). Except for illustrations of overall organs in "a"–"c", wild-type structures are schematically drawn at the top and those in *Celsr1* mutant or knockdown animals are at the *bottom*. Apical cell shape of epithelial cells (*black polygons*) and localized Celsr1 protein (*red*) are illustrated in individual tissues. (**c**) The oviduct consists of a simple columnar epithelium and a thin stromal layer surrounded by smooth muscle layers. See more details in the text

hair pattern and also loss of the asymmetry of gene expression within individual follicles (Devenport and Fuchs 2008; Ravni et al. 2009).

In addition to the above demonstration of the role of Celsr1 in global asymmetric morphogenesis of hair follicles, studies of basal epidermal stem cells are answering an important question at the molecular level: how PCP is maintained in highly proliferative tissues (Devenport et al. 2011; Shrestha et al. 2015). During mitosis when the cells are rounded up, asymmetrically distributed Celsr1 risks mislocalization or unequal inheritance, which could lead to profound perturbations of the long-range propagation of polarity. How is the polarized localization regained after cell division? Elegantly, Celsr1 proteins are internalized into endosomes during mitosis; and following mitosis, they are recycled back to the cell surface, where polarized localization is re-established with the help of the polarity of adjacent interphase cells (Devenport et al. 2011). This mechanism explains how extensive polarity disruption is minimized within tissues that must maintain function during rapid growth and/or turnover. A further proteomic approach shows that a key mitotic kinase, Polo-like kinase 1, phosphorylates conserved serine/threonine residues in the carboxyl tail of Celsr1, which promotes Celsr1 endocytosis during mitosis (Shrestha et al. 2015).

10.6 Celsr1 Orchestrates Neural Tube Formation by Way of Spatiotemporal Control of Actomyosin Contraction

The neural tube is a primordium of the brain and the spinal cord. Converting a wide flat neuroepithelial sheet—the neural plate—into the neural tube is a challenging feat (Fig. 10.5b). This is evident from clinical data, where common human birth defects are those affecting neural tube formation, including failure of neural tube closure, and from molecular genetic studies, where a variety of gene products, including Celsr1 and other PCP regulators, has been shown to be required for this tube formation in animal models and humans (Curtin et al. 2003; Doudney and Stanier 2005; Robinson et al. 2012; Allache et al. 2012; Juriloff and Harris 2012; Wallingford et al. 2013; Yamaguchi and Miura 2013). When apical cortical actin constricts locally within the epithelium, an indent forms in the sheet (St Johnston and Sanson 2011). However, if cells do not converge towards the midline, the sheet is incompletely rolled up along the A–P axis, resulting in a failure to close and the consequential inability to generate a tube. (Fig. 10.5b). How Celsr1 orchestrates other proteins and coordinates the two morphogenetic processes, apical constriction and midline conversion, has been best studied at the molecular level in chick embryos (Nishimura et al. 2012; Nishimura 2014).

Again, at the nexus of the known molecular machinery, there exists polarized localization of Celsr1. In this developmental context, Ceslr1 is enriched at anterior and posterior cell boundaries that are orthogonal to the A–P body axis in the bending neuroepithelium, and it cooperates with Dishevelled (one of the PCP core group) and other proteins to upregulate Rho kinase activity, causing shrinking of those boundaries through localized cytoskeletal reorganization (Fig. 10.5b). This planar-polarized acotomyosin contraction promotes midline convergence of neuroepithelium cells, leading to the bending of the neural plate along the A–P body axis. Knockdown of *Celsr1* and any of the other component genes of the machinery causes neural tube closure defects where the bending of the neural plate is insufficient (Fig. 10.5b). The role of Celsr1 in spatiotemporal regulation of cytoskeletal components such as phosphorylated myosin light chain and F-actin was also reported in otic placode invagination during inner ear morphogenesis (Sai et al. 2014).

10.7 Celsr1 Links Shaping of Individual Cells with Folding of Epithelial Sheets in the Oviduct

With respect to three-dimensional organ architecture, the oviduct is much more than a simple tube. In mice, the epithelium forms about 20 straight folds running in parallel with the ovary-to-uterus axis (Figs. 10.4g and 10.5c; Agduhr 1927). These folds are thought to increase the epithelial surface area and ensure the oocyte transport from the ovary to uterus, guided by coordinated ciliary beatings. In addition to this large-scale polarized structure, the apical surface of individual epithelial cells is elongated along the organ axis (Fig. 10.3e), and Celsr1 is localized at the shortened cellular junctions that are perpendicular to the organ axis (Figs. 10.4f, and 10.5c). Thus the wild-type oviduct generates polarity at multiple levels of single cell shape, directional ciliary beating on the plane, and three-dimensional tissue morphology; and all of these polarities are disrupted by loss of Celsr1 function (Shi et al. 2014). Compared to the wild-type, Celsr1-deficient cells are less elongated and less oriented along the organ axis (Fig. 10.3e), and the morphology of epithelial folds is disorganized, with the subsequent generation of ectopic and abnormal branches that run in randomized directions (Fig. 10.5c).

To address the cause-and-effect relationship between the cellular shape and the fold morphology, mosaic analyses were performed in mice, in which mutant clones of various size were generated in the wild-type background, and it was examined how the two features were affected. The results show that the cell shape is intrinsically regulated through Celsr1 and not by extrinsic cues; on the basis of this finding, it is hypothesized that the cell shape is primarily regulated by Celsr1 and as a consequence the epithelial folds are aligned. It requires future studies to unravel how Celsr1 drives polarization of cell shape, whether the cell elongation along the organ axis produces any anisotropic force, and whether such a force, if produced, leads to the parallel alignment of epithelial folds (Heisenberg and Bellaïche 2013; Sugimura and Ishihara 2013). The impact of this study is not limited to organogenesis of the oviduct, because folding of cellular sheets is observed throughout the animal body, including the sulcus in the brain (Striedter et al. 2015). It may be intriguing to re-examine directional relationships between cell shape and those folds.

Tissue-level polarity inside tubular organs is also seen in veins and lymphatic vessels, where one-way flow of blood and lymph is ensured by polarized multicellular structures, luminal valves (Fig. 10.5d). Morphogenesis of the lymphatic valve is initiated by Celsr1-dependent reorientation of endothelial cells perpendicular to the longitudinal organ axis, followed by migration of the cells into the luminal side to form a primordium of the valve (Shi et al. 2013; Tatin et al. 2013). Celsr1 and Vangl2 are localized to membrane protrusions and also recruited to cell–cell contacts during the reorientation of the endothelial cells. What the signal is for the reorientation, how Celsr1 drives this cell orientation, and what the link between localized Celsr1 and the cell migration is should be elucidated in the future.

This chapter has focused on just a handful of developmental processes where model organisms have provided some detailed knowledge of the molecular mechanisms, however, 7-TM cadherins play important roles in other developmental contexts in Drosophila and vertebrates that are not discussed in this chapter, including neuronal wiring (Takeichi 2007; Shima et al. 2007; Hakeda-Suzuki et al. 2011; Matsubara et al. 2011; Hirano and Takeichi 2012; Schwabe et al. 2013; Tissir and Goffinet 2013; Wang et al. 2014a), pancreatic β cell differentiation (Cortijo et al. 2012; Wang et al. 2014a), and maintenance of hematopoietic stem cells (Sugimura et al. 2012). We now turn our attention to Fat and Dachsous cadherins.

10.8 Functional Relationships Between 7-TM Cadherins and Fat and Dachsous Cadherins

The most outstanding feature of PCP would be its high fidelity to global axes, which is represented not only by the unidirectional beating of all cilia within individual cells, but also by the coordination between cells entirely along the tracheal or oviduct axis. One theoretical framework proposes that two distinct mechanisms can contribute to this long-range propagation of polarity (Abley et al. 2013): one is coupling between adjacent cells (designated as "cell–cell coupling"), generating local alignment, and the other is a concentration gradient of a signaling molecule across the tissue (designated as "tissue gradients"). Experimental data suggest that cell–cell coupling operates through the plasma membrane-spanning complexes comprising 7-TM cadherins and other core group members, whereas Dachsous that binds to Fat is implicated in the mechanism of tissue gradients as suggested by its graded expression along different axes in various tissues in Drosophila (Fig. 10.6a; Yang et al. 2002; Ma et al. 2003).

The functional relationship between the core group and the Ft–Ds system in PCP has been a target of intense investigations, and apparently there are numerous mechanistic variations among different tissues or even in the same tissue depending on the readouts (Goodrich and Strutt 2011; Lawrence and Casal 2013; Matis and Axelrod 2013; Carvajal-Gonzalez and Mlodzik 2014; Olofsson et al. 2014; Ayukawa et al. 2014; Merkel et al. 2014). In some tissues, the two systems work independently, whereas they interact in other tissues, as discussed at the end of this chapter. Much less is known about the relationship in vertebrate development, and one example in the context of collective cell migration is discussed below, where both CELSRs and the Fat4–Dchs1 system participate, but their roles are segregated.

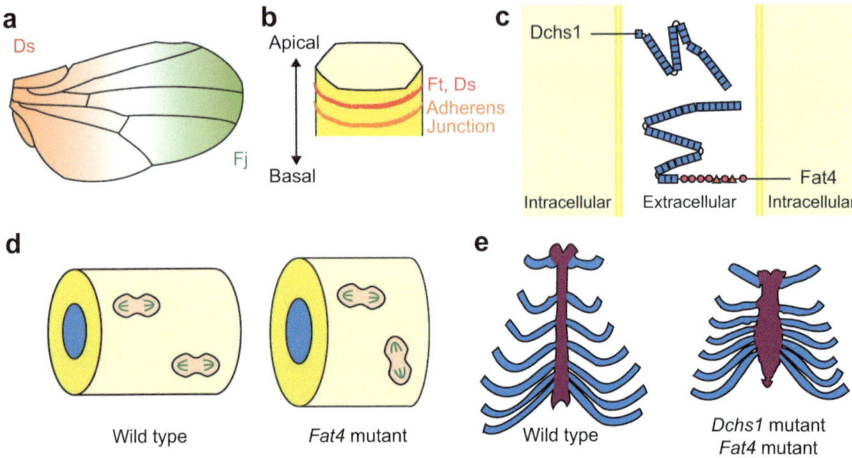

Fig. 10.6 Fat and Dachsous cadherins: expression at the tissue level, subcellular localization, protein conformations, and mutant phenotypes. (**a**) Dachsous (Ds, *orange*) and Four-jointed (Fj, *green*) are expressed in counter-gradients in the Drosophila wing. Fat (Ft) is uniformly expressed and not included in this diagram. (**b**) Ft and Ds are positioned more apically than the adherens junction where Fmi and E-cadherin are localized. (**c**) The model of conformations of Fat4 and Dchs1 ectodomains at the intercellular junction. (**d**) Fat4 is required for oriented cell division (OCD) in the kidney, and misorientation of cell division angles in the kidney in the mutant mice renders collecting ducts broader and shorter. (**e**) Illustrations of skeletal preparations of sternums (*purple*) of the wild-type and *Fat4* or *Dchs1* mutants. Sternums are wider and shorter in the mutants

10.9 Fat and Dachsous Cadherins: Control of the Heterophilic Binding by Ectodomain Phosphorylation and Protein Conformation at the Cell Interface

In Drosophila, Fat and Dachsous form a trio with Golgi kinase Four-jointed (Fj) and contribute to tissue patterning across the axis. Intriguingly, Fj phosphorylates the ectodomains of Ds and Ft to modulate their heterophilic bindings (Ishikawa et al. 2008; Sopko et al. 2009; Feng and Irvine 2009; Brittle et al. 2010; Simon et al. 2010). As discussed in the theoretical framework of PCP, Ds is thought to belong to the mechanism of "tissue gradients"; so is Fj, which is expressed in a counter-gradient fashion to Ds in multiple tissues (Fig. 10.6a; Villano and Katz 1995; Zeidler et al. 2000; Yang et al. 2002; Ma et al. 2003). It is this complementary expression pattern that controls Ft–Ds binding at cell boundaries differentially across the wing (Thomas and Strutt 2012; Mani et al. 2013; Jolly et al. 2014; Hale et al. 2015). Complementary expression patterns of *Dchs1* and the *Fj* homologue *Fjx1* and those of *Dchs1* and *Fat4* are found in mice (Rock et al. 2005) and their relevance to neuronal migration is described (Zakaria et al. 2014; and see below).

Along the apicobasal cell axis, Ft and Ds are localized in a cell–cell contact area positioned more apically than the adherens junction (AJ) in the Drosophila wing primordium (Fig. 10.6b; Ma et al. 2003); and Fat4 and Dchs1 show a similar localization in neural progenitor cells of the mouse cortex (Ishiuchi et al. 2009). The heterophilic binding between Fat4 and Dchs1 is shown by expressing the full-length molecules in cultured cells (Ishiuchi et al. 2009) and by surface plasmon resonance analysis using their entire ectodomains (Tsukasaki et al. 2014). At the organ level, the heterophilic binding is imaged in developing kidney, which consists of three cell layers: the epithelial ureteric bud, which forms the collecting ducts; the nephron progenitors (also known as the cap or condensing mesenchyme); and the stromal mesenchyme. For normal kidney development, it is necessary to balance self-renewal and a mesenchymal-to-epithelial transition of the progenitors. This control is achieved by signaling between Dchs1 in the progenitors and Fat4 in the stromal cells, and at least Dchs1 proteins are concentrated at the interface of these two layers (Bagherie-Lachidan et al. 2015; Mao et al. 2015).

Electron-microscopic observations verify how the large ectodomains of Fat4 and Dchs1, about 4500 and 3000 amino acids in length, respectively (Fig. 10.1) are fitted into limited intercellular spaces (Tsukasaki et al. 2014). Fat4 and Dchs1 ectodomains assume kinked conformations, in contrast to the linear configuration of the E-cadherin ectodomain (Fig. 10.6c). It was found that certain linker regions of Fat4 and Dchs1, which connect ECs (extracellular cadherin domains), lack the Ca^{2+}-binding motifs that are conserved in the E-cadherin ectodomain and are important for the linear conformation of E-cadherin (Tsukasaki et al. 2014). Due to these differences in amino acid sequences, the ectodomains of Fat4 and Dchs1 bend, thereby enabling them to fit in the confined intercellular spaces. Expansion of this structural approach may provide an explanation for how phosphorylation of the Ds and Ft ectodomains by Fj modulates their heterophilic binding.

10.10 Loss of Fat4 or Dchs1 Function Causes Various Morphogenetic Phenotypes

Fat4- or *Dchs1*-deficient mice show morphological defects in various organs, including size reductions of internal organs (intestine, lung, and kidney; Saburi et al. 2008; Mao et al. 2011). Which of these phenotypes, apart from the subtle misalignment of stereocilia of hair cells in cochlea (Saburi et al. 2008; Mao et al. 2011), can be explained by our current knowledge of PCP? The core group has been implicated in the regulation of oriented cell division (OCD) in various organs or embryos in Drosophila and vertebrates (Lu et al. 1999; Gong et al. 2004; Devenport 2014). The requirement of Fat4 for OCD was shown by misorientation of cell division angles in the kidney in KO mice, which renders collecting ducts broader and shorter, leading to cystic kidney disease (Fig. 10.6d; Saburi et al. 2008). In both the *Dchs1* mutant and the *Fat4* mutant mice, such a “compressed” morphological defect is also seen in the sternum (Fig. 10.6e; Mao et al. 2011), therefore elongation of this organ also might depend on Fat4–Dchs1 mediated OCD and/or cell intercalation.

Although Fat4 shows the highest sequence similarity to Drosophila Fat, genes of three other homologues are present in the mammalian genome, and the respective Fat proteins act both synergistically and antagonistically to affect morphogenesis of multiple organs including neural tube closure (Saburi et al. 2012; Badouel et al. 2015). Similarly, a partial redundancy between Dchs1 and its paralogue Dchs2 was shown (Bagherie-Lachidan et al. 2015). Molecular machineries downstream of Fat cadherins have been studied by hunting for binding proteins, and Fat1 and Fat4 bind to different sets of actin regulating and junctional proteins (Tanoue and Takeichi 2004; Ishiuchi et al. 2009; Sadeqzadeh et al. 2014; Badouel et al. 2015). Future studies will address the question of whether each Fat cadherin executes organogenesis in different contexts with a common set of binding partners (such as classic cadherins with catenins) or context-specific binding partners, or a combination of both. Drosophila Ds intracellular domain interacts with an unconventional myosin Dachs (D), and this binding is important for shaping the dorsal thorax epithelium (Bosveld et al. 2012). Although intracellular binders to vertebrate Dchs cadherins have not been reported, zebrafish Dchs1b regulates the actin and microtubule cytoskeleton, possibly independent of Fat in the unanticipated context of the single-celled embryo (Li-Villarreal et al. 2015).

10.11 "Intersection" of 7-TM Cadherins and the Fat4–Dchs1 System in the Guidance of Cell Migration

In Drosophila pupal development, one context of collective epidermal migration is dependent on Ds (Bischoff 2012), and the directional information is provided by an imbalance of the Ds level between the migrating cells (unpublished data of M.A. and T.U.). In vertebrate embryonic brains, the Fat4–Dchs1 system and 7-TM cadherins are shown to control migration of facial branchiomotor (FBM) neurons in the neuroepithelium (Fig. 10.7), but the roles of the two subfamilies are segregated along orthogonal axes: the A–P (anterior–posterior) axis and the mediolateral axis. FBM neurons arise within one compartment, rhombomere 4 (r4), of the hindbrain, undergo posterior migration towards r6 and then turn laterally in r6 (Fig. 10.7c; Tissir and Goffinet 2013). 7-TM cadherins control the former posterior migration, whereas the Fat4–Dchs1 system controls the latter. How each of the 7-TM cadherins contributes to the posterior migration is slightly different between Celsr1–Celsr3 and in different species (zebrafish and mice). In zebrafish, Celsr2 and Fz3a (one of the core group) act in the surrounding neuroepithelium to prevent the integration of the neurons into the neuroepithelium, thus restricting them to the correct A–P path (Wada et al. 2006; Wada and Okamoto 2009). A similar role of Celsr2 outside the neurons was reported in mice

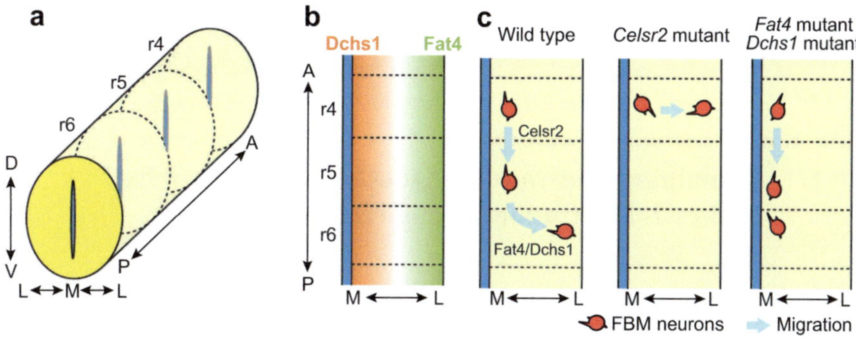

Fig. 10.7 Distinct roles of Celsr2 and Fat4 and Dchs1 in migration of facial branchiomotor neurons in the hindbrain. Diagrams of the hindbrain (**a**), of Fat4 and Dchs1 expression (**b**), and migration patterns of facial branchiomotor (FBM) neurons (**c**). The ventricular zones are highlighted in blue. (**a**) A portion of the hindbrain between rhombomere 4 (r4) and r6. The anterior–posterior (A–P) axis, the mediolateral (M–L) axis, and the dorsal–ventral (D–V) axis are indicated. (**b** and **c**) Dorsal views of lateral halves of longitudinal sections of the hindbrain. (**b**) Expression of Dchs1 is highest medially, whereas Fat4 expression is highest laterally. (**c**) Migration patterns of FBM neurons in the hindbrain neuroepithelium in the wild-type (WT) and the mutants are indicated. FBM neurons arise within one compartment, rhombomere 4 (r4), undergo posterior migrations towards r6 and then turn laterally in r5 and r6. The posterior migration depends on Celsr2, whereas the lateral migration depends on Fat4 and Dchs1

(Qu et al. 2010; Tissir and Goffinet 2013). However, the directional cue (anterior vs. posterior) is unknown.

In *Fat4* or *Dchs1* KO mice, the neurons fail to migrate laterally and stay medially positioned (Fig. 10.7c; Zakaria et al. 2014). A cell-level analysis showed that the neurons become polarized along their migratory path in normal development, on the basis of the localization of the Golgi apparatus relative to the nucleus and elongated cell shape, and that this polarization requires Fat4 and Dchs1. Curiously, at a specific stage of the neuronal migration, Dchs1 and Fat4 are expressed in opposing gradients along the mediolateral axis in the neuroepithelium, in such a way that the neurons migrate away from the medial region of high Dchs1 and towards the lateral region of high Fat4 (Fig. 10.7b). These expression patterns are reminiscent of the proximal–distal gradient of Ds in the Drosophila wing (Ma et al. 2003; Matakatsu and Blair 2004). Dchs1 and Fat4 are each required both within the migrating neurons and the neuroepithelial cells through which they migrate. Based on these results, along with the results of mosaic analyses, it is thought that the polarity of the neuron is established through the interpretation of the long-range gradients of Dchs1 and Fat4 across the epithelium and also by local communication between the neurons within the migration stream (Zakaria et al. 2014).

Cellular misalignment and defective migration are also observed in valvular interstitial cells of *Dchs1* mutant mice where cardiac valve formation is abnormal, and mutations in human *Dchs1* cause mitral valve prolapse, a common cardiac valve disease (Durst et al. 2015). Possibly, in multiple contexts of organogenesis, the Fat4–Dchs1 system plays a critical role in cell migration by providing directional cues.

10.12 Remaining Outstanding Questions Include What Makes 7-TM Cadherin Polarized

As discussed above, essentially all studies on 7-TM cadherins start with those proteins that already occupy selective cell boundaries along the tissue axis (Figs. 10.4d–g and 10.5a–c). In the microscopic sense, the assembly of the 7-TM cadherin-containing complex across one cell boundary could be made by a positive feedback mechanism; however, such a cell-by-cell polarity does not transform into the zigzag pattern throughout the tissue without an initial bias along the tissue axis. One possible cell-biological mechanism generating such a bias is polarized transport to particular cell boundaries (Fig. 10.4h). This hypothesis has been addressed experimentally by quantitative in vivo imaging and conventional as well as immunoelectron microscopy, using the Drosophila wing and its primordium (wing imaginal disc; Shimada et al. 2006; Harumoto et al. 2010; Matis et al. 2014; Olofsson et al. 2014). A current composite picture is as follows: (1) the Ft–Ds–Fj system interprets the information of the proximal–distal axis (see the explanation of "tissue gradients" above). (2) This system creates a subtle asymmetry of microtubules in the apical region of the cell, orienting along the tissue axis (planar MTs); that is, + end-distal MTs become slightly more abundant than + end-proximal MTs. (3) This asymmetrical MT organization allows + end motor proteins to transport vesicles containing 7-TM cadherins and other core-group members (Fz and Dsh) better to the distal cell boundary (Fig. 10.4h; see continued explanations in Fig. 10.4h legend). (4) This mechanism may contribute to establishing an initial bias in the proximal region where the imbalance of the Ds level is large (Matakatsu and Blair 2004; Harumoto et al. 2010), not necessarily throughout the wing. Open questions related to this hypothesis include:

- The vesicular transport has been imaged in pupal wings; however, the asymmetry of the core group emerges earlier, in growing imaginal discs in larvae (Classen et al. 2005; Aigouy et al. 2010; Sagner et al. 2012). Imaging vesicular transport in the growing disc is technically challenging (explained below).
- Although epithelia generally develop apicobasal MTs, planar MTs may not always exist in all tissues that are acquiring PCP (Devenport 2014). How does the asymmetry of 7-TM cadherins arise in such tissues?
- In tissues where the core group and the Ft–Ds system do not crosstalk, how is the asymmetrical MT organization generated? In such tissues, is the polarity generated by mechanisms other than the polarized transport?

Other approaches to investigating the polarity-generating mechanism include studies of how the localized domain is aligned between cells. The Drosophila orthologue of Wnt1, Wingless (Wg), has been long considered as an attractive candidate of the tissue-level polarity cue. One of the reasons for this is that Wg is one of the well-studied morphogens that are secreted from organizing centers in Drosophila, and it regulates the expression of *ds* and *fj* in opposite ways in the wing disc (*ds* high in the hinge and *fj* high in the wing blade; Cho and Irvine 2004; Zecca and Struhl 2010). One recent model proposes that gradients of Wg and Wnt4a are the long-range cues, which directly modulate Fz–PCP signaling (Wu et al. 2013). In contrast, an alternative model proposes that the Wg gradient is unlikely to act as a cue itself (Sagner et al. 2012). Data behind this model highlight that gradients of Wg and other morphogens do not play an instructive role in PCP. Instead, each organizing center reorients cells when the disc is small, and this "fixed" polarity is maintained as the tissue grows, and hence no longer depends on long-range biasing cues such as gradients throughout subsequent tissue growth (Aigouy et al. 2010; Sagner et al. 2012). Thus, the exact mechanisms by which the morphogens (or the organizing centers) regulate Fmi/Stan asymmetry early during tissue growth remain to be elucidated.

To address these remaining questions or to devise new models, it is preferable to perform in vivo time-lapse imaging of the growing tissue where the polarized pattern is emerging *de novo*. However, this approach often runs up against technical challenges due to folded smaller tissues that are located deep inside the body of younger animals, such as the Drosophila imaginal discs in immature larvae. One solution is the development of easy-access ex vivo tissue/organ or whole embryo cultures, such as the skin culture that provides a clue about a time window of the hypothetical directional cue and Xenopus embryos (Devenport and Fuchs 2008; Butler and Wallingford 2015). Another more drastic solution is the innovation of reconstitution systems, in which tissues are reconstituted from dissociated single cells, and they grow and acquire at least a local polarity. With this system in hand, we may be able to search for conditions to reproduce long-range polarity through genetic, chemical, and/or physical manipulations. This approach is promising, as illustrated by the airway epithelia that are reconstituted in primary cultures (Vladar et al. 2012; Vladar et al. 2015), and it is being applied for many other tissues/organs. With these new techniques, a detailed mechanistic understanding of how polarity is established may soon be within reach.

Acknowledgments This work was supported by a CREST grant and MEXT grands (Kakenhi) to T.U. and T.F,, and by a grant of The Mitsubishi Foundation and Takeda Science Foundation to T.U and a grant from NIBB to T.F. D.S. and M. A. were Research Fellows of the JSPS. Figure 10.4e is reprinted from "Flamingo, a Seven-Pass Transmembrane Cadherin, Regulates Planar Cell Polarity under the Control of Frizzled" by Usui et al., *Cell*, Volume 98, 585–595, 1999, with permission from Elsevier. We thank J. A. Hejna very much for his constructive suggestions and polishing the manuscript.

References

Abley K, De Reuille PB, Strutt D et al (2013) An intracellular partitioning-based framework for tissue cell polarity in plants and animals. Development 140:2061–2074. doi:10.1242/dev.062984

Adler PN (2012) The frizzled/stan pathway and planar cell polarity in the Drosophila wing. Curr Top Dev Biol 101:1–31. doi:10.1016/B978-0-12-394592-1.00001-6

Agduhr E (1927) Studies on the structure and development of the bursa ovarica and the tuba uterina in the mouse. Acta Zool 8:1–133. doi:10.1111/j.1463-6395.1927.tb00649.x

Aigouy B, Farhadifar R, Staple DB et al (2010) Cell flow reorients the axis of planar polarity in the wing epithelium of Drosophila. Cell 142:773–786. doi:10.1016/j.cell.2010.07.042

Allache R, De Marco P, Merello E et al (2012) Role of the planar cell polarity gene CELSR1 in neural tube defects and caudal agenesis. Birth Defects Res A Clin Mol Teratol 94:176–181. doi:10.1002/bdra.23002

Assheton R (1896) Notes on the ciliation of the ectoderm of the amphibian embryo.

Ayukawa T, Akiyama M, Mummery-Widmer JL et al (2014) Dachsous-dependent asymmetric localization of spiny-legs determines planar cell polarity orientation in Drosophila. Cell Rep 8:610–621. doi:10.1016/j.celrep.2014.06.009

Badouel C, Zander M, Liscio N et al (2015) Fat1 interacts with Fat4 to regulate neural tube closure, neural progenitor proliferation and apical constriction during mouse brain development. Development 142:2781–2791. doi:10.1242/dev.123539

Bagherie-Lachidan M, Reginensi A, Zaveri HP et al (2015) Stromal Fat4 acts non-autonomously with Dachsous1/2 to restrict the nephron progenitor pool. Development 142:2564–2573. doi:10.1242/dev.122648

Bastock R, Strutt H, Strutt D (2003) Strabismus is asymmetrically localised and binds to Prickle and Dishevelled during Drosophila planar polarity patterning. Development 130:3007–3014

Bischoff M (2012) Lamellipodia-based migrations of larval epithelial cells are required for normal closure of the adult epidermis of Drosophila. Dev Biol 363:179–190. doi:10.1016/j.ydbio.2011.12.033

Boisvieuxulrich E, Laine MC, Sandoz D (1985) The orientation of ciliary basal bodies in quail oviduct is related to the ciliary beating cycle commencement. Biol Cell 55:147–150

Bosveld F, Bonnet I, Guirao B et al (2012) Mechanical control of morphogenesis by Fat/Dachsous/Four-jointed planar cell polarity pathway. Science 336:724–727. doi:10.1126/science.1221071

Boutin C, Labedan P, Dimidschstein J et al (2014) A dual role for planar cell polarity genes in ciliated cells. Proc Natl Acad Sci U S A 111:E3129–E3138. doi:10.1073/pnas.1404988111

Brittle AL, Repiso A, Casal J et al (2010) Four-jointed modulates growth and planar polarity by reducing the affinity of dachsous for fat. Curr Biol 20:803–810. doi:10.1016/j.cub.2010.03.056

Brooks ER, Wallingford JB (2014) Multiciliated cells. Curr Biol 24:R973–R982. doi:10.1016/j.cub.2014.08.047

Butler MT, Wallingford JB (2015) Control of vertebrate core PCP protein localization and dynamics by Prickle2. Development. doi:10.1242/dev.121384

Carvajal-Gonzalez JM, Mlodzik M (2014) Mechanisms of planar cell polarity establishment in Drosophila. F1000Prime Rep 6:98. doi: 10.12703/P6-98

Chae J, Kim MJ, Goo JH et al (1999) The Drosophila tissue polarity gene starry night encodes a member of the protocadherin family. Development 126:5421–5429

Chen W-S, Antic D, Matis M et al (2008) Asymmetric Homotypic Interactions of the Atypical Cadherin Flamingo Mediate Intercellular Polarity Signaling. Cell 133:1093–1105. doi:10.1016/j.cell.2008.04.048

Cho E, Irvine KD (2004) Action of fat, four-jointed, dachsous and dachs in distal-to-proximal wing signaling. Development 131:4489–4500. doi:10.1242/dev.01315

Clark HF, Brentrup D, Schneitz K et al (1995) Dachsous encodes a member of the cadherin superfamily that controls imaginal disc morphogenesis in Drosophila. Genes Dev 9:1530–1542. doi:10.1101/gad.9.12.1530

Classen AK, Anderson KI, Marois E, Eaton S (2005) Hexagonal packing of Drosophila wing epithelial cells by the planar cell polarity pathway. Dev Cell 9:805–817. doi:10.1016/j.devcel.2005.10.016
Cortijo C, Gouzi M, Tissir F, Grapin-Botton A (2012) Planar cell polarity controls pancreatic beta cell differentiation and glucose homeostasis. Cell Rep 2:1593–1606. doi:10.1016/j.celrep.2012.10.016
Curtin JA, Quint E, Tsipouri V et al (2003) Mutation of Celsr1 disrupts planar polarity of inner ear hair cells and causes severe neural tube defects in the mouse. Curr Biol 13:1129–1133. doi:10.1016/s0960-9822(03)00374-9
Davies A, Formstone C, Mason I, Lewis J (2005) Planar polarity of hair cells in the chick inner ear is correlated with polarized distribution of c-flamingo-1 protein. Dev Dyn 233:998–1005. doi:10.1002/dvdy.20376
Del Bigio MR (1995) The ependyma: a protective barrier between brain and cerebrospinal fluid. Glia 14:1–13. doi:10.1002/glia.440140102
Devenport D (2014) Cell biology in development: The cell biology of planar cell polarity. J Cell Biol 207:171–179. doi:10.1083/jcb.201408039
Devenport D, Fuchs E (2008) Planar polarization in embryonic epidermis orchestrates global asymmetric morphogenesis of hair follicles. Nat Cell Biol 10:1257–1268. doi:10.1038/ncb1784
Devenport D, Oristian D, Heller E, Fuchs E (2011) Mitotic internalization of planar cell polarity proteins preserves tissue polarity. Nat Cell Biol 13:893–902. doi:10.1038/ncb2284
Doudney K, Stanier P (2005) Epithelial cell polarity genes are required for neural tube closure. Am J Med Genet C: Semin Med Genet 135C:42–47. doi:10.1002/ajmg.c.30052
Durst R, Sauls K, Peal DS et al (2015) Mutations in DCHS1 cause mitral valve prolapse. Nature. doi:10.1038/nature14670
Feng Y, Irvine KD (2009) Processing and phosphorylation of the Fat receptor. Proc Natl Acad Sci U S A 106:11989–11994. doi:10.1073/pnas.0811540106
Fuchs E (2007) Scratching the surface of skin development. Nature 445:834–842. doi:10.1038/nature05659
Gaengel K, Mlodzik M (2003) Egfr signaling regulates ommatidial rotation and cell motility in the Drosophila eye via MAPK/Pnt signaling and the Ras effector Canoe/AF6. Development 130:5413–5423. doi:10.1242/dev.00759
Gong Y, Mo C, Fraser SE (2004) Planar cell polarity signalling controls cell division orientation during zebrafish gastrulation. Nature 430:689–693. doi:10.1038/nature02796
Goodrich LV, Strutt D (2011) Principles of planar polarity in animal development. Development 138:1877–1892. doi:10.1242/dev.054080
Gray RS, Roszko I, Solnica-Krezel L (2011) Planar cell polarity: coordinating morphogenetic cell behaviors with embryonic polarity. Dev Cell 21:120–133. doi:10.1016/j.devcel.2011.06.011
Guirao B, Meunier A, Mortaud S et al (2010) Coupling between hydrodynamic forces and planar cell polarity orients mammalian motile cilia. Nat Cell Biol 12:341–350. doi:10.1038/ncb2040
Hadjantonakis AK, Sheward WJ, Harmar AJ et al (1997) Celsr1, a neural-specific gene encoding an unusual seven-pass transmembrane receptor, maps to mouse chromosome 15 and human chromosome 22qter. Genomics 45:97–104. doi:10.1006/geno.1997.4892
Hakeda-Suzuki S, Berger-Müller S, Tomasi T et al (2011) Golden Goal collaborates with Flamingo in conferring synaptic-layer specificity in the visual system. Nat Neurosci 14:314–323. doi:10.1038/nn.2756
Hale R, Brittle AL, Fisher KH et al (2015) Cellular interpretation of the long-range gradient of Four-jointed activity in the Drosophila wing. Elife. doi:10.7554/eLife.05789
Harumoto T, Ito M, Shimada Y et al (2010) Atypical cadherins Dachsous and Fat control dynamics of noncentrosomal microtubules in planar cell polarity. Dev Cell 19:389–401. doi:10.1016/j.devcel.2010.08.004
Heisenberg CP, Bellaïche Y (2013) Forces in Tissue Morphogenesis and Patterning. Cell 153:948–962. doi:10.1016/j.cell.2013.05.008

Hirano S, Takeichi M (2012) Cadherins in brain morphogenesis and wiring. Physiol Rev 92:597–634. doi:10.1152/physrev.00014.2011

Hirota Y, Meunier A, Huang S et al (2010) Planar polarity of multiciliated ependymal cells involves the anterior migration of basal bodies regulated by non-muscle myosin II. Development 137:3037–3046. doi:10.1242/dev.050120

Hoeng JC, Höng JC, Ivanov NV et al (2004) Identification of new human cadherin genes using a combination of protein motif search and gene finding methods. J Mol Biol 337:307–317. doi:10.1016/j.jmb.2004.01.026

Ishikawa HO, Takeuchi H, Haltiwanger RS, Irvine KD (2008) Four-jointed is a Golgi kinase that phosphorylates a subset of cadherin domains. Science 321:401–404. doi:10.1126/science.1158159

Ishiuchi T, Misaki K, Yonemura S et al (2009) Mammalian Fat and Dachsous cadherins regulate apical membrane organization in the embryonic cerebral cortex. J Cell Biol 185:959–967. doi:10.1083/jcb.200811030

Jolly MK, Rizvi MS, Kumar A, Sinha P (2014) Mathematical modeling of sub-cellular asymmetry of fat-dachsous heterodimer for generation of planar cell polarity. PLoS One 9:e97641. doi:10.1371/journal.pone.0097641

Juriloff DM, Harris MJ (2012) A consideration of the evidence that genetic defects in planar cell polarity contribute to the etiology of human neural tube defects. Birth Defects Res A Clin Mol Teratol 94:824–840. doi:10.1002/bdra.23079

Lawrence PA, Casal J (2013) The mechanisms of planar cell polarity, growth and the Hippo pathway: some known unknowns. Dev Biol 377:1–8. doi:10.1016/j.ydbio.2013.01.030

Li-Villarreal N, Forbes MM, Loza AJ et al (2015) Dachsous1b cadherin regulates actin and microtubule cytoskeleton during early zebrafish embryogenesis. Development 142:2704–2718. doi:10.1242/dev.119800

Lu B, Usui T, Uemura T et al (1999) Flamingo controls the planar polarity of sensory bristles and asymmetric division of sensory organ precursors in Drosophila. Curr Biol 9:1247–1250. doi:10.1016/S0960-9822(99)80505-3

Lu Q, Schafer DA, Adler PN (2015) The Drosophila planar polarity gene multiple wing hairs directly regulates the actin cytoskeleton. Development 142:2478–2486. doi:10.1242/dev.122119

Lucas AM, Douglas LC (1934) Principles underlying ciliary activity in the respiratory tract: II. A comparison of nasal clearance in man, monkey and other mammals. Arch Otolaryngol Head Neck Surg 20:518–541. doi:10.1001/archotol.1934.03600040074006

Ma D, Yang C, McNeill H et al (2003) Fidelity in planar cell polarity signalling. Nature 421:543–547. doi:10.1038/nature01366

Mahoney PA, Weber U, Onofrechuk P et al (1991) The fat tumor suppressor gene in Drosophila encodes a novel member of the cadherin gene superfamily. Cell 67:853–868. doi:10.1016/0092-8674(91)90359-7

Mani M, Goyal S, Irvine KD, Shraiman BI (2013) Collective polarization model for gradient sensing via Dachsous-Fat intercellular signaling. Proc Natl Acad Sci U S A 110:20420–20425. doi:10.1073/pnas.1307459110

Mao Y, Mulvaney J, Zakaria S et al (2011) Characterization of a Dchs1 mutant mouse reveals requirements for Dchs1-Fat4 signaling during mammalian development. Development 138:947–957. doi:10.1242/dev.057166

Mao Y, Francis-West P, Irvine KD (2015) A Fat4-Dchs1 signal between stromal and cap mesenchyme cells influences nephrogenesis and ureteric bud branching. Development 142:2574–2585. doi:10.1242/dev.122630

Matakatsu H, Blair SS (2004) Interactions between Fat and Dachsous and the regulation of planar cell polarity in the Drosophila wing. Development 131:3785–3794. doi:10.1242/dev.01254

Matis M, Axelrod JD (2013) Regulation of PCP by the Fat signaling pathway. Genes Dev 27:2207–2220. doi:10.1101/gad.228098.113

Matis M, Russler-Germain DA, Hu Q et al (2014) Microtubules provide directional information for core PCP function. Elife 3:e02893. doi:10.7554/eLife.02893

Matsubara D, Horiuchi S-Y, Shimono K et al (2011) The seven-pass transmembrane cadherin Flamingo controls dendritic self-avoidance via its binding to a LIM domain protein, Espinas, in Drosophila sensory neurons. Genes Dev 25:1982–1996. doi:10.1101/gad.16531611

May-Simera H, Kelley MW (2012) Planar cell polarity in the inner ear. Curr Top Dev Biol 101:111–140. doi:10.1016/B978-0-12-394592-1.00006-5

Merkel M, Sagner A, Gruber FS et al (2014) The balance of prickle/spiny-legs isoforms controls the amount of coupling between core and fat PCP systems. Curr Biol 24:2111–2123. doi:10.1016/j.cub.2014.08.005

Mirzadeh Z, Han Y-G, Soriano-Navarro M et al (2010) Cilia organize ependymal planar polarity. J Neurosci 30:2600–2610. doi:10.1523/JNEUROSCI.3744-09.2010

Nakajima D, Nakayama M, Kikuno R et al (2001) Identification of three novel non-classical cadherin genes through comprehensive analysis of large cDNAs. Mol Brain Res 94:85–95. doi:10.1016/S0169-328X(01)00218-2

Nakayama M, Nakajima D, Nagase T et al (1998) Identification of high-molecular-weight proteins with multiple EGF-like motifs by motif-trap screening. Genomics 51:27–34. doi:10.1006/geno.1998.5341

Nishimura T (2014) Making the Neural Plate to Fold into a Tube. In: Kondoh H, Kuroiwa A (eds) New Principles in Developmental Processes. Springer Japan, Tokyo, pp 123–136

Nishimura T, Honda H, Takeichi M (2012) Planar cell polarity links axes of spatial dynamics in neural-tube closure. Cell 149:1084–1097. doi:10.1016/j.cell.2012.04.021

Olofsson J, Sharp KA, Matis M et al (2014) Prickle/spiny-legs isoforms control the polarity of the apical microtubule network in planar cell polarity. Development 141:2866–2874. doi:10.1242/dev.105932

Qu Y, Glasco DM, Zhou L et al (2010) Atypical cadherins Celsr1-3 differentially regulate migration of facial branchiomotor neurons in mice. J Neurosci 30:9392–9401. doi:10.1523/JNEUROSCI.0124-10.2010

Ravni A, Qu Y, Goffinet AM, Tissir F (2009) Planar cell polarity cadherin Celsr1 regulates skin hair patterning in the mouse. J Invest Dermatol 129:2507–2509. doi:10.1038/jid.2009.84

Robinson A, Escuin S, Doudney K et al (2012) Mutations in the planar cell polarity genes CELSR1 and SCRIB are associated with the severe neural tube defect craniorachischisis. Hum Mutat 33:440–447. doi:10.1002/humu.21662

Rock R, Schrauth S, Gessler M (2005) Expression of mouse dchs1, fjx1, and fat-j suggests conservation of the planar cell polarity pathway identified in Drosophila. Dev Dyn 234:747–755. doi:10.1002/dvdy.20515

Saburi S, Hester I, Fischer E et al (2008) Loss of Fat4 disrupts PCP signaling and oriented cell division and leads to cystic kidney disease. Nat Genet 40:1010–1015. doi:10.1038/ng.179

Saburi S, Hester I, Goodrich L, McNeill H (2012) Functional interactions between Fat family cadherins in tissue morphogenesis and planar polarity. Development 139:1806–1820. doi:10.1242/dev.077461

Sadeqzadeh E, de Bock CE, Thorne RF (2014) Sleeping giants: emerging roles for the fat cadherins in health and disease. Med Res Rev 34:190–221. doi:10.1002/med.21286

Sagner A, Merkel M, Aigouy B et al (2012) Establishment of global patterns of planar polarity during growth of the Drosophila wing epithelium. Curr Biol 22:1296–1301. doi:10.1016/j.cub.2012.04.066

Sai X, Yonemura S, Ladher RK (2014) Junctionally restricted RhoA activity is necessary for apical constriction during phase 2 inner ear placode invagination. Dev Biol 394:206–216. doi:10.1016/j.ydbio.2014.08.022

Schwabe T, Neuert H, Clandinin TR (2013) A network of cadherin-mediated interactions polarizes growth cones to determine targeting specificity. Cell 154:351–364. doi:10.1016/j.cell.2013.06.011

Sharma P, McNeill H (2013) Regulation of long-range planar cell polarity by Fat-Dachsous signaling. Development 140:3869–3881. doi:10.1242/dev.094730

Shi D, Fujimori T, Uemura T (2013) Atypical cadherin negotiates a turn. Dev Cell 26:1–2

Shi D, Komatsu K, Hirao M et al (2014) Celsr1 is required for the generation of polarity at multiple levels of the mouse oviduct. Development 141:4558–4568. doi:10.1242/dev.115659
Shima Y, Kawaguchi S, Kosaka K et al (2007) Opposing roles in neurite growth control by two seven-pass transmembrane cadherins. Nat Neurosci 10:963–969. doi:10.1038/nn1933
Shimada Y, Yonemura S, Ohkura H et al (2006) Polarized transport of Frizzled along the planar microtubule arrays in Drosophila wing epithelium. Dev Cell 10:209–222. doi:10.1016/j.devcel.2005.11.016
Shrestha R, Little KA, Tamayo JV et al (2015) Mitotic Control of Planar Cell Polarity by Polo-like Kinase 1. Dev Cell 33:522–534. doi:10.1016/j.devcel.2015.03.024
Simon MA, Xu A, Ishikawa HO, Irvine KD (2010) Modulation of fat:dachsous binding by the cadherin domain kinase four-jointed. Curr Biol 20:811–817. doi:10.1016/j.cub.2010.04.016
Sopko R, Silva E, Clayton L et al (2009) Phosphorylation of the tumor suppressor fat is regulated by its ligand Dachsous and the kinase discs overgrown. Curr Biol 19:1112–1117. doi:10.1016/j.cub.2009.05.049
St Johnston D, Sanson B (2011) Epithelial polarity and morphogenesis. Curr Opin Cell Biol 23:540–546. doi:10.1016/j.ceb.2011.07.005
Striedter GF, Srinivasan S, Monuki ES (2015) Cortical folding: when, where, how, and why? Annu Rev Neurosci 38:291–307. doi:10.1146/annurev-neuro-071714-034128
Strutt DI (2001) Asymmetric localization of frizzled and the establishment of cell polarity in the Drosophila wing. Mol Cell 7:367–375
Strutt H, Strutt D (2002) Nonautonomous planar polarity patterning in Drosophila: dishevelled-independent functions of frizzled. Dev Cell 3:851–863
Strutt H, Strutt D (2008) Differential stability of flamingo protein complexes underlies the establishment of planar polarity. Curr Biol 18:1555–1564. doi:10.1016/j.cub.2008.08.063
Sugimura K, Ishihara S (2013) The mechanical anisotropy in a tissue promotes ordering in hexagonal cell packing. Development 140:4091–4101. doi:10.1242/dev.094060
Sugimura R, He XC, Venkatraman A et al (2012) Noncanonical Wnt signaling maintains hematopoietic stem cells in the niche. Cell 150:351–365. doi:10.1016/j.cell.2012.05.041
Sulik K, Dehart DB, Iangaki T et al (1994) Morphogenesis of the murine node and notochordal plate. Dev Dyn 201:260–278. doi:10.1002/aja.1002010309
Tada M, Heisenberg C-P (2012) Convergent extension: using collective cell migration and cell intercalation to shape embryos. Development 139:3897–3904. doi:10.1242/dev.073007
Tada M, Kai M (2012) Planar cell polarity in coordinated and directed movements. Curr Top Dev Biol 101:77–110. doi:10.1016/B978-0-12-394592-1.00004-1
Takeichi M (2007) The cadherin superfamily in neuronal connections and interactions. Nat Rev Neurosci 8:11–20. doi:10.1038/nrn2043
Tanoue T, Takeichi M (2004) Mammalian Fat1 cadherin regulates actin dynamics and cell-cell contact. J Cell Biol 165:517–528. doi:10.1083/jcb.200403006
Tatin F, Taddei A, Weston A et al (2013) Planar cell polarity protein celsr1 regulates endothelial adherens junctions and directed cell rearrangements during valve morphogenesis. Dev Cell 26:31–44. doi:10.1016/j.devcel.2013.05.015
Taylor J, Abramova N, Charlton J, Adler PN (1998) Van Gogh: a new Drosophila tissue polarity gene. Genetics 150:199–210
Thomas C, Strutt D (2012) The roles of the cadherins Fat and Dachsous in planar polarity specification in Drosophila. Dev Dyn 241:27–39. doi:10.1002/dvdy.22736
Tissir F, Goffinet AM (2013) Shaping the nervous system: role of the core planar cell polarity genes. Nat Rev Neurosci 14:525–535. doi:10.1038/nrn3525
Tissir F, Qu Y, Montcouquiol M et al (2010) Lack of cadherins Celsr2 and Celsr3 impairs ependymal ciliogenesis, leading to fatal hydrocephalus. Nat Neurosci 13:700–707. doi:10.1038/nn.2555
Tsukasaki Y, Miyazaki N, Matsumoto A et al (2014) Giant cadherins Fat and Dachsous self-bend to organize properly spaced intercellular junctions. Proc Natl Acad Sci U S A 111:16011–16016. doi:10.1073/pnas.1418990111

Usui T, Shima Y, Shimada Y et al (1999) Flamingo, a seven-pass transmembrane cadherin, regulates planar cell polarity under the control of frizzled. Cell 98:585–595. doi:10.1016/S0092-8674(00)80046-X

Vichas A, Zallen JA (2011) Translating cell polarity into tissue elongation. Semin Cell Dev Biol 22:858–864. doi:10.1016/j.semcdb.2011.09.013

Villano J, Katz F (1995) four-jointed is required for intermediate growth in the proximal-distal axis in Drosophila. Development 121:2767–2777

Vladar EK, Antic D, Axelrod JD (2009) Planar cell polarity signaling: the developing cell's compass. Cold Spring Harb Perspect Biol 1:a002964. doi:10.1101/cshperspect.a002964

Vladar EK, Bayly RD, Sangoram AM et al (2012) Microtubules enable the planar cell polarity of airway cilia. Curr Biol 22:2203–2212. doi:10.1016/j.cub.2012.09.046

Vladar EK, Lee YL, Stearns T, Axelrod JD (2015) Observing planar cell polarity in multiciliated mouse airway epithelial cells. Methods Cell Biol 127:37–54. doi:10.1016/bs.mcb.2015.01.016

Wada H, Okamoto H (2009) Roles of planar cell polarity pathway genes for neural migration and differentiation. Dev Growth Differ 51:233–240. doi:10.1111/j.1440-169X.2009.01092.x

Wada H, Tanaka H, Nakayama S et al (2006) Frizzled3a and Celsr2 function in the neuroepithelium to regulate migration of facial motor neurons in the developing zebrafish hindbrain. Development 133:4749–4759. doi:10.1242/dev.02665

Wallingford JB (2012) Planar cell polarity and the developmental control of cell behavior in vertebrate embryos. Annu Rev Cell Dev Biol 28:627–653. doi:10.1146/annurev-cellbio-092910-154208

Wallingford JB, Niswander LA, Shaw GM, Finnell RH (2013) The continuing challenge of understanding, preventing, and treating neural tube defects. Science 339:1222002. doi:10.1126/science.1222002

Wang X-J, Zhang D-L, Xu Z-G et al (2014a) Understanding cadherin EGF LAG seven-pass G-type receptors. J Neurochem 131:699–711. doi:10.1111/jnc.12955

Wang Y, Yan J, Lee H et al (2014b) The proteins encoded by the Drosophila Planar Polarity Effector genes inturned, fuzzy and fritz interact physically and can re-pattern the accumulation of "upstream" Planar Cell Polarity proteins. Dev Biol 394:156–169. doi:10.1016/j.ydbio.2014.07.013

Wolff T, Rubin GM (1998) Strabismus, a novel gene that regulates tissue polarity and cell fate decisions in Drosophila. Development 125:1149–1159

Wong LL, Adler PN (1993) Tissue polarity genes of Drosophila regulate the subcellular location for prehair initiation in pupal wing cells. J Cell Biol 123:209–221

Wu J, Roman A-C, Carvajal-Gonzalez JM, Mlodzik M (2013) Wg and Wnt4 provide long-range directional input to planar cell polarity orientation in Drosophila. Nat Cell Biol 15:1045–1055. doi:10.1038/ncb2806

Yamaguchi Y, Miura M (2013) How to form and close the brain: insight into the mechanism of cranial neural tube closure in mammals. Cell Mol Life Sci 70:3171–3186. doi:10.1007/s00018-012-1227-7

Yang Y (ed) (2012) Planar cell polarity during development, 1st edn. Academic, USA

Yang C, Axelrod JD, Simon MA (2002) Regulation of Frizzled by Fat-like Cadherins during Planar Polarity Signaling in the Drosophila Compound Eye. Cell 108:675–688. doi:10.1016/S0092-8674(02)00658-X

Zakaria S, Mao Y, Kuta A et al (2014) Regulation of neuronal migration by Dchs1-Fat4 planar cell polarity. Curr Biol 24:1620–1627. doi:10.1016/j.cub.2014.05.067

Zecca M, Struhl G (2010) A feed-forward circuit linking wingless, fat-dachsous signaling, and the warts-hippo pathway to Drosophila wing growth. PLoS Biol 8:e1000386. doi:10.1371/journal.pbio.1000386

Zeidler MP, Perrimon N, Strutt DI (2000) Multiple roles for four-jointed in planar polarity and limb patterning. Dev Biol 228:181–196. doi:10.1006/dbio.2000.9940

Chapter 11
Various Atypical Cadherins: T-Cadherin, RET, Calsyntenin, and 7D-Cadherin

Keiko Imai-Okano and Shinji Hirano

Abstract In addition to the various cadherins described in previous chapters, there are several atypical cadherins that have unique structures and functions. Because of their great diversity, some cadherins have unexpected roles beyond cell–cell adhesion. In this chapter, we review interesting features of T-cadherin, RET, calsyntenins, and 7D-cadherins that have been discovered in the past 20 years.

Keywords T-cadherin • 7D-cadherin • LI-cadherin • Ksp-cadherin • Cadherin-13 • Cadherin-16 • Cadherin-17 • RET • Calsyntenin • Alcadein

11.1 T-Cadherin (Cadherin-13/ V-Cadherin/ H-Cadherin)

T-cadherin is a unique cadherin that lacks transmembrane and cytoplasmic domains, and it is anchored to the cell membrane via a glycosilphosphoinositol (GPI) moiety (Ranscht and Dours-Zimmermann 1991). T-cadherin is named after its structural feature of "truncation" of the cytoplasmic domain, and it has also been called H-cadherin and V-cadherin due to high expression in heart cells and endothelial cells of blood vessels, respectively. Because it lacks a direct link to the cytoskeleton, it seems to act as a signaling receptor in recognition of environment, regulation of cell motility, and proliferation rather than as a mechanical adhesion molecule. In fact, T-cadherin can interact with specific ligands (lipoproteins and adiponectin), which is not a common feature of classical cadherins (see below). T-cadherin is highly expressed in neural tissue, skeletal muscle, and the cardiovascular system including heart, endothelial cells, smooth muscle cells, and pericytes (Ranscht and Bronner-Fraser 1991; Ivanov et al. 2001; Philippova et al. 2009; Andreeva and Kutuzov 2010). It is involved in various physiological processes and pathogenesis. Here, we briefly summarize current knowledge of T-cadherin. In addition, please see excellent recent reviews for further information (Philippova et al. 2009; Resink et al. 2009; Andreeva and Kutuzov 2010; Rivero et al. 2013).

K. Imai-Okano • S. Hirano (✉)
Department of Cell Biology, Kansai Medical University, 2-5-1 Shinmachi, Hirakata, Osaka 573-1010, Japan
e-mail: s-hirano@hirakata.kmu.ac.jp

S.T. Suzuki, S. Hirano (eds.), *The Cadherin Superfamily*,
DOI 10.1007/978-4-431-56033-3_11

11.1.1 T-Cadherin Gene

T-cadherin is encoded by a single gene in the vertebrate genome, with high homology among species (Philippova et al. 2009) It has appeared recently in evolution as a result of duplication of an ancestral gene of classic cadherins (Hulpiau and van Roy 2009); see Chap. 2. The gene structure is similar to those of classic cadherins; the extracellular (EC) region of T-cadherin is encoded by many exons in contrast to that of protocadherin (Philippova et al. 2009) Human T-cadherin is localized on chromosome 16q24, where several classic cadherin genes such as VE-cadherin, P-cadherin, E-cadherin, CDH8, and CDH11 exist (Rakha et al. 2006). Various regulatory elements are found in the 5’ flanking region of the T-cadherin gene, including androgen response element (ARE) and aryl hydrocarbon response (AHR) element (Philippova et al. 2009). Moreover, gene expression is regulated under various growth factors such as platelet-derived growth factor (PDGF-BB), epidermal growth factor (EGF), and insulin-like growth factor (IGF), leading to down-regulation of T-cadherin protein in the vascular smooth muscle (Kuzmenko et al. 1998).

11.1.2 Structure of T-Cadherin

The T-cadherin molecule has five cadherin domains, as do classic cadherins (Fig. 11.1A). The exact molecular weight of the protein has not been determined yet; among various molecular weights that have been reported, most are in the 90–130 kDa range (Philippova et al. 2009). The discrepancies between studies are probably due to alternative splicing, posttranslational modification, and different processing in different cell types, which would produce various protein sizes.

The amino acid sequence of T-cadherin is about 30 % identical to that of the EC region of classic cadherins (Dames et al. 2008). The calcium binding motifs LDRE and DXND, and the amino acid residue E11, which is important for calcium binding, are well conserved. In contrast, the amino acid residues that have important adhesive functions in classical cadherins are replaced in T-cadherin; these include W2, a key residue to form the strand dimer, and residues of the EC1 hydrophobic binding pocket in classic cadherins (Dames et al. 2008; Hulpiau and van Roy 2009). Thus, the homophilic binding mechanism of T-cadherin is somewhat different from that of classical cadherins; T-cadherin forms X-shaped dimers through an alternative non-swapped interface near the EC1–EC2 calcium binding site (Fig. 11.1B). In fact, mutations within this interface lost the ability of dimerization, and they also lost the ability to regulate neurite outgrowth from T-cadherin-expressing motor neurons in vitro (Ciatto et al. 2010) .

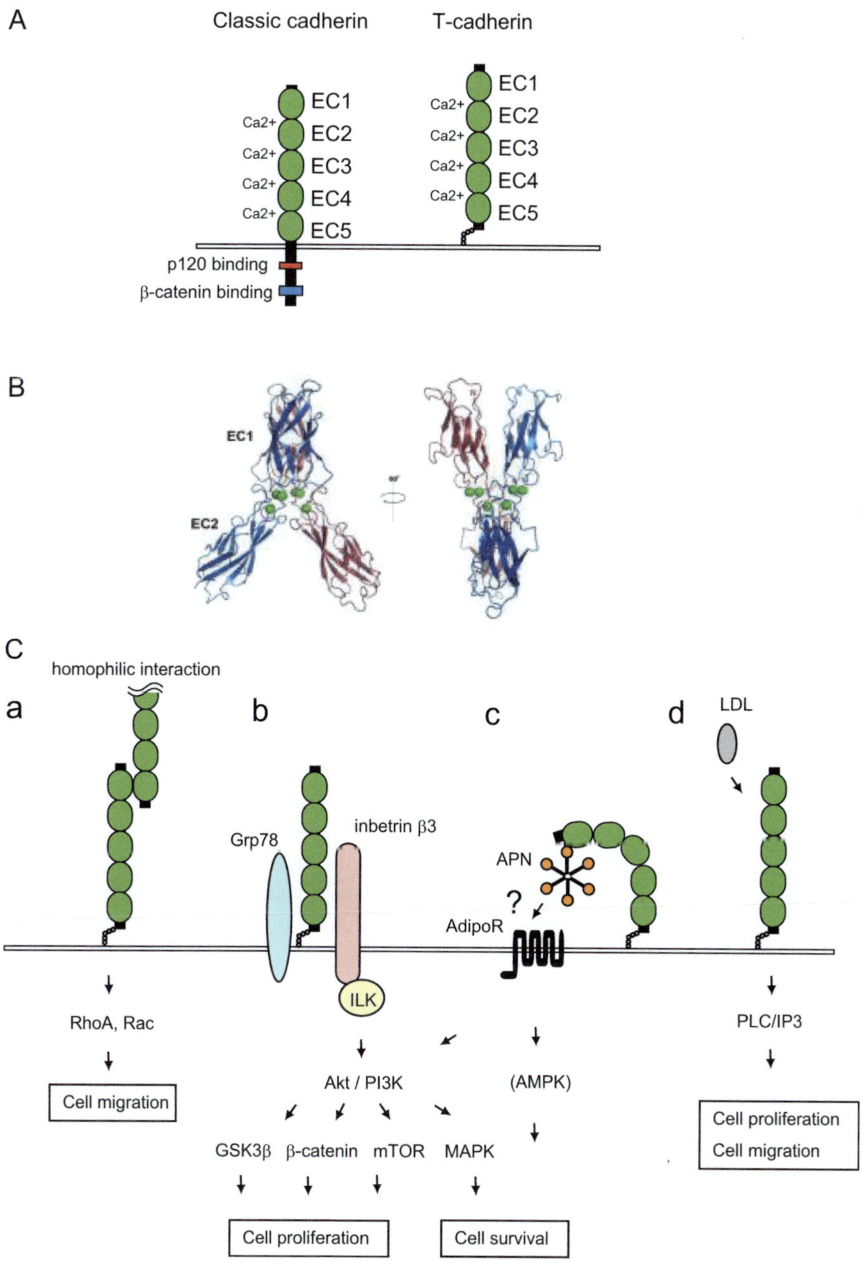

Fig.11.1 Structure and signaling of T-cadherin. (**A**) Structure of T-cadherin in comparison with that of a classic cadherin. T-cadherin is a GPI-anchored membrane protein with five cadherin domains that are stabilized with Ca^{2+} (EC1-5). (**B**) Structure of T-cadherin *trans*-dimer at EC1 and EC2. Note T-cadherin EC1-2 forms an X-shaped dimer by interaction near Ca^{2+} (green spheres)-binding sites. From Ciatto et al. 2010 by permission of Nature Publishing Group. See also the review by Shapiro in this book (Chap. 4). (**C**) T-cadherin–mediated signaling. Note information is accumulated from data with various cell types. (*a*) Homophilic interaction induces cell migration via RhoA and Rac. (*b*) T-cadherin–bound transmembrane proteins such as Grp78 and integrin β3

11.1.3 T-Cadherin Signaling

Several T-cadherin-mediated signaling pathways have been unveiled by studying T-cadherin effects on cell migration, adhesion, proliferation, and survival (Fig. 11.1C). GPI-anchored proteins acting as signaling molecules usually interact with transmembrane molecules that, in turn, give rise to intracellular signaling. Interactions between GPI-anchored proteins and transmembrane molecules are especially enhanced in specialized membrane components called lipid rafts (Simons and Toomre 2000). In fact, it has been reported that T-cadherin is localized in lipid rafts and interacts with some raft proteins including glucose-related protein Grp78, integrin ß3, and GABA-A receptor α1 in endothelial cells (Philippova et al. 1998; Doyle et al. 1998; Joshi et al. 2007; Philippova et al. 2008). Grp78 is a kind of molecular chaperone, and it mediates T-cadherin–dependent survival via the Akt pathway(s) in endothelial cells (Philippova et al. 2008). ß3 integrin is known to be important for angiogenesis via cell migration and survival in endothelial cells (Weis et al. 2007). β3 integrin is associated with integrin-linked kinase (ILK), and it has been reported that ILK is an essential mediator for T-cadherin–dependent effects on Akt. Akt mediates GSK3b and mTOR signaling for cell proliferation, and MAPK p38 for cell survival (Joshi et al. 2005; Joshi et al. 2007).

Moreover, other signaling pathways are also involved in T-cadherin–mediated cell proliferation and migration, although the exact mechanisms that produce intracellular signaling are not known. Homophilic interactions of T-cadherin molecules induce morphology change and cell migration that are mediated by small GTPases including Rho and Rac (Philippova et al. 2005). In addition, binding of LDL to T-cadherin gives rise to PLC/IP3 signaling, which is important for cell proliferation and migration (Tkachuk et al. 1998; Kipmen-Korgun et al. 2005; Rubina et al. 2005). LDL and adiponectin are closely related to vascular diseases, and dysfunction of T-cadherin may play important roles in the pathogenesis of vascular diseases (see next section). Further studies are needed to elucidate the complete mechanisms of T-cadherin–mediated signaling.

Fig.11.1 (continued) could give rise to intracellular signaling. At least ILK induces cell proliferation and cell survival via several pathways including GSK3b pathways, β-catenin–TCF/Lef pathway, mTOR–p70S6K pathway, and MAPKp38 /caspase3 pathway. (*c*) T-cadherin is a possible co-repressor of adiponectin hexamer (APN) with canonical adiponectin receptors (AdipoRs). AdiopR1 (but not AdipR2) induces the AMP-activated kinase (AMPK) signaling. Alternatively, T-cadherin may function as a low-affinity adiponectin receptor and produce signaling independently of adiponectin receptors. (*d*) T-cadherin functions as an LDL receptor, and it gives rise to PLC/IP3 signaling for cell proliferation and migration

11.1.4 T-Cadherin as an Adiponectin Receptor

Identification of T-cadherin as an adiponectin receptor has unveiled a new field of T-cadherin study (Hug et al. 2004). Adiponectin is an important adipose-derived hormone with insulin-sensitizing and anti-inflammatory activities, and lowered adiponectin plays critical roles in obesity-linked diseases such as type 2 diabetes and atherosclerosis (Ouchi et al. 2011; Yamauchi and Kadowaki 2013; Ghoshal and Bhattacharyya 2015). T-cadherin was identified as the third putative adiponectin receptor in addition to other canonical adiponectin receptors (AdipoR1, AdipoR2). T-cadherin binds to the hexameric and high-molecular–weight form of adiponectin, selectively (Hug et al. 2004). It is not yet known whether T- cadherin and associated transmembrane proteins such as integrin or Grp78 give rise to downstream signaling, or whether T-cadherin might be a co-receptor that presents adiponectin to canonical AdipoRs that, in turn, give rise to signaling (Parker-Duffen et al. 2013; Hebbard and Ranscht 2014; Fig. 11.1C). Adiponectin exists in both serum and cardiovascular tissue, but it mainly seems to be localized to cardiovascular tissue by T-cadherin (Parker-Duffen et al. 2013). In fact, T-cadherin and adiponectin were colocalized in some tissues such as aorta, heart, pancreas, and skeletal muscle at the cellular level (Denzel et al. 2010; Tyrberg et al. 2011; Parker-Duffen et al. 2013; Matsuda et al. 2015), and colocalization was abolished in T-cadherin–deficient mice (Hebbard et al. 2008; Denzel et al. 2010; Parker-Duffen et al. 2013). Thus, T-cadherin is essential for localization of adiponectin to cardiovascular tissue (Parker-Duffen et al. 2013; Parker-Duffen and Walsh 2014). It seems that the majority of adiponectin is localized to cardiovascular tissues by T-cadherin. Conversely, tissue T-cadherin levels are regulated by adiponectin via PI-PLC–mediated T-cadherin cleavage (Matsuda et al. 2015). This reciprocal regulation between adiponectin and T-cadherin forms a positive feedback possibly to enhance or stabilize binding of adiponectin to T-cadherin on the plasma membrane.

Expression of T-cadherin is up-regulated in endothelial cells under stress, suggesting its important roles in repairing damaged vessels (Joshi et al. 2005). In fact, T-cadherin–deficient mice showed similar phenotypes to those of adiponectin-deficient mice including impaired cardiac injury, blood flow recovery, and limb ischemia (Shibata et al. 2005; Shimano et al. 2010; Denzel et al. 2010; Parker-Duffen et al. 2013). Although administration of adiponectin rescues the phenotypes of stress-induced cardiac injuries and blood flow recovery for revascularization action in adiponectin-deficient mice, it does not improve recovery in T-cadherin–deficient mice (Denzel et al. 2010; Parker-Duffen et al. 2013). These studies showed that T-cadherin is essential for the cardiac-protective and vascular-protective actions of adiponectin.

11.1.5 Physiological Roles of T-Cadherin

In neural development, T-cadherin is known to be expressed in the caudal halves of the sclerotome, a region that neural crest cells and motor axons avoid, suggesting that T-cadherin may be a negative guidance cue in axon pathfinding and cell migration (Ranscht and Bronner-Fraser 1991). Similar avoidance was observed in the growing motor axons in hindlimb muscle (Fredette and Ranscht 1994). In fact, T-cadherin inhibits neurite outgrowth in culture of spinal motor neurons via homophilic interaction (Fredette et al. 1996). Interestingly, the expression pattern of T-cadherin is often complementary to that of N-cadherin, which is considered to be a permissive axon guidance cue (Fredette and Ranscht 1994; Miskevich et al. 1998); see also Hirano and Takeichi (2012) for review of N-cadherin. On the other hand, homophilic adhesion of T-cadherin molecules seems to contribute to axon guidance as a permissive regulator. It was reported recently that T-cadherin is expressed in the deep-layer cell axons projecting to subcortical structures but not in upper layer callosal axons in the brain (Hayano et al. 2014). Knock-down of T-cadherin induced an aberrant projection, suggesting that T-cadherin is involved in axonal pathway formation in the developing cortex (Hayano et al. 2014). Because T-cadherin is expressed in growing axons but not in surrounding brain regions, T-cadherin may be involved in axon–axon adhesion as a permissive adhesion molecule during development. If that is the case, then T-cadherin may be involved in axon guidance using different mechanisms in different contexts of neural development.

T-cadherin seems to be involved in the formation and maintenance of synapses (Fredette and Ranscht 1994; Paradis et al. 2007). An expression screening in cultured cells identified T-cadherin as a positive regulator that induces synapse formation along with cadherin-11 (Paradis et al. 2007). Knock-down of T-cadherin causes significant reduction of synaptic density with both pre- and postsynaptic markers, suggesting that T-cadherin is involved in early common steps for development of pre- and postsynaptic specialization (Paradis et al. 2007). However, molecular mechanisms of T-cadherin–mediated synapse formation remain elusive. For example, T-cadherin might act as a negative regulator of synapse formation at neuromuscular junctions because T-cadherin is specifically excluded from the junction (Fredette and Ranscht 1994). On the other hand, Akt is reported to be involved in controlling synaptic strength, suggesting possible involvement of T-cadherin signaling in synaptic plasticity (Wang et al. 2003). Because of its involvement in synapse development, T-cadherin dysfunction, as with dysfunction of other cadherin molecules, seems to be closely related to psychiatric disease (see Hirano and Takeichi 2012; Redies et al. 2012; Rivero et al. 2013 for review). In fact, recent genome-wide association studies showed that T-cadherin is associated with some neuropsychiatric disorders such as attention-deficit/hyperactivity disorder (ADHD), alcohol and drug dependence, and autistic spectrum disorders (ASDs; Lasky-Su et al. 2008; Lesch et al. 2008; Johnson et al. 2006; Treutlein et al. 2009;

Chapman et al. 2011). Please see a recent review on T-cadherin in the nervous system (Rivero et al. 2013).

As mentioned in the previous section, T-cadherin is known to be involved in various cellular processes including cell survival, proliferation, and migration in the cardiovascular system. T-cadherin promotes proliferation of endothelial cells and smooth muscle cells in vitro, although the effects of T-cadherin on proliferation are variable depending on the cell type (Ivanov et al. 2004a). Moreover, T-cadherin seems to act as a regulator of cell-cycle progression because a significant amount of T-cadherin is located in the nucleus and centrosomes, and aberrant expression of T-cadherin leads to disturbance in cytokinesis or centrosomal replication in endothelial cells (Andreeva et al. 2009). On the other hand, T-cadherin functions as a tumor suppressor that is down-regulated and/or silenced epigenetically in many tumors (see next section). In addition, T-cadherin seems to promote cell survival in damaged tissue. For example, it has been reported that expression of T-cadherin protects from oxidative stress-induced apoptosis and increases survival in endothelial cells (Joshi et al. 2005), and that expression of T-cadherin is significantly increased in atherosclerotic lesions from human aorta (Ivanov et al. 2001). T-cadherin is known to be essential for the vascular-protective actions of adiponectin (Parker-Duffen et al. 2013). Because of its importance in the cardiovascular system, T-cadherin has been identified as a susceptibility gene for hypertension and metabolic syndrome (Org et al. 2009; Monda et al. 2010).

T-cadherin is also involved in cell migration and regulation of cell behavior (Philippova et al. 2003; Ivanov et al. 2004b). It is mediated, at least in part, by small GTPases including Rho and Rac, as mentioned before. It is notable that in migrating endothelial cells, T-cadherin is concentrated at the leading edge that determines direction of cell migration during navigation (Philippova et al. 2003). Homophilic interaction of T-cadherin molecules serves as an antiadhesive signal and facilitates migration of cultured cells in vitro (Ivanov et al. 2004b). On the other hand, T-cadherin inhibits directional migration of endothelial cells during angiogenesis by contact inhibition with the surrounding cells (e.g., stroma cells), and this repulsive guidance cue allows a growing blood vessel to change direction (Philippova et al. 2006; Rubina et al. 2007; Hebbard et al. 2008). The repulsive guidance cue may be a common mechanism in blood vessel and peripheral nerve navigation, which show similar navigation patterns in vivo.

Another important role of T-cadherin is the promotion of angiogenesis. T-cadherin–mediated adhesion induces formation of capillary-like networks and sprout outgrowth of endothelial cells in vitro (Philippova et al. 2006). Overexpression of T-cadherin stimulates VEGF-mediated neovascularization in vivo (Philippova et al. 2006). Moreover, in a mouse mammary tumor model, T-cadherin promoted binding to adiponectin and angiogenesis (Hebbard et al. 2008). In fact, a similar effect was reported in a study using adiponectin-deficiency tumor model mice (Denzel et al. 2009). Similarly, T-cadherin and adiponectin were also shown to be critical in ischemia-induced revascularization in the limb (Parker-Duffen et al. 2013). However, the molecular mechanisms of T-cadherin/adiponectin-mediated angiogenesis are not well understood: the question of whether adiponectin exerts angiogenesis via AdipoR-dependent or AdipoR-independent signaling pathways remains unanswered.

Moreover, a novel role of T-cadherin in regulating endothelial permeability has recently been reported (Semina et al. 2014). T-cadherin overexpression leads to VE-cadherin phosphorylation on Y731 (β-catenin binding site), resulting in clathrin-mediated VE-cadherin endocytosis and degradation in lysosomes. This degradation of VE-cadherin abolishes endothelial barrier function. In addition, overexpression of T-cadherin results in activation of Rho GTPase, PAK1, and ROCK and induces reorganization of the actin cytoskeleton (Semina et al. 2014). These observations suggest that T-cadherin can regulate endothelial permeability via controlling dynamics of cell–cell adhesion.

T-cadherin seems to be related closely with glucose and lipid homeostasis, as it binds to LDL and adiponectin. A recent study using T-cadherin–deficient mice showed that T-cadherin is localized at insulin granules in β-cells and T-cadherin is shown to be important in second-phase insulin secretion (Tyrberg et al. 2011). However, this effect can occur independently of adiponectin binding. Moreover, there is cross-talk between T-cadherin signaling and insulin signaling (Philippova et al. 2012). The authors proposed a model of T-cadherin–insulin interaction in insulin resistance: oxidative stress, inflammation, and prolonged exposure to insulin lead to increased expression of T-cadherin and activation of its Akt signaling, which, in turn, inhibits insulin signaling as a negative feedback. As a result, endothelial cells lose their sensitivity to insulin.

11.1.6 T-Cadherin in Cancer

Down-regulation of T-cadherin expression has been reported in various human cancers and cancer cell lines (see review Andreeva and Kutuzov 2010). T-cadherin down-regulation in cancer was first reported in breast cancer tissue, and introduction of T-cadherin cDNA into these cells inhibited cellular growth and the infiltrative phenotype in a Matrigel outgrowth assay in vitro (Lee 1996). Re-expression of T-cadherin also reduced malignant properties of various cancers in human tissues (Andreeva and Kutuzov 2010). These observations suggest that down-regulation of T-cadherin in cancerous cells promotes tumor growth and invasiveness (Andreeva and Kutuzov 2010). However, it is interesting to note that T-cadherin functions as a tumor suppressor in various tissues where T-cadherin expression is normally relatively low as compared to neural tissue, skeletal muscle, and vessels.

The mechanisms of tumorigenesis mediated by T-cadherin are not fully understood. It should be noted that the cellular response of cancer cells is different from that of endothelial cells and smooth muscle cells. It is known that T-cadherin induces proliferation of endothelial cells, whereas loss of T-cadherin induces cell proliferation in tumors (Andreeva and Kutuzov 2010). It has also been reported that loss of T-cadherin suppresses apoptosis in melanomas (Bosserhoff et al. 2014).

Moreover, tumor progression depends on crosstalk between E-cadherin and epidermal growth factor receptor (EGFR), which is overexpressed in many carcinomas (Perrais et al. 2007). Down-regulation of T-cadherin increases EGFR phosphorylation and activates its signaling, resulting in promotion of cell proliferation (Mukoyama et al. 2007; Kyriakakis et al. 2012; Kyriakakis et al. 2013). In addition, EGFR activation induces endocytosis of β1 integrin, which seems to enhance cell motility (Mukoyama et al. 2007).

In cancer tissue where T-cadherin is down-regulated, metastases tend to increase. This has been confirmed by tumor models in T-cadherin–deficient mice (Hebbard et al. 2008). These observations suggest that T-cadherin plays a role in cell adhesion and antimigration. This situation stands in contrast to the case of vascular cells, neural crest cells, and motor axons, which use T-cadherin as a repellent in migration. Further study is needed to fully understand the role of T-cadherin in metastasis.

Neovascularization plays a critical role in tumor growth. Tumors grow in complex cellular environments and contexts in terms of molecular profile and cell–cell interactions. For example, T-cadherin is down-regulated in most tumor cells, but it is usually up-regulated in endothelial cells of tumor blood vessels (Wyder et al. 2000). As mentioned before, up-regulation of T-cadherin is thought to induce proliferation of endothelial cells and facilitate angiogenesis. In fact, intratumoral angiogenesis was potentiated by T-cadherin up-regulation in endothelial cells (Ghosh et al. 2007). In addition, using a tumor model of T-cadherin–deficient mice, T-cadherin was shown to be involved in adiponectin-mediated angiogenesis (Hebbard et al. 2008). On the other hand, down-regulation of T-cadherin in tumor cells seems to promote neovascularization by facilitation of endothelial cell invasion into tumor tissue (Andreeva and Kutuzov 2010). This is supported by the observation that, when transplanted into mice, cells expressing abundant T-cadherin suppressed neovascularization by inhibiting migration of endothelial cells (Rubina et al. 2007). Taken together, tumors seem to induce neovascularization partly by inducing proliferation of endothelial cells via up-regulation of T-cadherin in endothelial cells and partly by facilitating migration of endothelial cells via down-regulation of T-cadherin in tumor.

11.2 RET

RET is a unique member of the cadherin superfamily that has a tyrosine kinase domain in the cytoplasmic region (Fig. 11.2A). RET was first identified in humans as an oncogenic fusion protein via an in vitro transformation assay (Takahashi et al. 1985). The RET gene appeared at an early stage of evolution before separation of deuterostomes and protostomes; it is encoded by a single gene in the vertebrate genome, by two genes in *Drosophila,* but is missing in *C. elegans* (Hill et al. 2001; Fung et al. 2008; Hulpiau and van Roy 2009). The cytoplasmic region containing a kinase domain is highly conserved across species, whereas the extracellular region

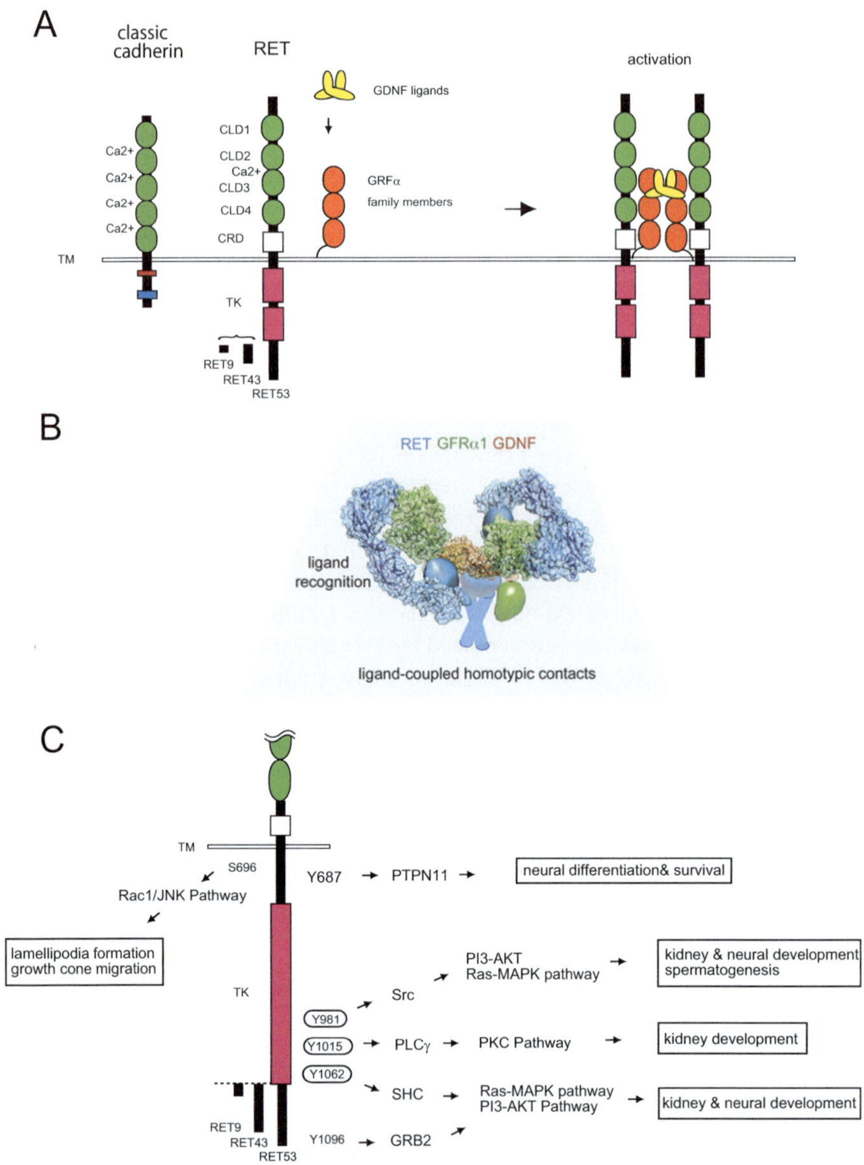

Fig.11.2 Structure and signaling of RET. (**A**) Structure of RET in comparison with that of a classic cadherin. RET is a receptor-type tyrosine kinase, and it has four cadherin-like domains (CLD 1–4) in the extracelluar region followed by a cysteine-rich domain (CRD). The CLD1 and CLD4 show lower homology with others, and Ca^{2+} binds only at the CLD2/3 region. RET can function as a coreceptor for GDNF ligands: GDNF ligands first bind the GFRα coreceptor, and the complex in turn binds to RET. Dimerization of the ternary complex induces activation of RET. (**B**) Extracellular architecture of RET–GFRα1–GDNF complex. From Goodman et al. 2014 by permission of Cell Press. (**C**) RET signaling via cytoplasmic regions. Activation of tyrosine kinase induces autophosphorylation of many tyrosine residues that serve as docking sites for various signaling effectors. Three tyrosine residues, Y981, Y1015, and Y1062, are major phosphorylation sites that activate PI3/AKT, Ras-MAPK pathway, and PKC pathways for various developmental

has diverged (Kjaer et al. 2010). According to the arrangements of cysteine residues in the extracellular domain, RET can be categorized into three phylogenic groups: mammalian-type, nonmammalian vertebrate-type, and invertebrate-type (Kjaer et al. 2010). Apparently, the mammalian RET evolved more recently, and it may have different features from the RETs of other animals. Because mammalian RET is the most well studied because of clinical interests, the description below is of the mammalian type unless otherwise stated. Please see excellent recent reviews on RET, especially for detailed signaling mechanisms and pathogenesis (Heanue and Pachnis 2007; Santoro and Carlomagno 2013; Ibanez 2013; Mulligan 2014; Davis et al. 2014).

11.2.1 RET Structure

RET protein has four cadherin-like domains (CLDs) followed by a cysteine-rich domain (CRD) in the extracellular (EC) region (Fig. 11.2A; Anders et al. 2001). The EC region has quite unique features, including non-adhesion activity and GDNF/GFRα-binding activity that are distinct from those of classic cadherins (Fig. 11.2A, B; Kjaer et al. 2010). RET lacks the adhesion motifs including HAV of classic cadherins in the CLD1 (Kjaer et al. 2010), but some cysteine residues exist instead (Anders et al. 2001). In addition, two expected Ca^{2+}-binding sites (i.e., between CLD1 and 2, CLD3 and 4) were lost over the course of evolution, resulting in only one Ca^{2+} binding site between CLD2 and CLD3 (Fig. 11.2A; Anders et al. 2001). This sole Ca^{2+}-binding site is critical for protein folding (van Weering et al. 1998; Anders et al. 2001; Kjaer and Ibanez 2003a). The linker region between CLD1 and CLD2 has a unique structure, and it is tethered by a stable disulfide bridge (Kjaer et al. 2010). The 3-D structure of CLD1 and CLD2 of mammalian RET forms a compact clam-shell arrangement that is distinct from that of classic cadherins but is reminiscent of that of T-cadherin (Fig. 11.2B; Kjaer et al. 2010).

Mammalian RET acquired binding capability for binary complex that includes glial cell line-derived neurotrophic factor (GDNF) family ligands (GFLs) and GDNF family co-receptor-α (GFRα; Fig. 11.2A, B; see next section, Abrescia et al. 2005). A recent study showed a composite binding site comprising four discrete contact sites in the extracellular domain (Goodman et al. 2014). The EC domains of RET envelop the dimeric ligand complex, forming a flower-shaped structure (Fig. 11.2B). Because there are only few direct contact sites for GFL, the EC region of RET can envelop four different GFL–GFRα complexes (Goodman et al. 2014).

Fig.11.2 (continued) processes. Phosphorylation of serine residue (S696) is also important for migration of cells. Please note that only some representative phosphorylation sites are shown here (See text for abbreviations)

The EC region is also important for stabilization of RET *cis*-dimers. FLF–GFRα binding promotes homotypic interaction between membrane-proximal regions of RET (Goodman et al. 2014). The cysteine-rich domain is important for homotypic *cis*-interactions, probably along with the TM domain (Goodman et al. 2014).

In the cytoplasmic region, there is a core kinase domain that contains an insertion similar to other receptor-type tyrosine kinases (RTKs). There are many tyrosine and serine residues, which are subject to phosphorylation, in the cytoplasimic region (Fig. 11.2C; see below).

In the C-terminal tail region, there are three isoforms with different length; RET9, RET43, and RET51 (Fig. 11.2C; see Santoro and Carlomagno 2013; Mulligan 2014). RET9 and RET51 exist in all vertebrates whereas RET 43 is specific and minor in primates (Mulligan 2014). Because Tyr1069, a major phosphorylation site, is present only in RET51, this isoform seems to have additional role(s). However, functional differences among isoforms remain controversial at the moment (Davis et al. 2014); it has been reported that mutations in RET 51 resulted in abnormal kidney development (Jain et al. 2006) whereas another report showed that RET9 could substitute for RET51 in kidney development (de Graaff et al. 2001).

11.2.2 RET Signaling

As mentioned before, mammalian RET acts as a coreceptor for GDNF family ligands including GDNF, neurturin, artemin, and persephin (Fig. 11.2A; see review Mulligan 2014). GDNF first binds to the GPI-anchored GFRα receptors, and the complex, in turn, binds to RET. There are four GFRαs (GFRα1-4) that bind GDNA, NRTN, ARTN, and PSPN preferentially, although there is some cross-binding. Because there are few direct contacts between GDNF and RET, RET can accommodate distinct ligand/coreceptor pairs (Goodman et al. 2014). In addition, it has also been reported that there is a soluble form of GFRα that acts in a non-cell–autonomous manner (Patel et al. 2012). Because there is no GDNF in *Drosophila*, GDNF–GFRα–RET interaction seems to be specific to the mammalian lineage, although the intracellular signaling mechanism via a kinase domain has some common pathways such as the Ras-ERK, Src, and JNK pathways in mammals and *Drosophila* (Fig. 11.2C; Read et al. 2005; Abrescia et al. 2005; Kallijarvi et al. 2012).

After their complex formation, GDNF–GFRα–RET complexes are recruited into lipid rafts where RET signaling occurs via dimerization (Mulligan 2014). As with other RTKs, homo-dimerization via the transmembrane region seems to be important for activation of RET (Ibanez 2013). Activation of RET induces autophosphorylation of tyrosine residues, which is important for recruitment of adaptor and signaling molecules. Among more than 14 possible phosphorylated Tyrosine residues, Y981, Y1015, and Y1062 are major sites for signaling (Fig. 11.2C; see reviews Ibanez 2013; Davis et al. 2014; Mulligan 2014). Phosphorylated Y981 recruits Src kinase and activates PI3K–AKT signaling, whereas phosphorylated Y1015 induces phospholipase Cγ (PLCγ)-protein kinase C (PKC) signaling. Phosphorylation of Y1062 recruits the Src homology 2 domain containing transforming protein (Shc), which in turn activates the Ras/MAPK pathway via growth factor receptor-bound protein (Grb2) and Sos. Moreover, phosphorylation of Y1062 induces activation of PI3 kinase pathway via Grb2 and GRB2-associated binding protein-2 (Gab2). In addition to these major sites, phosphorylation of Y752 and Y928 is involved in activation of Janus kinase (JAK) -signal transducer and activator of transcription (STAT) signaling. Y1096, which presents only in the RET51 isoform, is also important for recruiting GRB2 and its downstream signaling (Mulligan 2014). Thus, RET activates various signaling pathways including the Ras/MAP kinase pathway, the PI3 kinase/AKT pathway, and the PLCγ pathway (Ibanez 2013; Mulligan 2014).

In addition to tyrosine residues, phosphorylation of serine is also important; S696 in human RET, phosphorylated by protein kinase A (PKA), is a key for regulation of Rac1 and lamellipodia formation (Fukuda et al. 2002). Moreover, phosphorylation of the serine residue is involved in migration of enteric neural crest cells (Asai et al. 2006). Because phosphorylation of Y687 affects S696 signaling, the Y687 residue may be important in the integration of RET and PKA signaling (Ibanez 2013).

RET signaling has cross-talk with other signaling systems. For example, the expression of sonic hedgehog (Shh) and RET is reciprocally regulated (Gonzalez-Reyes et al. 2012). In development of the enteric nervous system, endothelin signaling regulates RET signaling (Heanue and Pachnis 2007). Moreover, the membrane-bound ectodomain of RET, which is produced via cleavage of the cytoplasmic region by caspase, can modulate N-cadherin–mediated adhesion by interacting p120 (Cabrera et al. 2011). In addition, there is a physical and functional interaction between RET and vascular endothelial growth factor receptor 2 (VEGFR2) in ureteric bud cells. RET signaling also controls adhesion and migration via integrin (Cockburn et al. 2010). Moreover, it has been reported that protocadherins (Pcdh-α and γ) form a heteromeric signaling complex with RET, and they stabilize RET signaling and vice versa in neural cells (Schalm et al. 2010). Furthermore, RET has been shown to be required for motor axon attraction mediated by *cis*-interaction with ephrin-A on the same cell (Bonanomi et al. 2012). Other interactions of RET with other systems are also known, including Robo2 and BMP4 via GDNF and Wnt 11 signaling (see review Davis et al. 2014). With harmony of these complex networks, RET is involved in various biological roles and pathogenesis. Please see more detailed reviews on RET signaling (e.g., Ibanez 2013; Mulligan 2014; Davis et al. 2014).

11.2.3 Biological Roles and Pathogenesis of RET

RET plays important roles in various developmental processes including formation of the enteric nervous system and other neural development, kidney development,

development of Payer's patches, and spermatogenesis (see Ibanez 2013; Davis et al. 2014; Mulligan 2014). Due to its critical roles, dysfunction of RET causes various human diseases such as cancer and Hirschsprung's disease, which have gotten much clinical attention in the past three decades.

11.2.4 RET in Cancers

Various RET gene mutations have been found in some neuroendocrine cancers such as papillary thyroid carcinoma, medullary thyroid carcinoma, and multiple endocrine neoplasia type 2A and type 2B (Ibanez 2013). It is notable that the frequency of germline mutations in these cancers is very high (up to 100 %; Mulligan 2014). Mutations can be found throughout the molecule in both the extracellular and cytoplasmic regions (Mulligan 2014). Most mutations are amino acid substitutions, but some chromosomal rearrangement leads to the production of chimeric proteins in somatic mutations of some cancers such as papillary thyroid carcinoma (PTC) and lung adenocarcinoma (Mulligan 2014). Basically, these mutations in cancers are gain-of-function mutations of the RET. For example, in FMTC and MEN2B, some mutations in the cytoplasmic domain lead to constitutive activation of RET kinase without binding to ligands, whereas some mutations in the EC domain lead to ligand-independent dimerization and activation. Activation of RET induces tumorigenesis via various cellular processes and steps (Mulligan 2014). As noted above, RET signaling is involved in the Ras/MAP kinase pathway, the PI3 kinase/AKT pathway, and the PLCγ pathway, which play major roles in cell proliferation. Activated RET also facilitates invasion and metastasis (Mulligan 2014). Moreover, mutated RET makes an inflammatory tumor environment by inducing secretion of cytokines and chemokines, and it also activates estrogen-responsive genes in the absence of estrogen in breast carcinomas (Mulligan 2014). Because the RET kinase domain is structurally similar to other tyrosine kinases, small-molecule tyrosine kinase inhibitors such as Vandetanib, Cabozantinib, and Sorafenib, which have been developed to target other RTKs, are being tested as cancer therapies. (Zuercher et al. 2010; Mulligan 2014; Grullich 2014). For detailed mechanisms of RET-mediated tumorigenesis, please see recent reviews (Santoro and Carlomagno 2013; Mulligan 2014).

11.2.5 RET in Enteric Nervous System and Hirschsprung's Disease

GDNF/RET signaling plays a critical role in the development of the enteric nervous system. The enteric nervous system derives from neural crest cells (the majority are from the vagal region but some from trunk and sacral regions), and it controls movements and secretions of the gut (Heanue and Pachnis 2007). RET and GFRα1 are expressed in enteric neural-crest–derived cells upon entry into the gut, whereas GDNF is expressed in the mesoderm of the gut (Heanue and Pachnis 2007). Endothelin 3 signaling is also important for development of neural crest cells, and there are interactions between GDNF/RET signaling and endothelin 3 signaling for development of the enteric nervous system (Heanue and Pachnis 2007).

Reduction of RET protein causes inhibition of migration of some enteric neural-crest–derived cells, resulting in symptoms of Hirschsprung's disease: a lack of ganglion cells in the enteric nervous systems and a megacolon phenotype (see Heanue and Pachnis 2007; Pan and Li 2012). In Hirschsprung's disease, many mutations can be found throughout the molecule (Manie et al. 2001). In contrast to gain-of-function of RET mutations in tumorigenesis, basically mutations in Hirschsprung's disease are loss-of-function mutations. Haploinsufficiency is the most possible mechanism for causing Hirshsprung's disease that shows dominant inheritance, although it shows incomplete penetrance (Amiel et al. 2008). Most of the mutations in Hirschsprung's prevent maturation of the RET protein in the endoplasmic reticulum (Ibanez 2013). The mutations lead to ubiquitination of the protein, which in turn reduces the protein level by the endoplasmic reticulum-associated degradation pathway (ERAD; Kjaer and Ibanez 2003b). In addition, mutations in the noncoding region are also found, and they are thought to affect RET expression (Emison et al. 2005).

11.2.6 Other Roles of RET in the Nervous System

RET is also involved in the development of various neuronal cell types. As in the case of enteric neurons, RET is essential for survival, migration, axonal growth, and axon guidance of neural-crest–derived sympathetic neurons (see, e.g., Enomoto et al. 2001; Encinas et al. 2008). RET is also involved in survival of spiral ganglia. It has been reported that impairments of Y1062 phosphorylation by mutation of RET kinase domains leads to loss of spiral ganglion neurons and hearing loss along with severe Hirschsprung's disease (Ohgami et al. 2010, 2012).

In motor neurons, GDNF/RET signaling is important not only for cell survival but also axonal branching when synaptic transmission is blocked (Simpson et al. 2013). Moreover, RET is involved in topographic projection of motor neurons (Kramer et al. 2006; Bonanomi et al. 2012). In the developing limb, motor axons make a dorsoventral choice. GDNF is expressed in part of the dorsal limb, and the dorsal branch of motor axons that express higher RET is attracted to GDNF. Thus, the GDNF/RET system functions as an instructive guidance signal for motor axons. The ephrinA/EphA4 system is also involved in this pathway choice, and these two systems work in a cooperative manner to enforce precise axon guidance.

Loss of midbrain dopaminergic (DA) neurons is critical in the pathogenesis of Parkinson's disease, and GDNF/RET signaling is involved in cell survival of these cells (Ibanez 2013). For example, in mice with regionally selective ablation of RET, progressive and adult-onset loss of DA neurons could be observed specifically in the substantia nigra pars compacta, resulting in degeneration of DA nerve terminals in the striatum (Kramer et al. 2007). Mechanisms of neural survival via GDNF/RET signaling are not fully understood, but recent study suggested that RET signaling could restore the activity of complex I of the mitochondrial electron transport chain in a *Drosophila* mutant of Pink1, a Parkinson's disease-associated protein (Klein et al. 2014). It is interesting to recall that mitochondrial dysfunction is considered an important factor in pathogenesis of Parkinson's disease (see McCoy and Cookson 2012).

11.2.7 RET in Kidney Development

RET is also critical in kidney development (see review Davis et al. 2014). During development, RET is widely expressed in the Wolffian duct at an early stage and becomes restricted to the ureteric bud (UB) during UB induction, and distal UB tips during branching morphogenesis, suggesting its roles in Wolffian duct patterning and branching morphogenesis (Davis et al. 2014). In fact, mutations of the RET lead to renal agenesis and hypodysplasia in addition to various defects in the urogenital system in the mouse (Davis et al. 2014). In humans, loss-of-function mutations of RET are associated with kidney agenesis, and congenital anomalies of the kidney and urinary tract (CAKUT) in 5–30 % and approximately 5 % of patients, respectively (Mulligan 2014; Davis et al. 2014). Because of its complex signaling pathways, different mutations of tyrosine residues lead to various effects on the urinary system. For example, inhibition of PLCγ signaling by loss of Y1015 causes renal dysplasia with multiplexed kidneys, megauretuer, vesicoureteral reflux (VUR), and so on, whereas loss of Y1062 causes MAPK and PI3K pathway disruptions that result in bilateral renal agenesis (Davis et al. 2014).

11.2.8 Roles of RET in Peyer's Patch Formation and Spermatogenesis

RET signaling is involved in Peyer's patch formation (Veiga-Fernandes et al. 2007; Fukuyama and Kiyono 2007). Peyer's patches are secondary lymphoid organs of the gut that are responsible for the early immune response. They consist of a framework of stroma cells and lymphocytes. Formation of Peyer's patch is initiated by interaction between haematopoietic cells and stroma cells. CD3-CD4-IL-7Ra-c-kit + CD11c + lymphoid tissue initiator (LTin) cells are recruited to the Peyer's patch anlagen and interact with mesenchymal lymphoid tissue organizer (LTo) cells. RET is expressed on LTin cells, but the identity of the cells that produce GFLs and GFRα has not yet been determined (Patel et al. 2012). Interestingly, the GFRα co-receptor is not expressed on LTin cells, and a soluble form of GFRα that binds ARTN interacts with RET in the *trans*-configuration (Patel et al. 2012).

RET is also required for male germ cell survival (Miles et al. 2012). A GDNF ligand is expressed by the somatic cells of the testis during development, whereas RET is expressed almost exclusively in male germ cells. In RET −/− male mice, germ cells undergo apoptosis. Interestingly, GDNF −/− mice did not show the same phenotype, suggesting a redundancy with GDNF-related genes, or that other ligand(s) are involved in germ cell survival. To date, the true ligands have not been determined because the expression profile of other GDNF ligands (artemin, neurturin, persephin) does not show significant correlation with germ cell differentiation in the developing testis.

11.3 Calsyntenin

The calsyntenin (independently named as alcadein) family consists of three members (calsytenin-1, calsyntein-2, and calysntenin-3) in mammals, and they are highly conserved during evolution because an orthologue exists not only in urochordates but also in *C. elegans* and *Drosophila* (Vogt et al. 2001; Hintsch et al. 2002; Hulpiau and van Roy 2009; see Chap. 2). Calsyntenins have two cadherin motifs in their extracellular regions in addition to an LG (laminin-globular)/LNS (laminin, neurexin, sex hormone-binding globulin) domain (Vogt et al. 2001; Ikeda et al. 2008; Fig. 11.3A). Although there are some structural differences among the three calsyntenins, the cytoplasmic domains contain an X11s binding motif (NP sequence), one or two kinesin-1 light chain-binding motif (s) (WD motifs; Araki et al. 2003; Konecna et al. 2006; Araki et al. 2007), and an acid region that can bind Ca^{2+} (Vogt et al. 2001). A significant amount of calsyntenins are shed from the membrane by proteolysis constitutively (Vogt et al. 2001; Pettem et al. 2013). Vertebrate calsyntenins can mediate Ca^{2+}-dependent adhesion in an in vitro beads assay, but adhesion specificity was not observed among members (Ortiz-Medina et al. 2015). Calsyntenins are mainly expressed in neural tissue; they are abundant in axons in postnatal neurons, whereas they are localized at postsynaptic membranes in adult neurons (Konecna et al. 2006).

Some unique functions of calsyntenin have been unveiled thus far. A well-known function of calsyntenins is vesicle transport in neurites. In vesicular transport along microtubules, calsyntenin-1 functions as a cargo-docking protein by the conserved WD motif of cytoplasmic domain that interacts with a light chain of kinesisn-1 (Konecna et al. 2006; Araki et al. 2007). Calsyntenin-1 is involved in at least two distinct transport pathways in neurites (Fig. 11.3B; Ludwig et al. 2009; Steuble et al. 2010, 2012). First, it is essential for segregation and concentration of APP (amyloid precursor protein) at the *trans*-Golgi network, and also subsequent anterograde transport to early endosomes (early endosome pathway). In addition,

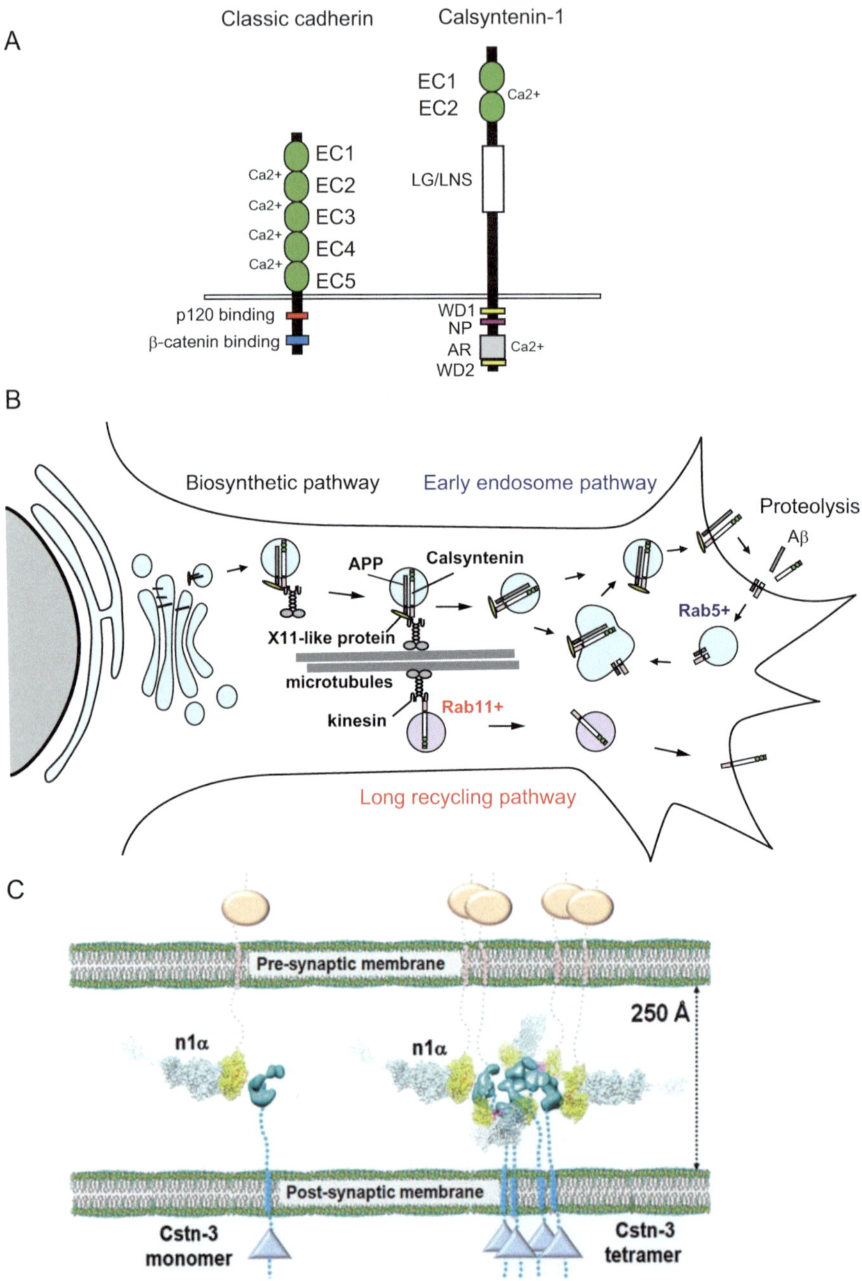

Fig.11.3 Structure and role of calsyntenin. (**A**) Structure of calsyntenin in comparison with that of a classic cadherin. Calsyntenin has two cadherin domains (EC1-2) followed by an LG (laminin-globular) /LNS (laminin, neurexin, sex hormone-binding globulin) domain in the extracellular region and WD1, 2 (kinesin light-chain–binding domain), NP (X11L binding site), and AR (acidic region) in the cytoplasmic region. (**B**) Calsyntenins in kinesin-1–dependent vesicle transport. Calsyntenins are involved in kinesin-1–dependent vesicle transport in distinct pathways, including the biosynthetic pathway, the APP and Rab5-positive early endosome pathway, and the Rab11-

calsyntenin-1 regulates trafficking of Rab5-containing endosomes, and it involves axon branching probably by directing trafficking of specific endosomes (Ponomareva et al. 2014). The second calsyntenin-dependent pathway is anterograde transport of APP-negative vesicles, which are mediated by Rab11/Rip11 (long recycling pathway).

In APP-positive pathways, calsyntenin is closely related to APP metabolism. In the vesicles, calsyntenin-1 and APP form a triple complex via X11L, and dissociation of this complex increases production of Aβ (Araki et al. 2003, 2007; Vagnoni et al. 2012; Takei et al. 2015). Calsyntenin-1 and APP show similar processing by a few secretases: the extracelluar domains are cleaved by α-secretase and the cytoplasmic domain is released by γ-secretase although APP is subject to additional processing by β-secretase (Araki et al. 2004). Interestingly, it was found that calsyntenin-1 accumulated with Aβ proteins in the brains of people with Alzheimer disease (Araki et al. 2003; Uchida et al. 2013). In addition, the cytoplasmic fragment of calsyntenin-1 regulates transport of Aβ (Vagnoni et al. 2012; Takei et al. 2015). Thus, coordinated metabolism of calsyntenin-1 and APP may be critical in the pathogenesis of Alzheimer disease.

The role of calsyntenins in synaptic plasticity and learning are beginning to be elucidated. By a genome-wide analysis, calsyntenin-2 was shown to be associated with memory performance in humans (Papassotiropoulos et al. 2006). In *C. elegans*, calsyntenin mutations showed defects in learning and memory, and the phenotype has been rescued by human calsyntenin-2 (Ikeda et al. 2008; Hoerndli et al. 2009).

Although calsyntenin protein function seems to be highly conserved, the underlying molecular mechanisms of calsyntenin-mediated learning and synaptic plasticity seem to be complex. In *C. elegans*, the ectodomain of calsyntenin produced by constitutive processing is important for learning because the ectodomain can rescue the phenotype of the mutants (Ikeda et al. 2008). On the other hand, a recent study showed that nematode calsyntenin is involved in learning via kinesin-dependent axonal transport of the insulin receptor isoform DAF-2c (Ohno et al. 2014). In the study, DAF-2c was shown to be involved in synaptic regulation via PI3 kinase signaling, and translocation of DAF-2c to synapse is crucial for learning. In addition, interaction of calsyntenin and kinesin is mediated by MAPK

Fig.11.3 (continued) positive, APP-negative recycling pathway. In the biosynthetic pathway, newly synthesized APP is included in the calsyntenin-containing vesicles in the *trans*-Golgi network. Calsyntenin is responsible for the recruitment of kinesin-1 to the vesicles. The vesicles are then transported in an anterograde direction along microtubules by kinesin-1. In endosomes, calsyntenins are involved in at least two distinct pathways: one is the APP-positive and Rab5-positive early endosome pathway, and the other is the APP-negative and Rab11-positive pathway. APP is processed in the early endosome pathway. **(C)** Model of the interaction between calsyntenin 3(Cstn-3) and neurexin1α at the synaptic cleft. A Cstn-3 monomer on the postsynaptic membrane interacts with neurexin1α on the presynaptic membrane, and a Cstn-3 tetramer can induce clustering of neurexin1α (From Lu et al. 2014 by permission of American Society for Biochemistry and Molecular Biology)

in response to starvation. Thus, calsyntenin mediates learning by controlling DAF-2c translocation in response to environmental stimuli.

Moreover, the role of calsyntenin in synaptic plasticity seems to be mediated by glutamate receptor transport, at least in part. It was first suggested by a genetic analysis of *C. elegans* in which gene dosage of a glutamate receptor gene rescued the behavioral defects of a calsyntenin mutant (Hoerndli et al. 2009). A recent study with mice showed that calsyntenin-1 is also involved in transport of NMDA receptor subunits (Ster et al. 2014). The two subunits of NMDA receptors (GluN2A and GluN2B) have functional differences; GluN2B-containing NMDA receptors are expressed in an earlier stage of development, and then replaced by GluN2A-containing NMDA receptors in the second postnatal week. Calsyntenin-1 is essential for transport of newly generated GluN2A and its switching from GluN2B to GluN2A. In fact, this switching did not occur in calsyntenin-1–deficent mice, resulting in persistence of GluN2B and premature synapses.

Furthermore, it was reported that mammalian calsyntenin-3 functions as a synaptic organizer in *trans*-synaptic interactions (Pettem et al. 2013). Postsynaptic calsyntenin-3 interacts with α-neurexins but not β-neurexins on the presynaptic membrane, and induces excitatory and inhibitory presynapse differentiation (see Fig. 11.3C). Calsyntenin-3 exists as a monomer and also forms tetramers, both of which can interact with α-neurexin, and the LNS domain of calsyntenin-3 is responsible for binding to α-neurexin (Lu et al. 2014). Ca^{2+} is involved not only in formation of calsyntenin tetramers but also in heterotypic interaction because the binding surface of the neurexin LNS domain contains a Ca^{2+}-binding site. At least half of calsyntenin-3 presents in a cleaved form in the brain, and interestingly, the shed ectodomain of calsyntenin-3 suppresses the ability of multiple α-neurexin partners including neuroligin 2 to induce presynapse differentiation in vitro (Lu et al. 2014).

Although three calsyntenins in mammals share structural similarities and show similar localization at the postsynaptic membrane, there seem to be functional differences among them. For example, calsyntenin-3 is involved in synapse induction but it has not been demonstrated in vesicle transport. In addition, calsyntenin-1 and calsyntenin-2 do not interact with α-neurexin and they do not have synaptogenic activity (Pettem et al. 2013). Thus, further studies are needed for a full understanding of the physiological roles of these three members of the calsyntenin family.

11.4 7D-Cadherins

The 7D-cadherin family consists of only two members: LI-cadherin (cadherin-17) and Ksp-cadherin (cadherin-16; Hulpiau and van Roy 2009; see Chap. 2). Their structures are unique in that they have seven cadherin motifs in the EC domain and a very short cytoplasmic tail (about 25a.a) that does not have any sequence homologies to that of classic cadherins (Fig. 11.4A). However, sequence analysis

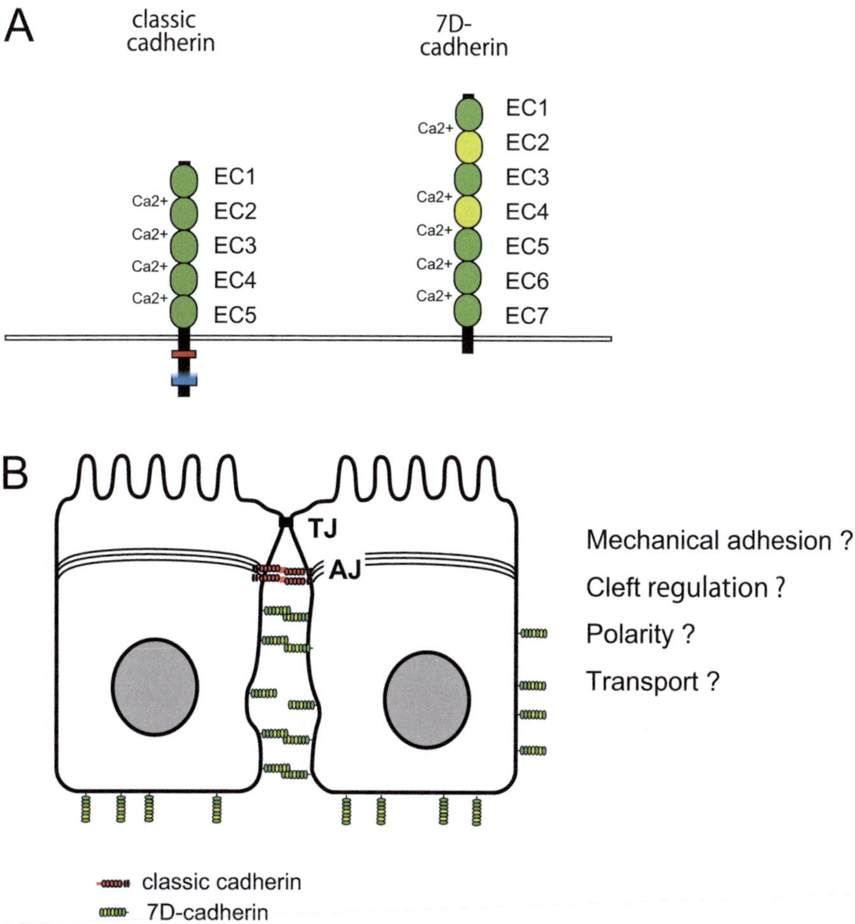

Fig.11.4 7D-cadherin. (**A**) Structure of 7D-cadherins in comparison with that of a classic cadherin. 7D-cadherin has 7 cadherin domains (EC1-7) and a very short cytoplasmic tail. EC2 and EC4 show lower homology with other cadherin domains. (**B**) Schematic diagram of localization and possible roles of 7D-cadherin. Classic cadherins are localized at adherens junctions whereas 7D-cadherins distribute on the basolateral membrane. *TJ*: tight junction, *AJ*: adherens junction

suggested that the additional EC domains evolved from a common ancestor of classic cadherins by partial duplication of EC1 and EC2 (Jung et al. 2004; Wendeler et al. 2006; Hulpiau and van Roy 2009).

11.4.1 LI-Cadherin (Cadherin 17)

LI-cadherin has Ca^{2+}-dependent adhesion activity, although the HAV sequence in the EC1 domain, which is thought to be important for adhesion, is replaced by an AAL sequence (Berndorff et al. 1994). The binding affinity of homophilic LI-cadherin-mediated adhesion is comparable to that of classic cadherin, but cells adhered by LI-cadherin are less tightly packed, probably because it is not closely associated with the cytoskeleton (Wendeler et al. 2007). The Ca^{2+} requirement of LI-cadherin can respond to small changes of extracellular Ca^{2+} concentration with a high degree of cooperativity (Wendeler et al. 2007). Due to this property, it is possible that LI-cadherin may function as a Ca^{2+}-regulated switch for adhesiveness. In addition to homophilic interactions, LI-cadherin can interact with E-cadherin in *trans* heterotypically, although the physiological importance is not clear (Baumgartner et al. 2008).

In addition, there is a RGD sequence in the sixth EC domain of human LI-cadherin but not in mouse LI-cadherin, and the RGD sequence can be used as a binding site by $\alpha2\beta1$ integrin. It has been reported that this interaction of LI-cadherin and integrin induces activation of integrin signaling including FAK and ERK1/2 activation, resulting in tumor growth and metastasis (Bartolmas et al. 2012, Bartolome et al. 2014a, b).

Because the cytoplasmic tail of LI-cadherin is very short and it does not have a catenin-binding site, LI-cadherin seems to be associated with cytoskeletons only weakly. In fact, LI-cadherins form *cis*-dimers and they diffuse freely on the plasma membrane. Interestingly, LI-cadherins are still highly diffusible at contact sites (Bartolmas et al. 2012). Thus, LI-cadherin distributes over the basolateral membrane whereas E-cadherin is localized at apical adherens junctions in simple epithelial cells (Berndorff et al. 1994). The cytoplasmic region of LI-cadherin seems to link different signaling pathways from those of classic cadherins. It has been reported that LI-cadherin is associated with Wnt signaling, Ras/MAPK signaling, and NFκB signaling in tumor cells, although the exact molecular interaction is not fully understood (Liu et al. 2009; Lin et al. 2014; Wang et al. 2013). In pancreatic ductal adenocarcinoma, the LI-cadherin cytoplasmic domain interacts with galectin-3, a member of the β-galactoside-binding proteins. Galectin-3 is known to be involved in apoptosis and immune response, and it also interacts with Wnt signaling (Takamura et al. 2003; Lee et al. 2010). Because the expression profiles of LI-cadherin and galectin-3 are inversely correlated in gastric cancer, they may be negatively regulated (Dong et al. 2008).

LI-cadherin was named after its specific expression in liver and intestine in rats (Berndorff et al. 1994). However, in humans and mice, expression is not observed in adult liver cells. It is expressed in mouse liver during embryogenesis, but its expression decreases in the postnatal stage (Zhu et al. 2010). In addition, it is also expressed in the B lymphocyte lineage, and is involved in B lymphocyte development (Ohnishi et al. 2005).

Expression of LI-cadherin is regulated by various molecules. For example, its expression is controlled by the intestine-specific transcription factor CDX2, the caudal-related homeobox transcription factor, and hepatic nuclear factor 1 (HNF1) (Hinoi et al. 2002; Zhu et al. 2010). In fact, expression of LI-cadherin and that of CDX2 is closely correlated in normal and cancer cells. CDX2 is, in turn, regulated by bone morphogenic protein 2 (BMP2) and BMP4 in gastric cell lines (Barros et al. 2008). In addition, expression of LI-cadherin is induced by both EGF and TGF-a, suggesting its involvement in intestinal differentiation (Sakamoto et al. 2012). Moreover, EphB6 can reduce LI-cadherin expression and inactivation of its downstream Wnt signaling (Bhushan et al. 2014).

The physiological role of LI-cadherin remains elusive (Fig. 11.4B). Because LI-cadherin is localized on the basolateral membrane of intestinal epithelial cells and hepatocytes (Berndorff et al. 1994), it seems to have some distinct function from E-cadherin that is localized at adherens junctions of epithelium. It is possible that LI-cadherin is involved in the initial phase of adhesion due to its high lateral mobility (Baumgartner 2013). Although LI-cadherin has adhesion activity in vitro, there is no direct evidence for its involvement in adhesion in vivo. It has been shown that LI-cadherin is enriched in cholesterol-rich membrane fractions where classic cadherins are absent, although the exact type of microdomain has not been determined for these membrane fractions (Baumgartner et al. 2008). Because some microdomains such as caveola and raft are known to be important for cellular signaling and vesicle transport, LI-cadherin may play important roles in those processes rather than cell–cell adhesion. On the other hand, LI-cadherin has been discovered independently as a human peptide transporter-1 (HPT-1) that is involved in intestinal peptide transport (Dantzig et al. 1994). However, its transport mechanism remains totally unknown because no study has been published since its first report. Thus, whether LI-cadherin itself has transport activity or it associates with an unknown transporter is not yet clear.

Recently, a hypothetical model of the role of 7D-cadherin was proposed from a mathematical perspective (Ahl et al. 2011). Because 7D-cadherins are expressed in the lateral membrane of epithelial cells of water-transporting tissue, the authors speculate that 7D-cadherins may be involved in water transport by regulating the width of the cell cleft. In the model, LI-cadherin molecules regulate the width of the intercellular cleft via engagement of adhesion in response to Ca^{2+} concentration that is coupled to the overall electrolyte concentration in the cleft. Inversely, cells can induce higher osmotic pressure easily in a narrower cleft by active transport of electrolytes. Because of this reciprocal regulation of width of the cleft and osmotic pressure in the cleft, change of the intercellular width can regulate efficiency of water influx through tight junctions between apical membranes by osmotic pressure. Although partial widening of the intercellular cleft was actually observed in gut epithelium, more evidence is needed to prove this model.

LI-cadherin seems to play important roles in cancer progression. In relation with LI-cadherin, gastric cancer is most studied but LI-cadherin also seems to be involved in pancreatic ductal adenocarcinoma, hepatocellular carcinoma, and intrahepatic cholangiocarcinoma (see review Takamura et al. 2013). Many studies have reported that LI-cadherin is a proto-oncogene that leads to tumors when it is overexpressed (Lee et al. 2010; Takamura et al. 2013). Because LI-cadherin is also abundantly expressed in embryonic tissue, LI-cadherin can be regarded as an oncofetal molecule (Lee et al. 2010; Zhu et al. 2010). However, expression patterns differ among different types of cancers and pathological stages, and observations are sometimes contradictory (Lee et al. 2010; Takamura et al. 2013). For example, LI-cadherin is highly expressed in well-differentiated gastric cancers, and highly up-regulated in advanced gastric cancer (Dong et al. 2007; Oue et al. 2004). Another report showed that there is a correlation between reduced expression of LI-cadherin and lymph node metastasis in gastric cancer (Park et al. 2007). Although there are some discrepancies, it often happens that up-regulation of LI-cadherins is associated with tumorigenesis in gastric and hepatic cancer where LI-cadherin is not expressed in normal adult tissue. On the other hand, reduced expression of LI-cadherin is associated with progression of cancers in many cases of colorectal cancer (see Takamura et al. 2013). Because of the complex processes of cancer initiation and tumor progression, further studies are needed to understand the molecular mechanisms of tumorigenisity by LI-cadherin.

11.4.2 Ksp-Cadherin (Cadherin-16)

Ksp-cadherin was named after its specific expression in kidney (Thomson et al. 1995). Ksp-cadherin shares some common features with LI-cadherin; both cadherins have seven cadherin motifs with a short cytoplasmic tail, they mediate cell–cell adhesion, and they are localized on the basolateral membrane of epithelium (Thomson et al. 1995; Thomson and Aronson 1999; Wendeler et al. 2004). It has been reported that the cytoplasmic tail of Ksp-cadherin can interact with αB-crystaline (Thedieck et al. 2008), which appears to link Ksp-cadherin to the actin cytoskeleton.

Ksp-cadherin is specifically expressed in kidney but it was later discovered that it is also expressed in lung, thyroid, and sex ducts (Wertz and Herrmann 1999; Boutet et al. 2006; de Cristofaro et al. 2012). Expression of Ksp-cadherin is regulated by hepatocyte nuclear factor-1 (HNF-1; Bai et al. 2002) and by Pax8 during thyroid development (de Cristofaro et al. 2012). Ksp-cadherin is suppressed by Snail, which is important in the epithelial–mesenchymal transition with repression of HNF-1B (Boutet et al. 2006). In fact, down-regulation of Ksp-cadherin was reported in renal cell carcinoma and thyroid carcinomas where the epithelial–mesenchymal transition seemed to occur (Thedieck et al. 2005; Cali et al. 2012).

The biological function of Ksp-cadherin is not known. It seems to be required in kidney development and/or function at a rather later stage of morphogenesis because onset of Ksp-cadherin expression is correlated with the onset of glomerular filtration and the acquisition of tubular epithelial cell polarity (Thomson and Aronson 1999). As mentioned in the LI-cadherin section, Ksp-cadherin may also play important roles in water transport via regulation of the width of the intercellular cleft of epithelium (Ahl et al. 2011). On the other hand, because expression of Ksp-cadherin is under regulation of Snail, which is a key regulator of the epithelial–mesenchymal transition, Ksp-cadherin may be involved in establishment and/or maintenance of cell polarity.

Acknowledgment We often cited review articles as references for simplicity and due to space limitations, and we are sorry for any omissions of original works from the references. We thank Drs. Lawrence Shapiro, Neil Q. McDonald, and Gabrielle Rudenko for their generous permissions to reproduce their original figures. This work was supported by grants from and the Kansai Medical University Project for Young Researchers (to K.I.), the JSPS KAKENHI Grant Number 25430037, RIKEN cooperative research, Takeda Science Foundation, Sumitomo Foundation, and MEXT-Supported Program for the Strategic Research Foundation at Private Universities (S1201038) (to S. H.).

References

Abrescia C, Sjostrand D, Kjaer S, Ibanez CF (2005) Drosophila RET contains an active tyrosine kinase and elicits neurotrophic activities in mammalian cells. FEBS Lett 579(17):3789–3796. doi: S0014-5793(05)00705-2 [pii] 10.1016/j.febslet.2005.05.075

Ahl M, Weth A, Walcher S, Baumgartner W (2011) The function of 7D-cadherins: a mathematical model predicts physiological importance for water transport through simple epithelia. Theor Biol Med Model 8:18. doi:1742-4682-8-18 [pii] 10.1186/1742-4682-8-18

Amiel J, Sproat-Emison E, Garcia-Barcelo M, Lantieri F, Burzynski G, Borrego S, Pelet A, Arnold S, Miao X, Griseri P, Brooks AS, Antinolo G, de Pontual L, Clement-Ziza M, Munnich A, Kashuk C, West K, Wong KK, Lyonnet S, Chakravarti A, Tam PK, Ceccherini I, Hofstra RM, Fernandez R (2008) Hirschsprung disease, associated syndromes and genetics: a review. J Med Genet 45(1):1–14. doi:jmg.2007.053959 [pii] 10.1136/jmg.2007 053959

Anders J, Kjar S, Ibanez CF (2001) Molecular modeling of the extracellular domain of the RET receptor tyrosine kinase reveals multiple cadherin-like domains and a calcium-binding site. J Biol Chem 276(38):35808–35817. doi:10.1074/jbc.M104968200 M104968200 [pii]

Andreeva AV, Kutuzov MA (2010) Cadherin 13 in cancer. Genes Chromosomes Cancer 49 (9):775–790. doi:10.1002/gcc.20787

Andreeva AV, Kutuzov MA, Tkachuk VA, Voyno-Yasenetskaya TA (2009) T-cadherin is located in the nucleus and centrosomes in endothelial cells. Am J Physiol Cell Physiol 297(5):C1168–C1177. doi:00237.2009 [pii] 10.1152/ajpcell.00237.2009

Araki Y, Tomita S, Yamaguchi H, Miyagi N, Sumioka A, Kirino Y, Suzuki T (2003) Novel cadherin-related membrane proteins, Alcadeins, enhance the X11-like protein-mediated stabilization of amyloid beta-protein precursor metabolism. J Biol Chem 278(49):49448–49458. doi:10.1074/jbc.M306024200 M306024200 [pii]

Araki Y, Miyagi N, Kato N, Yoshida T, Wada S, Nishimura M, Komano H, Yamamoto T, De Strooper B, Yamamoto K, Suzuki T (2004) Coordinated metabolism of Alcadein and amyloid beta-protein precursor regulates FE65-dependent gene transactivation. J Biol Chem 279 (23):24343–24354. doi:10.1074/jbc.M401925200 M401925200 [pii]

Araki Y, Kawano T, Taru H, Saito Y, Wada S, Miyamoto K, Kobayashi H, Ishikawa HO, Ohsugi Y, Yamamoto T, Matsuno K, Kinjo M, Suzuki T (2007) The novel cargo Alcadein induces vesicle association of kinesin-1 motor components and activates axonal transport. EMBO J 26(6):1475–1486. doi:10.1038/sj.emboj.7601609 7601609 [pii]

Asai N, Fukuda T, Wu Z, Enomoto A, Pachnis V, Takahashi M, Costantini F (2006) Targeted mutation of serine 697 in the Ret tyrosine kinase causes migration defect of enteric neural crest cells. Development 133(22):4507–4516. doi:dev.02616 [pii] 10.1242/dev.02616

Bai Y, Pontoglio M, Hiesberger T, Sinclair AM, Igarashi P (2002) Regulation of kidney-specific Ksp-cadherin gene promoter by hepatocyte nuclear factor-1beta. Am J Physiol Renal Physiol 283(4):F839–F851. doi:10.1152/ajprenal.00128.2002

Barros R, Pereira B, Duluc I, Azevedo M, Mendes N, Camilo V, Jacobs RJ, Paulo P, Santos-Silva F, van Seuningen I, van den Brink GR, David L, Freund JN, Almeida R (2008) Key elements of the BMP/SMAD pathway co-localize with CDX2 in intestinal metaplasia and regulate CDX2 expression in human gastric cell lines. J Pathol 215(4):411–420. doi:10.1002/path.2369

Bartolmas T, Hirschfeld-Ihlow C, Jonas S, Schaefer M, Gessner R (2012) LI-cadherin cis-dimerizes in the plasma membrane Ca(2+) independently and forms highly dynamic trans-contacts. Cell Mol Life Sci 69(22):3851–3862. doi:10.1007/s00018-012-1053-y

Bartolome RA, Barderas R, Torres S, Fernandez-Acenero MJ, Mendes M, Garcia-Foncillas J, Lopez-Lucendo M, Casal JI (2014a) Cadherin-17 interacts with alpha2beta1 integrin to regulate cell proliferation and adhesion in colorectal cancer cells causing liver metastasis. Oncogene 33(13):1658–1669. doi:onc2013117 [pii] 10.1038/onc.2013.117

Bartolome RA, Pelaez-Garcia A, Gomez I, Torres S, Fernandez-Acenero MJ, Escudero-Paniagua-B, Imbaud JI, Casal JI (2014b) An RGD motif present in cadherin 17 induces integrin activation and tumor growth. J Biol Chem 289(50):34801–34814. doi:M114.600502 [pii] 10.1074/jbc.M114.600502

Baumgartner W (2013) Possible roles of LI-Cadherin in the formation and maintenance of the intestinal epithelial barrier. Tissue Barriers 1(1):e23815. doi:10.4161/tisb.23815 2012TISSBARRIER007R [pii]

Baumgartner W, Wendeler MW, Weth A, Koob R, Drenckhahn D, Gessner R (2008) Heterotypic trans-interaction of LI- and E-cadherin and their localization in plasmalemmal microdomains. J Mol Biol 378(1):44–54. doi:S0022-2836(08)00207-6 [pii] 10.1016/j.jmb.2008.02.023

Berndorff D, Gessner R, Kreft B, Schnoy N, Lajous-Petter AM, Loch N, Reutter W, Hortsch M, Tauber R (1994) Liver-intestine cadherin: molecular cloning and characterization of a novel Ca(2+)-dependent cell adhesion molecule expressed in liver and intestine. J Cell Biol 125 (6):1353–1369

Bhushan L, Tavitian N, Dey D, Tumur Z, Parsa C, Kandpal RP (2014) Modulation of liver-intestine cadherin (Cadherin 17) expression, ERK phosphorylation and WNT signaling in EPHB6 receptor-expressing MDA-MB-231 cells. Cancer Genomics Proteomics 11 (5):239–249. doi:11/5/239 [pii]

Bonanomi D, Chivatakarn O, Bai G, Abdesselem H, Lettieri K, Marquardt T, Pierchala BA, Pfaff SL (2012) Ret is a multifunctional coreceptor that integrates diffusible- and contact-axon guidance signals. Cell 148(3):568–582. doi:S0092-8674(12)00090-6 [pii] 10.1016/j.cell.2012.01.024

Bosserhoff AK, Ellmann L, Quast AS, Eberle J, Boyle GM, Kuphal S (2014) Loss of T-cadherin (CDH-13) regulates AKT signaling and desensitizes cells to apoptosis in melanoma. Mol Carcinog 53(8):635–647. doi:10.1002/mc.22018

Boutet A, De Frutos CA, Maxwell PH, Mayol MJ, Romero J, Nieto MA (2006) Snail activation disrupts tissue homeostasis and induces fibrosis in the adult kidney. EMBO J 25 (23):5603–5613. doi:7601421 [pii] 10.1038/sj.emboj.7601421

Cabrera JR, Bouzas-Rodriguez J, Tauszig-Delamasure S, Mehlen P (2011) RET modulates cell adhesion via its cleavage by caspase in sympathetic neurons. J Biol Chem 286 (16):14628–14638. doi:M110.195461 [pii] 10.1074/jbc.M110.195461

Cali G, Gentile F, Mogavero S, Pallante P, Nitsch R, Ciancia G, Ferraro A, Fusco A, Nitsch L (2012) CDH16/Ksp-cadherin is expressed in the developing thyroid gland and is strongly down-regulated in thyroid carcinomas. Endocrinology 153(1):522–534. doi:en.2011-1572 [pii] 10.1210/en.2011-1572

Chapman NH, Estes A, Munson J, Bernier R, Webb SJ, Rothstein JH, Minshew NJ, Dawson G, Schellenberg GD, Wijsman EM (2011) Genome-scan for IQ discrepancy in autism: evidence for loci on chromosomes 10 and 16. Hum Genet 129(1):59–70. doi:10.1007/s00439-010-0899-z

Ciatto C, Bahna F, Zampieri N, VanSteenhouse HC, Katsamba PS, Ahlsen G, Harrison OJ, Brasch J, Jin X, Posy S, Vendome J, Ranscht B, Jessell TM, Honig B, Shapiro L (2010) T-cadherin structures reveal a novel adhesive binding mechanism. Nat Struct Mol Biol 17 (3):339–347. doi:nsmb.1781 [pii] 10.1038/nsmb.1781

Cockburn JG, Richardson DS, Gujral TS, Mulligan LM (2010) RET-mediated cell adhesion and migration require multiple integrin subunits. J Clin Endocrinol Metab 95(11):E342-346. doi: jc.2010-0771 [pii] 10.1210/jc.2010-0771

Dames SA, Bang E, Haussinger D, Ahrens T, Engel J, Grzesiek S (2008) Insights into the low adhesive capacity of human T-cadherin from the NMR structure of Its N-terminal extracellular domain. J Biol Chem 283(34):23485–23495. doi:M708335200 [pii] 10.1074/jbc.M708335200

Dantzig AH, Hoskins JA, Tabas LB, Bright S, Shepard RL, Jenkins IL, Duckworth DC, Sportsman JR, Mackensen D, Rosteck PR Jr et al (1994) Association of intestinal peptide transport with a protein related to the cadherin superfamily. Science 264(5157):430–433

Davis TK, Hoshi M, Jain S (2014) To bud or not to bud: the RET perspective in CAKUT. Pediatr Nephrol 29(4):597–608. doi:10.1007/s00467-013-2606-5

de Cristofaro T, Di Palma T, Fichera I, Lucci V, Parrillo L, De Felice M, Zannini M (2012) An essential role for Pax8 in the transcriptional regulation of cadherin-16 in thyroid cells. Mol Endocrinol 26(1):67–78. doi:me.2011-1090 [pii] 10.1210/me.2011-1090

de Graaff E, Srinivas S, Kilkenny C, D'Agati V, Mankoo BS, Costantini F, Pachnis V (2001) Differential activities of the RET tyrosine kinase receptor isoforms during mammalian embryogenesis. Genes Dev 15(18):2433–2444. doi:10.1101/gad.205001

Denzel MS, Hebbard LW, Shostak G, Shapiro L, Cardiff RD, Ranscht B (2009) Adiponectin deficiency limits tumor vascularization in the MMTV-PyV-mT mouse model of mammary cancer. Clin Cancer Res 15(10):3256–3264. doi:1078–0432.CCR-08-2661 [pii] 10.1158/1078-0432.CCR-08-2661

Denzel MS, Scimia MC, Zumstein PM, Walsh K, Ruiz-Lozano P, Ranscht B (2010) T-cadherin is critical for adiponectin-mediated cardioprotection in mice. J Clin Invest 120(12):4342–4352. doi:43464 [pii] 10.1172/JCI43464

Dong W, Yu Q, Xu Y (2007) Altered expression of a Li-cadherin in gastric cancer and intestinal metaplasia. Dig Dis Sci 52(2):536–542. doi:10.1007/s10620-006-9316-9

Dong WG, Yu QF, Xu Y, Fan LF (2008) Li-cadherin is inversely correlated with galectin-3 expression in gastric cancer. Dig Dis Sci 53(7):1811–1817. doi:10.1007/s10620-007-0080-2

Doyle DD, Goings GE, Upshaw-Earley J, Page E, Ranscht B, Palfrey HC (1998) T-cadherin is a major glycophosphoinositol-anchored protein associated with noncaveolar detergent-insoluble domains of the cardiac sarcolemma. J Biol Chem 273(12):6937–6943

Emison ES, McCallion AS, Kashuk CS, Bush RT, Grice E, Lin S, Portnoy ME, Cutler DJ, Green ED, Chakravarti A (2005) A common sex-dependent mutation in a RET enhancer underlies Hirschsprung disease risk. Nature 434(7035):857–863. doi:nature03467 [pii] 10.1038/nature03467

Encinas M, Rozen EJ, Dolcet X, Jain S, Comella JX, Milbrandt J, Johnson EM, Jr. (2008) Analysis of Ret knockin mice reveals a critical role for IKKs, but not PI 3-K, in neurotrophic factor-induced survival of sympathetic neurons. Cell Death Differ 15(9):1510–1521. doi:cdd200876 [pii] 10.1038/cdd.2008.76

Enomoto H, Crawford PA, Gorodinsky A, Heuckeroth RO, Johnson EM Jr, Milbrandt J (2001) RET signaling is essential for migration, axonal growth and axon guidance of developing sympathetic neurons. Development 128(20):3963–3974

Fredette BJ, Ranscht B (1994) T-cadherin expression delineates specific regions of the developing motor axon-hindlimb projection pathway. J Neurosci 14(12):7331–7346

Fredette BJ, Miller J, Ranscht B (1996) Inhibition of motor axon growth by T-cadherin substrata. Development 122(10):3163–3171

Fukuda T, Kiuchi K, Takahashi M (2002) Novel mechanism of regulation of Rac activity and lamellipodia formation by RET tyrosine kinase. J Biol Chem 277(21):19114–19121. doi:10.1074/jbc.M200643200 M200643200 [pii]

Fukuyama S, Kiyono H (2007) Neuroregulator RET initiates Peyer's-patch tissue genesis. Immunity 26(4):393–395. doi:S1074-7613(07)00220-8 [pii] 10.1016/j.immuni.2007.04.004

Fung S, Wang F, Chase M, Godt D, Hartenstein V (2008) Expression profile of the cadherin family in the developing Drosophila brain. J Comp Neurol 506(3):469–488. doi:10.1002/cne.21539

Ghosh S, Joshi MB, Ivanov D, Feder-Mengus C, Spagnoli GC, Martin I, Erne P, Resink TJ (2007) Use of multicellular tumor spheroids to dissect endothelial cell-tumor cell interactions: a role for T-cadherin in tumor angiogenesis. FEBS Lett 581(23):4523–4528. doi:S0014-5793(07)00914-3 [pii] 10.1016/j.febslet.2007.08.038

Ghoshal K, Bhattacharyya M (2015) Adiponectin: Probe of the molecular paradigm associating diabetes and obesity. World J Diab 6(1):151–166. doi:10.4239/wjd.v6.i1.151

Gonzalez-Reyes LE, Verbitsky M, Blesa J, Jackson-Lewis V, Paredes D, Tillack K, Phani S, Kramer ER, Przedborski S, Kottmann AH (2012) Sonic hedgehog maintains cellular and neurochemical homeostasis in the adult nigrostriatal circuit. Neuron 75(2):306–319. doi:S0896-6273(12)00484-9 [pii] 10.1016/j.neuron.2012.05.018

Goodman KM, Kjaer S, Beuron F, Knowles PP, Nawrotek A, Burns EM, Purkiss AG, George R, Santoro M, Morris EP, McDonald NQ (2014) RET recognition of GDNF-GFRalpha1 ligand by a composite binding site promotes membrane-proximal self-association. Cell Rep 8(6):1894–1904. doi:S2211-1247(14)00716-5 [pii] 10.1016/j.celrep.2014.08.040

Grullich C (2014) Cabozantinib: a MET, RET, and VEGFR2 tyrosine kinase inhibitor. Recent Results Cancer Res 201:207–214. doi:10.1007/978-3-642-54490-3_12

Hayano Y, Zhao H, Kobayashi H, Takeuchi K, Norioka S, Yamamoto N (2014) The role of T-cadherin in axonal pathway formation in neocortical circuits. Development 141(24):4784–4793. doi:141/24/4784 [pii] 10.1242/dev.108290

Heanue TA, Pachnis V (2007) Enteric nervous system development and Hirschsprung's disease: advances in genetic and stem cell studies. Nat Rev Neurosci 8(6):466–479. doi:nrn2137 [pii] 10.1038/nrn2137

Hebbard L, Ranscht B (2014) Multifaceted roles of adiponectin in cancer. Best Pract Res Clin Endocrinol Metab 28(1):59–69. doi:S1521-690X(13)00181-4 [pii] 10.1016/j.beem.2013.11.005

Hebbard LW, Garlatti M, Young LJ, Cardiff RD, Oshima RG, Ranscht B (2008) T-cadherin supports angiogenesis and adiponectin association with the vasculature in a mouse mammary tumor model. Cancer Res 68(5):1407–1416. doi:68/5/1407 [pii] 10.1158/0008-5472.CAN-07-2953

Hill E, Broadbent ID, Chothia C, Pettitt J (2001) Cadherin superfamily proteins in Caenorhabditis elegans and Drosophila melanogaster. J Mol Biol 305(5):1011–1024. doi:10.1006/jmbi.2000.4361 S0022-2836(00)94361-4 [pii]

Hinoi T, Lucas PC, Kuick R, Hanash S, Cho KR, Fearon ER (2002) CDX2 regulates liver intestine-cadherin expression in normal and malignant colon epithelium and intestinal metaplasia. Gastroenterology 123(5):1565–1577. doi:S0016508502002901 [pii]

Hintsch G, Zurlinden A, Meskenaite V, Steuble M, Fink-Widmer K, Kinter J, Sonderegger P (2002) The calsyntenins--a family of postsynaptic membrane proteins with distinct neuronal expression patterns. Mol Cell Neurosci 21(3):393–409. doi:S1044743102911814 [pii]

Hirano S, Takeichi M (2012) Cadherins in brain morphogenesis and wiring. Physiol Rev 92(2):597–634. doi:92/2/597 [pii] 10.1152/physrev.00014.2011

Hoerndli FJ, Walser M, Frohli Hoier E, de Quervain D, Papassotiropoulos A, Hajnal A (2009) A conserved function of C. elegans CASY-1 calsyntenin in associative learning. PLoS One 4(3):e4880. doi:10.1371/journal.pone.0004880

Hug C, Wang J, Ahmad NS, Bogan JS, Tsao TS, Lodish HF (2004) T-cadherin is a receptor for hexameric and high-molecular-weight forms of Acrp30/adiponectin. Proc Natl Acad Sci U S A 101(28):10308–10313. doi:10.1073/pnas.0403382101 0403382101 [pii]

Hulpiau P, van Roy F (2009) Molecular evolution of the cadherin superfamily. Int J Biochem Cell Biol 41(2):349–369. doi:S1357-2725(08)00404-4 [pii] 10.1016/j.biocel.2008.09.027

Ibanez CF (2013) Structure and physiology of the RET receptor tyrosine kinase. Cold Spring Harb Perspect Biol 5(2). doi:5/2/a009134 [pii] 10.1101/cshperspect.a009134

Ikeda DD, Duan Y, Matsuki M, Kunitomo H, Hutter H, Hedgecock EM, Iino Y (2008) CASY-1, an ortholog of calsyntenins/alcadeins, is essential for learning in *Caenorhabditis elegans*. Proc Natl Acad Sci U S A 105(13):5260–5265. doi:105/13/5260 [pii] 10.1073/pnas.0711894105

Ivanov D, Philippova M, Antropova J, Gubaeva F, Iljinskaya O, Tararak E, Bochkov V, Erne P, Resink T, Tkachuk V (2001) Expression of cell adhesion molecule T-cadherin in the human vasculature. Histochem Cell Biol 115(3):231–242

Ivanov D, Philippova M, Allenspach R, Erne P, Resink T (2004a) T-cadherin upregulation correlates with cell-cycle progression and promotes proliferation of vascular cells. Cardiovasc Res 64(1):132–143. doi:10.1016/j.cardiores.2004.06.010 S0008636304002524 [pii]

Ivanov D, Philippova M, Tkachuk V, Erne P, Resink T (2004b) Cell adhesion molecule T-cadherin regulates vascular cell adhesion, phenotype and motility. Exp Cell Res 293(2):207–218. doi:S0014482703005354 [pii]

Jain S, Encinas M, Johnson EM, Jr., Milbrandt J (2006) Critical and distinct roles for key RET tyrosine docking sites in renal development. Genes Dev 20(3):321–333. doi:20/3/321 [pii] 10.1101/gad.1387206

Johnson C, Drgon T, Liu QR, Walther D, Edenberg H, Rice J, Foroud T, Uhl GR (2006) Pooled association genome scanning for alcohol dependence using 104,268 SNPs: validation and use to identify alcoholism vulnerability loci in unrelated individuals from the collaborative study on the genetics of alcoholism. Am J Med Genet B Neuropsychiatr Genet 141B(8):844–853. doi:10.1002/ajmg.b.30346

Joshi MB, Philippova M, Ivanov D, Allenspach R, Erne P, Resink TJ (2005) T-cadherin protects endothelial cells from oxidative stress-induced apoptosis. FASEB J 19(12):1737–1739. doi:05-3834fje [pii] 10.1096/fj.05-3834fje

Joshi MB, Ivanov D, Philippova M, Erne P, Resink TJ (2007) Integrin-linked kinase is an essential mediator for T-cadherin-dependent signaling via Akt and GSK3beta in endothelial cells. FASEB J 21(12):3083–3095. doi:fj.06-7723com [pii] 10.1096/fj.06-7723com

Jung R, Wendeler MW, Danevad M, Himmelbauer H, Gessner R (2004) Phylogenetic origin of LI-cadherin revealed by protein and gene structure analysis. Cell Mol Life Sci 61 (10):1157–1166. doi:10.1007/s00018-004-3470-z

Kallijarvi J, Stratoulias V, Virtanen K, Hietakangas V, Heino TI, Saarma M (2012) Characterization of Drosophila GDNF receptor-like and evidence for its evolutionarily conserved interaction with neural cell adhesion molecule (NCAM)/FasII. PLoS One 7(12):e51997. doi:10.1371/journal.pone.0051997 PONE-D-12-23868 [pii]

Kipmen-Korgun D, Osibow K, Zoratti C, Schraml E, Greilberger J, Kostner GM, Jurgens G, Graier WF (2005) T-cadherin mediates low-density lipoprotein-initiated cell proliferation via the Ca(2+)-tyrosine kinase-Erk1/2 pathway. J Cardiovasc Pharmacol 45(5):418–430. doi:00005344-200505000-00006 [pii]

Kjaer S, Ibanez CF (2003a) Identification of a surface for binding to the GDNF-GFR alpha 1 complex in the first cadherin-like domain of RET. J Biol Chem 278(48):47898–47904. doi:10.1074/jbc.M309772200 M309772200 [pii]

Kjaer S, Ibanez CF (2003b) Intrinsic susceptibility to misfolding of a hot-spot for Hirschsprung disease mutations in the ectodomain of RET. Hum Mol Genet 12(17):2133–2144. doi:10.1093/hmg/ddg227 ddg227 [pii]

Kjaer S, Hanrahan S, Totty N, McDonald NQ (2010) Mammal-restricted elements predispose human RET to folding impairment by HSCR mutations. Nat Struct Mol Biol 17(6):726–731. doi:nsmb.1808 [pii] 10.1038/nsmb.1808

Klein P, Muller-Rischart AK, Motori E, Schonbauer C, Schnorrer F, Winklhofer KF, Klein R (2014) Ret rescues mitochondrial morphology and muscle degeneration of Drosophila Pink1 mutants. EMBO J 33(4):341–355. doi:embj.201284290 [pii] 10.1002/embj.201284290

Konecna A, Frischknecht R, Kinter J, Ludwig A, Steuble M, Meskenaite V, Indermuhle M, Engel M, Cen C, Mateos JM, Streit P, Sonderegger P (2006) Calsyntenin-1 docks vesicular cargo to kinesin-1. Mol Biol Cell 17(8):3651–3663. doi:E06-02-0112 [pii] 10.1091/mbc.E06-02-0112

Kramer ER, Knott L, Su F, Dessaud E, Krull CE, Helmbacher F, Klein R (2006) Cooperation between GDNF/Ret and ephrinA/EphA4 signals for motor-axon pathway selection in the limb. Neuron 50(1):35–47. doi:S0896-6273(06)00162-0 [pii] 10.1016/j.neuron.2006.02.020

Kramer ER, Aron L, Ramakers GM, Seitz S, Zhuang X, Beyer K, Smidt MP, Klein R (2007) Absence of Ret signaling in mice causes progressive and late degeneration of the nigrostriatal system. PLoS Biol 5(3):e39. doi:1544-9173-5-3-e39 [pii] 10.1371/journal.pbio.0050039

Kuzmenko YS, Kern F, Bochkov VN, Tkachuk VA, Resink TJ (1998) Density- and proliferation status-dependent expression of T-cadherin, a novel lipoprotein-binding glycoprotein: a function in negative regulation of smooth muscle cell growth? FEBS Lett 434(1–2):183–187. doi: S0014-5793(98)00977-6 [pii]

Kyriakakis E, Maslova K, Philippova M, Pfaff D, Joshi MB, Buechner SA, Erne P, Resink TJ (2012) T-Cadherin is an auxiliary negative regulator of EGFR pathway activity in cutaneous squamous cell carcinoma: impact on cell motility. J Invest Dermatol 132(9):2275–2285. doi: jid2012131 [pii] 10.1038/jid.2012.131

Kyriakakis E, Maslova K, Frachet A, Ferri N, Contini A, Pfaff D, Erne P, Resink TJ, Philippova M (2013) Cross-talk between EGFR and T-cadherin: EGFR activation promotes T-cadherin localization to intercellular contacts. Cell Signal 25(5):1044–1053. doi:S0898-6568(13)00037-5 [pii] 10.1016/j.cellsig.2013.02.001

Lasky-Su J, Anney RJ, Neale BM, Franke B, Zhou K, Maller JB, Vasquez AA, Chen W, Asherson P, Buitelaar J, Banaschewski T, Ebstein R, Gill M, Miranda A, Mulas F, Oades RD, Roeyers H, Rothenberger A, Sergeant J, Sonuga-Barke E, Steinhausen HC, Taylor E, Daly M, Laird N, Lange C, Faraone SV (2008) Genome-wide association scan of the time to onset of attention deficit hyperactivity disorder. Am J Med Genet B Neuropsychiatr Genet 147B(8):1355–1358. doi:10.1002/ajmg.b.30869

Lee SW (1996) H-cadherin, a novel cadherin with growth inhibitory functions and diminished expression in human breast cancer. Nat Med 2(7):776–782

Lee NP, Poon RT, Shek FH, Ng IO, Luk JM (2010) Role of cadherin-17 in oncogenesis and potential therapeutic implications in hepatocellular carcinoma. Biochim Biophys Acta 1806 (2):138–145. doi:S0304-419X(10)00034-X [pii] 10.1016/j.bbcan.2010.05.002

Lesch KP, Timmesfeld N, Renner TJ, Halperin R, Roser C, Nguyen TT, Craig DW, Romanos J, Heine M, Meyer J, Freitag C, Warnke A, Romanos M, Schafer H, Walitza S, Reif A, Stephan DA, Jacob C (2008) Molecular genetics of adult ADHD: converging evidence from genome-wide association and extended pedigree linkage studies. J Neural Transm 115(11):1573–1585. doi:10.1007/s00702-008-0119-3

Lin Z, Zhang C, Zhang M, Xu D, Fang Y, Zhou Z, Chen X, Qin N, Zhang X (2014) Targeting cadherin-17 inactivates Ras/Raf/MEK/ERK signaling and inhibits cell proliferation in gastric cancer. PLoS One 9(1):e85296. doi:10.1371/journal.pone.0085296 PONE-D-13-23947 [pii]

Liu LX, Lee NP, Chan VW, Xue W, Zender L, Zhang C, Mao M, Dai H, Wang XL, Xu MZ, Lee TK, Ng IO, Chen Y, Kung HF, Lowe SW, Poon RT, Wang JH, Luk JM (2009) Targeting cadherin-17 inactivates Wnt signaling and inhibits tumor growth in liver carcinoma. Hepatology 50(5):1453–1463. doi:10.1002/hep.23143

Lu Z, Wang Y, Chen F, Tong H, Reddy MV, Luo L, Seshadrinathan S, Zhang L, Holthauzen LM, Craig AM, Ren G, Rudenko G (2014) Calsyntenin-3 molecular architecture and interaction with neurexin 1alpha. J Biol Chem 289(50):34530–34542. doi:M114.606806 [pii] 10.1074/jbc.M114.606806

Ludwig A, Blume J, Diep TM, Yuan J, Mateos JM, Leuthauser K, Steuble M, Streit P, Sonderegger P (2009) Calsyntenins mediate TGN exit of APP in a kinesin-1-dependent manner. Traffic 10(5):572–589. doi:TRA886 [pii] 10.1111/j.1600-0854.2009.00886.x

Manie S, Santoro M, Fusco A, Billaud M (2001) The RET receptor: function in development and dysfunction in congenital malformation. Trends Genet 17(10):580–589. doi:S0168-9525(01) 02420-9 [pii]

Matsuda K, Fujishima Y, Maeda N, Mori T, Hirata A, Sekimoto R, Tsushima Y, Masuda S, Yamaoka M, Inoue K, Nishizawa H, Kita S, Ranscht B, Funahashi T, Shimomura I (2015) Positive feedback regulation between adiponectin and T-cadherin impacts adiponectin levels in tissue and plasma of male mice. Endocrinology 156(3):934–946. doi:10.1210/en.2014-1618

McCoy MK, Cookson MR (2012) Mitochondrial quality control and dynamics in Parkinson's disease. Antioxid Redox Signal 16(9):869–882. doi:10.1089/ars.2011.4019

Miles DC, van den Bergen JA, Wakeling SI, Anderson RB, Sinclair AH, Western PS (2012) The proto-oncogene Ret is required for male foetal germ cell survival. Dev Biol 365(1):101–109. doi:S0012-1606(12)00083-8 [pii] 10.1016/j.ydbio.2012.02.014

Miskevich F, Zhu Y, Ranscht B, Sanes JR (1998) Expression of multiple cadherins and catenins in the chick optic tectum. Mol Cell Neurosci 12(4–5):240–255. doi:S1044-7431(98)90718-7 [pii] 10.1006/mcne.1998.0718

Monda KL, North KE, Hunt SC, Rao DC, Province MA, Kraja AT (2010) The genetics of obesity and the metabolic syndrome. Endocr Metab Immune Disord Drug Targets 10(2):86–108. doi: EMID-DT-ABS-32 [pii]

Mukoyama Y, Utani A, Matsui S, Zhou S, Miyachi Y, Matsuyoshi N (2007) T-cadherin enhances cell-matrix adhesiveness by regulating beta1 integrin trafficking in cutaneous squamous carcinoma cells. Genes Cells 12(6):787–796. doi:GTC1092 [pii] 10.1111/j.1365-2443.2007. 01092.x

Mulligan LM (2014) RET revisited: expanding the oncogenic portfolio. Nat Rev Cancer 14 (3):173–186. doi:nrc3680 [pii] 10.1038/nrc3680

Ohgami N, Ida-Eto M, Shimotake T, Sakashita N, Sone M, Nakashima T, Tabuchi K, Hoshino T, Shimada A, Tsuzuki T, Yamamoto M, Sobue G, Jijiwa M, Asai N, Hara A, Takahashi M, Kato M (2010) c-Ret-mediated hearing loss in mice with Hirschsprung disease. Proc Natl Acad Sci U S A 107(29):13051–13056. doi:1004520107 [pii] 10.1073/pnas.1004520107

Ohgami N, Tamura H, Ohgami K, Iida M, Yajima I, Kumasaka MY, Goto Y, Sone M, Nakashima T, Kato M (2012) c-Ret-mediated hearing losses. Int J Clin Exp Pathol 5(1):23–28

Ohnishi K, Melchers F, Shimizu T (2005) Lymphocyte-expressed BILL-cadherin/cadherin-17 contributes to the development of B cells at two stages. Eur J Immunol 35(3):957–963. doi:10. 1002/eji.200425853

Ohno H, Kato S, Naito Y, Kunitomo H, Tomioka M, Iino Y (2014) Role of synaptic phosphatidylinositol 3-kinase in a behavioral learning response in C. elegans. Science 345 (6194):313–317. doi:345/6194/313 [pii] 10.1126/science.1250709

Org E, Eyheramendy S, Juhanson P, Gieger C, Lichtner P, Klopp N, Veldre G, Doring A, Viigimaa M, Sober S, Tomberg K, Eckstein G, Kelgo P, Rebane T, Shaw-Hawkins S, Howard P, Onipinla A, Dobson RJ, Newhouse SJ, Brown M, Dominiczak A, Connell J, Samani N, Farrall M, Caulfield MJ, Munroe PB, Illig T, Wichmann HE, Meitinger T, Laan M (2009) Genome-wide scan identifies CDH13 as a novel susceptibility locus contributing to blood pressure determination in two European populations. Hum Mol Genet 18 (12):2288–2296. doi:ddp135 [pii] 10.1093/hmg/ddp135

Ortiz-Medina H, Emond MR, Jontes JD (2015) Zebrafish calsyntenins mediate homophilic adhesion through their amino-terminal cadherin repeats. Neuroscience 286:87–96. doi: S0306-4522(14)00996-8 [pii] 10.1016/j.neuroscience.2014.11.030

Ouchi N, Parker JL, Lugus JJ, Walsh K (2011) Adipokines in inflammation and metabolic disease. Nat Rev Immunol 11(2):85–97. doi:nri2921 [pii] 10.1038/nri2921

Oue N, Hamai Y, Mitani Y, Matsumura S, Oshimo Y, Aung PP, Kuraoka K, Nakayama H, Yasui W (2004) Gene expression profile of gastric carcinoma: identification of genes and tags potentially involved in invasion, metastasis, and carcinogenesis by serial analysis of gene expression. Cancer Res 64(7):2397–2405

Pan ZW, Li JC (2012) Advances in molecular genetics of Hirschsprung's disease. Anat Rec (Hoboken) 295(10):1628–1638. doi:10.1002/ar.22538

Papassotiropoulos A, Stephan DA, Huentelman MJ, Hoerndli FJ, Craig DW, Pearson JV, Huynh KD, Brunner F, Corneveaux J, Osborne D, Wollmer MA, Aerni A, Coluccia D, Hanggi J, Mondadori CR, Buchmann A, Reiman EM, Caselli RJ, Henke K, de Quervain DJ (2006) Common Kibra alleles are associated with human memory performance. Science 314 (5798):475–478. doi:314/5798/475 [pii] 10.1126/science.1129837

Paradis S, Harrar DB, Lin Y, Koon AC, Hauser JL, Griffith EC, Zhu L, Brass LF, Chen C, Greenberg ME (2007) An RNAi-based approach identifies molecules required for glutamatergic and GABAergic synapse development. Neuron 53(2):217–232. doi:S0896-6273(06)00997-4 [pii] 10.1016/j.neuron.2006.12.012

Park SS, Kang SH, Park JM, Kim JH, Oh SC, Lee JH, Chae YS, Kim SJ, Kim CS, Mok YJ (2007) Expression of liver-intestine cadherin and its correlation with lymph node metastasis in gastric cancer: can it predict N stage preoperatively? Ann Surg Oncol 14(1):94–99. doi:10.1245/s10434-006-9114-9

Parker-Duffen JL, Walsh K (2014) Cardiometabolic effects of adiponectin. Best Pract Res Clin Endocrinol Metab 28(1):81–91. doi:S1521-690X(13)00134-6 [pii] 10.1016/j.beem.2013.09.001

Parker-Duffen JL, Nakamura K, Silver M, Kikuchi R, Tigges U, Yoshida S, Denzel MS, Ranscht B, Walsh K (2013) T-cadherin is essential for adiponectin-mediated revascularization. J Biol Chem 288(34):24886–24897. doi:M113.454835 [pii] 10.1074/jbc.M113.454835

Patel A, Harker N, Moreira-Santos L, Ferreira M, Alden K, Timmis J, Foster K, Garefalaki A, Pachnis P, Andrews P, Enomoto H, Milbrandt J, Pachnis V, Coles MC, Kioussis D, Veiga-Fernandes H (2012) Differential RET signaling pathways drive development of the enteric lymphoid and nervous systems. Sci Signal 5 (235):ra55. doi:5/235/ra55 [pii] 10.1126/scisignal.2002734

Perrais M, Chen X, Perez-Moreno M, Gumbiner BM (2007) E-cadherin homophilic ligation inhibits cell growth and epidermal growth factor receptor signaling independently of other cell interactions. Mol Biol Cell 18(6):2013–2025. doi:E06-04-0348 [pii] 10.1091/mbc.E06-04-0348

Pettem KL, Yokomaku D, Luo L, Linhoff MW, Prasad T, Connor SA, Siddiqui TJ, Kawabe H, Chen F, Zhang L, Rudenko G, Wang YT, Brose N, Craig AM (2013) The specific alpha-neurexin interactor calsyntenin-3 promotes excitatory and inhibitory synapse development. Neuron 80(1):113–128. doi:S0896-6273(13)00616-8 [pii] 10.1016/j.neuron.2013.07.016

Philippova MP, Bochkov VN, Stambolsky DV, Tkachuk VA, Resink TJ (1998) T-cadherin and signal-transducing molecules co-localize in caveolin-rich membrane domains of vascular smooth muscle cells. FEBS Lett 429(2):207–210. doi:S0014-5793(98)00598-5 [pii]

Philippova M, Ivanov D, Tkachuk V, Erne P, Resink TJ (2003) Polarisation of T-cadherin to the leading edge of migrating vascular cells in vitro: a function in vascular cell motility? Histochem Cell Biol 120(5):353–360. doi:10.1007/s00418-003-0584-6

Philippova M, Ivanov D, Allenspach R, Takuwa Y, Erne P, Resink T (2005) RhoA and Rac mediate endothelial cell polarization and detachment induced by T-cadherin. FASEB J 19 (6):588–590. doi:04-2430fje [pii] 10.1096/fj.04-2430fje

Philippova M, Banfi A, Ivanov D, Gianni-Barrera R, Allenspach R, Erne P, Resink T (2006) Atypical GPI-anchored T-cadherin stimulates angiogenesis in vitro and in vivo. Arterioscler Thromb Vasc Biol 26(10):2222–2230. doi:01.ATV.0000238356.20565.92 [pii] 10.1161/01.ATV.0000238356.20565.92

Philippova M, Ivanov D, Joshi MB, Kyriakakis E, Rupp K, Afonyushkin T, Bochkov V, Erne P, Resink TJ (2008) Identification of proteins associating with glycosylphosphatidylinositol-anchored T-cadherin on the surface of vascular endothelial cells: role for Grp78/BiP in T-cadherin-dependent cell survival. Mol Cell Biol 28(12):4004–4017. doi:MCB.00157-08 [pii] 10.1128/MCB.00157-08

Philippova M, Joshi MB, Kyriakakis E, Pfaff D, Erne P, Resink TJ (2009) A guide and guard: the many faces of T-cadherin. Cell Signal 21(7):1035–1044

Philippova M, Joshi MB, Pfaff D, Kyriakakis E, Maslova K, Erne P, Resink TJ (2012) T-cadherin attenuates insulin-dependent signalling, eNOS activation, and angiogenesis in vascular endothelial cells. Cardiovasc Res 93(3):498–507. doi:cvs004 [pii] 10.1093/cvr/cvs004

Ponomareva OY, Holmen IC, Sperry AJ, Eliceiri KW, Halloran MC (2014) Calsyntenin-1 regulates axon branching and endosomal trafficking during sensory neuron development in vivo. J Neurosci 34(28):9235–9248. doi:34/28/9235 [pii] 10.1523/JNEUROSCI.0561-14.2014

Rakha EA, Green AR, Powe DG, Roylance R, Ellis IO (2006) Chromosome 16 tumor-suppressor genes in breast cancer. Genes Chromosomes Cancer 45(6):527–535. doi:10.1002/gcc.20318

Ranscht B, Bronner-Fraser M (1991) T-cadherin expression alternates with migrating neural crest cells in the trunk of the avian embryo. Development 111(1):15–22

Ranscht B, Dours-Zimmermann MT (1991) T-cadherin, a novel cadherin cell adhesion molecule in the nervous system lacks the conserved cytoplasmic region. Neuron 7(3):391–402. doi:0896-6273(91)90291-7 [pii]

Read RD, Goodfellow PJ, Mardis ER, Novak N, Armstrong JR, Cagan RL (2005) A Drosophila model of multiple endocrine neoplasia type 2. Genetics 171(3):1057–1081. doi:genetics.104.038018 [pii] 10.1534/genetics.104.038018

Redies C, Hertel N, Hubner CA (2012) Cadherins and neuropsychiatric disorders. Brain Res 1470:130–144. doi:S0006-8993(12)01073-6 [pii] 10.1016/j.brainres.2012.06.020

Resink TJ, Philippova M, Joshi MB, Kyriakakis E, Erne P (2009) Cadherins and cardiovascular disease. Swiss Med Wkly 139(9–10):122–134. doi:smw-12429 [pii] smw-12429

Rivero O, Sich S, Popp S, Schmitt A, Franke B, Lesch KP (2013) Impact of the ADHD-susceptibility gene CDH13 on development and function of brain networks. Eur Neuropsychopharmacol 23(6):492–507. doi:S0924-977X(12)00161-7 [pii] 10.1016/j.euroneuro.2012.06.009

Rubina K, Talovskaya E, Cherenkov V, Ivanov D, Stambolsky D, Storozhevykh T, Pinelis V, Shevelev A, Parfyonova Y, Resink T, Erne P, Tkachuk V (2005) LDL induces intracellular signalling and cell migration via atypical LDL binding protein T-cadherin. Mol Cell Biochem 273(1–2):33–41

Rubina K, Kalinina N, Potekhina A, Efimenko A, Semina E, Poliakov A, Wilkinson DG, Parfyonova Y, Tkachuk V (2007) T-cadherin suppresses angiogenesis in vivo by inhibiting migration of endothelial cells. Angiogenesis 10(3):183–195. doi:10.1007/s10456-007-9072-2

Sakamoto N, Oue N, Sentani K, Anami K, Uraoka N, Naito Y, Oo HZ, Hinoi T, Ohdan H, Yanagihara K, Aoyagi K, Sasaki H, Yasui W (2012) Liver-intestine cadherin induction by epidermal growth factor receptor is associated with intestinal differentiation of gastric cancer. Cancer Sci 103(9):1744–1750. doi:10.1111/j.1349-7006.2012.02353.x

Santoro M, Carlomagno F (2013) Central role of RET in thyroid cancer. Cold Spring Harb Perspect Biol 5 (12):a009233. doi:5/12/a009233 [pii] 10.1101/cshperspect.a009233

Schalm SS, Ballif BA, Buchanan SM, Phillips GR, Maniatis T (2010) Phosphorylation of protocadherin proteins by the receptor tyrosine kinase Ret. Proc Natl Acad Sci U S A 107 (31):13894–13899. doi:1007182107 [pii] 10.1073/pnas.1007182107

Semina EV, Rubina KA, Sysoeva VY, Rutkevich PN, Kashirina NM, Tkachuk VA (2014) Novel mechanism regulating endothelial permeability via T-cadherin-dependent VE-cadherin phosphorylation and clathrin-mediated endocytosis. Mol Cell Biochem 387(1–2):39–53. doi:10.1007/s11010-013-1867-4

Shibata R, Sato K, Pimentel DR, Takemura Y, Kihara S, Ohashi K, Funahashi T, Ouchi N, Walsh K (2005) Adiponectin protects against myocardial ischemia-reperfusion injury through AMPK- and COX-2-dependent mechanisms. Nat Med 11(10):1096–1103. doi:nm1295 [pii] 10.1038/nm1295

Shimano M, Ouchi N, Shibata R, Ohashi K, Pimentel DR, Murohara T, Walsh K (2010) Adiponectin deficiency exacerbates cardiac dysfunction following pressure overload through disruption of an AMPK-dependent angiogenic response. J Mol Cell Cardiol 49(2):210–220. doi:S0022-2828(10)00082-9 [pii] 10.1016/j.yjmcc.2010.02.021

Simons K, Toomre D (2000) Lipid rafts and signal transduction. Nat Rev Mol Cell Biol 1 (1):31–39. doi:10.1038/35036052 35036052 [pii]

Simpson J, Keefe J, Nishi R (2013) Differential effects of RET and TRKB on axonal branching and survival of parasympathetic neurons. Dev Neurobiol 73(1):45–59. doi:10.1002/dneu. 22036

Ster J, Steuble M, Orlando C, Diep TM, Akhmedov A, Raineteau O, Pernet V, Sonderegger P, Gerber U (2014) Calsyntenin-1 regulates targeting of dendritic NMDA receptors and dendritic spine maturation in CA1 hippocampal pyramidal cells during postnatal development. J Neurosci 34(26):8716–8727. doi:34/26/8716 [pii] 10.1523/JNEUROSCI.0144-14.2014

Steuble M, Gerrits B, Ludwig A, Mateos JM, Diep TM, Tagaya M, Stephan A, Schatzle P, Kunz B, Streit P, Sonderegger P (2010) Molecular characterization of a trafficking organelle: dissecting the axonal paths of calsyntenin-1 transport vesicles. Proteomics 10(21):3775–3788. doi:10. 1002/pmic.201000384

Steuble M, Diep TM, Schatzle P, Ludwig A, Tagaya M, Kunz B, Sonderegger P (2012) Calsyntenin-1 shelters APP from proteolytic processing during anterograde axonal transport. Biol Open 1(8):761–774. doi:10.1242/bio.20121578 BIO20121578 [pii]

Takahashi M, Ritz J, Cooper GM (1985) Activation of a novel human transforming gene, ret, by DNA rearrangement. Cell 42(2):581–588. doi:0092-8674(85)90115-1 [pii]

Takamura M, Sakamoto M, Ino Y, Shimamura T, Ichida T, Asakura H, Hirohashi S (2003) Expression of liver-intestine cadherin and its possible interaction with galectin-3 in ductal adenocarcinoma of the pancreas. Cancer Sci 94(5):425–430

Takamura M, Yamagiwa S, Matsuda Y, Ichida T, Aoyagi Y (2013) Involvement of liver-intestine cadherin in cancer progression. Med Mol Morphol 46(1):1–7. doi:10.1007/s00795-012-0003-y

Takei N, Sobu Y, Kimura A, Urano S, Piao Y, Araki Y, Taru H, Yamamoto T, Hata S, Nakaya T, Suzuki T (2015) Cytoplasmic Fragment of Alcadein alpha Generated by Regulated Intramembrane Proteolysis Enhances Amyloid beta-Protein Precursor (APP) Transport into the Late Secretory Pathway and Facilitates APP Cleavage. J Biol Chem 290(2):987–995. doi: M114.599852 [pii] 10.1074/jbc.M114.599852

Thedieck C, Kuczyk M, Klingel K, Steiert I, Muller CA, Klein G (2005) Expression of Ksp-cadherin during kidney development and in renal cell carcinoma. Br J Cancer 92 (11):2010–2017. doi:6602597 [pii] 10.1038/sj.bjc.6602597

Thedieck C, Kalbacher H, Kratzer U, Lammers R, Stevanovic S, Klein G (2008) alpha B-crystallin is a cytoplasmic interaction partner of the kidney-specific cadherin-16. J Mol Biol 378 (1):145–153. doi:S0022-2836(08)00174-5 [pii] 10.1016/j.jmb.2008.02.008

Thomson RB, Aronson PS (1999) Immunolocalization of Ksp-cadherin in the adult and developing rabbit kidney. Am J Physiol 277(1 Pt 2):F146–F156

Thomson RB, Igarashi P, Biemesderfer D, Kim R, Abu-Alfa A, Soleimani M, Aronson PS (1995) Isolation and cDNA cloning of Ksp-cadherin, a novel kidney-specific member of the cadherin multigene family. J Biol Chem 270(29):17594–17601

Tkachuk VA, Bochkov VN, Philippova MP, Stambolsky DV, Kuzmenko ES, Sidorova MV, Molokoedov AS, Spirov VG, Resink TJ (1998) Identification of an atypical lipoprotein-binding protein from human aortic smooth muscle as T-cadherin. FEBS Lett 421 (3):208–212. doi:S0014-5793(97)01562-7 [pii]

Treutlein J, Cichon S, Ridinger M, Wodarz N, Soyka M, Zill P, Maier W, Moessner R, Gaebel W, Dahmen N, Fehr C, Scherbaum N, Steffens M, Ludwig KU, Frank J, Wichmann HE, Schreiber S, Dragano N, Sommer WH, Leonardi-Essmann F, Lourdusamy A, Gebicke-Haerter P, Wienker TF, Sullivan PF, Nothen MM, Kiefer F, Spanagel R, Mann K, Rietschel M (2009) Genome-wide association study of alcohol dependence. Arch Gen Psychiatry 66 (7):773–784. doi:66/7/773 [pii] 10.1001/archgenpsychiatry.2009.83

Tyrberg B, Miles P, Azizian KT, Denzel MS, Nieves ML, Monosov EZ, Levine F, Ranscht B (2011) T-cadherin (Cdh13) in association with pancreatic beta-cell granules contributes to second phase insulin secretion. Islets 3(6):327–337. doi:17705 [pii] 10.4161/isl.3.6.17705

Uchida Y, Gomi F, Murayama S, Takahashi H (2013) Calsyntenin-3 C-terminal fragment accumulates in dystrophic neurites surrounding abeta plaques in tg2576 mouse and Alzheimer disease brains: its neurotoxic role in mediating dystrophic neurite formation. Am J Pathol 182 (5):1718–1726. doi:S0002-9440(13)00098-9 [pii] 10.1016/j.ajpath.2013.01.014

Vagnoni A, Perkinton MS, Gray EH, Francis PT, Noble W, Miller CC (2012) Calsyntenin-1 mediates axonal transport of the amyloid precursor protein and regulates Abeta production. Hum Mol Genet 21(13):2845–2854. doi:dds109 [pii] 10.1093/hmg/dds109

van Weering DH, Moen TC, Braakman I, Baas PD, Bos JL (1998) Expression of the receptor tyrosine kinase Ret on the plasma membrane is dependent on calcium. J Biol Chem 273 (20):12077–12081

Veiga-Fernandes H, Coles MC, Foster KE, Patel A, Williams A, Natarajan D, Barlow A, Pachnis V, Kioussis D (2007) Tyrosine kinase receptor RET is a key regulator of Peyer's patch organogenesis. Nature 446(7135):547–551. doi:nature05597 [pii] 10.1038/nature05597

Vogt L, Schrimpf SP, Meskenaite V, Frischknecht R, Kinter J, Leone DP, Ziegler U, Sonderegger P (2001) Calsyntenin-1, a proteolytically processed postsynaptic membrane protein with a cytoplasmic calcium-binding domain. Mol Cell Neurosci 17(1):151–166. doi:10.1006/mcne.2000.0937 S1044-7431(00)90937-0 [pii]

Wang Q, Liu L, Pei L, Ju W, Ahmadian G, Lu J, Wang Y, Liu F, Wang YT (2003) Control of synaptic strength, a novel function of Akt. Neuron 38(6):915–928. doi:S0896627303003568 [pii]

Wang Y, Shek FH, Wong KF, Liu LX, Zhang XQ, Yuan Y, Khin E, Hu MY, Wang JH, Poon RT, Hong W, Lee NP, Luk JM (2013) Anti-cadherin-17 antibody modulates beta-catenin signaling and tumorigenicity of hepatocellular carcinoma. PLoS One 8(9):e72386. doi:10.1371/journal.pone.0072386 PONE-D-13-11881 [pii]

Weis SM, Lindquist JN, Barnes LA, Lutu-Fuga KM, Cui J, Wood MR, Cheresh DA (2007) Cooperation between VEGF and beta3 integrin during cardiac vascular development. Blood 109(5):1962–1970. doi:blood-2005-10-038893 [pii] 10.1182/blood-2005-10-038893

Wendeler MW, Praus M, Jung R, Hecking M, Metzig C, Gessner R (2004) Ksp-cadherin is a functional cell-cell adhesion molecule related to LI-cadherin. Exp Cell Res 294(2):345–355. doi:10.1016/j.yexcr.2003.11.022 S0014482703006098 [pii]

Wendeler MW, Jung R, Himmelbauer H, Gessner R (2006) Unique gene structure and paralogy define the 7D-cadherin family. Cell Mol Life Sci 63(13):1564–1573. doi:10.1007/s00018-006-6014-x

Wendeler MW, Drenckhahn D, Gessner R, Baumgartner W (2007) Intestinal LI-cadherin acts as a Ca^{2+-} dependent adhesion switch. J Mol Biol 370(2):220–230. doi:S0022-2836(07)00550-5 [pii] 10.1016/j.jmb.2007.04.062

Wertz K, Herrmann BG (1999) Kidney-specific cadherin (cdh16) is expressed in embryonic kidney, lung, and sex ducts. Mech Dev 84(1–2):185–188. doi:S0925-4773(99)00074-X [pii]

Wyder L, Vitaliti A, Schneider H, Hebbard LW, Moritz DR, Wittmer M, Ajmo M, Klemenz R (2000) Increased expression of H/T-cadherin in tumor-penetrating blood vessels. Cancer Res 60(17):4682–4688

Yamauchi T, Kadowaki T (2013) Adiponectin receptor as a key player in healthy longevity and obesity-related diseases. Cell Metab 17(2):185–196. doi:S1550-4131(13)00006-5 [pii] 10.1016/j.cmet.2013.01.001

Zhu R, Wong KF, Lee NP, Lee KF, Luk JM (2010) HNF1alpha and CDX2 transcriptional factors bind to cadherin-17 (CDH17) gene promoter and modulate its expression in hepatocellular carcinoma. J Cell Biochem 111(3):618–626. doi:10.1002/jcb.22742

Zuercher WJ, Turunen BJ, Lackey KE (2010) Current review of small molecule Ret kinase inhibitors. Mini Rev Med Chem 10(2):138–146. doi:MRMC Epub-037 [pii]

Part IV
Cadherins in Tissues and Diseases

Chapter 12
Cadherins in Neural Development

Lewis L. Brayshaw and Stephen R. Price

Abstract Cadherins play many diverse roles in the development of the nervous system of vertebrates. Far from being simple adhesion molecules, they also orchestrate cell generation, cell movements, and cell morphogenesis. Cadherins also regulate specificity of cell-to-cell interactions during neuronal circuit formation and function. Cadherin expression during neural development is also dynamic and highly regulated. At each phase of embryo development, cadherins emerge as key molecular determinants of neural function through their many diverse binding partners. Additionally, they play important roles in the plasticity of the nervous system, a key feature believed to underpin the ability of the brain to function. Many neurodevelopmental disorders also have cadherin disfunction at their heart indicating that cadherin-based therapies may emerge as future treatments for these devastating conditions.

Keywords Cadherin • Neuron • Progenitor cell • Morphogenesis • Synapse • Mental disorders • Dendrite • Axon • Nucleogenesis

12.1 Introduction

The nervous system of vertebrates is a functioning structure of awe-inspiring complexity. Many billions of neurons connect and communicate with one another via many trillions of structures known as synapses. Synapses are dynamic and can be strengthened or weakened, created or destroyed depending on the needs of the given neuronal circuit for information storage or processing. But these circuits are not born fully formed. During embryo development many thousands of different specialised subtypes of cells are generated which have to migrate to where they will assemble into a circuit. Concomitantly, neurons elaborate an axonal process that will grow towards and make contact with the other neurons in the circuit which may be some distance away from the neuron cell body. Cadherins are found expressed

L.L. Brayshaw • S.R. Price (✉)
Research Department of Cell and Developmental Biology, University College London, Gower Street, London WC1E 6BT, UK
e-mail: Stephen.price@ucl.ac.uk

S.T. Suzuki, S. Hirano (eds.), *The Cadherin Superfamily*,
DOI 10.1007/978-4-431-56033-3_12

differentially in most areas of the developing nervous system and have been shown to play key roles in the assembly and functional plasticity of neuronal circuits. For neurons, cadherins are more than just cell adhesion proteins; they are essential to the formation and function of the entire brain. As the cadherin family consists of over 100 different subtypes, a whole textbook could be written on the role of cadherins in neural development alone. Consequently, we have decided to focus our attention on the roles of members of the classical cadherin family in vertebrate nervous system development. We discuss current understanding of cadherin function in neural development by first discussing their role in neural tissue morphogenesis. We then discuss cadherin function in the formation of individual postmitotic neurons with their characteristic morphology and finally their role in coordinating cell-to-cell assemblies which, when they go wrong, could underpin neurodevelopmental disorders. This latter field of endeavour is still in its infancy and much work remains to uncover the many varied roles that cadherins play in the lifetime of the nervous system.

12.2 Cadherins in Neural Tissue Morphogenesis

Tissue morphogenesis during neural development requires coordinated changes in cell shape, adhesion, and movement. As tissue morphogenesis involves the collective movement of cells together, cadherins have an obvious function in maintaining cell–cell adhesions throughout gross changes in embryo structure. However, this function is far from trivial, as the expression of multiple cadherin subtypes must be tightly regulated in time and space, and the dynamic assembly and disassembly of cadherin-mediated interaction accurately orchestrated in order to permit change without loss of tissue integrity. In this section, the involvement of cadherins in several processes of early embryo and neural tissue morphogenesis is outlined, and the role of cadherins in processes beyond simply cell adhesion is discussed.

12.2.1 Gastrulation

Cadherins play a central role in one of the earliest of morphogenetic processes in embryos, gastrulation. Gastrulation involves large-scale cell movements to reorganize the embryo from the blastula, a simple single-layered sphere of cells, into a trilayered structure known as the gastrula (Fig. 12.1a; Solnica-Krezel and Sepich 2012; Ozair et al. 2013). The endoderm, mesoderm, and ectoderm are the primary germ layers formed during gastrulation and will, amongst other tissues, give rise to the digestive system, muscles, and nervous system, respectively. Cadherins have been shown to be crucial mediators of cell–cell adhesions during the morphogenetic movements of gastrulation in vertebrates (Tepass et al. 2000; Nakaya and Sheng 2008; Nishimura and Takeichi 2009). In zebrafish, E-cadherin facilitates adhesion

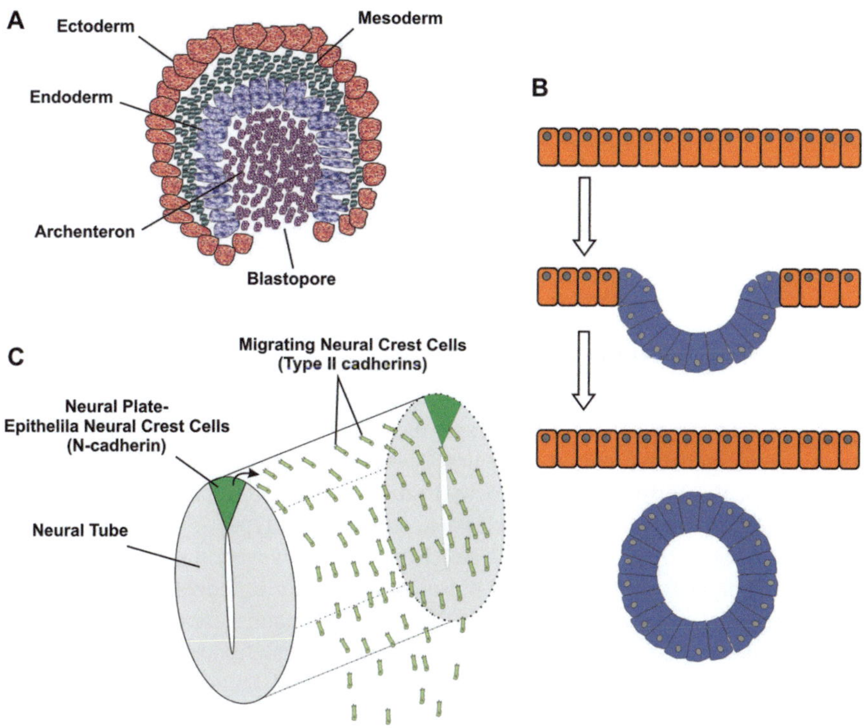

Fig. 12.1 Cadherin roles in neural tissue morphogenesis. Cadherin cell–cell adhesions are important in maintaining tissue integrity of morphogenic structures, such as the gastrula (**a**). Dynamic regulation of cadherins is required for gross cellular rearrangements such as neurulation, where E-cadherin is replaced by N cadherin in the invaginating neural plate (**b**). Cadherin subtype switching facilitates EMT during neural crest cell migration by permitting key changes in a cell's adhesive interactions and phenotype (**c**)

between the enveloping layer and deep cells, two cellular domains in the zebrafish blastula, and inhibition of E-cadherin expression significantly disrupts gastrulation processes such as epiboly movement and the thinning and spreading of the ectoderm (Babb and Marrs 2004; Shimizu et al. 2005; Lepage and Bruce 2010). Similarly, in *Xenopus* embryos C-cadherin adhesions are crucial for gastrulation movements as expression of dominant-negative C-cadherin results in failure to close the blastopore and impaired involution (Lee and Gumbiner 1995). Whereas cadherin-mediated adhesions are important for maintaining structural integrity of the tissue and facilitating collective cell migration, adhesions must also be downregulated in order to permit movement and changes in the tissue by promoting epithelial-to-mesenchymal transition (EMT). For example, C-cadherin must be downregulated by mesoderm-inducing factor activin in order to permit convergent extension in *Xenopus* embryos, which is the anterior–posterior extension of the embryo as cells move towards and intercalate at the dorsal midline (Zhong et al. 1999). In zebrafish and mice, FGF signalling promotes EMT during gastrulation by Snail-mediated transcriptional downregulation of E-cadherin, and the mesoderm in mice deficient in Snail activity are unable to lose epithelial morphology and apicobasal cell polarity (Ciruna and Rossant 2001; Carver et al. 2001). Furthermore, disassembly of cadherin adhesions must occur rapidly in order to correlate with the gross movements of gastrulation, therefore cadherins are also regulated at the protein level. For example, EPB4.1 L5, p38 interacting protein and p38-MAP kinase all downregulate E-cadherin during EMT in gastrulation (Hirano et al. 2008; Zohn et al. 2006).

12.2.2 Neurulation

Cadherin subtypes display a distinct spatiotemporal expression pattern throughout many morphogenetic processes in neural development (Hirano et al. 2012). During neurulation, the formation of the neural tube, E-cadherin expression is replaced with N-cadherin as well as other classical cadherin subtypes in the dorsal neural ectoderm (Fig. 12.1b; Hatta and Takeichi 1986). However, the purpose of this cadherin subtype switching and its correlation to the morphogenetic movements during neurulation is under debate. In N-cadherin mutant zebrafish, key cellular rearrangements such as convergent extension and intercalation are impaired during neurulation (Hong and Brewster 2006). However, in N-cadherin knockout mice, neural tube formation and closure occur normally with only some slight malformations in the tissue organization (Radice et al. 1997). Furthermore, close analysis of cadherin expression patterns during early morphogenesis in chick embryos revealed that the kinetics of E-to-N switching do not appear to be synchronised with the movements of neurulation (Dady et al. 2012). Instead, based on the fact that the transcriptional regulators involved are distinct from those in epithelial-to-mesenchymal transition (EMT), it is suggested that the switch from E-cadherin to N-cadherin during neurulation is more a reflection of the segregation of the neuroectoderm into its three main populations: ectoderm, neural crest, and neural tube (Dady et al. 2012). This is an interesting example where the loss of E-cadherin and gain of N-cadherin do not result in EMT, unlike during tumourigenesis and cancer metastasis (Hazan et al. 2004).

12.2.3 Neural Crest Cell Migration

Epithelial-to-mesenchymal transition (EMT) is an extraordinary process in which cells undergo changes in cell shape and adhesion to transform from an epithelial phenotype into a migratory one (Thiery et al. 2009). EMT is required for multiple tissue morphogenic movements in neural development and cadherins play a major role in facilitating EMT, as cadherin subtype switching is required for key changes in a cell's adhesive interactions and phenotype (Theveneau and Mayor 2012).

Neural crest cells (NCCs) are a neural stem cell population located at the neural plate border that give rise to craniofacial structures, smooth muscles, cells of the cardiac system, and most of the neurons and glial of the peripheral nervous system (Dupin et al. 2006; Hall 2008). In a process called delamination, NCCs undergo EMT and detach from neighbouring neuroepithelial cells in the neural plate in order to migrate to various destinations in the embryo and differentiate (Fig. 12.1c; Thiery et al. 2009; Theveneau and Mayor 2012). During EMT, NCCs typically undergo a switch in cadherin expression, downregulating N-cadherin and upregulating Type II cadherins 6/7/11 (Nakagawa and Takeichi 1995; Vallin et al. 1998). At the initiation of EMT, N-cadherin expression is downregulated posttranslationally by the activation of metalloprotease ADAM10 by BMP/Wnt signalling in NCCs (Hall and Erickson 2003). Cleavage of N-cadherin aids NCC delamination firstly by loosening cell–cell adhesions, and secondly by the cytosolic cleavage-product of N-cadherin inducing transcription of cyclin-D1, which results in the activation of β-catenin signalling, an important promoter of NCC EMT (Shoval et al. 2007). Prior to delamination, premigratory NCCs express cadherin-6 (formally cadherin-6B in chick) and the expression of this cadherin is believed to play a role in segregating this population from other cells in the neuroepithelium, which do not express cadherin-6 (Coles et al. 2007; Theveneau and Mayor 2012). Following emigration from the neural tube, all populations of NCCs lack cadherin-6 expression. However, differences in the timing of downregulation suggest that cadherin-6 adhesions mediate different functions in the delamination of cranial and trunk NCC populations (Clay and Halloran 2014). As cranial NCCs undergo EMT, their cadherin-6 levels are rapidly reduced transcriptionally by Snail2 and posttranslationally via proteolytic cleavage by ADAM10, ADAM9, and γ-secretase (Taneyhill et al. 2007; Schiffmacher et al. 2014). Furthermore, evidence shows that this loss of cadherin-6 adhesion is critical for the transformation of cranial NCCs to the migratory state. *In ovo* knockout of cadherin-6 in chick embryos increases cranial NCC emigration from the neural plate and in vitro results support the conclusion that loss of cadherin-6 adhesions play a critical role in regulating the timing of cranial NCC delamination (Coles 2007; Schiffmacher et al. 2014). Trunk NCCs, on the other hand, maintain cadherin-6 expression throughout EMT and downregulation is only observed in chick and zebrafish embryos following delamination (Clay and Halloran 2014; Park and Gumbiner 2010). Furthermore, the current evidence actually points towards a pro-EMT role for cadherin-6 adhesions in trunk NCCs (Taneyhill and Schiffmacher 2013). Cadherin-6–mediated bone morphogenetic signalling has been shown to promote de-epithelialisation in chick trunk NCCs and in vivo live-cell imaging of zebrafish reveals a novel role for cadherin-6 in promoting apical detachment of NCCs by regulating F-actin dynamics (Park and Gumbiner 2010; Park and Gumbiner 2012; Clay and Halloran 2014). The involvement of cadherins in NCC development is complex, but the benefits of our understanding are great due to the number of NCC-related developmental disorders, neurocristopathies, and due to the remarkable similarity between the mechanisms used to regulate cadherin adhesions in NCCs and tumour cells during EMT (Theveneau 2012; Mayor and Theveneau 2013).

12.2.4 Retinal Morphogenesis

Retinal morphogenesis is a highly complex process in neural development that results in the astonishing organ that is the eye. Cadherin-mediated adhesions have been demonstrated to be important for retinal development in mice, chick, and zebrafish and in particular are crucial for appropriate patterning of the retina (Matsunaga et al. 1988b; Rungger-Brändle et al. 2010; Chen et al. 2012). Expression patterns of N-cadherin and R-cadherin are regulated by pax6 in the retina and zebrafish cadherin-mutants have severe disorders in retinal lamination (Rungger-Brändle et al. 2010). Additionally, cadherin adhesions have been shown to be important for optic fissure closure in both zebrafish and mice (Masai 2003, Chen 2012). Optic fissure closure is an important epithelial fusion event in eye development, and occurs when the growing edges of the optic cup at the optic fissure margins come together and fuse in order to form a continuous optic cup (Chow and Lang 2001). Cadherins are believed to play an important role in coordinating cell morphology changes in the optic fissure margins to ensure proper alignment and closure (Chen 2012). Furthermore, cadherin-defective mice develop coloboma, a congenital disease leading to childhood blindness. Therefore, there is hope that a better understanding of cadherin functions in optic fissure closure will help to pin down the cellular mechanisms responsible for coloboma development in humans (Chang et al. 2006).

12.2.5 Summary

There has been significant progress in elucidating the complex cadherin expression patterns seen throughout tissue morphogenesis, and the development of live-cell microscopy techniques to follow cadherin dynamics in vivo has been invaluable. However, in understanding the roles of cadherin in a tissue morphogenetic process, having an accurate spatiotemporal map of cadherin expression is just the first step. It is clear that the resulting effect of a specific cadherin's expression is strongly dependent on cellular context and cadherins can carry out multiple roles during neural development. Thus, as well as identifying the downstream effects of cadherin expression, it is equally important to elucidate the integration of upstream signals which regulate cadherin functions in a given morphogenetic process. Although challenging, obtaining an appreciation of the precise and various functions of cadherins in tissue morphogenesis will be crucial in advancing our knowledge of related neurodevelopmental disorders and cancer pathogenesis.

12.3 Cadherins in Neural Progenitor Cell Maintenance and Differentiation

During neural development, neural progenitor cells (NPCs) give rise to all of the neuronal cells that will populate the entire adult nervous system (Temple 2001). In this process called neurogenesis, it is essential that these self-renewing progenitor cells are sustained as well as undergo differentiation at the appropriate time and place in the embryo (Götz and Huttner 2005; Doe 2008). Cadherin molecules play a central role in this balance of NPC maintenance and differentiation, as many processes hinge on the appropriate assembly and disassembly of cadherin-mediated adhesions (Halbleib and Nelson 2006). For the maintenance of NPCs, cadherins have functions in organization, regulation of NPC proliferation, and preservation of NPC identity. Consequently, dynamic downregulation of cadherin-mediated adhesions is also necessary for NPC differentiation and migration during neurogenesis. This section provides an overview of cadherin roles in NPC maintenance and differentiation, and discusses how regulation of adhesion, signalling, and cell polarity is vital in eliciting these roles.

12.3.1 Cadherins in Organising the NPC Microenvironment

The appropriate structure and organization of the NPC microenvironment, known as the ventricular zone, is both essential to the maintenance of NPCs and to the process of neurogenesis (Fig. 12.2). This includes the controlled residence of NPCs, as removal of NPCs from the ventricular zone results in their exit from the cell cycle and terminal differentiation (Temple 2001). Cadherins facilitate many of the adhesions that are critical to the positioning of NPCs and maintenance of the NPC microenvironment. Early in development, cadherins in adherens junctions physically link neuroepithelial progenitors to each other and to the ventricular surface of the neuroepithelium (Chenn et al. 1998). Unlike in adult stem cell microenvironments, the neuroepithelium is made up entirely of precursor cells with no supporting cells present, thus the cell–cell adhesions mediated by cadherins are essential to maintaining the integrity of the NPC microenvironment (Takahashi et al. 1993). Further in development, as radial glial cells become the predominant NPC, cadherin adhesions also play an important role in organising radial glial cells and providing the architecture for neurogenesis. N-cadherin is responsible for the anchoring of radial glial cells to the ventricular surface of the developing cortex and loss of N-cadherin results in the failure of radial glial cells to extend processes from the apical surface to the basal lamina of the cortical layer, which provide the migrational track for newly formed neuronal cells during neurogenesis (Kadowaki et al. 2007). As well as being necessary for the formation of these processes, N-cadherin is also required for neuronal attachment to radial glial processes during migration (Shikanai et al. 2011). Therefore, deregulation of cadherin molecules has

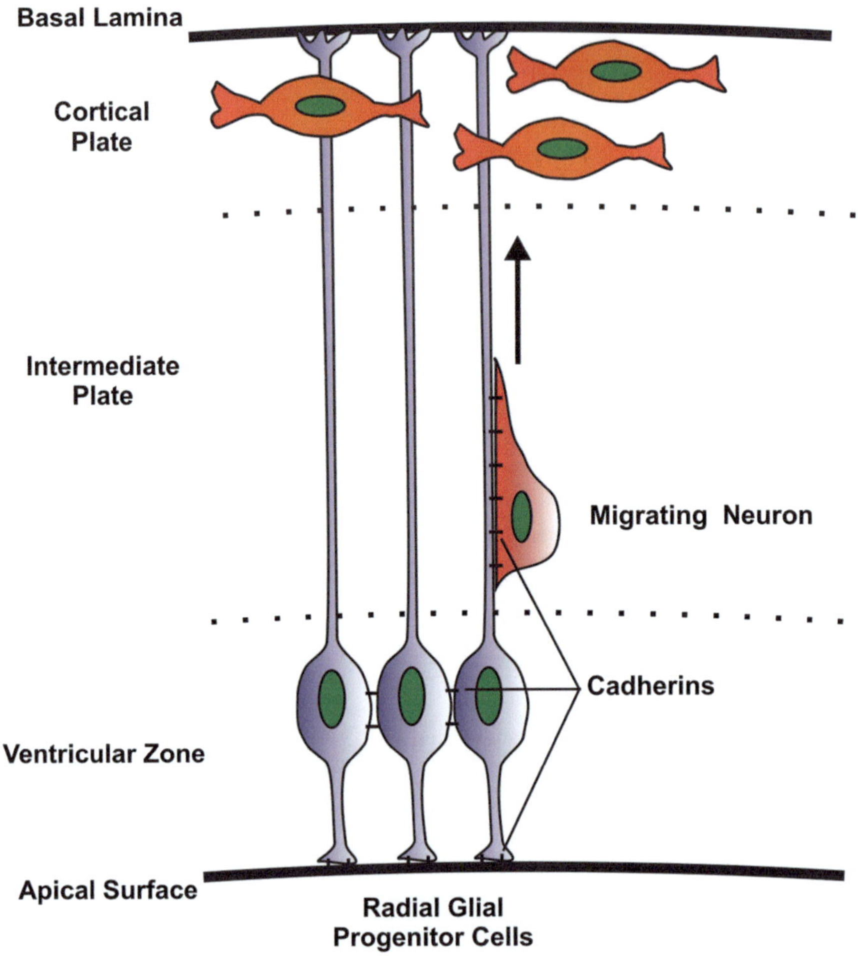

Fig. 12.2 Neurogenesis and the organisation of the developing cortex. Radial glial neural progenitor cells reside in the ventricular zone and extend processes to the basal lamina which form the migrational track for nascent neurons. Cadherin adhesions are important in providing the architecture for neurogenesis, attaching radial glial cells to each other and to the apical surface as well as facilitating neuronal attachment to radial glial processes

dire consequence for neurogenesis and results in improper population and layering of the cerebral cortex (Marín et al. 2010; Kadowaki et al. 2007).

12.3.2 Cadherins in NPC Proliferation

Even subtle changes to NPC populations can result in dire developmental consequences. Insufficient proliferation can result in microcephaly, abnormally reduced brain size (Gilmore and Walsh 2013), whereas uncontrolled growth of NPCs has been linked to brain tumours such as astrocytomas (Perego et al. 2002) and medulloblastomas (Swartling et al. 2012). Cadherin adhesions within adherens junctions have been demonstrated to play an important role in the regulation of NPC proliferation during development. For example, mislocalisation of N-cadherin in the developing cortex and spinal cord results in hyperproliferation of NPCs and the formation of tumour-like rosettes (Teng et al. 2005).

The classic link between cadherin adhesion and cell proliferation has always been β-catenin. β-catenin connects cadherin molecules to the cell's actin cytoskeleton and is a central signalling molecule in the Wnt-pathway, which is responsible for NPC growth and cell cycling (Chenn and Walsh 2003; Junghans et al. 2005). It is believed that cadherin adhesion sequesters β-catenin to the cell membrane, inhibiting its activity and thus cell proliferation pathways (Nelson and Nusse 2004). Some evidence for this mechanism in NPCs has been found in vitro, however, there is a growing body of in vivo evidence supporting a positive regulatory role of cadherin adhesion on β-catenin signalling and subsequent proliferation in neural progenitor cells (Noles and Chenn 2007; Zhang et al. 2010; Zhang et al. 2013). Most recently, N-cadherin was shown to be necessary for proper proliferation and cell cycling of NPCs in developing mice cortex by activating β-catenin signalling in a Wnt-mediated manner (Zhang et al. 2013). Furthermore, these findings also revealed N-cadherin positively regulates AKT activity, an inhibitor of NPC exit from the VZ and apoptosis, demonstrating an additional mechanism by which cadherin regulates NPC proliferation during development.

It is clear that further understanding of the complex relationship between cadherins and β-catenin in the context of NPC proliferation is required. Although, what is generally accepted is that it is the cell autonomous changes in β-catenin signalling and not changes in cadherin cell adhesion which primarily regulate NPC proliferation (Farkas and Huttner 2008). There is some suggestion, however, for a possible non-cell autonomous role for cadherins in controlling precursor behaviour in neural development. Earlier studies have shown a N-cadherin–dependent increase in β-catenin transcriptional activation when neural precursors are cultured at high density, suggesting a possible cell–cell contact ('outside–in') regulation mechanism (Zhang et al. 2010). However, recent attempts to test this non-cell autonomous regulation of cadherin adhesion on β-catenin and NPC proliferation in vitro were inconclusive (Zhang et al. 2013). Additional investigation will be required to evaluate a role for cadherins in transducing extracellular signals and will help to further understand the interplay of cadherin adhesion and signalling functions in regulating the proliferation of NPCs during development.

12.3.3 Cadherins in the Maintenance of NPC Identity

In order to give rise to the millions of cells in the nervous system, NPCs must be maintained in the undifferentiated, continuously dividing state during neural development. Premature differentiation of NPCs and loss of their stem cell-like identity will result in the depletion of the progenitor pool and underdevelopment of the nervous system (Zhu et al. 2006; Hatakeyama et al. 2004). Central to the identity of NPCs is the maintenance of their epithelial apicobasal cell polarity. This is demonstrated by the disruption of cell polarity complexes in NPCs, which leads to loss of neuroepithelial markers and premature differentiation (Cappello et al. 2006; Costa et al. 2008). Cadherins in adherens junctions facilitate apicobasal polarity by positioning important determinants and adhering processes of radial glial and neuroepithelial cells to the ventricular surface and basal lamina (Götz and Huttner 2005; Martin-Belmonte and Perez-Moreno 2011). Indeed, multiple groups have demonstrated that the disruption of cadherin adhesions leads to the loss of apicobasal polarity in NPCs and subsequent premature differentiation (Zhadanov et al. 1999; Kadowaki et al. 2007; Lein et al. 2006). Cadherins also function to maintain the undifferentiated NPC population by influencing outcomes of individual mitotic divisions; promoting self-renewing divisions, and inhibiting terminally differentiating divisions which would deplete the progenitor pool (Götz and Huttner 2005; Noles and Chenn 2007). Additionally, cadherins have also been shown to maintain NPC identity by facilitating communication between NPCs and differentiating cells in an 'outside–in' regulation mechanism (Hatakeyama et al. 2014). In vitro and in vivo evidence in chick and mice embryos demonstrates that cadherin-mediated adhesions in adherens junctions of the apical end-feet of differentiating cells keep Notch signalling active in neighbouring NPCs, preventing premature differentiation in a non-cell autonomous manner. It is often believed that adherens junctions simply mediate physical contact between cells, but discoveries such as the one above have led to a growing appreciation for cadherin-mediated adherens junctions as sites for intercellular signalling, which have important roles in regulating spatiotemporal maintenance and differentiation of NPCs.

12.3.4 Cadherins in NPC Differentiation and Migration of Differentiated Cells

Cadherin adhesions are crucial in maintaining the self-renewing NPC population, and consequently dynamic disassembly of their adhesive contacts is important for the eventual differentiation of NPCs and detachment from the ventricular zone (Doe 2008). However, the loss of cadherin adhesions must be tightly regulated in order not to disrupt the careful balance between NPC maintenance and differentiation, as aberrant disruption of cadherin adhesions has dire consequences for the NPC population. Although loss of cadherin adhesions does not appear to affect the ability of differentiated cells to arise in N-cadherin–deficient mice, it is becoming clear that the precise regulation of cadherin adhesions is important for the successful formation of differentiated cells during neural development (Kadowaki et al. 2007; Kostetskii et al. 2001). At the end of the neurogenic period, retinal ganglion cells undergo changes in cell polarity and adhesive contacts in order to differentiate into the required cell types (Götz and Huttner 2005). Some RGCs downregulate adherens junctions and lose apical contacts to differentiate into multipolar parenchymal astrocytes, whereas others maintain adherens junctions and retract basal processes to form the ependymal lining of ventricles (Rakic 2003; Schmid et al. 2003). Downregulation of N-cadherin is also required for apical abscission, the process where differentiated NPCs detach and migrate away from the ventricular surface during neurogenesis. High-resolution live-cell imaging in chick neural tubes reveals disassembly of cadherin adhesions is essential for the retraction of apical processes during apical abscission, likely by loosening cell–cell junctions and actin–myosin tension (Das and Storey 2014). Recent work has focused on understanding the signalling networks which regulate cadherin adhesions in order to control the balance between NPC self-renewal and differentiation. Numb and Numb-like, regulators of Notch signalling, are required for the maintenance of adherens junctions in cortical progenitor cells in mice and consequently dictate NPC cell fate and polarity in a cadherin-dependent manner (Rasin et al. 2007). Additionally, a transcription factor network involving Sox2 and two Forkhead proteins (Foxp2 and Foxp4) has been identified, which regulates the expression of N-cadherin in order to control the balance of NPC self-renewal and differentiation in the developing neuroepithelium (Rousso et al. 2013). Foxp2 and Foxp4 are potent suppressors of N-cadherin expression, and disruption of the Foxp proteins inhibits NPC differentiation and migration from the VZ in the spinal cords of chicks and mice (Rousso et al. 2013). Sox2 acts in opposition to activate N-cadherin expression, and together with the Foxp proteins it helps to establish the level of cadherin expression in the developing nervous system in order to regulate NPC self-renewal and differentiation. Further work such as those outlined above will give us a complete understanding of the genetic circuits that dictate cadherin expression, and how the embryo is able to regulate the behaviour of NPCs spatiotemporally with exquisite precision.

12.3.5 Summary

Cadherins sit at the centre of the balance between NPC maintenance and differentiation during neural development. In vitro investigations and invaluable embryo models have revealed the diverse functions of cadherin molecules in maintaining NPCs and the pathways involved in their regulation for mediating neurogenesis. In the future, there is hope to elucidate fully the network of interactions dictating cadherin adhesions and this will set the foundation for advancements in treating neural developmental diseases and utilising stem cell technology to its full potential.

12.4 Cadherins in Neuronal Form and Function

Once generated from a progenitor cell population, postmitotic cells of the nervous system still undergo many phases of development to become integral parts of neural circuits. This includes their taking on the characteristic neuronal morphology of having a single axon and an elaboration of dendrites, the growth and patterning of those processes, assembly of the neuronal soma into their mature position, and formation of synapses between axons and dendrites of neurons within the circuit.

12.4.1 Cadherins in Axon Formation

A key first step in postmitotic neuronal development is to break symmetry of a new-born cell to generate the characteristic neuronal polarity of a long thin axon emanating from the neuronal cell body (Fig. 12.3; Dotti et al. 1988). This asymmetry is generated around the time of the differentiation of the neuron where the localisation of the centrosome predicts the location of the first neurite process that

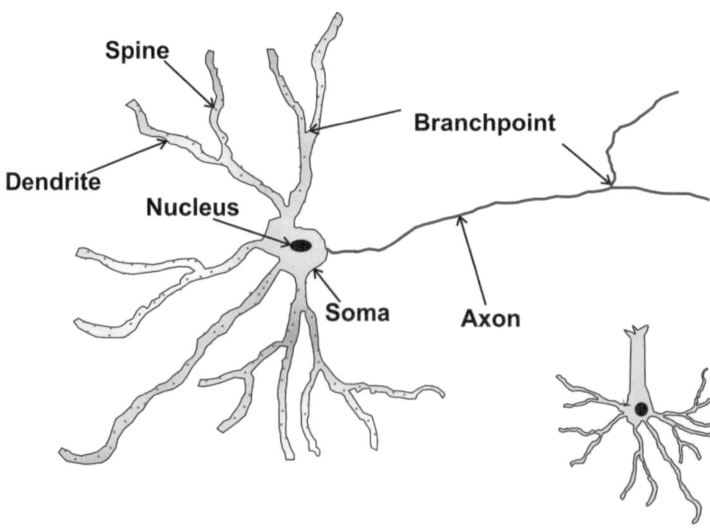

Fig. 12.3 Neuron structure and cadherin function. Neurons consist of a cell soma with a single axon and multiple dendrites. Each of these structures branches to varying degrees. Dendritic spines are found on excitatory neurons. Many different neuronal morphologies can be observed in the central nervous system. Two examples are shown here

will become the axon (de Anda et al. 2005, 2008; Bradke and Dotti 2000; Powell et al. 1997). Clusters of cadherins, most notably N-cadherin and E-cadherin are required for the positioning of the centrosome (Gärtner et al. 2012a, b, 2014a, b; Pollarolo et al 2011). Interestingly, in vitro studies suggest that this asymmetry of cadherins is cell autonomous, in other words an intrinsic function of the cell and thus not related to the binding of the cadherin to an extracellular substrate. It may be that cadherin clustering from the plane of progenitor division in the last mitosis that generates the postmitotic neuron may define the neurite precursor of the axon. In Xenopus retinal ganglion cells, expression of an N-cadherin construct lacking its extracellular region inhibits axon elongation (Riehl et al. 1996). It is thought that cytoplasmic interactions of the cadherin with members of the armadillo family of catenins, β-catenin, or p120 regulate cytosolic levels of the catenins. Each catenin has different roles in activating or inactivating members of the Rho GTPase family that may be critical for the cell autonomous regulation of axon elongation (Hirano and Takeichi 2012; Gärtner et al. 2014a, b).

For excitatory cells of the cerebral cortex, the definition of the axon is predictive of the orientation of the first dendrite processes. The centrosome moves around from the axon to the opposite side of the neuron and a dendrite is elaborated (Kadowaki et al. 2007; de Anda et al. 2010; Bellion et al. 2005; Gregory and Edmondson 1988; Higginbotham and Gleeson 2007; Solecki et al. 2004; Tanaka et al. 2004; Zmuda and Rivas 1998). This orientation then subsequently predicts the direction of migration of the cortical neuron along radial glial cell processes towards the pial surface of the developing brain. Thus, for excitatory cortical neurons, the asymmetry of axon and dendrite formation and direction of initial migration are hardwired and depend on cadherin localisation within the new-born neuron.

12.4.2 Cadherins in Axon Patterning

The contact and subsequent formation of synapses with other cells requires the axon to grow towards its synaptic target and for dendrites to elaborate ready for the axon–dendrite contact which will generate the beginnings of a neuronal circuit. Cadherin function is also implicated in aspects of axon growth and branching (Bixby et al. 1988; Matsunaga et al. 1988a; Tomaselli et al. 1988; Masai et al. 2003; Riehl et al. 1996; Tanabe et al. 2006; Andrews and Mastick 2003; Barnes et al. 2010; Borchers et al. 2001; Oblander et al. 2007, Oblander and Brady-Kalnay 2010; Redies and Takeichi 1993) as well as in the function and plasticity of synapses and the specificity of neuronal circuit formation.

Within cranial motor neurons, temporal differential cadherin expression has been shown to regulate axon outgrowth or branching (Barnes et al. 2010). For example, cadherin-7 is expressed in the motor neurons early in their development and cadherin-7 interactions are important for the growth of the axon from the neuron cell body. In contrast, cadherin-6b, which is expressed later in motor neuron development, is important for regulating the branching of the cranial motor neurons. This later branching is important for the subsequent arborisation of the motor axons when they reach their muscle target. Of note is that both cadherin-7 and cadherin-6b actions seem to require binding to substrate cadherins. In other words, in contrast to the cell autonomous role for cadherins in the initial specification of the axon, subsequent phases of axon development require extracellular cadherin–cadherin interactions. Also of note is that the effect of cadherin-6b on branching requires the PI3Kinase/AKT pathway. Both β-catenin and γ-catenin can bind to the PI3Kinase and so it seems likely that the effect of the cadherin is, intracellularly, through catenin binding.

These data suggest the differential actions of cadherin-7 and cadherin-6b on different aspects of cranial motor axonogenesis presumably through cadherin action at different times. As a family, classical cadherins have also been shown to be differentially expressed throughout the developing nervous system (Hirano and Takeich 2012; Matsunaga et al. 2013; Bekirov et al. 2008; Tsuchiya et al. 2006; Redies et al. 1993; Inuzuka et al. 1991; Takeichi et al. 1990). There are two likely ways that these expression patterns could operate in nervous system development. In one scheme, different functions of the cadherins could operate in different neurons, necessitating differential expression. In a related scheme, different combinations of cadherin expression could further refine the actions of the cadherins, particularly through specificity of cadherin function within a defined subset of neurons.

A recent example of the first scheme shows that differential expression of cadherin-8 and cadherin-9 in mouse retinal bipolar cells controls connectivity in different types of direction-selective visual circuits (Duan et al. 2014). Cadherin-8 and cadherin-9 are expressed in different classes of bipolar cells and each cadherin directs specificity of axonal lamina targeting in the inner plexiform layer of the retina. In the absence of either cadherin, the retinal cells' axonal arbours target both the correct and incorrect lamina. The inappropriate targeting of these axons disrupted the visual responses of the neurons with the synapses formed being highly attenuated in their synaptic transmission. This suggests that differential cadherin function in the retina targets axonal arborisation in the correct lamina and is important for the function of synapses. Interestingly, the functions of cadherin-8 and cadherin-9 appear to act heterophilically as introduction of either cadherin sparsely into each respective mutant mouse was sufficient to rescue the lamina targeting of the retinal bipolar cells. Should each cadherin be acting through cadherin–cadherin interactions then the presumption must be that they act through binding to additional cadherins. Cadherin-6 has been shown to be expressed in the retina and it may be that heterophilic cadherin interactions are mediated by that family member (Kay et al. 2011).

12.4.3 Cadherins in Neuronal Clustering

Differential expression of multiple cadherins within neuronal subsets has also been described (Liu et al. 2004). For example, within spinal and cranial motor neurons up to four different cadherins are expressed in defined functional groupings of motor neurons (Price et al. 2002; Demireva et al. 2011; Bello et al. 2012; Astick et al. 2014). Within the spinal cord, the motor neurons that project axons to an individual muscle in the limb cluster in groupings known as motor neuron pools. Different motor pools segregate from one another with little mixing of neurons of different pools. In order to form these pools, motor neurons pass through a migratory phase followed by a pool coalescence phase (Fig. 12.4). Each motor neuron pool expresses a different combination of cadherins and this combinatorial expression is instructive for the clustering of the motor neurons into pools. For example, the adductor motor neuron pool expresses cadherins-6b, -8, -13 and -20 (also called MN-cadherin) whereas the femorotibialis motor pool expresses cadherin-6b, -8, and -13 (Fig. 12.4c). Expression of cadherin-20 in the femorotibialis motor neurons results in their mixing with the adductor motor neurons. Additionally, removal of cadherin-20 function by expression of a dominant negative also causes mixing of

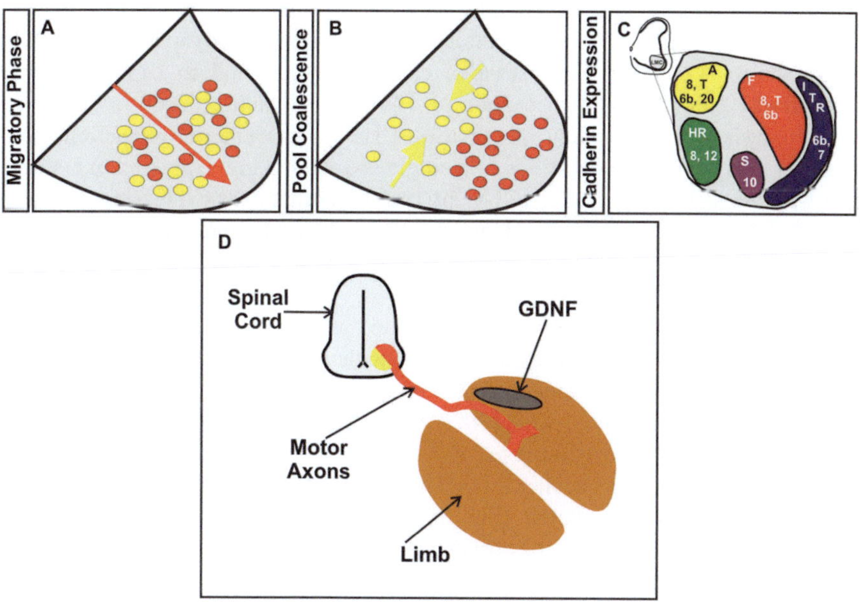

Fig. 12.4 Motor neuron pool formation. Motor neurons pass through an early migratory phase (**a**) which coincides with a pan-motor neuron expression of cadherins. Following this, motor pool coalescence occurs (**b**). Differential cadherin expression is found in motor neuron pools and is instructive in motor pool coalescence (**c**). The refinement of cadherin expression occurs through neurotrophic factor expression in the limb which is read out by motor neurons, presumably via their axons (**d**)

the femorotibialis and adductor motor neurons. Expression of other cadherins not predicted to equalise expression between the two motor pools had no effect on pool segregation. These data argue that the specificity of motor pool segregation and motor neuron coalescence is driven by the specific nature of the combination of cadherins expressed in the motor neurons. A similar combinatorial code also operates in the segregation of cranial motor nuclei. Again, each cluster of cranial motor neurons expresses a different combination of cadherins and this drives specificity of coalescence and segregation of the motor neurons during development. Interestingly, with the notable exception of cadherin-13, all of the cadherins that drive motor pool segregation are members of the type II subfamily of cadherins. Furthermore, with the exception of cadherin-5, all type II cadherins are expressed differentially in motor neurons. This suggests that possibility that combinatorial type II cadherin expression is a major driver of specificity of cell-to-cell recognition within the developing nervous system. Combinatorial expression of cell adhesion molecules is an attractive and rather elegant mechanism for generating diversity to drive specificity of intercellular interactions. With relatively few different family members a large number of different combinations can be achieved. For example, with just 6 family members 462 different combinations are possible. However, the molecular nature of the display of different combinations of cadherins in an individual cell is currently not known. Additionally, considerable heterophilic interactions between different type II cadherin family members have been observed (Ahrens et al. 2002; Shimoyama et al. 2000; Katsamba et al. 2009). For example, cadherin-8 can bind to cadherin-9 and cadherin-11 can also bind to cadherin-8. A note of caution needs to be raised with analysis of these binding specificities. Classically, cadherin interaction specificity is assayed under conditions of a single cadherin being expressed in a single cell with that cell being challenged to interact with other cells expressing the same or different individual cadherins. How specificity of cadherin interaction manifests itself when multiple cadherins are expressed within a given cell has not been studied.

Cadherin expression is also highly dynamic during motor neuron development. During spinal motor pool and cranial motor nucleus formation, the motor neurons pass through a phase that appears to have no differential cadherin expression. In other words, initially, motor neurons seem to express the same combination of cadherins with this expression being refined during the period of cell sorting. For spinal motor neurons, this cadherin refinement depends on a limb-derived source of the neurotrophic factor GDNF (Fig. 12.4d; Livet et al. 2002). In the GDNF knockout mutant mouse or in the absence of its cognate receptor, GFRα1, normally expressed within motor neurons, motor pool coalescence is perturbed (Haase et al. 2002). Cadherin expression is also perturbed in these mutant mice consistent with the role for cadherin expression in driving pool coalescence. The GDNF signals to motor neurons to express members of the ETS family of transcription factors in a pool-specific manner and it is this ETS expression that appears to drive the refinement of cadherin expression in the motor neurons. Interestingly, this GDNF signal is permissive for ETS and cadherin expression as the receptor is expressed in a pool-specific manner prior to the motor axons encountering the GDNF source in the limb.

The initial, pan-motor neuron, expression of cadherins plays a role in the migration of newly born motor neurons from the ventricular zone into the ventral horn of the developing spinal cord (Bello et al. 2012). This migration occurs on spinal radial glia, that act as guides for the motor neurons as they migrate. The cadherin expression in motor neurons during their migration might act to anchor the migration machinery within the cell providing traction for retrograde flow of actin to be used to force cell movement.

12.5 Cadherins in Synapses

One of the defining, and last, parts of neural development is the formation of functional synapses between postmitotic neurons. One could argue that this aspect of development continues throughout life as synaptic plasticity, the strengthening or weakening of synapses in response to circuit activity is a key feature of the functioning of the nervous system. Additionally, the genetic basis of many mental disorders can be traced to proteins whose function is predominantly in synapse function. Cadherins are implicated in many of the processes of synapse formation and plasticity (Arikkath and Reichardt 2008; Brigidi and Bamji 2011; Suzuki and Takeichi 2008; Tai et al. 2008; Takeichi 2007) and cadherin perturbations also underlie many psychiatric disorders.

12.5.1 Cadherins in Synapse Formation and Function

Synapses are small structures formed by axonal contact with the dendrites or soma of another neuron. They have a so-called presynaptic part, termed the active zone, which contains synaptic vesicles loaded with a neurotransmitter and a postsynaptic part that contains the receptors for the neurotransmitter. The receptors are anchored to the so-called postsynaptic density. For excitatory neurons, synapses are localised on a specialised dendritic structure known as a spine. Classical cadherins span the pre- and postsynaptic structures and are located at the outer edges of the active zone in mature synapses (Fannon and Colman 1996; Uchida et al. 1996). For example, N-cadherin and its associated catenins are found at synaptic junctions (Uchida et al. 1996). Their expression along with that of neuroligin has been shown to cooperate to regulate synapse formation (Aiga et al. 2011). Additionally, cadherins-11 and -13 can regulate the formation of both inhibitory and excitatory synapses (Paradis et al. 2007). General blockade of cadherin function using a blocking antibody results in smaller synapses with impaired function (Bozdagi et al. 2004, 2010). One of the major roles for cadherins, however, appears to be in dendritic spine morphogenesis. Cadherins are required for the formation of spines. Inhibition of cadherin function results in abnormal shapes of spines, such as their length and spine loss (Abe et al 2004; Mysore et al. 2007; Togashi et al 2002). These functions of cadherins require the cytoplasmic binding partners of cadherins such as αN-catenin and p120 catenin. p120 catenin is also important in the maturation of dendritic spines and this requires the Rho family of GTPases (Elia et al. 2006). Cadherins are also involved in the more general structure of synapses. For example, cadherins recruit PSD95 and synapsin to spines (Togashi et al. 2002) and bind to an AMPA receptor thus regulating its localisation in the synapse (Dunah et al. 2005; Nuriya and Huganir 2006; Saglietti et al. 2007). Spines are highly dynamic structures, this presumably being important for strengthening or weakening of synapses in response to neuronal activity. A major mechanism of synapse strengthening is a process called long term potentiation (LTP) whereby, following prolonged stimulation, a given input elicits a larger synaptic output (Bliss and Lomo 1973). LTP is associated with enlargement of spines which requires N-cadherin protein. The dynamic interplay between structural and functional plasticity at spines is also illustrated in the changes to the strength of cadherin adhesion related to the activity-dependent concentration of calcium ions at the synaptic cleft (Tai et al. 2008). Cadherin activation can also influence intracellular levels of calcium ions (Bixby et al. 1994; Chadborn et al. 2002; Marrs et al. 2009; Sheng et al. 2013). These phenomena could indicate that cadherins could act as activity sensors at the synapse and thus be intimate players in regulating the scaling of synaptic responses to prolonged activity (Thalhammer and Cingolani 2014). In addition to the biophysical changes to cadherin function at synapses, the recruitment and retention of cadherins at the synapse is also regulated by activity. Cadherins can be cleaved by proteases, for example, N-cadherin is processed by both ADAM10 and PS1/γ-secretase in a manner that depends on NMDA receptor activity (Monea et al. 2006; Reiss et al. 2005; Uemura et al. 2006; Malinverno et al. 2010). Interestingly, one of the cytoplasmic fragments of N-cadherin generated by proteolysis (N-cad/CTF2) induces the destruction of CREB-binding proteins. CREB-dependent gene expression is critical to synapse plasticity offering a transcriptional link between activity-dependent cadherin function and longer term changes to synapses (Marambaud et al. 2003; Uemura et al. 2006; Alberini 2009; Lonze and Ginty 2002). NMDA Receptor activation also reduces the rate of endocytosis of N-cadherin at the synapse (Tai et al. 2007). Thus, multiple mechanisms of activity-dependent changes to cadherin function feed into changes in synapse function.

12.5.2 Cadherins in Neural Disorders

Synaptic disfunction is believed to play a role in some disorders of the nervous system. A growing body of evidence suggests that at least some of the phenotypes found in these disorders may have a genetic basis linked to cadherin loci (Bhalla et al. 2008; Rose et al. 1995; Singh et al. 2010; Wang et al. 2009). Mutations in cadherin genes have been found in a wide spectrum of different disorders including autism spectrum disorders (Crepel et al. 2014), schizophrenia and bipolar disorder, and addiction-related disorders. For example, cadherin-13 has been linked to autism spectrum disorder, attention-deficit and hyperactivity disorder, schizophrenia, and addiction disorders (Børglum et al. 2014; Chapman et al. 2011; Johnson et al. 2006; Lasky-Su et al. 2008; Lesch et al. 2008; Treutlein et al. 2009). Members of the catenin family of cadherin-interacting proteins are also linked to neural disorders with αN-catenin and δ-catenin mutations found in schizophrenia and severe intellectual disability (Chu and Liu 2010; Medina et al. 2000). Exactly how cadherin/catenin mutations are involved in these disorders is not currently clear. It seems likely that synaptic functions of cadherins underpin their role in neural disorders but other functions of cadherins in circuit formation in general may also play a role (Gleeson 2001).

12.6 Summary

Cadherins play key roles throughout the development of the nervous system. The expression of cadherins is highly dynamic and regulated at both the transcriptional and posttranslational level. Far from being relatively simple homophilic molecular adhesives, the multiple binding partners of cadherins indicate that they play an important part in orchestrating many aspects of neural development both in the embryo as well as throughout life. There is still much to learn about the roles of cadherins in disorders of the nervous system. Additionally, it seems clear that we have only begun to scratch the surface of the diversity of functions that cadherins play in neural development.

References

Abe K, Chisaka O, Van Roy F, Takeichi M (2004) Stability of dendritic spines and synaptic contacts is controlled by alpha N-catenin. Nat Neurosci 7:357–363

Ahrens T, Pertz O, Haussinger D et al (2002) Analysis of heterophilic and homophilic interactions of cadherins using the c-Jun/c-Fos dimerization domains. J Biol Chem 277:19455–19460

Aiga M, Levinson JN, Bamji SX (2011) N-cadherin and neuroligins cooperate to regulate synapse formation in hippocampal cultures. J Biol Chem 286:851–858

Alberini CM (2009) Transcription factors in long-term memory and synaptic plasticity. Physiol Rev 89:121–145

Andrews GL, Mastick GS (2003) R-cadherin is a Pax6-regulated, growth-promoting cue for pioneer axons. J Neurosci 23:9873–9880

Arikkath J, Reichardt LF (2008) Cadherins and catenins at synapses: roles in synaptogenesis and synaptic plasticity. Trends Neurosci 31:487–494

Astick M, Tubby K, Mubarak WM et al (2014) Central topography of cranial motor nuclei controlled by differential cadherin expression. Curr Biol 24:2541–2547

Babb SG, Marrs J (2004) E-cadherin regulates cell movements and tissue formation in early zebrafish embryos. Dev Dyn 230:263–277

Barnes SH, Price SR, Wentzel C, Guthrie SC (2010) Cadherin-7 and cadherin-6B differentially regulate the growth, branching and guidance of cranial motor axons. Development 137:805–814

Bekirov IH, Nagy V, Svoronos A et al (2008) Cadherin-8 and N cadherin differentially regulate pre- and postsynaptic development of the hippocampal mossy fiber pathway. Hippocampus 18:349–363

Bellion A, Baudoin J-P, Alvarez C (2005) Nucleokinesis in tangentially migrating neurons comprises two alternating phases: forward migration of the Golgi/centrosome associated with centrosome splitting and myosin contraction at the rear. J Neurosci 25:5691–5699

Bello SM, Millo H, Rajebhosale M, Price SR (2012) Catenin-dependent cadherin function drives divisional segregation of spinal motor neurons. J Neurosci 32:490–505

Bhalla K, LuoY BT et al (2008) Alterations in CDH15 and KIRREL3 in patients with mild to severe intellectual disability. Am J Hum Genet 83:703–713

Bixby JL, Lilien J, Reichardt LF (1988) Identification of the major proteins that promote neuronal process outgrowth on Schwann cells in vitro. J Cell Biol 107:353–361

Bixby JL, Grunwald GB, Bookman RJ (1994) Ca^{2+} influx and neurite growth in response to purified N-cadherin and laminin. J Cell Biol 127:1461–1475

Bliss TV, Lomo T (1973) Long-lasting potentiation of synaptic transmission in the dentate area of the anaesthetized rabbit following stimulation of the perforant path. J Physiol 232:331–356

Borchers A, David R, Wedlich D (2001) Xenopus cadherin-11 restrains cranial neural crest migration and influences neural crest specification. Development 128:3049–3060

Børglum AD, Demontis D, Grove J et al (2014) Genome-wide study of association and interaction with maternal cytomegalovirus infection suggests new schizophrenia loci. Mol Psychiatry 19:325–333

Bozdagi O, Valcin M, Poskanzer K et al (2004) Temporally distinct demands for classic cadherins in synapse formation and maturation. Mol Cell Neurosci 27:509–521

Bozdagi O, Wang XB, Nikitczuk JS et al (2010) Persistence of coordinated long-term potentiation and dendritic spine enlargement at mature hippocampal CA1 synapses requires N-cadherin. J Neurosci 30:9984–9989

Bradke F, Dotti CG (2000) Establishment of neuronal polarity: lessons from cultured hippocampal neurons. Curr Opin Neurobiol 10:574–581

Brigidi GS, Bamji SX (2011) Cadherin-catenin adhesion complexes at the synapse. Curr Opin Neurobiol 21:208–214

Cappello S, Attardo A, Wu X et al (2006) The Rho-GTPase cdc42 regulates neural progenitor fate at the apical surface. Nat Neurosci 9:1099–1107. doi:10.1038/nn1744

Carver EA, Jiang R, Lan Y et al (2001) The mouse snail gene encodes a key regulator of the epithelial-mesenchymal transition the mouse snail gene encodes a key regulator of the epithelial-mesenchymal transition. Mol Cell Biol 21:8184–8188. doi:10.1128/MCB.21.23.8184

Chadborn N, Eickholt B, Doherty P, Bolsover S (2002) Direct measurement of local raised subplasmalemmal calcium concentrations in growth cones advancing on an N-cadherin substrate. Eur J Neurosci 15:1891–1898

Chang L, Blain D, Bertuzzi S, Brooks BP (2006) Uveal coloboma: clinical and basic science update. Curr Opin Ophthalmol 17:447–470. doi:10.1097/01.icu.0000243020.82380.f6

Chapman NH, Estes A, Munson J et al (2011) Genome-scan for IQ discrepancy in au- tism: evidence for loci on chromosomes 10 and 16. Hum Genet 129:59–70

Chen S, Lewis B, Moran A, Xie T (2012) Cadherin-mediated cell adhesion is critical for the closing of the mouse optic fissure. PloS One 7:1–8. doi:10.1371/journal.pone.0051705

Chenn A, Walsh C (2003) Increased neuronal production, enlarged forebrains and cytoarchitectural distortions in beta-catenin overexpressing transgenic mice. Cereb Cortex 13:599–606. doi:10.1093/cercor/13.6.599

Chenn A, Zhang YA, Chang BT, McConnell SK (1998) Intrinsic polarity of mammalian neuroepithelial cells. Mol Cell Neurosci 11:183–193. doi:10.1006/mcne.1998.0680

Chow RL, Lang RA (2001) Early eye development in vertebrates. Annu Rev Cell Dev Biol 17:255–296

Chu TT, Liu Y (2010) An integrated genomic analysis of gene function correlation on schizophrenia susceptibility genes. J Hum Genet 55:285–292

Ciruna B, Rossant J (2001) FGF signalling regulates mesoderm cell Fate specification and morphogenetic movement at the primitive streak. Dev Cell 1:37–49. doi:10.1016/S1534-5807(01)00017-X

Clay MR, Halloran MC (2014) Cadherin 6 promotes neural crest cell detachment via F-actin regulation and influences active Rho distribution during epithelial-to-mesenchymal transition. Development 141:2506–2515. doi:10.1242/dev.105551

Coles EG, Taneyhill LA, Bronner-Fraser M (2007) A critical role for Cadherin6B in regulating avian neural crest emigration. Dev Biol 312:533–544. doi:10.1016/j.ydbio.2007.09.056

Costa MR, Wen G, Lepier A et al (2008) Par-complex proteins promote proliferative progenitor divisions in the developing mouse cerebral cortex. Development 135:11–22. doi:10.1242/dev.009951

Crepel A, De Wolf V, Brison N et al (2014) Association of CDH11 with non-syndromic ASD. Am J Med Genet B Neuropsychiatr Genet 165:391–398

Dady A, Blavet C, Duband JL (2012) Timing and kinetics of E- to N-cadherin switch during neurulation in the avian embryo. Dev Dyn 241:1333–1349. doi:10.1002/dvdy.23813

Das RM, Storey KG (2014) Apical abscission alters cell polarity and dismantles the primary cilium during neurogenesis. Science 343:200–204. doi:10.1126/science.1247521

de Anda FC, Pollarolo G, Da Silva JS et al (2005) Centrosome localization determines neuronal polarity. Nature 436:704–708

de Anda F, Gartner A, Tsai LH, Dotti CG (2008) Pyramidal neuron polarity axis is defined at the bipolar stage. J Cell Sci 121:178–185

de Anda FC, Meletis K, Ge X et al (2010) Centrosome motility is essential for initial axon formation in the neocortex. J Neurosci 30:10391–10406

Demireva EY, Shapiro LS, Jessell TM, Zampieri N (2011) Motor neuron position and topographic order imposed by β - and γ-catenin activities. Cell 147:641–52

Doe CQ (2008) Neural stem cells: balancing self-renewal with differentiation. Development 135:1575–1587. doi:10.1242/dev.014977

Dotti CG, Sullivan CA, Banker GA (1988) The establishment of polarity by hippocampal neurons in culture. J Neurosci 8:1454–1468

Duan X, Krishnaswamy A, De la Huerta I, Sanes JR (2014) Type II cadherins guide assembly of a direction-selective retinal circuit. Cell 158:793–807

Dunah AW, Hueske E, Wyszynski M (2005) LAR receptor protein tyrosine phosphatases in the development and maintenance of excitatory synapses. Nat Neurosci 8:458–467

Dupin E, Creuzet S, Le Douarin NM (2006) The contribution of the neural crest to the vertebrate body. Adv Exp Med Biol 589:96–119

Elia LP, Yamamoto M, Zang K, Reichardt LF (2006) p120 catenin regulates dendritic spine and synapse development through Rho-family GTPases and cadherins. Neuron 51:43–56

Fannon AM, Colman DR (1996) A model for central synaptic junctional complex formation based on the differential adhesive specificities of the cadherins. Neuron 17:423–434

Farkas LM, Huttner WB (2008) The cell biology of neural stem and progenitor cells and its significance for their proliferation versus differentiation during mammalian brain development. Curr Opin Cell Biol 20:707–715. doi:10.1016/j.ceb.2008.09.008

Gartner A, Fornasiero EF, Dotti CG (2012a) N-cadherin: a new player in neuronal polarity. Cell Cycle 11:2223–2224

Gartner A, Fornasiero EF, Munck S et al (2012b) N-cadherin specifies first asymmetry in developing neurons. EMBO J 31:1893–1903

Gärtner A, Fornasiero EF, Dotti CG (2014a) Cadherins as regulators of neuronal polarity. Cell Adh Migr 14:1–8

Gärtner A, Fornasiero EF, Valtorta F, Dotti CG (2014b) Distinct temporal hierarchies in membrane and cytoskeleton dynamics precede the morphological polarization of developing neurons. J Cell Sci 127:4409–4419

Gilmore EC, Walsh CA (2013) Genetic causes of microcephaly and lessons for neuronal development. Wiley Interdiscp Rev Dev Biol 2:461–478. doi:10.1002/wdev.89

Gleeson JG (2001) Neuronal migration disorders. Ment Retard Dev Disabil Res Rev 7:167–171

Götz M, Huttner WB (2005) The cell biology of neurogenesis. Nat Rev Mol Cell Biol 6:777–788. doi:10.1038/nrm1739

Gregory W, Edmondson J (1988) Cytology and neuron-glial apposition of migrating cerebellar granule cells in vitro. J Neurosci 8:1728–1738

Haase G, Dessaud E, Garcès A (2002) GDNF acts through PEA3 to regulate cell body positioning and muscle innervation of specific motor neuron pools. Neuron 35:893–905

Halbleib JM, Nelson WJ (2006) Cadherins in development: cell adhesion, sorting, and tissue morphogenesis. Genes Dev 20:3199–3214. doi:10.1101/gad.1486806

Hall BK (2008) The neural crest and neural crest cells in vertebrate development and evolution. TripleC. doi:10.1007/128

Hall R, Erickson C (2003) ADAM 10: an active metalloprotease expressed during avian epithelial morphogenesis. Dev Biol 256:146–159

Hatakeyama J, Bessho Y, Katoh K et al (2004) Hes genes regulate size, shape and histogenesis of the nervous system by control of the timing of neural stem cell differentiation. Development 131:5539–5550. doi:10.1242/dev.01436

Hatakeyama J, Wakamatsu Y, Nagafuchi A et al (2014) Cadherin-based adhesions in the apical endfoot are required for active Notch signaling to control neurogenesis in vertebrates. Development 141:1671–1682. doi:10.1242/dev.102988

Hatta K, Takeichi M (1986) Expression of N-cadherin adhesion molecules associated with early morphogenetic events in chick development. Nature 320:447–449. doi:10.1038/320447a0

Hazan RB, Qiao R, Keren R et al (2004) Cadherin switch in tumor progression. Ann N Y Acad Sci 1014:155–163. doi:10.1196/annals.1294.016

Higginbotham HR, Gleeson JG (2007) The centrosome in neuronal development. Trends Neurosci 30:276–283

Hirano S, Takeichi M (2012) Cadherins in brain morphogenesis and wiring. Physiol Rev 92:597–634

Hirano M, Hashimoto S, Yonemura S et al (2008) EPB41L5 functions to post-transcriptionally regulate cadherin and integrin during epithelial-mesenchymal transition. J Cell Biol 182:1217–1230. doi:10.1083/jcb.200712086

Hong E, Brewster R (2006) N-cadherin is required for the polarized cell behaviours that drive neurulation in the zebrafish. Development 133:3895–3905. doi:10.1242/dev.02560

Inuzuka H, Redies C, Takeichi M (1991) Differential expression of R- and N-cadherin in neural and mesodermal tissues during early chicken development. Development 113:959–967

Johnson C, Drgon T, Liu QR et al (2006) Pooled association genome scanning for alcohol dependence using 104,268 SNPs: validation and use to identify alcoholism vulnerability loci in unrelated individuals from the collaborative study on the genetics of alcoholism. Am J Med Genet B Neuropsychiatr Genet 141:844–853

Junghans D, Hack I, Frotscher M et al (2005) Beta-catenin-mediated cell-adhesion is vital for embryonic forebrain development. Dev Dyn 233:528–539. doi:10.1002/dvdy.20365

Kadowaki M, Nakamura S, Machon O et al (2007) N-cadherin mediates cortical organization in the mouse brain. Dev Biol 304:22–33

Katsamba P, Carroll K, Ahlsen G et al (2009) Linking molecular affinity and cellular specificity in cadherin-mediated adhesion. Proc Natl Acad Sci U S A 106:11594–11599

Kay JN, De la Huerta I, Kim IJ (2011) Retinal ganglion cells with distinct directional preferences differ in molecular identity, structure, and central projections. J Neurosci 31:7753–7762

Kostetskii I, Moore R, Kemler R, Radice GL (2001) Differential adhesion leads to segregation and exclusion of N-cadherin-deficient cells in chimeric embryos. Dev Biol 234:72–79. doi:10.1006/dbio.2001.0250

Lasky-Su J, Neale BM, Franke B et al (2008) Genome-wide association scan of quantitative traits for attention deficit hyperactivity disorder identifies novel associations and confirms candidate gene associations. Am J Med Genet B Neuropsychiatr Genet 147:1345–1354

Lee CH, Gumbiner BM (1995) Disruption of gastrulation movements in Xenopus by a dominant-negative mutant for C-cadherin. Dev Biol 171:363–373. doi:10.1006/dbio.1995.1288

Lein W-H, Klezovitch O, Fernandez TE, et al (2006) αE-catenin controls cerebral cortical size by regulating hedgehog signalling pathway. 311:1609–1612. doi:10.1126/science.1121449.

Lepage SE, Bruce AEE (2010) Zebrafish epiboly: mechanics and mechanisms. Int J Dev Biol 54:1213–1228. doi:10.1387/ijdb.093028sl

Lesch KP, Timmesfeld N, Renner TJ et al (2008) Molecular genetics of adult ADHD: converging evidence from genome-wide association and extended pedigree linkage studies. J Neural Transm 115:1573–1585

Liu Q, Marrs JA, Azodi E et al (2004) Differential expression of cadherins in the developing and adult zebrafish olfactory system. J Comp Neurol 478:269–281

Livet J, Sigrist M, Stroebel S et al (2002) ETS gene Pea3 controls the central position and terminal arborization of specific motor neuron pools. Neuron 35:877–892

Lonze BE, Ginty DD (2002) Function and regulation of CREB family transcription factors in the nervous system. Neuron 35:605–623

Malinverno M, Carta M, Epis R et al (2010) Synaptic localization and activity of ADAM10 regulate excit- atory synapses through N-cadherin cleavage. J Neurosci 30:16343–16355

Marambaud P, Wen PH, Dutt A et al (2003) A CBP binding transcriptional repressor produced by the PS1/epsilon-cleavage of N-cadherin is inhibited by PS1 FAD mutations. Cell 114:635–645

Marín O, Valiente M, Ge X, Tsai L-H (2010) Guiding neuronal cell migrations. Cold Spring Harb Perspect Biol 2:1–21. doi:10.1101/cshperspect.a001834

Marrs GS, Theisen CS, Bruses JL (2009) N-cadherin modulates voltage activated calcium influx via RhoA, p120-catenin, and myosin-actin interaction. Mol Cell Neurosci 40:390–400

Martin-Belmonte F, Perez-Moreno M (2011) Epithelial cell polarity, stem cells and cancer. Nat Rev Cancer 12:23–38. doi:10.1038/nrc3169

Masai I, Lele Z, Yamaguchi M (2003) N-cadherin mediates retinal lamination, maintenance of forebrain compartments and patterning of retinal neurites. Development 130:2479–2494

Matsunaga M, Hatta K, Nagafuchi A, Takeichi M (1988a) Guidance of optic nerve fibres by N-cadherin adhesion molecules. Nature 334:62–64

Matsunaga M, Hatta K, Takeichi M (1988b) Role of N-cadherin cell adhesion molecules in the histogenesis of neural retina. Neuron 1:289–295. doi:10.1016/0896-6273(88)90077-3

Matsunaga E, Nambu S, Oka M, Iriki A (2013) Differential cadherin expression in the developing postnatal telencephalon of a New World monkey. J Comp Neurol 521:4027–4060

Mayor R, Theveneau E (2013) The neural crest. Development 140:2247–2251. doi:10.1242/dev.091751

Medina M, Marinescu RC, Overhauser J, Kosik KS (2000) Hemizygosity of delta-catenin (CTNND2) is associated with severe mental retardation in cri-du-chat syndrome. Genomics 63:157–164

Monea S, Jordan BA, Srivastava S et al (2006) Membrane localization of membrane type 5 matrix metalloproteinase by AMPA receptor binding protein and cleavage of cadherins. J Neurosci 26:2300–2312

Mysore SP, Tai CY, Schuman EM (2007) Effects of N-cadherin disruption on spine morphological dynamics. Front Cell Neurosci 1:1. doi:10.3389/neuro.03.001.2007

Nakagawa S, Takeichi M (1995) Neural crest cell-cell adhesion controlled by sequential and subpopulation-specific expression of novel cadherins. Development 121:1321–1332. doi:10.1016/0168-9525(95)90550-2

Nakaya Y, Sheng G (2008) Epithelial to mesenchymal transition during gastrulation: an embryological view. Dev Growth Differ 50:755–766. doi:10.1111/j.1440-169X.2008.01070.x

Nelson J, Nusse R (2004) Convergence of Wnt, β-Catenin, and Cadherin pathways. Science (80–) 303:1483–1487. doi:10.1126/science.1094291

Nishimura T, Takeichi M (2009) Remodeling of the Adherens Junctions During Morphogenesis. In: Lecuit T (ed) Tissue remodeling and epithelial morphogenesis, 1st edn. Elsevier, San Diego, pp 33–49

Noles SR, Chenn A (2007) Cadherin inhibition of β -catenin signaling regulates the proliferation and differentiation of neural precursor cells. Mol Cell Neurosci 35:549–558. doi:10.1016/j.mcn.2007.04.012

Nuriya M, Huganir RL (2006) Regulation of AMPA receptor trafficking by N-cadherin. J Neurochem 97:652–661

Oblander SA, Brady-Kalnay SM (2010) Distinct PTPmu-associated signaling molecules differentially regulate neurite outgrowth on E-, N-, and R-cadherin. Mol Cell Neurosci 44:78–93

Oblander SA, Ensslen-Craig SE, Longo FM, Brady-Kalnay SM (2007) E-cadherin promotes retinal ganglion cell neurite outgrowth in a protein tyrosine phosphatase mu dependent manner. Mol Cell Neurosci 34:481–492

Ozair MZ, Kintner C, Brivanlou AH (2013) Neural induction and early patterning in vertebrates. Wiley Interdiscip Rev Dev Biol 2:479–498. doi:10.1002/wdev.90

Paradis S, Harrar DB, Lin Y et al (2007) An RNAi-based approach identifies molecules required for gluta- matergic and GABAergic synapse development. Neuron 53:217–232

Park K-S, Gumbiner BM (2010) Cadherin 6B induces BMP signaling and de-epithelialization during the epithelial mesenchymal transition of the neural crest. Development 137:2691–2701. doi:10.1242/dev.050096

Park K-S, Gumbiner BM (2012) Cadherin-6B stimulates an epithelial mesenchymal transition and the delamination of cells from the neural ectoderm via LIMK/ cofilin mediated non-canonical BMP receptor signaling. Dev Biol 366:232–243. doi:10.1016/j.biotechadv.2011.08.021. Secreted

Perego C, Vanoni C, Massari S et al (2002) Invasive behaviour of glioblastoma cell lines is associated with altered organisation of the cadherin-catenin adhesion system. J Cell Sci 115:3331–3340

Pollarolo G, Schulz JG, Munck S, Dotti CG (2011) Cytokinesis remnants define first neuronal asymmetry in vivo. Nat Neurosci 14:1525–1533

Powell SK, Rivas RJ, Rodriguez-Boulan E, Hatten ME (1997) Development of polarity in cerebellar granule neurons. J Neurobiol 32:223–236

Price SR, De Marco GNV, Ranscht B, Jessell TM (2002) Regulation of motor neuron pool sorting by differential expression of type II cadherins. Cell 109:205–16

Radice GL, Rayburn H, Matsunami H et al (1997) Developmental defects in mouse embryos lacking N-cadherin. Dev Biol 181:64–78. doi:10.1006/dbio.1996.8443

Rakic P (2003) Developmental and evolutionary adaptations of cortical radial glia. Cereb Cortex 13:541–549. doi:10.1093/cercor/13.6.541

Rasin M-R, Gazula V-R, Breunig JJ et al (2007) Numb and Numbl are required for maintenance of cadherin-based adhesion and polarity of neural progenitors. Nat Neurosci 10:819–827. doi:10.1038/nn1924

Redies C, Takeichi M (1993) N-and R-cadherin expression in the optic nerve of the chicken embryo. Glia 8:161–171

Redies C, Engelhart K, Takeichi M (1993) Differential expression of N- and R-cadherin in functional neuronal systems and other structures of the developing chicken brain. J Comp Neurol 333:398–416

Reiss K, Maretzky T, Ludwig A et al (2005) ADAM10 cleavage of N-cadherin and regulation of cell-cell adhesion and beta-catenin nuclear signalling. EMBO J 24:742–752

Riehl R, Johnson K, Bradley R et al (1996) Cadherin function is required for axon outgrowth in retinal ganglion cells in vivo. Neuron 17:837–848

Rose O, Grund C, Reinhardt S et al (1995) Contactus adherens, a special type of plaque-bearing adhering junction containing M-cadherin, in the granule cell layer of the cerebellar glomerulus. Proc Natl Acad Sci U S A 92:6022–6026

Rousso DL, Pearson CA, Gaber ZB et al (2013) Foxp-mediated suppression of N-cadherin regulates neuroepithelial character and progenitor maintenance in the CNS. Neuron 74:314–330. doi:10.1038/nature11130.Reduced

Rungger-Brändle E, Ripperger JA, Steiner K et al (2010) Retinal patterning by Pax6-dependent cell adhesion molecules. Dev Neurobiol 70:764–780. doi:10.1002/dneu.20816

Saglietti L, Dequidt C, Kamieniarz K et al (2007) Extracellular interactions between GluR2 and N-cadherin in spine regulation. Neuron 54:461–477

Schiffmacher AT, Padmanabhan R, Jhingory S, Taneyhill LA (2014) Cadherin-6B is proteolytically processed during epithelial-to-mesenchymal transitions of the cranial neural crest. Mol Biol Cell 25:41–54. doi:10.1091/mbc.E13-08-0459

Schmid RS, McGrath B, Berechid BE et al (2003) Neuregulin 1-erbB2 signaling is required for the establishment of radial glia and their transformation into astrocytes in cerebral cortex. Proc Natl Acad Sci U S A 100:4251–4256. doi:10.1073/pnas.0630496100

Sheng L, Leshchyns'ka I, Sytnyk V (2013) Cell adhesion and intracellular calcium signaling in neurons. Cell Commun Signal 11:94. doi:10.1186/1478-811X-11-94

Shikanai M, Nakajima K, Kawauchi T (2011) N-Cadherin regulates radial glial fiber-dependent migration of cortical locomoting neurons. Commun Integr Biol 4:326–330. doi:10.4161/cib.4.3.14886

Shimizu T, Yabe T, Muraoka O et al (2005) E-cadherin is required for gastrulation cell movements in zebrafish. Mech Dev 122:747–763. doi:10.1016/j.mod.2005.03.008

Shimoyama Y, Tsujimoto G, Kitajima M, Natori M (2000) Identification of three human type-II classic cadherins and frequent heterophilic interactions between different subclasses of type-II classic cadherins. Biochem J 349:159–167

Shoval I, Ludwig A, Kalcheim C (2007) Antagonistic roles of full-length N-cadherin and its soluble BMP cleavage product in neural crest delamination. Development 134:491–501. doi:10.1242/dev.02742

Singh SM, Castellani C, O'Reilly R (2010) Autism meets schizophrenia via cadherin pathway. Schizophr Res 116:293–294

Solecki DJ, Model L, Gaetz J (2004) Par6alpha signaling controls glial-guided neuronal migration. Nat Neurosci 7:1195–1203

Solnica-Krezel L, Sepich DS (2012) Gastrulation: making and shaping germ layers. Annu Rev Cell Dev Biol 28:687–717

Suzuki SC, Takeichi M (2008) Cadherins in neuronal morphogenesis and function. Dev Growth Differ 1:S119–S130

Swartling FJ, Savov V, Persson AI et al (2012) Distinct neural stem cell populations give rise to disparate brain tumors in response to N-MYC. Cancer Cell 21:601–613. doi:10.1016/j.ccr.2012.04.012

Tai CY, Mysore SP, Chiu C, Schuman EM (2007) Activity-regulated N-cadherin endocytosis. Neuron 54:771–785

Tai CY, Kim SA, Schuman EM (2008) Cadherins and synaptic plasticity. Curr Opin Cell Biol 20:567–575

Takahashi T, Nowakowski RS, Caviness VS (1993) Cell cycle parameters and patterns of nuclear movement in the neocortical proliferative zone of the fetal mouse. J Neurosci 13:820–833

Takeichi M (2007) The cadherin superfamily in neuronal connections and interactions. Nat Rev Neurosci 8:11–20

Takeichi M, Inuzuka H, Shimamura K et al (1990) Cadherin subclasses: differential expression and their roles in neural morphogenesis. Cold Spring Harb Symp Quant Biol 55:319–325

Tanabe K, Takahashi Y, Sato Y et al (2006) Cadherin is required for dendritic morphogenesis and synaptic terminal organization of retinal horizontal cells. Development 133:4085–4096

Tanaka T, Serneo FF, Higgins C et al (2004) Lis1 and doublecortin function with dynein to mediate coupling of the nucleus to the centrosome in neuronal migration. J Cell Biol 165:709–721

Taneyhill LA, Schiffmacher AT (2013) Chapter 3: The Cell Biology of Neural Crest Cell Delamination and EMT. In: Trainor P (ed) Neural Crest Cells: Evolution, Development and Disease. Elsevier Inc, Taramani, pp 1–22

Taneyhill LA, Coles EG, Bronner-Fraser M (2007) Snail2 directly represses cadherin6B during epithelial-to-mesenchymal transitions of the neural crest. Development 134:1481–1490. doi:10.1242/dev.02834

Temple S (2001) The development of neural stem cells. Nature 414:112–117. doi:10.1038/35102174

Teng J, Rai T, Tanaka Y et al (2005) The KIF3 motor transports N-cadherin and organizes the developing neuroepithelium. Nat Cell Biol 7:474–482. doi:10.1038/ncb1249

Tepass U, Truong K, Godt D et al (2000) Cadherins in embryonic and neural morphogenesis. Nat Rev Mol Cell Biol 1:91–100. doi:10.1038/35040042

Thalhammer A, Cingolani LA (2014) Cell adhesion and homeostatic synaptic plasticity. Neuropharmacology 78:23–30

Theveneau E, Mayor R (2012) Neural crest delamination and migration: from epithelium-to-mesenchyme transition to collective cell migration. Dev Biol 366:34–54. doi:10.1016/j.ydbio.2011.12.041

Thiery JP, Acloque H, Huang RYJ, Nieto MA (2009) Epithelial-Mesenchymal Transitions in Development and Disease. Cell 139:871–890. doi:10.1016/j.cell.2009.11.007

Togashi H, Abe K, Mizoguchi A et al (2002) Cadherin regulates dendritic spine morphogenesis. Neuron 35:77–89

Tomaselli KJ, Neugebauer KM, Bixby JL et al (1988) N-cadherin and integrins: two receptor systems that mediate neuronal process outgrowth on astrocyte surfaces. Neuron 1:33–43

Treutlein J, Cichon S, Ridinger M et al (2009) Genome-wide association study of alcohol dependence. Arch Gen Psychiatry 66:773–784

Tsuchiya B, Sato Y, Kameya T et al (2006) Differential expression of N-cadherin and E-cadherin in normal human tissues. Arch Histol Cytol 69:135–145

Uchida N, Honjo Y, Johnson KR et al (1996) The catenin/cadherin adhesion system is localized in synaptic junctions bordering transmitter release zones. J Cell Biol 135:767–779

Uemura K, Kihara T, Kuzuya A et al (2006) Characterization of sequential N-cadherin cleavage by ADAM10 and PS1. Neurosci Lett 402:278–283

Vallin J, Girault JM, Thiery JP, Broders F (1998) Xenopus cadherin-11 is expressed in different populations of migrating neural crest cells. Mech Dev 75:171–174. doi:10.1016/S0925-4773(98)00099-9

Wang K, Zhang H, Ma D et al (2009) Common genetic variants on 5p14.1 associate with autism spectrum dis- orders. Nature 459:528–533

Zhadanov AB, Provance DW, Speer CA et al (1999) Absence of the tight junctional protein AF-6 disrupts epithelial cell-cell junctions and cell polarity during mouse development. Curr Biol 9:880–888. doi:10.1016/S0960-9822(99)80392-3

Zhang J, Woodhead GJ, Swaminathan SK et al (2010) Cortical neural precursors inhibit their own differentiation via N- cadherin maintenance of beta-catenin signaling. Dev Cell 18:472–479. doi:10.1016/j.devcel.2009.12.025.Cortical

Zhang J, Shemezis JR, Mcquinn ER et al (2013) AKT activation by N-cadherin regulates beta-catenin signaling and neuronal differentiation during cortical development. Neural Dev 8:1. doi:10.1186/1749-8104-8-7

Zhong Y, Brieher WM, Gumbiner BM (1999) Analysis of C-cadherin regulation during tissue morphogenesis with an activating antibody. J Cell Biol 144:351–359. doi:10.1083/jcb.144.2.351

Zhu X, Zhang J, Tollkuhn J et al (2006) Sustained Notch signaling in progenitors is required for sequential emergence of distinct cell lineages during organogenesis. Genes Dev 20:2739–2753. doi:10.1101/gad.1444706

Zmuda JF, Rivas RJ (1998) The Golgi apparatus and the centrosome are localized to the sites of newly emerging axons in cerebellar granule neurons in vitro. Cell Motil Cytoskel 41:18–38

Zohn IE, Li Y, Skolnik EY et al (2006) p38 and a p38-Interacting Protein Are Critical for Downregulation of E-Cadherin during Mouse Gastrulation. Cell 125:957–969. doi:10.1016/j.cell.2006.03.048

Chapter 13
Cadherins in the Auditory Sensory Organ

Aziz El-Amraoui and Christine Petit

Abstract The exquisite sensitivity and frequency tuning of hearing depend on the correct structure and functioning of the auditory sensory hair cells, the neighbouring supporting cells, and the homeostasis of their ionic environment. The increasing number of adhesion proteins identified as causing hearing impairment in humans and mice when defective is consistent with a critical role for cell–cell junctions between neighbouring epithelial cells of the cochlea, and of fibrous links within the hair bundle, the sensory hair cell structure responsible for sound reception. Classical cadherins and/or associated adherens-junction proteins, such as p120-catenin or nectin 3, have been shown to be essential for establishment of the regular mosaic cellular pattern of the auditory sensory epithelium. Two cadherin-related proteins, protocadherin-15 and cadherin-23, are key components of both lateral links and tip-links in hair bundles; they are essential components of the mechanoelectrical transduction machinery. Studies of the role of these adhesion proteins and of the pathogenesis of the forms of deafness caused by defects of these proteins have provided considerable insight into the development and functioning of the auditory sensory epithelium, and of the hair cells in particular.

A. El-Amraoui (✉)
Unité de Génétique et Physiologie de l'Audition, Département de Neuroscience, Institut Pasteur, 25 rue Dr Roux, 75015 Paris, France

Unité Mixte de Recherche, UMR-S 1120, Institut National de la Santé et de la Recherche Médicale (INSERM), 75015 Paris, France

Sorbonne Universités, UPMC Univ Paris 06, Complexité du vivant, 4 place Jussieu, 75005 Paris, France
e-mail: aziz.el-amraoui@pasteur.fr

C. Petit (✉)
Unité de Génétique et Physiologie de l'Audition, Département de Neuroscience, Institut Pasteur, 25 rue Dr Roux, 75015 Paris, France

Unité Mixte de Recherche, UMR-S 1120, Institut National de la Santé et de la Recherche Médicale (INSERM), 75015 Paris, France

Sorbonne Universités, UPMC Univ Paris 06, Complexité du vivant, 4 place Jussieu, 75005 Paris, France

Collège de France, 75005 Paris, France
e-mail: christine.petit@pasteur.fr

S.T. Suzuki, S. Hirano (eds.), *The Cadherin Superfamily*,
DOI 10.1007/978-4-431-56033-3_13

Keywords Cadherins • Usher syndrome • Deafness • Convergent extension • Planar cell polarity • Cell-cell contacts • Tip-link • Stereocilia

13.1 Introduction

Our ability to discriminate sounds in noisy environments and to understand speech involves some unique properties of the mechanoreceptive sensory cells of the auditory sense organ (the cochlea), the outer hair cells (OHCs), which amplify the sound stimulus, and the inner hair cells (IHCs), the genuine sensory cells responsible for transmitting the sound-induced electrical signal to the primary auditory neurons (Fig. 13.1a–c). In mammals, hair cells are organised in a regular mosaic pattern running from the base to the apex of the snail-shaped cochlea. The requirement for strict control over ion homeostasis, and for the adhesive strength and stability of cell–cell contacts renders the cochlea very sensitive to defects of adhesion-related proteins (see El-Amraoui and Petit 2010; Richardson et al. 2011). Defects of several genes encoding adhesion proteins have been found to be responsible for hearing impairment in humans (see Table 13.1) and/or mice (http://hearingimpairment.jax.org/). The genes implicated in hearing impairment in humans include those encoding cadherin-23 (*CDH23*) and protocadherin-15

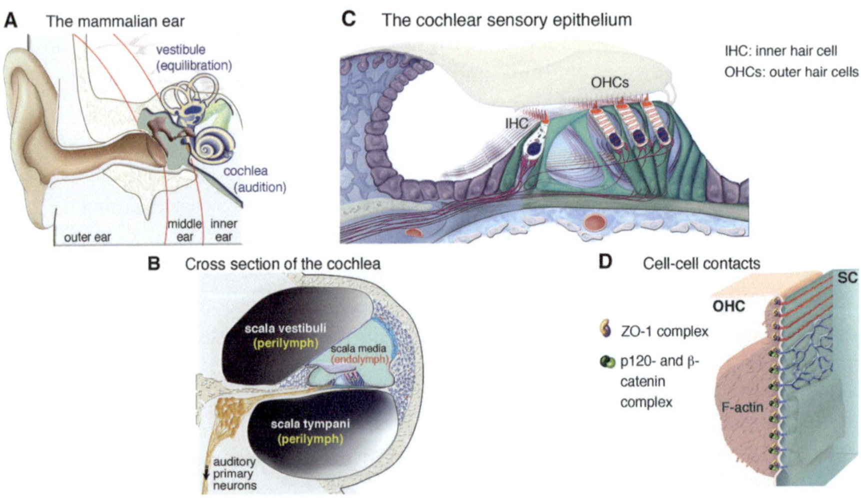

Fig. 13.1 The mammalian ear, the inner ear, and the hair cells, the sensory transducers. (**a**) The inner ear contains three semi-circular canals, the utricle and the saccule (balance organs) and the cochlea (auditory organ). (**b**) A cross-section through the cochlea showing the three fluid-filled compartments: scala vestibuli, scala tympani, and scala media. (**c**) The scala media house the auditory sensory epithelium, the organ of Corti (OC), consisting of one row of inner hair cells (IHCs), three rows of OHCs, and various types of supporting cells. (**d**) At the atypical cell–cell junctions between OHCs and Deiters cell (DC), catenin complexes are distributed together with claudin-containing domains (Adapted from Nunes et al. 2006)

Table 13.1 Cadherins and cell–cell junction proteins involved in inherited human diseases

Proteins present in cell-cell contacts
***CLDN14*, claudin-14** (OMIM 605608)
Sensorineural recessive deafness, DFNB29 (OMIM 614035)
TRIC, MARVELD2 (OMIM 610153)
Sensorineural recessive deafness, DFNB49 (OMIM 610572)
***ILDR1*, immunoglobulin-like domain containing receptor 1** (OMIM 609739)
Sensorineural recessive deafness, DFNB42 (OMIM 609646)
TJP2, tight-junction protein 2 (OMIM 607709)
Sensorineural dominant deafness, DFNA51 (OMIM 6613558)
Proteins present in hair bundles
***CDH23*, cadherin-23** (OMIM 605516)
Sensorineural recessive deafness, DFNB12 (OMIM 601386) and Usher syndrome type 1D (OMIM 601067)
***PCDH15*, protocadherin-15** (OMIM 605514)
Sensorineural recessive deafness, DFNB23 (OMIM 609533) and Usher syndrome type 1F (OMIM 602083)
Proteins present in cell-cell contacts & hair bundles
***VEZT*, vezatin, an adherens junction protein**
Sensorineural recessive deafness, DFNBi (Bahloul et al. 2010)

The OMIM numbers contain the links between the indicated molecules and the related diseases

(*PCDH15*), which are responsible for syndromic (Usher syndrome) and isolated forms of deafness (Ahmed et al. 2001; Alagramam et al. 2001b; Bolz et al. 2001; Bork et al. 2001); vezatin (*VEZT)*, which causes an autosomal recessive form of deafness (DFNBi; Bahloul et al. 2009); and several tight junction proteins, such as claudin-14 (*CLDN14)*, tricellulin (*TRIC*), and immunoglobulin-like domain-containing receptor 1 (*ILDR1*), responsible for the DFNB29 (Wilcox et al. 2001), DFNB49 (Riazuddin et al. 2006), and DFNB48 (Borck et al. 2011) recessive forms of deafness, respectively. In addition, mutations of the gene encoding tight-junction protein 2 (*TJP2)* cause the DFNA51 dominant form of deafness (Walsh et al. 2010).

The Ca^{2+}-dependent adhesion molecules of the cadherin superfamily play a key role in cell–cell adhesion, acting in concert with other intercellular or matrix-associated proteins. About 114 cadherins and cadherin-related molecules have been identified in humans (Hulpiau et al. 2013). During development and at mature stages, epithelial and neuronal cells express a "cadherin code" involved in cell fate specification, morphogenetic and cellular rearrangements, the maintenance of tissue integrity, and intracellular signal transduction (Nelson 2008; Harris and Tepass 2010; Takeichi 2011; Hirano and Takeichi 2012; El-Amraoui and Petit 2013). This review focuses on recent advances towards understanding the contribution of cadherins to the development and functioning of the auditory sensory organ. The detailed molecular mechanisms underlying the role of cadherins in the early steps of otic organogenesis, differentiation of the inner ear sensory epithelia, and cochlear compartment morphogenesis are not well understood. By contrast, large amounts of data have been accumulated concerning the role of two cadherin-related proteins, cadherin-23 and protocadherin-15, in developing and mature hair cells.

13.2 Structure of the Mammalian Auditory Organ: Focus on Membrane–Membrane Contacts

The mammalian inner ear consists of various sense organs: the utricle, the saccule, and three semi-circular canals (forming the balance organ), which detect linear and angular accelerations, and the cochlea, which detects sound waves (Fig. 13.1a). The cochlea consists of three fluid-filled compartments of different ionic compositions: the scala vestibuli and the scala tympani, both filled with perilymph, and the scala media (cochlear duct) filled with endolymph (Fig. 13.1b). Endolymph is an unusual fluid with a high K^+ concentrations and a low Na^+ concentration (Fig. 13.1b, c), whereas perilymph has an ion composition more typical of extracellular fluids (high Na^+ concentration and low K^+ concentration) (Wangemann 2006). The apical surface of the hair cells and supporting cells is bathed in endolymph, whereas the cell bodies of these cells are surrounded by perilymph. The absence of paracellular permeability between the endolymphatic and perilymphatic compartments of the cochlea is crucial for hearing. The cells of the sensory epithelium experience prolonged mechanical stress due to the sound-induced motion of this epithelium. A resilient epithelial barrier is required to withstand such mechanical stress and to prevent the leakage of fluid between adjacent cells of the auditory sensory epithelium. Evolution has led to adaptive changes in structure at the junctions between OHCs and their neighbouring supporting cells (Deiters cells) (Togashi et al. 2011; Fukuda et al. 2014; Fig. 13.1d), to maintain the basoapical axial forces involved in OHC electromotility, membrane potential-driven changes in cell length mediating the frequency-tuned cochlear amplification (Dallos et al. 2008). Specialised junctions connect OHCs with their four neighbouring supporting cells. These large hybrid junctions (about 3–5 μm high) are unique in that they combine the ultrastructural features of tight junctions and adherens junctions, hence their name of 'tight-adherens junctions' (see Fig. 13.1d). At these junctions, claudins are arranged into claudin-14 and claudin-9/6 subdomain strands. At least three adherens junction proteins, p120-catenin, and the α- and β-catenins, colocalise with the claudin-9/6 subdomain and recruit a dense cytoskeletal actin network (Nunes et al. 2006).

The integrity of cadherin-mediated extracellular contacts between opposing membranes of the same cell also are essential for normal hearing. Indeed, fibrous extracellular links connect the cell apical membranes in a highly organised structure within the hair cell, the hair bundle. The hair bundles are the site at which sound waves are converted into electrical signals by a process of mechanoelectrical transduction (MET) (Corey and Hudspeth 1983) (Fig. 13.2a, b). The hair bundle, a V-shaped structure that crowns the apical surface of the hair cell, consists of 20–300 actin-filled, stiff microvilli, the stereocilia, organised into three rows with a

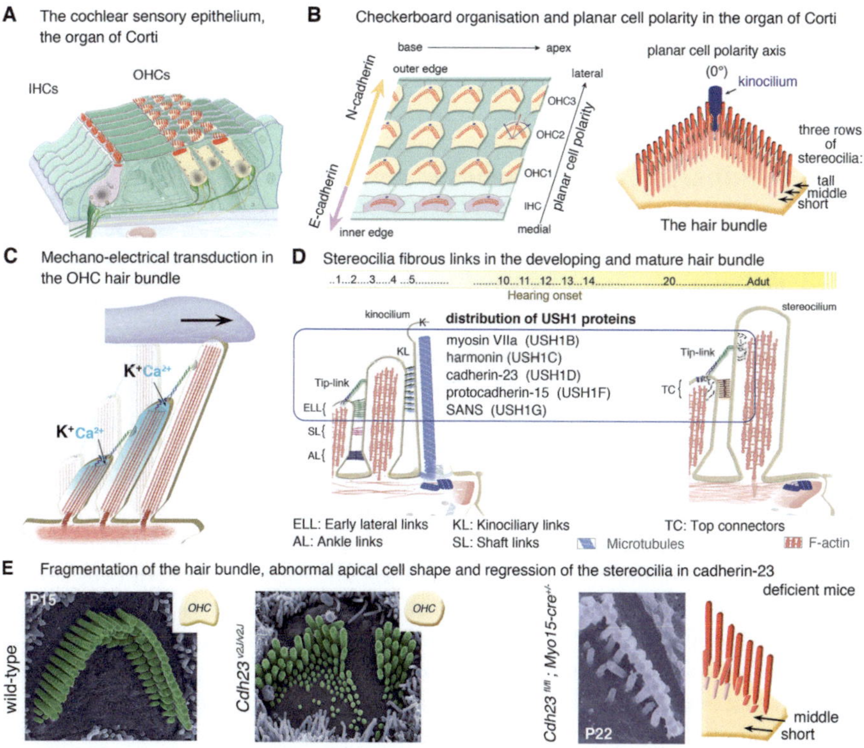

Fig. 13.2 The planar cell polarity of the auditory sensory epithelium, and hair bundle structure. (**a**) A top view of the organ of Corti showing the typical checkerboard-like organisation of the auditory sensory epithelium. (**b**) Positioning of the kinocilium, which directs the staircase pattern of the hair bundle, and determines the planar cell polarity axis. (**c**) Deflection of the hair bundle in the direction of the tall stereocilia leads to MET channel opening, K^+ influx and cell depolarisation. (**d**) Within the hair bundle, the stereocilia are connected by different types of transient and persistent extracellular links. The early transient lateral links (*ELL*), shaft links (*SL*), ankle links (*AL*), and kinociliary links (*KL*) are progressively replaced by the apical top connectors (*TC*) found in mature outer hair cells. The tip-links appear during development and persist in mature hair cells. (**e**) *Left* and *middle panels*: hair bundle disorganisation in a cadherin-23 deficient mouse (*Cdh23*$^{v2j/v2j}$). The *insets* are *top* views of the cell apical circumference, illustrating the loss of shape transition in *Cdh23*$^{v2j/v2j}$ mice: the cadherin-23-deficient OHC is round rather than having a convex (typically hexagonal) circumference. *Right panels*: in a mouse with postnatal loss of cadherin-23 (conditional knockout mouse, *Cdh23*$^{fl/fl}$*Myo15*-Cre$^{+/-}$), the tip-links disappear, and the stereocilia of the *short* and *middle rows* have regressed by P22

gradient in height, yielding a staircase-like structure. A single primary cilium, the kinocilium, and its basal body are located at the vertex of the V-shaped hair bundle (Fig. 13.2b). The kinocilium is thought to serve as a 'guidepost', determining the orientation and morphology of the hair bundle. Along the longitudinal axis of the cochlea, all the vertices of the hair bundles point in the same direction, towards the lateral wall of the scala media. This establishes a uniform polarity in the auditory sensory epithelium, the organ of Corti (Jones and Chen 2008; Kelly and Chen 2009;

Ezan and Montcouquiol 2013) (Fig. 13.2b). The kinocilium has been lost from the mature hair bundles of mammalian auditory hair cells (see Fig. 13.2c, d).

13.3 Cadherins, Organogenesis, and Planar Cell Polarity in the Cochlea

The auditory sensory epithelium has, over the years, proved very useful as a model system for studying planar cell polarity (PCP). Several excellent reviews have been published on hair cell polarity, PCP signalling pathways, and the relationships between these processes and the concomitant remodelling of apical junction complexes in the hair cells and the supporting cells (Ezan and Montcouquiol 2013; Jones and Chen 2008; Kelly and Chen 2009; Montcouquiol et al. 2006). The organ of Corti acquires a checkerboard-like pattern due to the arrangement of hair cells into a single row of IHCs and three rows of OHCs, interspersed with several types of supporting cells: the inner and outer pillar cells, Deiters, and phalangeal cells (see Fig. 13.2a). In the mouse, this organisation into different rows of cells occurs via a convergent extension process (Ezan and Montcouquiol 2013; Jones and Chen 2008; Kelly and Chen 2009; Montcouquiol et al. 2006). On embryonic day 14.5 (E14.5), all postmitotic hair cells and supporting cells are tightly packed and confined to a small region at the base of the cochlear duct. From E14.5 to E18.5, these differentiating cells undergo extensive morphogenetic movements along the elongating cochlear duct, leading to a thinning and lengthening of the epithelium due to cell intercalations, and the remodelling of cell–cell junctions (Keller et al. 2000; Kelly and Chen 2007; Etournay et al. 2010; Lepelletier et al. 2013).

The integrity and strength of the physical coupling between hair cells and supporting cells have been shown to affect hair cell polarity and the morphology of the hair bundles (Choi and Peifer 2011; Fukuda et al. 2014). A lack of nectin 3, an adherens junction protein of supporting cells, has been shown to cause cellular mispatterning in the organ of Corti, to alter the positioning of kinocilia and to result in an abnormal orientation and structure of the hair bundles (Fukuda et al. 2014). The role and spatiotemporal pattern of expression of the classical cadherins, E-cadherin and N-cadherin, have been studied in the developing organ of Corti and at early postnatal stages (Leonova and Raphael 1997; Mahendrasingam et al. 1997; Nunes et al. 2006; Simonneau et al. 2003; Whitlon 1993). Further investigations of the cellular and subcellular distribution of these cadherins showed that E- and N-cadherins had complementary expression patterns, delineating a sharp boundary between specific cell populations within the auditory sensory epithelium (see Fig. 13.2b) (Chacon-Heszele et al. 2012; Etournay et al. 2010). Indeed, E-cadherin has been detected in the outer pillar cells, OHCs, Deiters cells, and Hensen cells, whereas N-cadherin is restricted to cells in the medial IHC region, including the inner pillar cells, IHCs, and inner border cells (Etournay et al. 2010). Based on the differences in their properties, it has been suggested

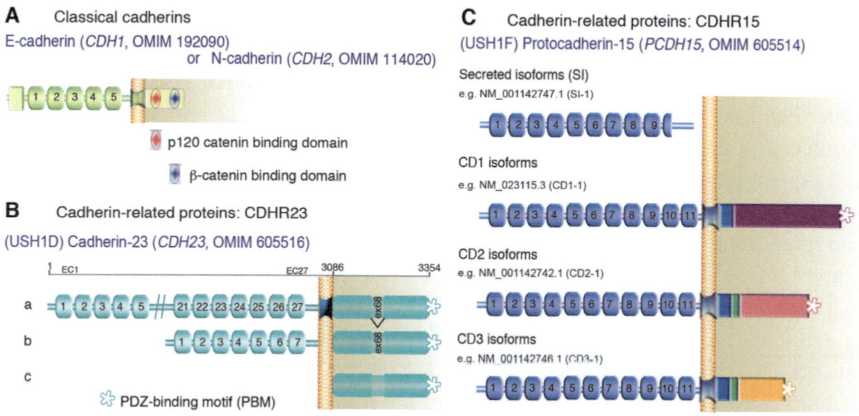

Fig. 13.3 Inner ear classical cadherins and cadherin-related proteins (**a**) All cadherins have an extracellular region consisting of extracellular cadherin (EC) domains, which are ~110 amino acids in length. (**a**) E- and N-cadherins are classical cadherins consisting of an extracellular region with five EC domains, a transmembrane domain (TM), and a cytoplasmic domain capable of binding two catenins, p120-catenin and β-catenin. (**b**, **c**) Cadherin-23 and protocadherin-15 both have an unusually large number of EC domains, and their cytodomains display no sequence similarity either to each other or to those of classical cadherins; in particular, they lack the consensus motif for binding to p120- or to β-catenins. (**b**) Three classes of cadherin-23 isoforms (*a*, *b*, and *c*) have been described. (**c**) The protocadherin-15 splice isoforms are grouped into four distinct classes: integral membrane isoforms with different cytodomains (CD), CD1, CD2, CD3, and secreted isoforms (SI)

that N-cadherin plays a major role in the early steps of cochlear extension, favouring cell mobility, whereas E-cadherin is involved in the establishment of stable cell–cell contacts (Chacon-Heszele et al. 2012). Evidence for cadherin involvement in the rapid changes in cell–cell contact occurring during cochlear extension was provided by the altered convergent extension process due to low N- and E-cadherin cell contents in the absence of p120-catenin (Chacon-Heszele et al. 2012). Classical cadherins interact principally with two catenins: p120-catenin, and β-catenin (see Fig. 13.3a). β-catenin anchors cadherins to the actin cytoskeleton by interacting with α-catenin, whereas p120-catenin is required for cadherin stabilisation at the plasma membrane (see Kourtidis et al. 2013; Nelson 2008).

A lack of p120-catenin has been shown to cause impaired cellular patterning, but does not lead to the misorientation of hair cells (Chacon-Heszele et al. 2012). Thus, different mechanisms underlie cochlear extension and hair cell polarity. It was recently shown that the establishment of hair cell planar polarity is influenced by kinocilium/basal body positioning and by junctional complex remodelling at the contacts between hair cells and their supporting cells (see Ezan and Montcouquiol 2013; Jones and Chen 2008; Kelly and Chen 2009). Indeed, core PCP signalling is required to establish the complementary compartmentalisation at the apical surface of the hair cell of two polarised protein complexes, the Pins (Partner Inscutable)/

Gαi (G-protein alpha i) and aPKC/PAR6/PAR3 complexes (Ezan et al. 2013; Tarchini et al. 2013). Pins//Gαi has been found to be restricted to a stereocilium-free lateral region also known as the 'bare zone', whereas aPKC occupies a medial zone that does not overlap with the core PCP protein Vangl2 region. These well-defined and delimited zones guide kinocilium positioning at the vertex, and the subsequent shaping and orientation of the hair bundle. Video microscopy studies monitoring movements of the kinocilium basal body (mother centriole) and its daughter centriole ex vivo established that the gradual displacement of the kinocilium to the cell periphery occurs at the same time as the convergent cochlear extension process (Lepelletier et al. 2013). The structural constraints ensuring kinocilium confinement to the vertices of the hair cells and the cytoskeletal mechanisms underlying kinocilium migration are not understood.

During early postnatal development in the mouse, a further refinement of hair bundle orientation and of the establishment of V-shaped OHC hair bundles has been shown to be correlated with a change in the shape of the apical circumference of these cells, which are initially round, but then develop a convex (typically hexagonal) circumference (Etournay et al. 2010). This shape transition is impaired in the absence of some hair bundle proteins, such as cadherin-23 (see insets in Fig. 13.2e, and Etournay et al. 2010) in mice. Several proteins of the apical junctional complex, including myosin II, myosin VIIa, shroom2, and F-actin, have been shown to be asymmetrically redistributed during the remodelling of the OHC apical circumference (Etournay et al. 2010). Consistent with the importance of actin- and myosin-based mechanisms in epithelial remodelling (see Takeichi 2014), a study of convergent extension in myosin II-deficient mice revealed that this type of myosin was required for cell patterning and alignment within the cochlear sensory epithelium (Yamamoto et al. 2009). Furthermore, the inhibition of myosin II activity in the auditory sensory epithelium has been shown to deform cells, changing their apical shape, and to lead to an overall expansion of the epithelium (Ebrahim et al. 2013). Along the apical junctional complex, nonmuscle myosin II sarcomeres are arranged in precise pairs across the junctional barrier between hair cells and adjacent supporting cells, forming an integrated, transcellular contractile network across the apical epithelial surface (Ebrahim et al. 2013). Connections between stereociliary rootlets and the apical junctions have been described (Etournay et al. 2010), but additional studies are required to clarify the mechanisms by which crosstalk between the protein complexes and the signalling pathways in the hair bundle and at the tight and adherens apical junctions between hair cells and supporting cells ensure correct planar cell polarity.

13.4 Cadherins in the Differentiating and Mature Auditory Hair Cells

13.4.1 The Sound-Receptive Structure of Hair Cells, the Hair Bundle

The transformation of sound waves into electrical signals (mechanoelectrical transduction; Fig. 13.2c), takes place in the hair bundle of cochlear hair cells (Corey and Hudspeth 1983). In developing hair cells, the stereocilia of the hair bundle are connected to each other, and some are connected to the kinocilium. Five different types of extracellular lateral links have been identified: early transient lateral links, kinociliary links, shaft links, ankle links, and the tip-link (Goodyear et al. 2005; El-Amraoui and Petit 2014; Pepermans and Petit 2015; Michalski and Petit 2015; Fig. 13.2d). In the mature hair bundle, two types of links persist: the tip-link and apical links, referred to as top connectors in OHCs (Fig. 13.2d). The tip-link, which is 5 nm in diameter and 125–250 nm long, is an oblique intertwined filament that connects each stereocilium of the short and middle rows to the side of the adjacent taller stereocilium (Pickles et al. 1984). Ultrastructural analyses have shown that there are electron-dense regions at the upper and lower insertion points of the tip-link (Corey and Hudspeth 1983; Furness and Hackney 1985; Furness et al. 2008) (see Fig. 13.2c).

According to the 'gating-spring' model of MET (Corey and Hudspeth 1983), a positive displacement of the bundle (towards the tall stereocilia) increases the tension in the tip-links, thereby increasing the open probability of the MET channels located at tips of stereocilia. The resulting influx of cations (K^+ and Ca^{2+}) sets up a depolarising receptor potential in OHCs and IHCs, eventually leading to neurotransmitter release by IHCs and signalling to the brain (Howard and Hudspeth 1988; Hudspeth et al. 2000). The steady-state tension at the tip-link has been estimated at ~8 pN in frogs (Jaramillo and Hudspeth 1993), but is unknown in mammals. Indirect estimates suggest that, during hair-cell stimulation, the tip-link tension probably lies in the range of 10–40 pN, although higher values cannot be excluded (Jaramillo and Hudspeth 1993).

13.4.2 Two Cadherin-Related Proteins Form the Tip-Link, a Key Component of the MET Machinery

Studies of a form of human syndromic deafness, Usher syndrome type I (USH1), have provided considerable insight into the development and functioning of the hair bundle (see Caberlotto et al. 2011a; Petit and Richardson 2009; Richardson et al. 2011; Pepermans and Petit 2015). Usher syndrome has an estimated prevalence of 1/10,000, and seems to account for about 50 % of all cases of monogenic deafness/blindness. USH1, the most severe form, is characterised by severe to profound congenital deafness, constant vestibular dysfunction, and retinitis pigmentosa beginning before puberty, eventually leading to blindness (reviewed by El-Amraoui and Petit 2014; Mathur and Yang 2015; Pepermans and Petit 2015). Six *USH1* genes have been identified; they encode the actin-based motor protein myosin VIIa (USH1B), the PDZ-domain–containing submembrane protein harmonin (USH1C), the scaffold protein Sans (USH1G), the integrin and calcium-binding protein CIB2 (USH1J), and the two Ca^{2+}-dependent adhesion proteins, cadherin-23 (USH1D) and protocadherin-15 (USH1F; reviewed by Petit and Richardson 2009; Pepermans and Petit 2015).

Animal USH1 models mimicking the abnormal hearing phenotype are available for all USH1 genes. Cadherin-23 is defective in the deaf *waltzer* mouse (Di Palma et al. 2001), and in the *sputnik* zebrafish (Sollner et al. 2004), whereas protocadherin-15 is defective in the *Ames waltzer* mouse (Alagramam et al. 2001a) and in the *orbiter* zebrafish (Seiler et al. 2005). These mutants have fragmented and disorganised hair bundles (see Fig. 13.2e) (Ahmed et al. 2003; Boeda et al. 2002; Lagziel et al. 2005; Michel et al. 2005; Senften et al. 2006; Siemens et al. 2002). A similar abnormal hair bundle phenotype has been observed in mutant mice defective for the other USH1 proteins (Johnson et al. 2003; Kikkawa et al. 2003; Self et al. 1998). This led us to suggest that USH1 proteins function together to establish the cohesiveness of the developing hair bundle (Lefevre et al. 2008; see Fig. 13.2e). An interdisciplinary study of these animal models, including an in-depth analysis of USH1 proteins, led to the suggestion that the physical coupling of cadherin-23– and protocadherin-15–mediated early transient interstereociliary and kinociliary links is required for the formation of individual, correctly shaped hair bundles (Boeda et al. 2002; Lagziel et al. 2005; Michel et al. 2005; Siemens et al. 2002, reviewed by El-Amraoui and Petit 2005; Mathur and Yang 2015; Richardson et al. 2011; Pepermans and Petit 2015). The tip-link, which has a helical structure (Kachar et al. 2000), consists of at least two cadherin heterodimers: *cis*-homodimers of cadherin-23 interact in trans with *cis*-homodimers of protocadherin-15 (Kazmierczak et al. 2007), to form the upper and lower parts of this link (Fig. 13.4). Crystallographic studies have shown that the two aminoterminal cadherin repeats (extracellular cadherin (EC) repeats 1 and 2) of protocadherin-15 and cadherin-23 interact in an 'extended handshake' manner to form an overlapping antiparallel heterodimer (Sotomayor et al. 2012). Molecular dynamics simulations and binding experiments have indicated that the bond formed by protocadherin-15 and cadherin-23 is mechanically strong enough to withstand the forces applied to the tip-link during hair bundle deflection (Sotomayor et al. 2012). The three-dimensional (3D) structures of the cadherin-23 and protocadherin-15 EC1 domains are similar to those of other cadherins (three Ca^{2+} -binding sites (1, 2, and 3) at the linker between EC1 and EC2), but with several unusual features, such as an elongated N-terminus stabilised at the tip by Ca^{2+}-binding site 0 (Elledge et al. 2010; Sotomayor et al. 2010). As in classical cadherins, Ca^{2+}-ion binding probably renders the cadherin extracellular domains more rigid, promoting *trans*-junctional interactions. The local Ca^{2+} concentration in the

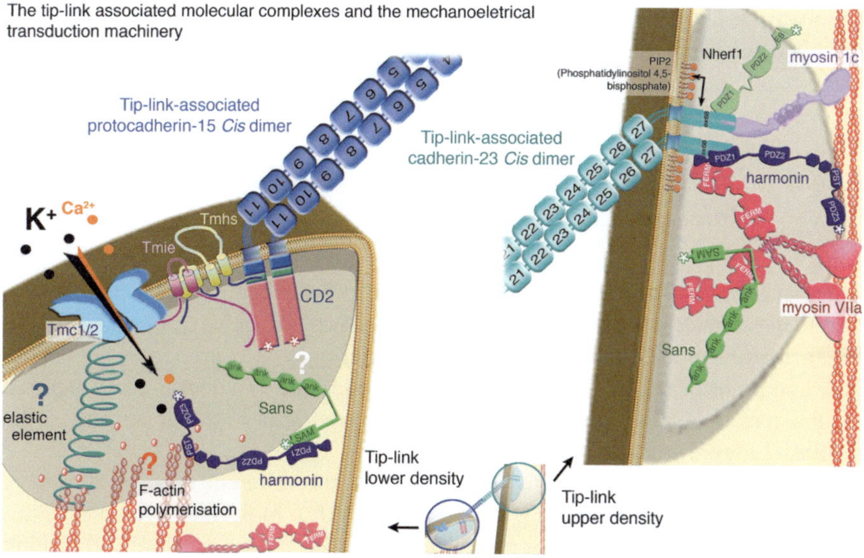

Fig. 13.4 In the mature hair bundle, large extracellular regions of the transmembrane proteins of cadherin-23 and protocadherin-15 CD2 form the tip-link. At the upper extremity of the tip-link, cadherin-23 is connected to the actin core of the stereocilia through interactions with USH1 proteins: harmonin b, and/or myosin VIIa. Myosin 1c and Nhref1 have also been shown to interact with cadherin-23 in vitro. In mature hair cells, the protocadherin-15 CD2 isoform forms the lower extremity of the tip-link. Interactions of this isoform with the transmembrane proteins Tmhs, Tmie, and Tmc1/2 channels have been described in vitro. The identity of the elastic element underlying the gating spring model and the way in which the tip-link and Ca^{2+} influx through the MET channel are coupled to F-actin polymerisation remain unclear

endolymph close to the tip-links is probably crucial for their formation and stability, particularly given the low concentration of Ca^{2+}, about 20–40 μM, in the endolymph (Bosher and Warren 1978).

According to the 'gating spring' model, mechanoelectrical transduction in vertebrate hair cells involves an elastic element (see Fig. 13.4) that pulls on the transducer channels to open these channels upon hair bundle deflection towards the tallest stereocilia (Howard and Hudspeth 1988). Until recently, the tip-link was thought to act as the gating spring and was assumed to have a working extension of 10–20 nm (Corey and Hudspeth 1983; Howard and Hudspeth 1988), to accommodate large sound stimuli (see Sotomayor et al. 2005). Whether cadherin homo- and hetero-dimers, which so far are thought to form a stiff element, also contribute to the gating spring is yet unclear. Other components of the stereocilia may also perform this role in conjunction with the tip-link. The extension and stiffness of large ankyrin-repeat–containing proteins (i.e., TRPA1, TRPN1) are thought to be similar to those predicted by the gating-spring model (Sotomayor et al. 2005), but there is still no evidence to suggest that these proteins are present in the hair bundles of vertebrates. The asymmetric nature of the tip-link suggests that its two ends have different biophysical and biochemical properties. Investigations are currently underway to determine the role of the various proteins in the functioning of the mechanoelectrical transduction machinery.

13.4.3 Cadherin-23 at the Core of the Upper Tip-Link Molecular Complex

Three classes of cadherin 23 isoforms have been described (Lagziel et al. 2005). The longest splice isoform of cadherin-23 consists of 27 EC domains (Fig. 13.3a, b). In mature mouse auditory hair cells, cadherin-23 is restricted to the upper insertion point of the tip-link (Kazmierczak et al. 2007). Cadherin-23 is required for the correct localisation of harmonin and myosin VIIa to the upper end of the tip-link (Lefevre et al. 2008; Bahloul et al. 2010).

Little is known about the respective roles of the different splice isoforms of cadherin-23 (Lagziel et al. 2005; Lagziel et al. 2009). Molecular dynamics simulations of the composition of the tip-link, based on its crystal structure and tip-link size, combined with the use of antibodies specific for class-a isoforms have led to suggestions that the tip-link contains cadherin-23a (the longest isoform) (Kazmierczak et al. 2007). The roles of the smaller isoforms of cadherin-23, cadherin-23b (smaller extracellular domain), or cadherin-23c (the cytoplasmic splice isoform) are currently unknown.

13.4.4 Protocadherin-15 at the Core of the Lower Tip-Link Molecular Complex

There are about 24 different protocadherin-15 transcripts (Ahmed et al. 2008; Pepermans and Petit 2015) classified as CD1, CD2, CD3, and SI on the basis of the predicted encoded proteins (see Fig. 13.3c). The expression of the secreted forms of protocadherin-15 (SI) by the hair cells and the roles of these forms have yet to be described. The three integral membrane isoforms consist of 11 EC repeats, a single transmembrane domain, and a variable cytoplasmic domain carrying different C-terminal PDZ-binding motifs (PBMs). The various cytoplasmic regions of these protocadherin-15 transmembrane splice isoforms are encoded by different final coding exons: exon 35 for CD1, exon 38 for CD2, and exon 39 for CD3 (Ahmed et al. 2008; Pepermans and Petit 2015). The different cytoplasmic domains probably impart functional specificity by conferring on the protein an ability to interact with different binding partners. Establishment of the distribution of each isoform in the developing and mature hair bundles, and assessments of their contribution to tip-link formation are thus important. Specific mutants of each isoform have been generated to address these issues, and the distribution of each isoform in the hair bundle has been determined (Alagramam et al. 2011; Webb et al. 2011; Pepermans et al. 2014).

The lower part of the mechanoelectrical transduction machinery has been shown to undergo molecular maturation at about the time of hearing onset in mice. This maturation involves a switch from the initial functional redundancy between the three Pcdh15 isoforms to a situation in which only the protocadherin-15 CD2 isoform is essential for tip-link formation, at mature stages. The CD2 isoform has been localised to the lower insertion point of the tip-link in mature hair bundles, in studies with an antibody specific for this isoform. Moreover, the postnatal deletion of CD2, which preserves the formation and differentiation of the hair bundles, has been shown to abolish tip-links and mechanoelectrical transduction currents. Finally, mutation analysis in patients with an isolated (nonsyndromic) form of profound deafness revealed frameshift mutations in exon 38 of *PCDH15*. This exon is specific to CD2 isoforms (Pepermans et al. 2014). Knockout mice defective for any of the first five USH1 proteins – myosin VIIa, harmonin, cadherin-23, protocadherin-15, and Sans – displayed a loss of the stereocilia in the short and middle rows (Lefevre et al. 2008). Subsequent studies of conditional knockout mice with postnatal inactivation of USH1 proteins (i.e., Sans, cadherin-23, or protocadherin-15 CD2 isoforms; Caberlotto et al. 2011b; Pepermans et al. 2014) provided further evidence for a connection between the mechanoelectrical transduction machinery and the F-actin polymerisation machinery, by showing a progressive regression of the short and middle-row stereocilia following the loss of tip-links. This connection may be constitutive or functional, possibly involving the tip-link tension and/or the magnitude of Ca^{2+} influx through the MET channels as a regulator of F-actin polymerisation (Lefevre et al. 2008; Caberlotto et al. 2011a; Pepermans et al. 2014).

13.5 Constructing the Tip-Link Interactome

In all cells, cadherins have diverse roles, involving interactions with multiple cytoplasmic proteins, including adaptor proteins, actin- and/or microtubule-binding proteins, protein kinases and phosphatases, and transcription regulators (Nelson 2008; Harris and Tepass 2010; Takeichi 2011; Hirano and Takeichi 2012). The tip region of the stereocilia may be considered to be a mechanoresponsive adhesion site, a focal adhesion point at which membrane-associated complexes interact with extracellular components. Studies of the 'integrin adhesome', a key component of focal adhesions, have revealed an associated network of more than 150 linked components potentially modified by 690 identified interactions (Zaidel-Bar et al. 2007). It seems likely that the lower and upper tip-link regions of high electron density also contain a large number of proteins governing the structural and signalling activities of the mechanoelectrical transduction machinery.

At the upper insertion point of the tip-link, cadherin-23 has been shown to form a ternary complex with harmonin and myosin VIIa (Bahloul et al. 2010). The cytoplasmic region of cadherin-23 and harmonin, both independently and as a binary complex, also binds phosphatidylinositol 4,5-bisphosphate (PIP2), which may account for the role of this membrane phospholipid in mechanoelectrical transduction current adaptation (Bahloul et al. 2010). Harmonin-b (the largest harmonin isoform) binds to actin filaments (Boeda et al. 2002), myosin VIIa and cadherin-23 (Boeda et al. 2002; Siemens et al. 2002). Harmonin b has thus been identified as a putative intracellular membrane-cytoskeleton crosslinker, coupling the cadherin-23 cytodomain with the underlying cytoskeleton (Boeda et al. 2002). It is involved in MET current adaptation, probably by recruiting adaptation motors to the upper insertion point of the tip-link (Michalski et al. 2009). It has been suggested that this coupling optimises the sensitivity of the hair bundle to sound-induced motion (Grillet et al. 2009; Michalski et al. 2009). This tip-link connection to the cytoskeleton may also be mediated by myosin 1c, an adaptation motor protein in the hair bundle (Stauffer et al. 2005) that has been shown to coimmunoprecipitate with cadherin-23 (Siemens et al. 2004). Its levels in the stereocilia are decreased in the absence of cadherin-23 (Phillips et al. 2006). The PDZ-domain-containing protein Nherf1 (Kamiya et al. 2014), and Magi-1, a membrane-associated guanylate kinase (Xu et al. 2008), have both been found to bind cadherin-23 in vitro, and may also be considered possible components of the upper mechanoelectrical transduction machinery. Finally, the USH1G protein, Sans, has been identified as a critical component of the tip-link complex (Caberlotto et al. 2011b; Grati and Kachar 2011), because postnatal depletion of this protein leads to a loss of the tip-links, followed by a progressive decrease in the length of the short and middle rows of stereocilia, until their complete disappearance (Caberlotto et al. 2011b).

At the lower insertion point, the Tmc1 (transmembrane channel protein 1), and Tmc2 proteins have been identified as possible subunits of the mechanoelectrical transduction channel (Kawashima et al. 2011). However, it remains unclear whether Tmc1, which persists in the mature hair bundle, unlike Tmc2, is a component of the pore channel (see Pepermans and Petit 2015). Two other transmembrane proteins located in the apical region of the stereocilia, and the absence of which leads to deafness, have been identified as key proteins of the MET machinery: Tmhs (tetraspan membrane protein of hair cell stereocilia; Xiong et al. 2012) and Tmie (transmembrane protein of inner ear hair cells; Zhao et al. 2014). It remains unknown how the mechanoelectrical transduction channel is connected to the lower part of the tip-link. Interactions between the tip-link protein protocadherin-15 and the other components of the lower MET machinery have been investigated (see Fig. 13.4). Yeast two-hybrid and coimmunoprecipitation experiments have shown that the CD1 and CD3 isoforms of protocadherin-15 interact directly with Tmc1 and Tmie (Maeda et al. 2014). All three isoforms of protocadherin-15 were found to coimmunoprecipitate with both TMC1 and TMC2 (Beurg et al. 2015). In another study, Zhao and colleagues showed that the CD2 isoform bound Tmie directly, but that the CD1 and CD3 isoforms bound Tmie only indirectly, via Tmhs (Zhao et al. 2014; see Fig. 13.4). These interactions among multiple splice isoforms suggest that there may be flexibility in the potential interactions between protocadherin-15 isoforms and Tmc1/2 channels. They also highlight the need to determine precisely the temporal distribution of the components of the mechanoelectrical transduction machinery, in both developing and mature hair cells, to ensure the correct interpretation of their potential interactions in situ. The possibility that these complexes differ along the tonotopic axis of the cochlea cannot be excluded.

Two recent studies based on multi-isotope imaging mass spectrometry and live imaging of single differentiating hair cells to capture stereociliogenesis (Drummond et al. 2015), and on the monitoring of β-actin-GFP incorporation into the stereocilia of adult mouse hair cells in vivo and ex vivo (Narayanan et al. 2015), demonstrated that the actin cores of the stereocilia were stable structures and that new F-actin incorporation occurred very slowly, and almost exclusively at the barbed ends of the filament near the distal tips of the stereocilia. In mutant mice either lacking ADF (actin depolymerising factor) or expressing a mutated form of AIP1 (actin interacting protein 1), two proteins involved in the severing and disassembly of F-actin, the stereocilia of the short and middle rows were found to be shortened, and some were missing (Narayanan et al. 2015), although they did not disappear entirely as they do in USH1 mutant mice (Caberlotto et al. 2011a; Lefevre et al. 2008; Pepermans et al. 2014). This suggests that two as yet uncharacterised mechanisms control the length of stereocilia. These processes probably involve various classes of actin-binding proteins, including nucleators of actin polymerisation, actin-capping proteins, actin-severing proteins, actin-crosslinking proteins, and molecular motors (reviewed by Caberlotto et al. 2011a; Michalski and Petit 2015).

13.6 Concluding Remarks

As described above, many questions about the roles played by cadherin and cadherin-related molecules in the hair cells remain unresolved. In particular, further studies are required to determine how the two extremities of the tip-link are coupled to the F-actin cytoskeleton at the tips of the stereocilia. The diversity of cochlear phenotypes observed in mutant mice lacking the various protocadherin-15 splice isoforms suggests that these isoforms contribute to the formation of different types of links associated with different molecular complexes with temporally and spatially specific functions. It has been suggested that cadherin-23 and protocadherin-15 connect the light-sensitive outer segment to the inner segment of photoreceptor cells and the adjacent calyceal processes, microvillus-like structures similar to the stereocilia of the hair cells (Sahly et al. 2012). Parallel studies of cadherin-mediated pathways in the photoreceptor cells will probably shed light on possible associations between the USH1-associated mechanosensory functions and the calyceal processes.

Acknowledgements We thank Jean-Pierre Hardelin for critical reading of the manuscript, and Jacques Boutet de Monvel for useful comments. We apologise for omitting to cite certain references, due to space constraints. The work of the authors is supported by Institut Pasteur,

INSERM, the European Union Seventh Framework Programme, under grant agreement HEALTH-F2-2010-242013 (TREATRUSH), LHW-Stiftung, Fondation Raymonde & Guy Strittmatter, Fighting Blindness, FAUN Stiftung (Suchert Foundation), Conny Maeva Charitable Foundation, Fondation Orange, European Research Council (ERC) advanced grant "Hair bundle" (ERC-2011-AdG 294570), LABEX Lifesenses [ANR-10-LABX-65], the French National Research Agency (ANR) as part of the second "Investissements d'Avenir" programme (ANR-15-RHUS-0001), Retina France, and the Fondation Voir et Entendre.

References

Ahmed ZM, Riazuddin S, Bernstein SL, Ahmed Z, Khan S, Griffith AJ, Morell RJ, Friedman TB, Riazuddin S, Wilcox ER (2001) Mutations of the protocadherin gene *PCDH15* cause Usher syndrome type 1F. Am J Hum Genet 69:25–34

Ahmed ZM, Riazuddin S, Ahmad J, Bernstein SL, Guo Y, Sabar MF, Sieving P, Griffith AJ, Friedman TB, Belyantseva IA, Wilcox ER (2003) PCDH15 is expressed in the neurosensory epithelium of the eye and ear and mutant alleles are responsible for both USH1F and DFNB23. Hum Mol Genet 12:3215–3223

Ahmed ZM, Riazuddin S, Aye S, Ali RA, Venselaar H, Anwar S, Belyantseva PP, Qasim M, Friedman TB (2008) Gene structure and mutant alleles of PCDH15: nonsyndromic deafness DFNB23 and type 1 Usher syndrome. Hum Genet 124:215–223

Alagramam KN, Murcia CL, Kwon HY, Pawlowski KS, Wright CG, Woychik RP (2001a) The mouse Ames waltzer hearing-loss mutant is caused by mutation of *Pcdh15*, a novel protocadherin gene. Nat Genet 27:99–102

Alagramam KN, Yuan H, Kuehn MH, Murcia CL, Wayne S, Srisailpathy CR, Lowry RB, Knaus R, Van Laer L, Bernier FP, Schwartz S, Lee C, Morton CC, Mullins RF, Ramesh A, Van Camp G, Hageman GS, Woychik RP, Smith RJ (2001b) Mutations in the novel protocadherin PCDH15 cause Usher syndrome type 1F. Hum Mol Genet 10:1709–1718

Alagramam KN, Goodyear RJ, Geng R, Furness DN, van Aken AF, Marcotti W, Kros CJ, Richardson GP (2011) Mutations in protocadherin 15 and cadherin 23 affect tip links and mechanotransduction in mammalian sensory hair cells. PLoS One 6:e19183

Bahloul A, Simmler MC, Michel V, Leibovici M, Perfettini I, Roux I, Weil D, Nouaille S, Zuo J, Zadro C, Licastro D, Gasparini P, Avan P, Hardelin JP, Petit C (2009) Vezatin, an integral membrane protein of adherens junctions, is required for the sound resilience of cochlear hair cells. EMBO Mol Med 1:125–138

Bahloul A, Michel V, Hardelin J-P, Nouaille S, Hoos S, Houdusse A, England P, Petit C (2010) Cadherin-23, myosin VIIa and harmonin, encoded by Usher syndrome type I genes, form a ternary complex and interact with membrane phospholipids. Hum Mol Genet 19:3557–3565

Beurg M, Xiong W, Zhao B, Muller U, Fettiplace R (2015) Subunit determination of the conductance of hair-cell mechanotransducer channels. Proc Natl Acad Sci U S A 112:1589–1594

Boeda B, El-Amraoui A, Bahloul A, Goodyear R, Daviet L, Blanchard S, Perfettini I, Fath KR, Shorte S, Reiners J, Houdusse A, Legrain P, Wolfrum U, Richardson G, Petit C (2002) Myosin VIIa, harmonin and cadherin 23, three Usher I gene products that cooperate to shape the sensory hair cell bundle. EMBO J 21:6689–6699

Bolz H, von Brederlow B, Ramirez A, Bryda EC, Kutsche K, Nothwang HG, Seeliger M, del C-Salcedó CM, Vila MC, Molina OP, Gal A, Kubisch C (2001) Mutation of *CDH23*, encoding a new member of the cadherin gene family, causes Usher syndrome type 1D. Nat Genet 27:108–112

Borck G, Ur Rehman A, Lee K, Pogoda HM, Kakar N, von Ameln S, Grillet N, Hildebrand MS, Ahmed ZM, Nurnberg G, Ansar M, Basit S, Javed Q, Morell RJ, Nasreen N, Shearer AE,

Ahmad A, Kahrizi K, Shaikh RS, Ali RA, Khan SN, Goebel I, Meyer NC, Kimberling WJ, Webster JA, Stephan DA, Schiller MR, Bahlo M, Najmabadi H, Gillespie PG, Nurnberg P, Wollnik B, Riazuddin S, Smith RJ, Ahmad W, Muller U, Hammerschmidt M, Friedman TB, Leal SM, Ahmad J, Kubisch C (2011) Loss-of-function mutations of ILDR1 cause autosomal-recessive hearing impairment DFNB42. Am J Hum Genet 88:127–137

Bork JM, Peters LM, Riazuddin S, Bernstein SL, Ahmed ZM, Ness SL, Polomeno R, Ramesh A, Schloss M, Srisailpathy CRS, Wayne S, Bellman S, Desmukh D, Ahmed Z, Khan SN, Der Kaloustian VM, Li XC, Lalwani A, Riazuddin S, Bitner-Glindzicz M, Nance WE, Liu X-Z, Wistow G, Smith RJH, Griffith AJ, Wilcox ER, Friedman TB, Morell RJ (2001) Usher syndrome 1D and nonsyndromic autosomal recessive deafness DFNB12 are caused by allelic mutations of the novel cadherin-like gene *CDH23*. Am J Hum Genet 68:26–37

Bosher SK, Warren RL (1978) Very low calcium content of cochlear endolymph, an extracellular fluid. Nature 273:377–378

Caberlotto E, Michel V, de Monvel JB, Petit C (2011a) Coupling of the mechanotransduction machinery and F-actin polymerization in the cochlear hair bundles. Biogeosciences 1:169–174

Caberlotto E, Michel V, Foucher I, Bahloul A, Goodyear RJ, Pepermans E, Michalski N, Perfettini I, Alegria-Prevot O, Chardenoux S, Do Cruzeiro M, Hardelin JP, Richardson GP, Avan P, Weil D, Petit C (2011b) Usher type 1G protein sans is a critical component of the tip-link complex, a structure controlling actin polymerization in stereocilia. Proc Natl Acad Sci U S A 108:5825–5830

Chacon-Heszele MF, Ren D, Reynolds AB, Chi F, Chen P (2012) Regulation of cochlear convergent extension by the vertebrate planar cell polarity pathway is dependent on p120-catenin. Development 139:968–978

Choi W, Peifer M (2011) Cell biology. Arranging a cellular checkerboard. Science 333:1099–1100

Corey DP, Hudspeth AJ (1983) Kinetics of the receptor current in bullfrog saccular hair cells. J Neurosci 3:962–976

Dallos P, Wu X, Cheatham MA, Gao J, Zheng J, Anderson CT, Jia S, Wang X, Cheng WH, Sengupta S, He DZ, Zuo J (2008) Prestin-based outer hair cell motility is necessary for mammalian cochlear amplification. Neuron 58:333–339

Di Palma F, Holme RH, Bryda EC, Belyantseva IA, Pellegrino R, Kachar B, Steel KP, Noben-Trauth K (2001) Mutations in *Cdh23*, encoding a new type of cadherin, cause stereocilia disorganization in waltzer, the mouse model for Usher syndrome type 1D. Nat Genet 27:103–107

Drummond MC, Barzik M, Bird JE, Zhang DS, Lechene CP, Corey DP, Cunningham LL, Friedman TB (2015) Live-cell imaging of actin dynamics reveals mechanisms of stereocilia length regulation in the inner ear. Nat Commun 6:6873

Ebrahim S, Fujita T, Millis BA, Kozin E, Ma X, Kawamoto S, Baird MA, Davidson M, Yonemura S, Hisa Y, Conti MA, Adelstein RS, Sakaguchi H, Kachar B (2013) NMII forms a contractile transcellular sarcomeric network to regulate apical cell junctions and tissue geometry. Curr Biol 23:731–736

El-Amraoui A, Petit C (2005) Usher I syndrome: unravelling the mechanisms that underlie the cohesion of the growing hair bundle in inner ear sensory cells. J Cell Sci 118:4593–4603

El-Amraoui A, Petit C (2010) Cadherins as targets for genetic diseases. Cold Spring Harb Perspect Biol 2:a003095

El-Amraoui A, Petit C (2013) Cadherin defects in inherited human diseases. Prog Mol Biol Transl Sci 116C:361–384

El-Amraoui A, Petit C (2014) The retinal phenotype of Usher syndrome: pathophysiological insights from animal models. C R Biol 337:167–177

Elledge HM, Kazmierczak P, Clark P, Joseph JS, Kolatkar A, Kuhn P, Muller U (2010) Structure of the N terminus of cadherin 23 reveals a new adhesion mechanism for a subset of cadherin superfamily members. Proc Natl Acad Sci U S A 107:10708–10712

Etournay R, Lepelletier L, Boutet de Monvel J, Michel V, Cayet N, Leibovici M, Weil D, Foucher I, Hardelin JP, Petit C (2010) Cochlear outer hair cells undergo an apical circumference remodeling constrained by the hair bundle shape. Development 137:1373–1383
Ezan J, Montcouquiol M (2013) Revisiting planar cell polarity in the inner ear. Semin Cell Dev Biol 24:499–506
Ezan J, Lasvaux L, Gezer A, Novakovic A, May-Simera H, Belotti E, Lhoumeau AC, Birnbaumer L, Beer-Hammer S, Borg JP, Le Bivic A, Nurnberg B, Sans N, Montcouquiol M (2013) Primary cilium migration depends on G-protein signalling control of subapical cytoskeleton. Nat Cell Biol 15:1107–1115
Fukuda T, Kominami K, Wang S, Togashi H, Hirata K, Mizoguchi A, Rikitake Y, Takai Y (2014) Aberrant cochlear hair cell attachments caused by Nectin-3 deficiency result in hair bundle abnormalities. Development 141:399–409
Furness DN, Hackney CM (1985) Cross-links between stereocilia in the guinea pig cochlea. Hear Res 18:177–188
Furness DN, Katori Y, Nirmal Kumar B, Hackney CM (2008) The dimensions and structural attachments of tip links in mammalian cochlear hair cells and the effects of exposure to different levels of extracellular calcium. Neuroscience 154:10–21
Goodyear RJ, Marcotti W, Kros CJ, Richardson GP (2005) Development and properties of stereociliary link types in hair cells of the mouse cochlea. J Comp Neurol 485:75–85
Grati M, Kachar B (2011) Myosin VIIa and sans localization at stereocilia upper tip-link density implicates these Usher syndrome proteins in mechanotransduction. Proc Natl Acad Sci U S A 108:11476–11481
Grillet N, Xiong W, Reynolds A, Kazmierczak P, Sato T, Lillo C, Dumont RA, Hintermann E, Sczaniecka A, Schwander M, Williams D, Kachar B, Gillespie PG, Muller U (2009) Harmonin mutations cause mechanotransduction defects in cochlear hair cells. Neuron 62:375–387
Harris TJ, Tepass U (2010) Adherens junctions: from molecules to morphogenesis. Nat Rev Mol Cell Biol 11:502–514
Hirano S, Takeichi M (2012) Cadherins in brain morphogenesis and wiring. Physiol Rev 92:597–634
Howard J, Hudspeth AJ (1988) Compliance of the hair bundle associated with gating of mechanoelectrical transduction channels in the bullfrog's saccular hair cell. Neuron 1:189–199
Hudspeth AJ, Choe Y, Mehta AD, Martin P (2000) Putting ion channels to work: mechanoelectrical transduction, adaptation, and amplification by hair cells. Proc Natl Acad Sci U S A 97:11765–11772
Hulpiau P, Gul IS, van Roy F (2013) New insights into the evolution of metazoan cadherins and catenins. Prog Mol Biol Transl Sci 116:71–94
Jaramillo F, Hudspeth AJ (1993) Displacement-clamp measurement of the forces exerted by gating springs in the hair bundle. Proc Natl Acad Sci U S A 90:1330–1334
Johnson KR, Gagnon LH, Webb LS, Peters LL, Hawes NL, Chang B, Zheng QY (2003) Mouse models of USH1C and DFNB18: phenotypic and molecular analyses of two new spontaneous mutations of the *Ush1c* gene. Hum Mol Genet 12:3075–3086
Jones C, Chen P (2008) Primary cilia in planar cell polarity regulation of the inner ear. Curr Top Dev Biol 85:197–224
Kachar B, Parakkal M, Kurc M, Zhao Y, Gillespie PG (2000) High-resolution structure of hair-cell tip links. Proc Natl Acad Sci U S A 97:13336–13341
Kamiya K, Michel V, Giraudet F, Riederer B, Foucher I, Papal S, Perfettini I, Le Gal S, Verpy E, Xia W, Seidler U, Georgescu MM, Avan P, El-Amraoui A, Petit C (2014) An unusually powerful mode of low-frequency sound interference due to defective hair bundles of the auditory outer hair cells. Proc Natl Acad Sci U S A 111:9307–9312
Kawashima Y, Geleoc GS, Kurima K, Labay V, Lelli A, Asai Y, Makishima T, Wu DK, Della Santina CC, Holt JR, Griffith AJ (2011) Mechanotransduction in mouse inner ear hair cells requires transmembrane channel-like genes. J Clin Invest 121:4796–4809

Kazmierczak P, Sakaguchi H, Tokita J, Wilson-Kubalek EM, Milligan RA, Muller U, Kachar B (2007) Cadherin 23 and protocadherin 15 interact to form tip-link filaments in sensory hair cells. Nature 449:87–91

Keller R, Davidson L, Edlund A, Elul T, Ezin M, Shook D, Skoglund P (2000) Mechanisms of convergence and extension by cell intercalation. Philos Trans R Soc Lond B Biol Sci 355:897–922

Kelly M, Chen P (2007) Shaping the mammalian auditory sensory organ by the planar cell polarity pathway. Int J Dev Biol 51:535–547

Kelly MC, Chen P (2009) Development of form and function in the mammalian cochlea. Curr Opin Neurobiol 19:395–401

Kikkawa Y, Shitara H, Wakana S, Kohara Y, Takada T, Okamoto M, Taya C, Kamiya K, Yoshikawa Y, Tokano H, Kitamura K, Shimizu K, Wakabayashi Y, Shiroishi T, Kominami R, Yonekawa H (2003) Mutations in a new scaffold protein Sans cause deafness in Jackson shaker mice. Hum Mol Genet 12:453–461

Kourtidis A, Ngok SP, Anastasiadis PZ (2013) p120 catenin: an essential regulator of cadherin stability, adhesion-induced signalling, and cancer progression. Prog Mol Biol Transl Sci 116:409–432

Lagziel A, Ahmed ZM, Schultz JM, Morell RJ, Belyantseva IA, Friedman TB (2005) Spatiotemporal pattern and isoforms of cadherin 23 in wild type and waltzer mice during inner ear hair cell development. Dev Dyn 205:295–306

Lagziel A, Overlack N, Bernstein SL, Morell RJ, Wolfrum U, Friedman TB (2009) Expression of cadherin 23 isoforms is not conserved: implications for a mouse model of Usher syndrome type 1D. Mol Vis 15:1843–1857

Lefevre G, Michel V, Weil D, Lepelletier L, Bizard E, Wolfrum U, Hardelin JP, Petit C (2008) A core cochlear phenotype in USH1 mouse mutants implicates fibrous links of the hair bundle in its cohesion, orientation and differential growth. Development 135:1427–1437

Leonova EV, Raphael Y (1997) Organization of cell junctions and cytoskeleton in the reticular lamina in normal and ototoxically damaged organ of Corti. Hear Res 113:14–28

Lepelletier L, de Monvel JB, Buisson J, Desdouets C, Petit C (2013) Auditory hair cell centrioles undergo confined Brownian motion throughout the developmental migration of the kinocilium. Biophys J 105:48–58

Maeda R, Kindt KS, Mo W, Morgan CP, Erickson T, Zhao H, Clemens-Grisham R, Barr-Gillespie PG, Nicolson T (2014) Tip-link protein protocadherin 15 interacts with transmembrane channel-like proteins TMC1 and TMC2. Proc Natl Acad Sci U S A 111:12907–12912

Mahendrasingam S, Katori Y, Furness DN, Hackney CM (1997) Ultrastructural localization of cadherin in the adult guinea-pig organ of Corti. Hear Res 111:85–92

Mathur P, Yang J (2015) Usher syndrome: Hearing loss, retinal degeneration and associated abnormalities. Biochim Biophys Acta 1852:406–420

Michalski N, Petit C (2015) Genetics of auditory mechano-electrical transduction. Pflugers Arch 467:49–72

Michalski N, Michel V, Caberlotto E, Lefevre GM, van Aken AF, Tinevez JY, Bizard E, Houbron C, Weil D, Hardelin JP, Richardson GP, Kros CJ, Martin P, Petit C (2009) Harmonin-b, an actin-binding scaffold protein, is involved in the adaptation of mechanoelectrical transduction by sensory hair cells. Pflugers Arch 459:115–130

Michel V, Goodyear RJ, Weil D, Marcotti W, Perfettini I, Wolfrum U, Kros C, Richardson GP, Petit C (2005) Cadherin 23 is a component of the transient lateral links in the developing hair bundles of cochlear sensory cells. Dev Biol 280:281–294

Montcouquiol M, Crenshaw EB 3rd, Kelley MW (2006) Noncanonical Wnt signaling and neural polarity. Annu Rev Neurosci 29:363–386

Narayanan P, Chatterton P, Ikeda A, Ikeda S, Corey DP, Ervasti JM, Perrin BJ (2015) Length regulation of mechanosensitive stereocilia depends on very slow actin dynamics and filament-severing proteins. Nat Commun 6:6855

Nelson WJ (2008) Regulation of cell-cell adhesion by the cadherin-catenin complex. Biochem Soc Trans 36:149–155

Nunes FD, Lopez LN, Lin HW, Davies C, Azevedo RB, Gow A, Kachar B (2006) Distinct subdomain organization and molecular composition of a tight junction with adherens junction features. J Cell Sci 119:4819–4827

Pepermans E, Petit C (2015) The tip-link molecular complex of the auditory mechano-electrical transduction machinery. Hear Res 330(1):10–17

Pepermans E, Michel V, Goodyear R, Bonnet C, Abdi S, Dupont T, Gherbi S, Holder M, Makrelouf M, Hardelin JP, Marlin S, Zenati A, Richardson G, Avan P, Bahloul A, Petit C (2014) The CD2 isoform of protocadherin-15 is an essential component of the tip-link complex in mature auditory hair cells. EMBO Mol Med 6:984–992

Petit C, Richardson G (2009) Linking deafness genes to hair-bundle development and function. Nat Neurosci 12:703–710

Phillips KR, Tong S, Goodyear R, Richardson GP, Cyr JL (2006) Stereociliary myosin-1c receptors are sensitive to calcium chelation and absent from cadherin 23 mutant mice. J Neurosci 26:10777–10788

Pickles JO, Comis SD, Osborne MP (1984) Cross-links between stereocilia in the guinea pig organ of Corti, and their possible relation to sensory transduction. Hear Res 15:103–112

Riazuddin S, Ahmed ZM, Fanning AS, Lagziel A, Kitajiri S, Ramzan K, Khan SN, Chattaraj P, Friedman PL, Anderson JM, Belyantseva IA, Forge A, Friedman TB (2006) Tricellulin is a tight-junction protein necessary for hearing. Am J Hum Genet 79:1040–1051

Richardson GP, Boutet de Monvel J, Petit C (2011) How the genetics of deafness illuminates auditory physiology. Annu Rev Physiol 73:311–334

Sahly I, Dufour E, Schietroma C, Michel V, Bahloul A, Perfettini I, Pepermans E, Estivalet A, Carette D, Aghaie A, Ebermann I, Lelli A, Iribarne M, Hardelin JP, Weil D, Sahel JA, El-Amraoui A, Petit C (2012) Localization of Usher 1 proteins to the photoreceptor calyceal processes, which are absent from mice. J Cell Biol 199:381–399

Seiler C, Finger-Baier KC, Rinner O, Makhankov YV, Schwarz H, Neuhauss SC, Nicolson T (2005) Duplicated genes with split functions: independent roles of protocadherin15 orthologues in zebrafish hearing and vision. Development 132:615–623

Self T, Mahony M, Fleming J, Walsh J, Brown SD, Steel KP (1998) Shaker-1 mutations reveal roles for myosin VIIA in both development and function of cochlear hair cells. Development 125:557–566

Senften M, Schwander M, Kazmierczak P, Lillo C, Shin JB, Hasson T, Geleoc GS, Gillespie PG, Williams D, Holt JR, Muller U (2006) Physical and functional interaction between protocadherin 15 and myosin VIIa in mechanosensory hair cells. J Neurosci 26:2060–2071

Siemens J, Kazmierczak P, Reynolds A, Sticker M, Littlewood-Evans A, Muller U (2002) The Usher syndrome proteins cadherin 23 and harmonin form a complex by means of PDZ-domain interactions. Proc Natl Acad Sci U S A 99:14946–14951

Siemens J, Lillo C, Dumont RA, Reynolds A, Williams DS, Gillespie PG, Muller U (2004) Cadherin 23 is a component of the tip link in hair-cell stereocilia. Nature 428:950–955

Simonneau L, Gallego M, Pujol R (2003) Comparative expression patterns of T-, N-, E-cadherins, beta-catenin, and polysialic acid neural cell adhesion molecule in rat cochlea during development: implications for the nature of Kolliker's organ. J Comp Neurol 459:113–126

Sollner C, Rauch GJ, Siemens J, Geisler R, Schuster SC, Muller U, Nicolson T (2004) Mutations in cadherin 23 affect tip links in zebrafish sensory hair cells. Nature 428:955–959

Sotomayor M, Corey DP, Schulten K (2005) In search of the hair-cell gating spring elastic properties of ankyrin and cadherin repeats. Structure 13:669–682

Sotomayor M, Weihofen WA, Gaudet R, Corey DP (2010) Structural determinants of cadherin-23 function in hearing and deafness. Neuron 66:85–100

Sotomayor M, Weihofen WA, Gaudet R, Corey DP (2012) Structure of a force-conveying cadherin bond essential for inner-ear mechanotransduction. Nature 492:128–132

Stauffer EA, Scarborough JD, Hirono M, Miller ED, Shah K, Mercer JA, Holt JR, Gillespie PG (2005) Fast adaptation in vestibular hair cells requires myosin-1c activity. Neuron 47:541–553

Takeichi M (2011) Self-organization of animal tissues: cadherin-mediated processes. Dev Cell 21:24–26

Takeichi M (2014) Dynamic contacts: rearranging adherens junctions to drive epithelial remodelling. Nat Rev Mol Cell Biol 15:397–410

Tarchini B, Jolicoeur C, Cayouette M (2013) A molecular blueprint at the apical surface establishes planar asymmetry in cochlear hair cells. Dev Cell 27:88–102

Togashi H, Kominami K, Waseda M, Komura H, Miyoshi J, Takeichi M, Takai Y (2011) Nectins establish a checkerboard-like cellular pattern in the auditory epithelium. Science 333:1144–1147

Walsh T, Pierce SB, Lenz DR, Brownstein Z, Dagan-Rosenfeld O, Shahin H, Roeb W, McCarthy S, Nord AS, Gordon CR, Ben-Neriah Z, Sebat J, Kanaan M, Lee MK, Frydman M, King MC, Avraham KB (2010) Genomic duplication and overexpression of TJP2/ZO-2 leads to altered expression of apoptosis genes in progressive nonsyndromic hearing loss DFNA51. Am J Hum Genet 87:101–109

Wangemann P (2006) Supporting sensory transduction: cochlear fluid homeostasis and the endocochlear potential. J Physiol 576:11–21

Webb SW, Grillet N, Andrade LR, Xiong W, Swarthout L, Della Santina CC, Kachar B, Muller U (2011) Regulation of PCDH15 function in mechanosensory hair cells by alternative splicing of the cytoplasmic domain. Development 138:1607–1617

Whitlon DS (1993) E-cadherin in the mature and developing organ of Corti of the mouse. J Neurocytol 22:1030–1038

Wilcox ER, Burton QL, Naz S, Riazuddin S, Smith TN, Ploplis B, Belyantseva I, Ben-Yosef T, Liburd NA, Morell RJ, Kachar B, Wu DK, Griffith AJ, Friedman TB (2001) Mutations in the gene encoding tight junction claudin-14 cause autosomal recessive deafness DFNB29. Cell 104:165–172

Xiong W, Grillet N, Elledge HM, Wagner TF, Zhao B, Johnson KR, Kazmierczak P, Muller U (2012) TMHS is an integral component of the mechanotransduction machinery of cochlear hair cells. Cell 151:1283–1295

Xu Z, Peng AW, Oshima K, Heller S (2008) MAGI-1, a candidate stereociliary scaffolding protein, associates with the tip-link component cadherin 23. J Neurosci 28:11269–11276

Yamamoto N, Okano T, Ma X, Adelstein RS, Kelley MW (2009) Myosin II regulates extension, growth and patterning in the mammalian cochlear duct. Development 136:1977–1986

Zaidel-Bar R, Itzkovitz S, Ma'ayan A, Iyengar R, Geiger B (2007) Functional atlas of the integrin adhesome. Nat Cell Biol 9:858–867

Zhao B, Wu Z, Grillet N, Yan L, Xiong W, Harkins-Perry S, Muller U (2014) TMIE is an essential component of the mechanotransduction machinery of cochlear hair cells. Neuron 84:954–967

Chapter 14
Cadherins in Cancer

Lauren V. Albrecht, Kathleen J. Green, and Adi D. Dubash

Abstract Despite decades of research, cancer remains one of the leading causes of death worldwide. Progression of cancer includes the breakdown or loss of normal tissue structure, which closely depends on the proper expression and regulation of numerous cell–cell adhesion molecules. Not surprisingly, the multifunctional cadherin cell–cell adhesion protein family members have emerged as critical regulators of tumorigenesis. The maintenance of cell–cell junctions and adhesion-mediated signaling pathways are tightly regulated by cadherin expression in a tissue-specific manner. In addition to their adhesive functions, cadherins integrate diverse cellular inputs (from cell–cell adhesion to mechanical forces or receptor tyrosine kinase activity) and translate these cues into biochemical intracellular signaling events involved in cell proliferation, motility, survival, and tissue homeostasis. Alterations in cadherin function can lead to cancer progression through a variety of molecular mechanisms including cadherin switching/EMT and the misregulation of different signaling mediators, including Rho GTPases, Ras/MAPK, Hippo/YAP, PI3K/Akt, and other pathways that have been implicated in tumor progression. Furthermore, cadherins have been recently implicated in mechanotransduction and cancer stem cell signaling. In this chapter, we report both fundamental findings and novel insights that define the roles of cadherins in human cancer and discuss how changes in the expression and regulation of these molecules contribute to cancer progression.

L.V. Albrecht
Department of Pathology, Northwestern University Feinberg School of Medicine, 303 E. Chicago Ave, Chicago, IL 60611, USA

K.J. Green, Ph.D.
Department of Pathology, Northwestern University Feinberg School of Medicine, 303 E. Chicago Ave, Chicago, IL 60611, USA

Department of Dermatology, Northwestern University Feinberg School of Medicine, 303 E. Chicago Ave, Chicago, IL 60611, USA

A.D. Dubash, Ph.D. (✉)
Department of Pathology, Northwestern University Feinberg School of Medicine, 303 E. Chicago Ave, Chicago, IL 60611, USA

Department of Biology, Furman University, 3300 Poinsett Highway, Greenville, SC 29613
e-mail: adi.dubash@furman.edu

S.T. Suzuki, S. Hirano (eds.), *The Cadherin Superfamily*,
DOI 10.1007/978-4-431-56033-3_14

Keywords Cadherins • Cancer progression • Metastasis • E-cadherin • Cadherin switching • EMT • EGFR • MAP Kinase • YAP • VE-cadherin • Angiogenesis • Desmosomal cadherins

14.1 Introduction

Oncogenesis is a complex process driven by epigenetic changes, genetic factors, alterations in signaling pathways, and different environmental conditions (Gurevich-Panigrahi et al. 2009; Virani et al. 2012; Gasparini and Longo 2012; Farahani et al. 2014). Metastasis is responsible for 90 % of cancer-related mortalities, a process that normally requires loss of intercellular adhesion between tumor cells to promote an invasive and migratory phenotype. Over decades of study in the field of cancer research, our understanding of how aberrant regulation of intercellular adhesion leads to tumorigenesis has continued to evolve.

Cadherins are emerging as key regulators of cancer progression. Representing one of the major families of cell adhesion molecules, cadherins mediate cell–cell interactions through homophilic calcium-dependent interactions of their extracellular cadherin (EC) domains. Based on sequence comparison, members of the cadherin family can be divided into either type I (E-, N-, P-) classical cadherins, type II (VE-, OB-) classical cadherins, desmosomal cadherins (desmogleins and desmocollins), and truncated (T-) cadherin (Fig. 14.1). In normal tissues, cadherins play an essential role in the maintenance of cell–cell adhesion. In addition to this central function, cadherins coordinate a broad spectrum of cellular functions, which includes cell proliferation, apoptosis, movement, differentiation, and tissue morphogenesis. The mechanisms by which these cell adhesion molecules regulate these different biological processes involve the regulation of a wide array of signaling pathways. In particular, crosstalk between cadherins and receptor tyrosine kinases are responsible for control of signaling mediators such as Rho and Ras GTPases, PI3-kinase/Akt, Src kinase, as well as transcription factors including MAP kinases, SMADs, YAP, and TCF. The coordinated control of these signaling pathways and their biological outcomes is tightly regulated via tissue-specific regulation of cadherin expression, which is compromised during the progression of different cancers and contributes to the mechanisms by which tumor cells invade and metastasize to distant sites (Berx and van Roy 2009).

Cadherin function is also well studied during critical prometastatic processes such as epithelial-to-mesenchymal transitions (EMT), cell migration, and invasion (Makrilia et al. 2009). The goal of this chapter is to review the literature that highlights key mechanistic roles for cadherins in the process of cancer formation and metastasis, and to elucidate the biggest questions that still remain. We conclude the chapter with a discussion of recent clinical work targeting cadherin adhesion and signaling for therapeutic intervention in cancer.

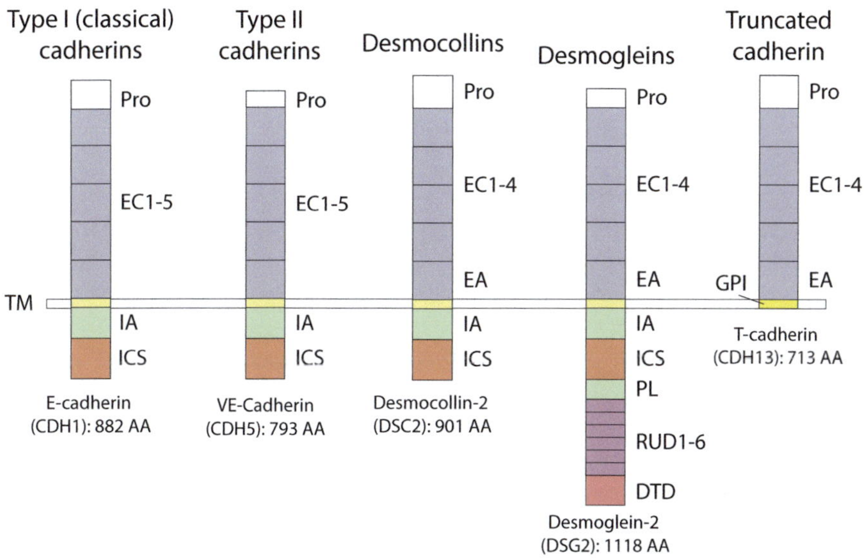

Fig. 14.1 Representative members of the cadherin superfamily. Schematic models of the structure of human members of the cadherin family of proteins involved in cancer (drawn to scale). Type I cadherins, represented by epithelial (E)-cadherin (CDH1), are characterized by short ectodomains that contains five extracellular cadherin (EC) repeats, a single transmembrane domain, and a cytoplasmic domain. Type II cadherins are represented by vascular endothelial (VE)-cadherin (CDH5). Desmosomal cadherins shown include desmocollin-2 (DSC2) and desmoglein-2 (DSG2). As the only member of its subfamily, truncated (T-) cadherin (CDH13) lacks a cytoplasmic domain and is connected to the plasma membrane by a GPI anchor. Not all cadherin subtypes are depicted. Protein lengths are depicted below each protein with the number of amino acids (AA). The domains shown here include: Pro domain, extracellular cadherin (EC) domain, extracellular anchor (EA) domain, transmembrane region (TM), intracellular anchor (IA) domain, intracellular cadherin sequence (ICS) domain, intracellular proline rich linker (PL), repeating unit domain (RUD), desmoglein terminal domain (DTD), and glycosylphosphatidylinositol (GPI) anchor

14.2 E-Cadherin

Derived from epithelial tissues, carcinomas constitute more than 70 % of tumors and are responsible for 80 % of cancer-related deaths in the United States (Jemal et al. 2011; Farahani et al. 2014). The classical, prototypic epithelial cadherin, E-cadherin provides adhesive properties to epithelial cells and mediates the morphogenesis and development of epithelial tissues. A key component of adherens junctions (AJs), the E-cadherin molecule has five extracellular cadherin repeats, a single transmembrane domain, and a cytoplasmic domain (Fig. 14.1). The cytoplasmic domain of E-cadherin interacts with catenin family members β-catenin, α-catenin, p120-catenin, and γ-catenin/plakoglobin (Pg) and coordinates the organization of the actin microfilament network at the plasma membrane (McEwen, Escobar, and Gottardi 2012).

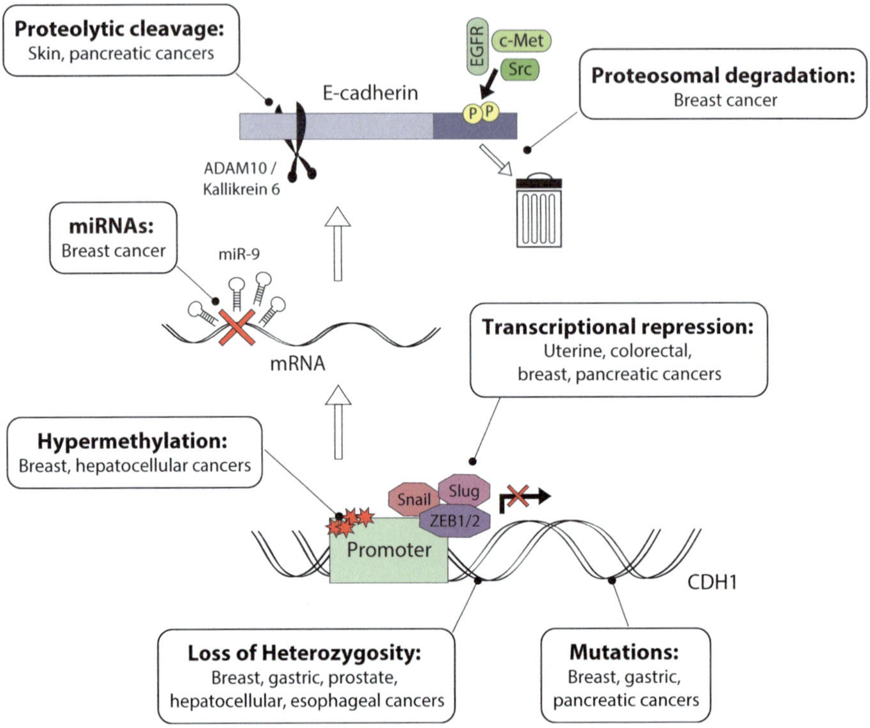

Fig. 14.2 Mechanisms of E-cadherin downregulation. A hallmark of cancer progression is the downregulation of E-cadherin (CDH1). Different mechanisms for downregulation of E-cadherin are depicted here, and their identification in different types of cancer (Sect. 14.2.1). Genetic mutations and loss of heterozygosity affect E-cadherin expression at the level of the CDH1 gene, and are found in many different types of cancer. Hypermethylation of the promoter region of E-cadherin causes a loss of transcriptional activation of the CDH1 gene, and occurs in breast and hepatocellular cancers. Transcription factors Snail, Slug, and ZEB1/2 repress E-cadherin transcription and are overexpressed in uterine, colorectal, breast, and pancreatic cancers. miR-9, is overexpressed in breast cancer and suppresses E-cadherin at the mRNA level. Overexpression of proteinases such as kallikrein 6 cleave E-cadherin and promote aberrant protein degradation. Overactivation of different oncogenic proteins (EGFR, c-Met, Src) in cancer leads to E-cadherin phosphorylation that signals for protein degradation

E-cadherin represents the most well-studied cadherin in cancer as it is commonly downregulated during the progression of epithelial tumors, leading to tumor growth and invasion. Downregulation or loss of E-cadherin has been documented in breast, nasopharynx, gastrointestinal tract, pancreas, lung, stomach, kidney, prostate, and esophageal primary cancers (Brooks et al. 2010). The mechanisms through which E-cadherin expression is lost during cancer progression varies depending on the stage and type of tumor, and includes loss of heterozygosity, genetic mutations, promoter hypermethylation, transcriptional silencing, and defects in protein processing (Fig. 14.2) (Berx and van Roy 2009).

14.2.1 Downregulation of E-Cadherin Gene Expression

14.2.1.1 LOH and Genetic Mutations

Breast cancer is the major cause of cancer-related deaths among women worldwide with greater than 40,000 breast cancer fatalities in the United States alone (Jemal et al. 2011). Loss of heterozygosity (LOH) of the E-cadherin gene occurs in 50 % of ductal carcinomas of breast cancer (Cleton-Jansen et al. 2001) and is frequently observed in lobular breast cancer (Fig. 14.2) (Berx et al. 1996). LOH has also been identified in gastric, prostate, hepatocellular, and esophageal carcinomas, correlating with malignant progression of these cancers (Strathdee 2002).

In addition to LOH, genetic mutations also lead to the downregulation of E-cadherin and have been identified in breast and gastric cancer (Fig. 14.2; Becker et al. 1994; Gayther et al. 1998; Becker et al. 1999; Van Marck et al. 2005; Carvalho et al. 2012). Inactivating mutations were first identified in diffuse gastric cancer where E-cadherin mutants lack the calcium-binding extracellular repeats (Becker et al. 1993; Becker et al. 1994; Becker et al. 1999; Carneiro et al. 1999). Additionally, somatic mutations of E-cadherin have also been identified in breast, gastric, and pancreatic cancers and correlate with high growth patterns for breast cancers (Berx et al. 1998). Genomic and proteomic studies have been performed to identify the different gene expression profiles that are associated with specific tumor cell types (Minn et al. 2005). With the recent advances in human genome sequencing techniques, the identification of E-cadherin mutations could prove to be a powerful tool as a clinical marker of tumor stage in cancer patients.

14.2.1.2 Promoter Methylation and Transcriptional Silencing

The most well-studied and frequent mechanism of downregulation is hypermethylation of the E-cadherin promoter, having been identified in at least eight human carcinoma types (Fig. 14.2; (Graff et al. 1995; Graff et al. 1998; Graff et al. 2000; Machado et al. 2001; Oki and Issa 2010). Aberrant E-cadherin promoter methylation increases during malignant progression of breast and hepatocellular carcinomas (Kanai et al. 2000; Nass et al. 2000).

E-cadherin expression is also tightly regulated on the transcriptional level (Fig. 14.2). Transcription factors Snail, Slug, ZEB1, and ZEB2 have all been demonstrated to decrease E-cadherin expression by binding directly to the E-box sequence of the gene promoter. ZEB1 is overexpressed in highly aggressive uterine cancers and localizes to the tumor border of advanced colorectal carcinomas (Comijn et al. 2001; Spoelstra et al. 2006; Spaderna et al. 2006; Shamir et al. 2014). Inversely correlated to differentiation grade of the tumors, Snail is overexpressed in human breast carcinomas with infiltrating ductal carcinomas (Blanco et al. 2002). Compelling studies of the molecular mechanisms underlying Snail activity demonstrated that Snail recruits the mSin3A corepressor and histone deacetylase (HDAC1/2) to the E-cadherin promoter in metastatic pancreatic cancer (Peinado et al. 2004; von Burstin et al. 2009). Furthermore, Snail silencing also depends on Polycomb repressive complex 2 (PRC2) (Herranz et al. 2008). Thus, transcriptional silencing is a widespread mechanism for E-cadherin downregulation in different cancer types.

14.2.1.3 microRNAs

Recently, microRNAs (miRNAs) have emerged as modulators of E-cadherin gene expression (Fig. 14.2). miR-9, miR-101, and specific members of the miR-200 family regulate E-cadherin expression through different mechanisms (Ma et al. 2010; Tan et al. 2010). By directly targeting E-cadherin, the overexpression of miR-9 in breast cancer cells promotes metastasis where miR-9–mediated suppression of E-cadherin leads to β-catenin mislocalization and activation, promoting the induction of tumor angiogenic gene expression (Ma et al. 2010). Furthermore, miR-9 sensitizes tumor cells to metastatic signals from the surrounding tissue and tumor micoenvironment indicating that epithelial cells crosstalk with their environment to promote cancer through modulation of cadherins (Ma et al. 2010). In contrast, miRNAs from the miR-200 family enforce an epithelial phenotype of tissues by increasing E-cadherin expression through the direct targeting of ZEB1/ΔEF1 and ZEB2/SIP1 mRNA and repression of ZEB1 and ZEB2 protein expression (Hurteau et al. 2007; Park et al. 2008; Gregory et al. 2008; Korpal et al. 2008). The misregulation of microRNA–mRNA feedback loops in cancer exemplify the delicate balance of cadherin expression required for normal biological processes.

14.2.1.4 Posttranslational Processing

Defects in E-cadherin protein processing are also associated with cancer progression (Fig. 14.2). Missense mutations of E-cadherin associated with hereditary diffuse gastric cancer (HDGC) are regulated by endoplasmic reticulum associated degradation (ERAD) and the mutations at these sites lead to premature proteasomal degradation (Simões-Correia et al. 2008). Proteolyic processing of E-cadherin is mediated by matrix metalloproteinases that cleave E-cadherin's ectodomain near the plasma membrane (Lochter et al. 1997; Noë et al. 2001; Davies et al. 2001; Covington et al. 2006; Symowicz et al. 2007). Cadherin fragments have been found in the serum isolated from cancer patients (De Wever et al. 2007). Furthermore, global gene expression analyses of tumor samples identified that serine proteinases such as kallikrein 6 (Klk6) are aberrantly expressed in human squamous cell carcinomas (SCCs) and pancreatic carcinomas (Johnson et al. 2007; Klucky et al. 2007). Klk6 overexpression promotes the proteolytic activity of a disintegrin and metalloproteinase domain-containing protein 10 (ADAM10), a protease that mediates E-cadherin ectodomain shedding, and results in solubilized E-cadherin fragments, disrupted cell–cell junctions, and increased metastatic dissemination (Fig. 14.2; Klucky et al. 2007).

Tyrosine phosphorylation of E-cadherin's cytoplasmic domain leads to its internalization and ubiquitin-mediated degradation (Fig. 14.2; van Roy and Berx 2008). E-cadherin endocytosis occurs through clathrin-mediated vesicles and nonclathrin pathways, which include caveolae-endocytosis and macropinocytosis (de Beco et al. 2012). Overactivation of proto-oncogenes, EGFR, c-Met, and Src, lead to increased phosphorylation of E-cadherin and aberrant E-cadherin degradation in breast cancer and other carcinomas (Fujita et al. 2002; Shen et al. 2008). Additionally, anomalous glycosylation has recently been proposed as a mechanism leading to downregulation of E-cadherin (Pinho et al. 2009; Pinho et al. 2011). Intriguingly, the deregulation of endocytic pathways has been classified as a new hallmark of cancer as cadherin adhesion and growth factor receptors are both compromised by aberrant vesicular transport (Mosesson et al. 2008). Collectively, the levels of complexity in the regulation of E-cadherin underscore the importance of maintaining E-cadherin expression as a cellular break on the road to cancer.

14.2.2 Epithelial-to-Mesenchymal Transitions (EMT)

Tumor-suppressing roles of E-cadherin are often interpreted in the context of epithelial-to-mesenchymal transitions (EMT), a dynamic process that leads to metastasis. EMT is characterized by genetic, biochemical, and ultimately phenotypic alterations of an epithelial cell type to a fibroblastic cell type (Bolós et al. 2003). The loss of E-cadherin is a hallmark in the initial stages of EMT.

14.2.2.1 Loss of E-Cadherin in EMT

There are multiple signaling pathways that converge to induce the downregulation of E-cadherin during EMT such as the Wnt, Notch/delta, and TGF-β pathways (Fig. 14.3; Grotegut et al. 2006; Leroy and Mostov 2007; Lee et al. 2008; Ding et al. 2010). In intestinal and mammary epithelial cells, increases in integrin-linked kinase (ILK) activity triggers Wnt signaling and transcription factors Snail and Slug, leading to downregulation of E-cadherin and induction of tumorigenic characteristics (Novak et al. 1998; Oloumi et al. 2004; Serrano et al. 2013). Under hypoxic conditions, the Notch signaling pathway is activated and also leads to the activation of E-cadherin repressors Snail and Slug (Leong et al. 2007; Sahlgren et al. 2008; Niessen et al. 2008). Additionally, transforming growth factor beta (TGF-β), a pivotal regulator of diverse cellular processes such as proliferation, migration, and extracellular matrix remodeling, also functions as a prominent inducer of EMT by enhancing Snail activity that subsequently leads to the loss of E-cadherin (Zeisberg et al. 2003; Kalluri and Weinberg 2009). Together, the

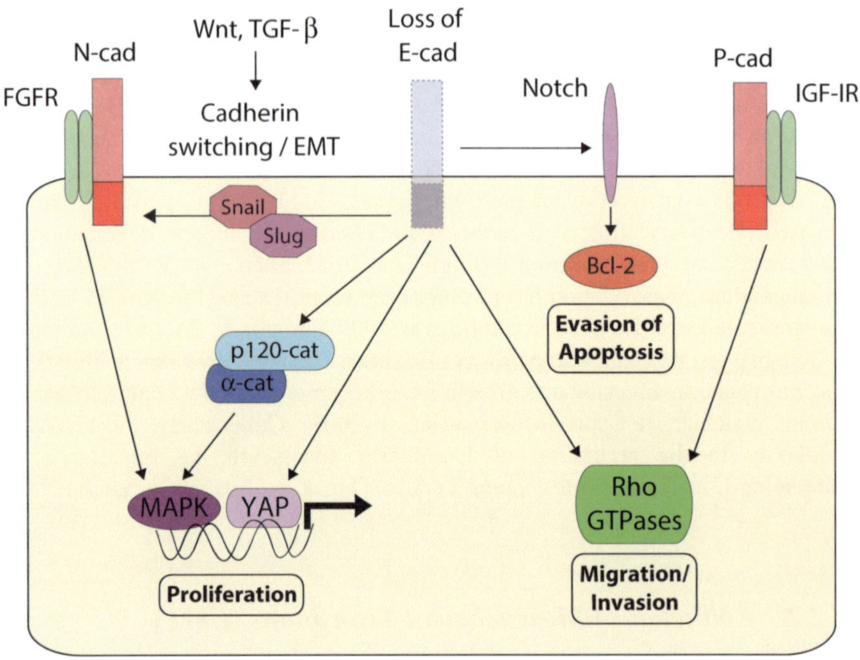

Fig. 14.3 Signaling pathways of epithelial-to-mesenchymal transition. During EMT, the loss of epithelial-cadherin (E-cad) leads to the activation of oncogenic signaling pathways (Sects. 14.2.2 and 14.3). Transcription factors Snail and Slug coordinate a "cadherin switch" by downregulating E-cadherin and inducing the expression of mesenchymal neuronal-cadherin (N-cad). Different signaling pathways (including Wnt and TGF-β) have been shown to activate Snail and Slug. In association with fibroblast growth factor receptor 1 (FGFR1), N-cad promotes cell proliferation through mitogen activated protein kinase (MAPK) transcriptional pathways. The loss of E-cad can also result in the mislocalization of α-catenin and p120 catenin, which leads to the activation of oncogenic Ras-MAPK signaling pathways. By controlling the subcellular localization and activation of Yap, E-cad acts as a direct regulator of Hippo signaling, which promotes EMT and cell proliferation. Somatic mutations of E-cad leading to its downregulation disrupt normal signaling to Rho GTPases (Rac1 and RhoA), which leads to tumor cell migration and invasion. Decreased E-cad expression also leads to the deregulated activation of Notch, inducing a resistance to apoptosis through the Notch1-dependent upregulation of Bcl-2, an anti-apoptotic protein. Placental-cadherin (P-cad) upregulation activates the insulin-like growth factor 1 receptor (IGF-1R) signaling pathway, which also leads to Rho GTPase signaling that promotes migration and invasion. For the sake of clarity, context-dependent tumor-suppressive functions of mesenchymal cadherins are not depicted here

activation of transcriptional repressors of E-cadherin represents a common downstream target for multiple signaling pathways implicated in EMT progression.

14.2.2.2 E-Cadherin Signaling in EMT

The loss of E-cadherin during EMT weakens cellular adhesion to promote invasion and migration. In addition to its roles in adhesion, E-cadherin also functions as a repressor of the ligand activation of many receptor tyrosine kinases (RTKs) in healthy tissues (Fig. 14.3; Thiery 2002; Qian et al. 2004; van Roy and Berx 2008). Thus, the loss of E-cadherin during EMT also contributes to tumor progression through the activation of pro-oncogenic signaling pathways. In primary gastric cancers, decreased E-cadherin expression leads to the deregulated activation of Notch and human epidermal growth factor receptors (HER), inducing a resistance to apoptosis through the Notch1-dependent upregulation of Bcl-2, an anti-apoptotic protein (Fig. 14.3; Ferreira et al. 2012). Also contributing to EMT progression, the loss of E-cadherin results in the mislocalization of AJ components such as α-catenin and p120 catenin, which leads to the activation of the oncogenic Ras and Rac1-MAPK signaling pathways, respectively (Fig. 14.3; Jamora et al. 2005; Soto et al. 2008).

The Hippo tumor suppressor signaling pathway controls organ size by promoting apoptosis and inhibiting cell proliferation (Kim et al. 2011). Many cancers express higher levels of Yap, a transcriptional factor regulated by the Hippo pathway that promotes EMT and cell proliferation, while inhibiting apoptosis (Overholtzer et al. 2006; Lei et al. 2008; Zhao et al. 2008; Kim et al. 2011). Intriguingly, recent reports demonstrate that E-cadherin also functions as an upstream mediator of the Hippo pathway (Nishioka et al. 2009; Kim et al. 2011; Hirate et al. 2013). By controlling the subcellular localization and activation of Yap, E-cadherin acts as a direct regulator of Hippo signaling to control cell proliferation (Fig. 14.3; Kim et al. 2011). Together, these findings demonstrate that, in addition to weakened cellular adhesion, the loss of E-cadherin also leads to aberrant signaling that endows cells with tumorigenic properties.

14.2.3 Tumor-Promoting Activities of E-Cadherin

Although expression of E-cadherin is generally anti-tumorigenic, E-cadherin displays tumor-promoting activities in a few cancer types, where it is responsible for promoting collective cell invasion and pro-survival EGFR signaling (van Roy and Berx 2008; Rodriguez et al. 2012; Oliveira et al. 2013). Interestingly, E-cadherin expression is consistently upregulated in ovarian tumors, a cancer that not does undergo EMT during tumorigenesis (Auersperg et al. 1999). In inflammatory breast cancer, E-cadherin plays an important role in promoting tumor intravasation, the process by which tumor cells gain access to vascular channels (van Zijl et al. 2011; Kim et al. 2011). Nevertheless, aside from these few cases, E-cadherin downregulation usually represents a critical step in promoting the initial stages of tumor development.

14.2.4 Mechanotransduction and E-Cadherin Signaling in Tumor Development

Mechanotransduction, the process whereby mechanical cues are translated into biochemical signals, has recently been implicated in tumor development. In the earliest stages of tumor development, alteration of the biomechanics of cancer cells leads to the activation of EMT signaling pathways that contribute to the downregulation of E-cadherin (Alessandri et al. 2013). Increasing the stiffness of the microenvironment surrounding the primary tumor leads to the activation of TGF-β signaling and tumor cell invasion in vivo in both glioma and epithelial mammary cancers. These findings raise the possibility that this is a general mechanism in malignant tumor progression (Suyama et al. 2002; Gaggioli et al. 2007; Goetz et al. 2011; Leight et al. 2012). As TGF-β activation contributes to the downregulation of E-cadherin during EMT, uncovering the connections between mechanotransduction signaling pathways and cadherin function could reveal novel insights into our understanding of tumor development.

Cadherins can also act as mechanosensors by responding to different mechanical stresses and triggering intracellular signaling pathways that coordinate actin cytoskeletal rearrangement, cell migration, and invasion (Smutny et al. 2010; Engl et al. 2014). These dynamic processes require the coordination of polarized activity of the small Rho GTPase protein family. Iconic members of the Rho family GTPases primarily mediate actin cytoskeletal dynamics: Cdc42 induces filopodial formation, Rac promotes lamellipodia formation at the leading edge of migrating cells, and Rho regulates cellular contractility and cell–matrix adhesion (Mateus et al. 2007; Mateus et al. 2009). In normal tissues, E-cadherin activates Rac1 (Kim et al. 2000; Nakagawa et al. 2001; Kraemer et al. 2007) and inhibits Rho activity (Noren et al. 2001; Arthur et al. 2002). However, E-cadherin's regulation of Rac1 and RhoA is disrupted by the presence of somatic E-cadherin mutations, which leads to tumor cell migration and invasion in hereditary diffuse-type gastric carcinoma (Fig. 14.3; Deplazes et al. 2009).

The molecular machinery that couples mechanical forces to metastatic properties such as cell migration has recently been investigated (Tambe et al. 2011; Trepat and Fredberg 2011). Collective cell migration requires cadherin-mediated AJs where the distribution of AJs at the leading edge regulates the direction and rate of migration (Peglion et al. 2014). In an exciting report, Das et al. demonstrated a link between cadherin-mediated, intercellular mechanical forces and collective cell motions in a moving epithelial monolayer. In this report, the tumor suppressor protein, Merlin, acted as a mechanotransducer by converting cadherin cell–cell adhesive forces into polarized Rac1 activation (Das et al. 2015). It is well established that Merlin localizes to E-cadherin to facilitate AJ formation while also coordinating AJ-mediated cellular adhesion with cell polarity (Gladden et al. 2010). As Merlin is a bona fide tumor suppressor, the tantalizing links underlying Merlin's mechanosensitive regulation of cytoskeleton-anchorage and of cadherin-modulated signal transduction in cancer development warrant further research.

14.3 Cadherin-Switching

A growing collection of data demonstrate that other members of the cadherin superfamily also play vital roles in tumorigenesis. "Cadherin switching" is the process during EMT where the loss of E-cadherin expression is coordinated with a subsequent induction of the expression of mesenchymal cadherins (Wheelock et al. 2008). The "cadherin switch" can influence the behavior of cells through a variety of mechanisms such as the modulation of nuclear signaling or growth factor receptor signaling, or through the activation of GTPases that promote cell migration and invasion (Wheelock et al. 2008).

14.3.1 N-Cadherin

Normally expressed in nonepithelial tissues, neuronal-cadherin (N-cadherin) plays important roles in tumor progression, where a cadherin switch to N-cadherin expression is associated with enhanced migration (Thiery et al. 2009; De Craene and Berx 2013). Additionally, transcriptional repressors of E-cadherin induce N-cadherin expression during EMT, implicating the "cadherin switch" as being a key factor in the progression of EMT (Fig. 14.3; Cano et al. 2000; Vandewalle et al. 2005; Sarrio et al. 2008). In epithelial breast tumor cells and various other cancer cells in nude mice, the overexpression of N-cadherin increases motility, invasiveness, and metastatic capacity by enhancing the capture of tumor cells in circulation in the vasculature (Nieman et al. 1999; Hazan et al. 2004; Qian et al. 2004). The mechanism through which N-cadherin expression promotes EMT is not completely understood, however, it has also been proposed that N-cadherin AJ junctions are much weaker than E-cadherin AJs and that the weakened integrity of these tissues leads to enhanced migration (Chu et al. 2006).

A fundamental role of N-cadherin in tumor progression is promoting collective cell migration. An in-depth study recently investigated the fundamental question of how cadherins can simultaneously provide adhesive strength to tissues while also promoting the cellular rearrangements that occur during cell migration, using primary astrocytes expressing fluorescently labeled N-cadherin. Peglion et al. demonstrated in a wound-healing assay that N-cadherin is recycled during migration, supporting a continuous "treadmilling" of AJs along the lateral sides of adjacent cells. The results of this study provide critical details of the molecular mechanisms underlying N-cadherin function in the collective invasion of cancer cells and could have implications for how other cadherin family members dynamics are regulated (Peglion et al. 2014).

In addition, N-cadherin also promotes tumorigenesis through the activation of oncogenic signaling pathways. N-cadherin physically interacts with fibroblast growth factor receptor 1 (FGFR1) and stimulates the continuous activity of the extracellular regulated MAP kinase (ERK) signaling pathway (Fig. 14.3; Williams et al. 1994; Suyama et al. 2002; Sanchez-Heras et al. 2006). In melanoma and many carcinomas, N-cadherin promotes FGFR signaling and the induction of several genes that are implicated in tumor progression (Qian et al. 2004; Hulit et al. 2007; Shintani et al. 2008; Augustine et al. 2008; Tanaka et al. 2010). In contrast, N-cadherin has also been shown to act as a tumor suppressor in neuroblastoma, highlighting the relevance of evaluating the relationship between cadherins and tumor progression in the context of specific cancer types (Suyama et al. 2002; Sanchez-Heras et al. 2006; Libusova et al. 2010; Kotb et al. 2011; Lammens et al. 2012).

14.3.2 P-Cadherin

Normally expressed in the basal layers of the epithelium, placental-cadherin (P-cadherin) expression in cancer cells is commonly associated with increased invasion and migration. P-cadherin tumor-promoting activities have been documented in colon cancer metastasis, alveolar rhabdomyosarcomas, bladder cancer, and numerous invasive carcinomas (Taniuchi et al. 2005; Paredes et al. 2005). Furthermore, P-cadherin was reported to be a potential prognostic marker for breast cancer where a tissue array demonstrated that the expression of P-cadherin correlates with occurrence of HER2 positivity and breast carcinoma subtypes of the basal B phenotype (May et al. 2011; Turashvili et al. 2011). The mechanisms underlying P-cadherin's tumor-promoting activities have been linked to the activation of the insulin-like growth factor 1 receptor (IGF1R) signaling pathway (Fig. 14.3). Additionally, it has been reported that P-cadherin expression can lead to the distortion of E-cadherin function and the downregulation of its expression (Paredes et al. 2008; Cheung et al. 2010; Van Marck et al. 2011; Jacobs et al. 2011; Cheung et al. 2011; Sun et al. 2011; Vlahova et al. 2012; Ribeiro et al. 2013; Thuault et al. 2013).

In contrast, loss of P-cadherin has been associated with higher tumor grade of colorectal adenocarcinomas (Koehler et al. 2004). Furthermore, reports have also been published regarding skin and colon cancer where P-cadherin expression is associated with an anti-invasive phenotype. In these cases, it was proposed that P-cadherin might act as a backup molecule in the absence of E-cadherin (Van Marck et al. 2005; Werling et al. 2011). However, the molecular mechanisms underlying how P-cadherin can rescue E-cadherin function in some cancers while promoting the tumorigenesis of others is unknown.

14.3.3 OB-Cadherin

Similar to N- and P-cadherin, OB-cadherin expression has been both positively and negatively correlated with cancer progression in different contexts. Normally expressed in mesoderm-derived tissues, osteoblast-cadherin (OB-cadherin) is inactivated by promoter methylation in several cancer cell lines of hepatocellular, colorectal, and breast origin (Li et al. 2012). An interesting report applied a DNA methylation microarray platform to compare OB-cadherin promoter hypermethylation of primary tumors with lymph node metastatic cell lines from the same patient (Carmona et al. 2012). Intriguingly, hypermethylation of OB-cadherin only occurred in the metastases of melanoma and head and neck cancers but not the primary tumors raising the possibility that epigenetic regulation of OB-cadherin occurs in a metastasis-specific manner. These data support the idea that the tumor microenvironment plays a critical role in metastasis by epigenetically regulating cadherin expression.

Additionally, the loss of OB-cadherin has also been linked to the activation of oncogenic signaling pathways. In retinoblastoma, OB-cadherin reduces Wnt-mediated anti-apoptotic signaling by stabilizing β-catenin at the plasma membrane (Nyberg et al. 2005; Marchong et al. 2010). In contrast, OB-cadherin is upregulated in prostate and brain cancer and is also strongly expressed in invasive breast cancer cell lines (Hoffmann and Balling 1995; Devaud et al. 2014). It has been proposed that OB-cadherin promotes the malignancy of prostate and brain tumors through the engagement of homophilic interactions between OB-cadherin on tumor cells and bone tissues (Tomita et al. 2000; Nakajima et al. 2008; Tamura et al. 2008; Chu et al. 2008; Floor et al. 2011; Kaur et al. 2012).

14.4 Cadherins in the Tumor Microenvironment

The modes in which cancer cells migrate and invade vary depending on the environment surrounding the tumor cells. Growth of a tumor and colonization of distant sites is dependent on its adaptation to the tumor microenvironment, or the surrounding environmental conditions of a tissue (Klonisch et al. 2008; Faurobert et al. 2015). A key process that dictates the microenvironment of a tumor is angiogenesis, defined as the growth and development of endothelial cells into new capillaries from an initial vascular network (Cavallaro et al. 2006). Critical for tumor development, angiogenesis provides the growing tumor with access to oxygen and essential nutrients for further growth and subsequent metastasis.

14.4.1 VE-Cadherin

Whereas E-cadherin is the primary constituent of epithelial adherens junctions, vascular endothelial- (VE-) cadherin is the adhesive molecule in endothelial adherens junctions (Fig. 14.1). Many angiogenic and inflammatory mediators released into the tumor microenvironment influence the function of VE-cadherin in the tumor endothelium, which exhibits higher proliferation and vascularization (Liao et al. 2000; Eastham et al. 2007). VE-cadherin is essential in promoting endothelial junctions, blood vessel integrity, and regulating angiogenesis (Francavilla et al. 2009). During the development of vascular tumors, the downregulation of VE-cadherin is associated with increased endothelial tumor growth and hemorrhagic complications (Zanetta et al. 2005). In addition to its barrier function, VE-cadherin also mediates contact growth inhibition and regulates downstream intracellular signaling pathways through its interactions with growth factor receptors. When bound to β-catenin, VE-cadherin associates with the vascular endothelial growth factor receptor-2 (VEGF-R2) to inhibit downstream proliferation signals (Lampugnani et al. 2006). Importantly, tyrosine phosphorylation of VE-cadherin disrupts interactions with β-catenin and leads to the weakening of endothelial junctions to promote diapedesis of invasive breast cancer cells (Fig. 14.4; Haidari et al. 2012). VE-cadherin has also been shown to be responsible for regulating VEGF-R2 dependent Src and PI3-Kinase/Akt signaling, leading to cell survival and angiogenesis (Fig. 14.4; Ha et al. 2008; Jiang and Liu 2008). Additionally, VE-cadherin interacts with the TGF-β receptor complex to promote endothelial growth and motility through downstream Smad signaling (Fig. 14.4; Cavallaro et al. 2006; Rudini et al. 2008; Giampietro et al. 2012).

VE-cadherin is normally only expressed by endothelial cells, however, the expression of VE-cadherin in other cell types is typically associated with tumor-promoting activity and is specific to the cancer environment. VE-cadherin overexpression has been reported in uveal melanoma and aggressive melanoma (Hendrix et al. 2001). In breast cancer, the loss of E-cadherin during EMT led to the induction of VE-cadherin expression at the cell surface of cancer cells where VE-cadherin stimulated protumorigenic TGF-β signaling that enhanced cell proliferation and invasion (Labelle et al. 2008; Berx and van Roy 2009; Rezaei et al. 2012; Breier et al. 2014). Understanding the signaling that contributes to VE-cadherin adhesion remodeling could facilitate the development of therapeutic strategies that prevent the endothelial barrier breakdown in metastasis.

In a recent report, Bentley et al. reported that pathological angiogenesis arises from a specific distribution of VE-cadherin patterns in a synchronized large group of cells where a "dynamic state", with high turnover of VE-cadherin, promotes cell motility (Bentley et al. 2014). In addition, VEGFR2 and Notch signaling pathways converge to differentially regulate the "state" of VE-cadherin to drive functional endothelial cell dynamics during angiogenic sprouting. As Notch and VEGFR signaling pathways are implicated in tumor progression, how this model of VE-cadherin regulation correlates to the pathological angiogenesis of different

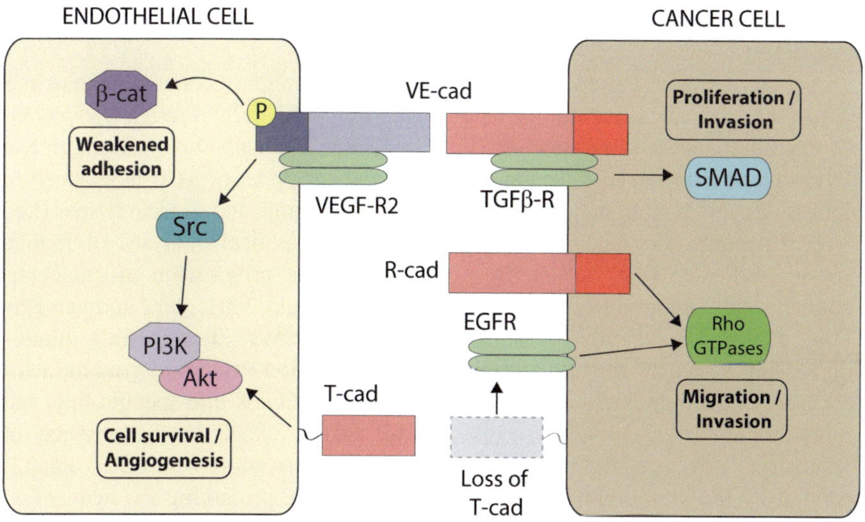

Fig. 14.4 Cadherins in the tumor microenvironment. Cadherin-mediated signaling between cancer cells and endothelial cells in the tumor microenvironment control tumor growth and dissemination (Sect. 14.4). In endothelial cells (*left*), tyrosine phosphorylation of vascular endothelial-cadherin (VE-cad) disrupts interactions with β-catenin and weakens endothelial junctions to promote diapedesis of invasive breast cancer cells. VE-cadherin also regulates vascular endothelial growth factor receptor 2 (VEGF-R2) dependent activation of Src and PI3-Kinase/Akt signaling, leading to increased proliferation and angiogenesis. Hypoxia-driven abnormal induction of truncated-cadherin (T-cad) activates PI3-Kinase/Akt signaling, which leads to increased cell proliferation, angiogenesis, and tumor growth and dissemination. In cancer cells (*right*), the induction of VE-cad expression at the cell surface stimulates protumorigenic TGF-β and Smad signaling that enhances cell proliferation and invasion. In the absence of T-cad, cells are highly sensitized to epidermal growth factor receptor (EGFR) activation from EGF ligands secreted by the surrounding tumor microenvironment, promoting the acquisition of migratory and invasive phenotypes through Rho GTPase activity. Retinal-cadherin (R-cad) promotes rhabdomysarcomas tumorigenesis through Rac1 activation. For the sake of clarity, not all tumorigenic signaling pathways in the tumor microenvironment are depicted here

cancers warrants further investigation and may aid in the development of novel therapeutic strategies.

14.4.2 R-Cadherin and T-Cadherin

Endothelial cells also express truncated-cadherin (T-cadherin) and retinal-cadherin (R-cadherin). Expressed in myoblast-derived rhabdomyosarcomas, R-cadherin is downregulated in gastrointestinal and colorectal cancer where the promoter of the R-cadherin gene is hypermethylated in early tumor progression (Fig. 14.4; Miotto et al. 2004; Agiostratidou et al. 2009; Berx and van Roy 2009). However, R-cadherin has also been shown to promote rhabdomysarcomas tumorigenesis through Rac1 activation (Charrasse et al. 2004; Kucharczak et al. 2008; Makrilia et al. 2009).

Lacking a cytoplasmic domain, T-cadherin is a unique cadherin in that it is localized to the apical membrane in polarized epithelial cells through a glycosylphosphatidylinositol (GPI) anchor (Fig. 14.1; Ranscht and Dours-Zimmermann 1991). T-cadherin is typically considered a tumor suppressor as it is silenced in melanomas and in malignancies of breast, pancreatic, lung, and ovarian tissues (Lee 1996; Takeuchi and Ohtsuki 2001). Immunohistochemical analyses of human biopsies indicated that T-cadherin is lost during the progression of cutaneous squamous cell carcinomas (SCCs; Takeuchi and Ohtsuki 2001; Berx and van Roy 2009; Philippova et al. 2013). Independent of EMT, T-cadherin's tumor-suppressing activities are instead linked to the inhibition epidermal growth factor receptor (EGFR)-mediated signaling by sequestering EGFR into specific lipid raft compartments (Fig. 14.4; Kyriakakis et al. 2012). Thus, in the absence of T-cadherin, cells are highly sensitized to EGFR activation from EGF ligands secreted by the surrounding tumor microenvironment, promoting the acquisition of migratory phenotypes. Furthermore, the loss of T-cadherin enhances stable adhesion between tumor cells and the endothelium during extravasation, thereby promoting the formation of metastases in distant tissues (Philippova et al. 2013).

In contrast, the hypoxia-driven abnormal induction of T-cadherin activates Akt signaling, which leads to increased cell proliferation and angiogenesis and tumor growth and dissemination (Fig. 14.4; Berx et al. 1996; Andreeva and Kutuzov 2010). Additionally, T-cadherin can bind to adiponectin, the adipocyte-secreted hormone, which exhibits cardioprotective activity and promotes tumor angiogenesis (Denzel et al. 2009; Denzel et al. 2010). Collectively, the roles of endothelial cadherins in cancer progression are still a large area of research that requires further investigation.

14.5 Desmosomal Cadherins

14.5.1 Desmosomes

Desmosomes are intercellular junctions that provide structural integrity to tissues that undergo mechanical stress such as the skin and the heart (Kimura et al. 2007; Brooke et al. 2012). The adhesive strength of desmosomes is mediated by desmosomal cadherins, desmogleins (Dsg1-4), and desmocollins (Dsc1-3; Fig. 14.1). Desmosomal cadherins make direct contacts between neighboring cells and span the plasma membrane, attaching with their cytoplasmic tails to armadillo proteins plakoglobin (Pg) and plakophilin (Pkp1-3) and the cytoskeletal linker protein, desmoplakin. Through direct interactions with desmoplakin, desmosomes anchor the intermediate filament cytoskeleton (keratin in epithelial tissues, vimentin in fibroblasts and endothelial cells, and desmin in cardiac tissues) to sites of cell–cell contact (Ruhrberg and Watt 1997; Sonnenberg and Liem 2007; Kowalczyk and Green 2013). Although desmosomes are crucial for providing strong cellular adhesion, a function that is often lost during cancer progression, the roles of desmosomes in carcinogenesis have only recently been investigated (Dusek and Attardi 2011).

14.5.2 Desmosome Expression in Cancer

14.5.2.1 Tumor-Promoting Properties

Protein expression patterns of desmosomes are altered in a variety of epithelial-derived tumors, where almost all desmosomal components have been reported to be misregulated in different forms of cancer (Stahley and Kowalczyk 2015). O'Shea et al. recently reported that acantholytic squamous cell carcinomas display reduced expression of desmosomal components in 90 % of tumor sections (O'Shea et al. 2014). The mechanisms by which desmosomal cadherins contribute to cancer are still being uncovered. Expressed in a differentiation-dependent pattern in the epidermis, desmoglein 2 (Dsg2) expression is increased in malignant skin carcinomas (Kurzen et al. 2003; Schmitt et al. 2007; Brennan and Mahoney 2009) and prostate cancer (Trojan et al. 2005). Additionally, Dsg2 was also reported to be overexpressed in suprabasal layers of skin, a phenotype that led to hyperplasia and benign tumor formation (Brennan et al. 2007). Desmoglein 3 (Dsg 3) overexpression is associated with poor clinicopathological status for head and neck squamous cancers (Chen et al. 2007). Furthermore, overexpression of Dsg3 increased migration, invasion, and growth of head and neck cancer cell lines (Chen et al. 2007).

14.5.2.2 Tumor-Suppressing Properties

Desmosomal cadherins also display expression patterns that are consistent with tumor-suppressing activities. Dsg2 expression is reduced in familial gastric cancer whereas Dsc2 and Dsc3 are reduced in pancreatic and breast cancers (Oshiro et al. 2005; Biedermann et al. 2005; Yashiro et al. 2006; Hamidov et al. 2011). Intriguingly, upregulation of the Slug transcription factor has been reported to be associated with reduced expression of Dsg3, Dsc2, and Pkp1 in oral squamous cell carcinomas (Katafiasz et al. 2011). Additionally, DNA methylation has been reported to play a role in the downregulation of Dsc1, Dsc2, and Dsc3 in lung cancer (Cui et al. 2012a, b). During epidermal differentiation, kallikreins (KLKs) contribute to the proteolytic cleavage of Dsg1. Elevated expression of serine proteinases KLK5, 7, 8, and 10 is correlated with the formation of more aggressive tumors in human oral squamous cell carcinoma (OSCC). In a recent report, Jiang et al. provide evidence to support the idea that KLK5-mediated cleavage of Dsg1

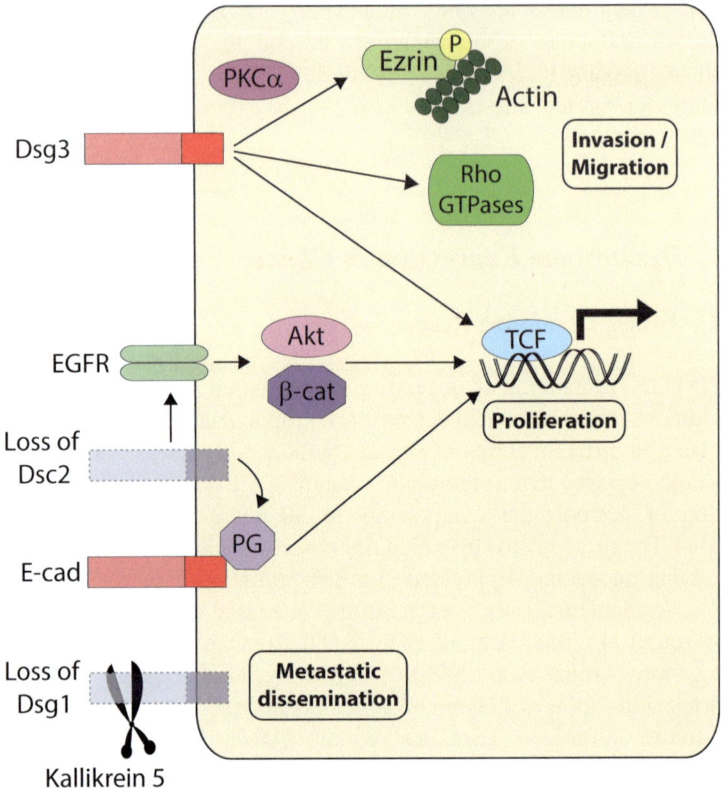

Fig. 14.5 Desmosome signaling in cancer. Classically been regarded as static "spot-welds", recent reports demonstrate that desmosome intercellular junctions are also dynamic scaffolds that mediate intracellular signaling. Desmoglein 3 (Dsg3) has been shown to form a complex with Ezrin and facilitates PKCα-mediated phosphorylation of Ezrin, which, in turn, leads to cell protrusions, cell spreading, and increased invasion via regulation of the actin cytoskeleton. Dsg3 can also regulate actin dynamics and membrane protrusion via activation of Rac and Cdc42 GTPases. Lastly, expression of Dsg3 was shown to promote cell growth/invasion and initiation of TCF/LEF target genes. In the absence of desmocollin 2 (Dsc2), activation of Akt/β-catenin signaling via EGFR promoted cell proliferation and tumor growth of colorectal cancer cells. Loss of Dsc2 has also been shown to promote β-cat/TCF-dependent transcription due to an increase in free PG regulating E-cadherin containing adherens junctions. Elevated expression of Kallikrein 5 (Klk5), a serine proteinase that cleaves desmoglein 1 (Dsg1), was shown to promote metastatic dissemination by inducing loss of junctional integrity

promoted metastatic dissemination of OSCC by promoting loss of junctional integrity (Fig. 14.5; Jiang et al. 2011). Thus, in the cases where desmosomal cadherins exhibit tumor-suppressing activities, it is possible that downregulation would lead to weakened intercellular adhesion to promote invasive phenotypes of tumors. The reduction of Dsc2 is also associated with enhanced tumor metastasis in oesophageal squamous cell carcinoma (ESCC) where restoring Dsc2 decreased cell migration and invasion both in vitro and in vivo, likely via its regulation of

E-cadherin containing adherens junctions (Fig. 14.5). Fang et al. found that Dsc2 was a downstream target of miR-25 where enhanced miR-25 promoted ESCC cell invasiveness. (Fang et al. 2013). Together, these reports provide novel insights into the mechanistic contributions of miRNA-mediated regulation of desmosomal cadherins in tumorigenesis.

14.5.3 Desmosome Signaling in Cancer

Although desmosomes have classically been regarded as static "spot-welds", recent reports demonstrate that desmosomes are also dynamic scaffolds that mediate intracellular signaling. There have been multiple recent studies that have advanced our understanding of the crosstalk between signaling molecules and desmosomal cadherins in tumor progression and invasion. Kolegraft et al. investigated the molecular mechanisms by which the decreased expression of Dsc2 (observed in colorectal carcinomas) contributes to cancer progression. This report demonstrates that in the absence of Dsc2, Akt/β-catenin signaling is activated and promotes cell proliferation and tumor growth of colorectal cancer cells in vivo. Together, this report unveils a novel mechanism for desmosomal cadherin-mediated β-catenin signaling, which contributes to tumor progression (Fig. 14.5; Kolegraff et al. 2011).

Several studies report a correlation between Dsg3 upregulation and increasing clinical stages of malignancy in SCC of head and neck, lung, skin, and oesophagus (Brown and Wan 2015). Dsg3 expression was shown to promote cell growth/invasion and expression of TCF/LEF target genes in a murine tumor model (Fig. 14.5; Chen et al. 2013). In a recent study, Brown et al. discovered a link between Dsg3 overexpression in cancer and the activation of c-Jun/AP-1, a transcription factor that plays a pivotal role in cancer metastasis. Additionally, this report also identified a novel interaction between Dsg3 and Ezrin, which form in a complex at the plasma membrane to promote Ezrin interactions with F-actin and CD44 (Fig. 14.5; Brown et al. 2014). As Dsg3 is upregulated in squamous cell carcinoma (SCC), Dsg3 functioned in this protein complex to facilitate PKC-mediated phosphorylation of Ezrin, which, in turn, leads to cell protrusions, cell spreading, and increased invasion. A separate report demonstrated that Dsg3 modulates the activity of members of the Rho family GTPases, Rac1 and Cdc42, to mediate actin dynamics and that the overexpression of Dsg3 promoted cell migration and membrane protrusions (Fig. 14.5; Tsang et al. 2012; 2010). Collectively, these data demonstrate that the crosstalk between desmosomal cadherins and signaling scaffold proteins such as Ezrin could be responsible for promoting the hallmarks of cancer progression.

Additional signaling mechanisms underlying the tumor-suppressing activities of desmosomal cadherins could be linked to the mislocalization of desmosomal binding partners. Similar to β-catenin, plakoglobin can also localize to the nucleus and initiate LEF/TCF target genes (McCrea et al. 1991; Butz et al. 1992; Näthke et al. 1994; Zhurinsky et al. 2000; Conacci-Sorrell et al. 2002; Klucky et al. 2007).

Thus, the absence of desmosomal cadherins could promote the modulation of oncogenic Wnt-β-catenin signaling through mislocalized Pg and PKPs. An additional reported component of the desmosome is the tetraspan membrane protein, p53 apoptosis effector related to PMP-22 (PERP). Linking the p53 family of transcriptional regulators to cell–cell adhesion, PERP is activated by the p53 tumor suppressor during DNA damage-induced apoptosis (Ihrie et al. 2005). Furthermore, as p53 is inactivated in at least 50 % of all human cancers, PERP represents tumor suppressor activities for the desmosome. PERP is also activated during development of the stratified epithelia by the transcription factor p63, a tumor suppressor in specific cancers (Flores et al. 2005; Carroll et al. 2007). Finally, Harmon et al. identified that Erbin, a known ERK regulator, is recruited to Dsg1 to suppress MAPK signaling and promote keratinocyte differentiation. Intriguingly, malignant skin and basal cell carcinomas have been reported to have decreased expression of Dsg1 and cytoplasmic localization of Erbin raising the possibility that the aberrant activity of Erbin's downstream targets contributes to tumor progression (Hammers and Stanley 2013; Harmon et al. 2013; Lebeau et al. 2005; Tada et al. 2000). Collectively, putative mechanisms for tumor-promoting activities of desmosomal cadherins include the stimulation of proliferation, invasion, and the inhibition of apoptosis.

14.6 Future Clinical Perspectives

14.6.1 Pharmaceutical Restoration of E-Cadherin Expression

As loss of E-cadherin clearly represents an early step in many different cancers, pharmaceutical intervention targeting the mechanisms of E-cadherin downregulation provides many opportunities for the development of cancer therapeutics. Currently, therapeutics targeting the hypermethylation of E-cadherin are being developed using a pleotropic epigenetic drug that inhibits histone deacetylases (HDAC) and DNA methyltransferases to restore E-cadherin gene expression (van Roy and Berx 2008; Gregory et al. 2012). Furthermore, targeting the signaling pathways that promote the loss of E-cadherin during EMT represent another avenue for therapeutic intervention. Recent studies have identified that BMP-7 is a potent inhibitor of TGF-β-induced EMT and reverses the loss of E-cadherin (Zohn et al. 2006).

14.6.2 Targeting Cadherin-Mediated Oncogenic Signaling

An additional area of cancer therapeutics is focused on targeting cadherin-dependent signaling pathways that promote the development of tumors. For example, the Hippo signaling pathway restricts cell growth and is commonly deregulated in cancer progression. Targeting the Hippo pathway, verteprofin is a compound that specifically reduces the transcriptional activity of Yap and has been shown to reduce tumor growth in mice (Kim et al. 2011; Liu-Chittenden et al. 2012). Additionally, dobutamine, a β-adrenergic GPCR agonist that is commonly used clinically in the treatment of heart failure, was also recently shown to promote the phosphorylation and inactivation of Yap in cell culture, reducing tumorigenesis (Overholtzer et al. 2006; Lei et al. 2008; Zhao et al. 2008; Bao et al. 2011a; Kim et al. 2011; Bao et al. 2011b). Finally, studies have recently identified a novel role for AMPK, a kinase known as the energy sensor of the cell, in regulating the Hippo signaling pathway and Yap activity (Nishioka et al. 2009; Kim et al. 2011; Hirate et al. 2013; Mo et al. 2015; Wang et al. 2015). Excitingly, these findings demonstrate that the use of a well-known clinically used compound metformin is sufficient to suppress Yap-driven proliferation in mice. As E-cadherin plays a critical role in regulating Hippo signaling and Yap activation, future studies investigating the links between AMPK and cadherin function during tumor progression could provide novel approaches for the treatment of cancer using pharmaceuticals that are already being used in a clinical setting.

14.6.3 Immunotherapy

Promising new avenues of cancer therapeutics have been developed using cancer immunotherapy. A novel and highly selective monoclonal antibody against P-cadherin, PF-03732010, has demonstrated anti-tumor and anti-metastatic activities in mice overexpressing P-cadherin (Zhang et al. 2010). Additional studies have focused on the contributions of the tumor microenvironment to the responses of tumors to immunotherapy (van Roy and Berx 2008; Devaud et al. 2014). Blocking angiogenesis, a key component of the tumor microenvironment, promotes tumor dormancy through the secretion of angiostatin and endostatin (Hanahan and Folkman 1996; Nyberg et al. 2005). In experimental tumors, treatment with a VE-cadherin specific antibody was sufficient to block angiogenesis and tumor growth (Liao et al. 2000; Corada et al. 2002). Additionally, targeting the signaling downstream of VE-cadherin with the use of a monoclonal antibody for VEGF, Bevacizumab, was also sufficient to decrease angiogenesis (Nyberg et al. 2005). Furthermore, Manning et al. developed an immunotherapeutic strategy that directly altered the microenvironment of the tumor by blocking vessel formation, to increase hypoxic conditions, using an anti-VEGF-R2 antibody, which ultimately led to apoptosis and necrosis. Future goals to improve the efficacy of immunotherapeutic approaches will aim specifically to reduce the immune suppressive responses within a tumor microenvironment (Devaud et al. 2014). Additionally, the efficacy of combining immunotherapy with chemotherapeutic and radiotherapy strategies has been examined in mouse models and represents a promising method to test in clinical trials (Devaud et al. 2013; Kershaw et al. 2013).

14.6.4 Stem Cell Research

The way that we treat cancer patients and design cancer therapeutics has been revolutionized by the novel concept that tumors are composed of heterogeneous populations of cells and that not all cancer cells have equal tumor growth-supporting potential (Klonisch et al. 2008; Floor et al. 2011). A promising new avenue of cancer therapeutic research is the field of regenerative-medicine–based treatment strategies using stem cells. Cancer stem cells (CSCs) are tumorigenic stem cell populations that have undergone genetic alterations and a change in the microenvironment of the stem cell niche (Smith 2001; Klonisch et al. 2008). Possessing the unique capacity for unlimited divisions, cancer stem cells display resistance to radiotherapy and chemotherapy. Stem cell research aims to develop therapies that would specifically target the initiating population of cancer stem cells that give rise to the tumors. It has been proposed that CSCs represent the origin of the disease; therefore, their elimination could mean eradication of cancer within a patient.

Recent studies demonstrate that E-cadherin plays a critical role in regulating homeostasis of embryonic stem cells (ESCs), where differentiation of the stem cell (SC) niche is controlled by cadherin expression (Soncin et al. 2011). Chou et al. demonstrated that E-cadherin is a regulator of pluripotency, as loss of E-cadherin leads to rapid ESC differentiation. N-cadherin replaces E-cadherin during the differentiation of ESCs, which leads to the upregulation of E-cadherin repressors Snail and Slug, gelatinase activity of matrix metalloproteinases MMP-2 and MMP-9, and increased cellular motility (Spencer et al. 2007; Eastham et al. 2007). Understanding the function of cadherins in cancer stem cell differentiation will undoubtedly provide key insights into the development of therapeutic strategies to enhance the effects of chemotherapies and to improve clinical markers for diagnosis.

Although cancer stem cells have been identified in many cancers associated with cadherin-deficiencies such as breast cancer, prostate cancer, ovarian cancer, and brain tumors, biomarkers for cancer stem cells are extremely limited and a universal marker for cancer stem cells does not currently exist (Schulenburg et al. 2006). Importantly, the assessment of secreted factors by tumors can also be used as a marker for cancer progression. Identifying markers for cancer stem cells would allow for the detection and diagnosis of cancer at its earliest stage.

Exhibiting the highest levels of recurrence, tumor progression of squamous cell carcinomas (SCC) is tightly associated with cadherin expression (Fuchs and

Raghavan 2002). Progress has been made towards our understanding of the molecular mechanisms promoting the cancer stem cell niche in tumor development where gene expression signatures have been reported for squamous cell carcinoma cancer stem cells and allowed for the identification of genes in squamous skin cells (Schober and Fuchs 2011; Beronja et al. 2013). In a recent report, Oshimori et al. devised an in vivo system to monitor and manipulate the pro-EMT TGF-β signaling pathway, and found that TGF-β slows proliferation while aiding malignancy of squamous cancer cells. As the activation of TGF-β signaling is a critical promoter of E-cadherin downregulation during EMT, future investigations of the link between SCC cancer stem cell progression and cadherin function could provide critical details in our understanding of tumorigenesis and tumor recurrence in patients.

In addition to TFG-β, both cancer stem cell populations and cadherin function in tumor progression are regulated by the Wnt, Notch, Hedgehog, and tyrosine kinase receptor signaling pathways (Li and Laterra 2012). Intriguingly, our understanding of the Notch signaling in cancer stem cells has provided a useful treatment option for breast cancer, where the loss of E-cadherin plays a critical step in the early stages of EMT, through the inhibition of the γ-secretase enzyme in the Notch pathway (Wicha et al. 2006; Thiery et al. 2009; De Craene and Berx 2013). Further investigations into the links between cadherin function and cancer stem cells hold promise for significantly enhancing our understanding of the signaling pathways that promote tumor progression.

14.7 Concluding Remarks

The body of research studying cadherins in cancer clearly demonstrates that cadherins play critical roles in tumorigenesis and can function as either oncogenic promoters or tumor suppressors. The identity of the cadherin, type of tissue, and tumor microenvironment are all factors that contribute to the often contradictory and context-dependent roles for cadherins in both the primary tumor and disseminated tumor cells (Nieman et al. 1999; Hazan et al. 2004; Qian et al. 2004; Berx and van Roy 2009). A comprehensive understanding of how each cadherin member differentially influences the expression of other cadherins will provide important insights into elucidating cadherin-based signatures that will be useful for prognosis. In addition, future studies focused on identifying the context-dependent control of oncogenic signaling pathways by cadherins will provide much needed clarity into the functions of these cell adhesion molecules in cancer.

References

Agiostratidou G, Li M, Suyama K et al (2009) Loss of retinal cadherin facilitates mammary tumor progression and metastasis. Cancer Res 69:5030–5038. doi:10.1158/0008-5472.CAN-08-4007

Alessandri K, Sarangi BR, Gurchenkov VV et al (2013) Cellular capsules as a tool for multicellular spheroid production and for investigating the mechanics of tumor progression in vitro. Proc Natl Acad Sci U S A 110:14843–14848. doi:10.1073/pnas.1309482110

Andreeva AV, Kutuzov MA (2010) Cadherin 13 in cancer. Genes Chromosomes Cancer 49:775–790. doi:10.1002/gcc.20787

Arthur WT, Noren NK, Burridge K (2002) Regulation of Rho family GTPases by cell-cell and cell-matrix adhesion. Biol Res 35:239–246

Auersperg N, Pan J, Grove BD et al (1999) E-cadherin induces mesenchymal-to-epithelial transition in human ovarian surface epithelium. Proc Natl Acad Sci U S A 96:6249–6254

Augustine CK, Yoshimoto Y, Gupta M et al (2008) Targeting N-cadherin enhances antitumor activity of cytotoxic therapies in melanoma treatment. Cancer Res 68:3777–3784. doi:10.1158/0008-5472.CAN-07-5949

Bao Y, Hata Y, Ikeda M, Withanage K (2011a) Mammalian Hippo pathway: from development to cancer and beyond. J Biochem 149:361–379. doi:10.1093/jb/mvr021

Bao Y, Nakagawa K, Yang Z et al (2011b) A cell-based assay to screen stimulators of the Hippo pathway reveals the inhibitory effect of dobutamine on the YAP-dependent gene transcription. J Biochem 150:199–208. doi:10.1093/jb/mvr063

Becker KF, Atkinson MJ, Reich U et al (1993) Exon skipping in the E-cadherin gene transcript in metastatic human gastric carcinomas. Hum Mol Genet 2:803–804

Becker KF, Atkinson MJ, Reich U et al (1994) E-cadherin gene mutations provide clues to diffuse type gastric carcinomas. Cancer Res 54:3845–3852

Becker KF, Kremmer E, Eulitz M et al (1999) Analysis of E-cadherin in diffuse-type gastric cancer using a mutation-specific monoclonal antibody. Am J Pathol 155:1803–1809. doi:10.1016/S0002-9440(10)65497-1

Bentley K, Franco CA, Philippides A et al (2014) The role of differential VE-cadherin dynamics in cell rearrangement during angiogenesis. Nat Cell Biol 16:309–321. doi:10.1038/ncb2926

Beronja S, Janki P, Heller E et al (2013) RNAi screens in mice identify physiological regulators of oncogenic growth. Nature 501:185–190. doi:10.1038/nature12464

Berx G, van Roy F (2009) Involvement of members of the cadherin superfamily in cancer. Cold Spring Harb Perspect Biol 1:a003129. doi:10.1101/cshperspect.a003129

Berx G, Cleton-Jansen AM, Strumane K et al (1996) E-cadherin is inactivated in a majority of invasive human lobular breast cancers by truncation mutations throughout its extracellular domain. Oncogene 13:1919–1925

Berx G, Becker KF, Höfler H, van Roy F (1998) Mutations of the human E-cadherin (CDH1) gene. Hum Mutat 12:226–237. doi:10.1002/(SICI)1098-1004(1998)12:4<226::AID-HUMU2>3.0.CO;2-D

Biedermann K, Vogelsang H, Becker I et al (2005) Desmoglein 2 is expressed abnormally rather than mutated in familial and sporadic gastric cancer. J Pathol 207:199–206. doi:10.1002/path.1821

Blanco MJ, Moreno-Bueno G, Sarrio D et al (2002) Correlation of Snail expression with histological grade and lymph node status in breast carcinomas. Oncogene 21:3241–3246. doi:10.1038/sj.onc.1205416

Bolós V, Peinado H, Pérez-Moreno MA et al (2003) The transcription factor Slug represses E-cadherin expression and induces epithelial to mesenchymal transitions: a comparison with Snail and E47 repressors. J Cell Sci 116:499–511

Breier G, Grosser M, Rezaei M (2014) Endothelial cadherins in cancer. Cell Tissue Res 355:523–527. doi:10.1007/s00441-014-1851-7

Brennan D, Mahoney MG (2009) Increased expression of Dsg2 in malignant skin carcinomas: A tissue-microarray based study. Cell Adh Migr 3:148–154

Brennan D, Hu Y, Joubeh S et al (2007) Suprabasal Dsg2 expression in transgenic mouse skin confers a hyperproliferative and apoptosis-resistant phenotype to keratinocytes. J Cell Sci 120:758–771. doi:10.1242/jcs.03392

Brooke MA, Nitoiu D, Kelsell DP (2012) Cell-cell connectivity: desmosomes and disease. J Pathol 226:158–171. doi:10.1002/path.3027

Brooks SA, Lomax-Browne HJ, Carter TM et al (2010) Molecular interactions in cancer cell metastasis. Acta Histochem 112:3–25. doi:10.1016/j.acthis.2008.11.022

Brown L, Wan H (2015) Desmoglein 3: a help or a hindrance in cancer progression? Cancers 7 (1):266–286. doi:10.3390/cancers7010266

Brown L, Waseem A, Cruz IN et al (2014) Desmoglein 3 promotes cancer cell migration and invasion by regulating activator protein 1 and protein kinase C-dependent-Ezrin activation. Oncogene 33:2363–2374. doi:10.1038/onc.2013.186

Butz S, Stappert J, Weissig H, Kemler R (1992) Plakoglobin and beta-catenin: distinct but closely related. Science 257:1142–1144

Cano A, Pérez-Moreno MA, Rodrigo I et al (2000) The transcription factor snail controls epithelial-mesenchymal transitions by repressing E-cadherin expression. Nat Cell Biol 2:76–83. doi:10.1038/35000025

Carmona FJ, Villanueva A, Vidal A et al (2012) Epigenetic disruption of cadherin-11 in human cancer metastasis. J Pathol 228:230–240. doi:10.1002/path.4011

Carneiro F, Machado JC, Seruca R, Sobrinho-Simões M (1999) E-cadherin changes in gastric carcimona. Histopathology 35:477–478

Carroll DK, Brugge JS, Attardi LD (2007) p63, cell adhesion and survival. Cell Cycle 6:255–261

Carvalho J, van Grieken NC, Pereira PM et al (2012) Lack of microRNA-101 causes E-cadherin functional deregulation through EZH2 up-regulation in intestinal gastric cancer. J Pathol 228:31–44. doi:10.1002/path.4032

Cavallaro U, Liebner S, Dejana E (2006) Endothelial cadherins and tumor angiogenesis. Exp Cell Res 312:659–667. doi:10.1016/j.yexcr.2005.09.019

Charrasse S, Comunale F, Gilbert E et al (2004) Variation in cadherins and catenins expression is linked to both proliferation and transformation of Rhabdomyosarcoma. Oncogene 23:2420–2430. doi:10.1038/sj.onc.1207382

Chen Y-J, Chang JT, Lee L et al (2007) DSG3 is overexpressed in head neck cancer and is a potential molecular target for inhibition of oncogenesis. Oncogene 26:467–476. doi:10.1038/sj.onc.1209802

Chen Y-J, Lee L-Y, Chao Y-K et al (2013) DSG3 facilitates cancer cell growth and invasion through the DSG3-plakoglobin-TCF/LEF-Myc/cyclin D1/MMP signaling pathway. PLoS One 8:e64088. doi:10.1371/journal.pone.0064088

Cheung LWT, Leung PCK, Wong AST (2010) Cadherin switching and activation of p120 catenin signaling are mediators of gonadotropin-releasing hormone to promote tumor cell migration and invasion in ovarian cancer. Oncogene 29:2427–2440. doi:10.1038/onc.2009.523

Cheung LWT, Mak ASC, Cheung ANY et al (2011) P-cadherin cooperates with insulin-like growth factor-1 receptor to promote metastatic signaling of gonadotropin-releasing hormone in ovarian cancer via p120 catenin. Oncogene 30:2964–2974. doi:10.1038/onc.2011.7

Chu K, Cheng C-J, Ye X et al (2008) Cadherin-11 promotes the metastasis of prostate cancer cells to bone. Mol Cancer Res 6:1259–1267. doi:10.1158/1541-7786.MCR-08-0077

Cleton-Jansen AM, Callen DF, Seshadri R et al (2001) Loss of heterozygosity mapping at chromosome arm 16q in 712 breast tumors reveals factors that influence delineation of candidate regions. Cancer Res 61:1171–1177

Comijn J, Berx G, Vermassen P et al (2001) The two-handed E box binding zinc finger protein SIP1 downregulates E-cadherin and induces invasion. Mol Cell 7:1267–1278

Conacci-Sorrell ME, Ben-Yedidia T, Shtutman M et al (2002) Nr-CAM is a target gene of the beta-catenin/LEF-1 pathway in melanoma and colon cancer and its expression enhances motility and confers tumorigenesis. Genes Dev 16:2058–2072. doi:10.1101/gad.227502

Corada M, Zanetta L, Orsenigo F et al (2002) A monoclonal antibody to vascular endothelial-cadherin inhibits tumor angiogenesis without side effects on endothelial permeability. Blood 100:905–911

Covington MD, Burghardt RC, Parrish AR (2006) Ischemia-induced cleavage of cadherins in NRK cells requires MT1-MMP (MMP-14). Am J Physiol Renal Physiol 290:F43–F51. doi:10.1152/ajprenal.00179.2005

Cui T, Chen Y, Yang L et al (2012a) The p53 target gene desmocollin 3 acts as a novel tumor suppressor through inhibiting EGFR/ERK pathway in human lung cancer. Carcinogenesis 33:2326–2333. doi:10.1093/carcin/bgs273

Cui T, Chen Y, Yang L et al (2012b) Diagnostic and prognostic impact of desmocollins in human lung cancer. J Clin Pathol 65:1100–1106. doi:10.1136/jclinpath-2011-200630

Das T, Safferling K, Rausch S, Grabe N, Boehm H, Spatz JP (2015) A molecular mechanotransduction pathway regulates collective migration of epithelial cells. Nat Cell Biol 17(3):276–287. doi:10.1038/ncb3115

Davies G, Jiang WG, Mason MD (2001) Matrilysin mediates extracellular cleavage of E-cadherin from prostate cancer cells: a key mechanism in hepatocyte growth factor/scatter factor-induced cell-cell dissociation and in vitro invasion. Clin Cancer Res 7:3289–3297

de Beco S, Amblard F, Coscoy S (2012) New insights into the regulation of E-cadherin distribution by endocytosis. Int Rev Cell Mol Biol 295:63–108. doi:10.1016/B978-0-12-394306-4.00008-3

De Craene B, Berx G (2013) Regulatory networks defining EMT during cancer initiation and progression. Nat Rev Cancer 13:97–110. doi:10.1038/nrc3447

De Wever O, Derycke L, Hendrix A et al (2007) Soluble cadherins as cancer biomarkers. Clin Exp Metastasis 24:685–697. doi:10.1007/s10585-007-9104-8

Denzel MS, Hebbard LW, Shostak G et al (2009) Adiponectin deficiency limits tumor vascularization in the MMTV-PyV-mT mouse model of mammary cancer. Clin Cancer Res 15:3256–3264. doi:10.1158/1078-0432.CCR-08-2661

Denzel MS, Scimia M-C, Zumstein PM et al (2010) T-cadherin is critical for adiponectin-mediated cardioprotection in mice. J Clin Invest 120:4342–4352. doi:10.1172/JCI43464

Deplazes J, Fuchs M, Rauser S et al (2009) Rac1 and Rho contribute to the migratory and invasive phenotype associated with somatic E-cadherin mutation. Hum Mol Genet 18:3632–3644. doi:10.1093/hmg/ddp312

Devaud C, John LB, Westwood JA et al (2013) Immune modulation of the tumor microenvironment for enhancing cancer immunotherapy. Oncoimmunology 2:e25961. doi:10.4161/onci.25961

Devaud C, Westwood JA, John LB et al (2014) Tissues in different anatomical sites can sculpt and vary the tumor microenvironment to affect responses to therapy. Mol Ther 22:18–27. doi:10.1038/mt.2013.219

Ding W, You H, Dang H et al (2010) Epithelial-to-mesenchymal transition of murine liver tumor cells promotes invasion. Hepatology 52:945–953. doi:10.1002/hep.23748

Dusek RL, Attardi LD (2011) Desmosomes: new perpetrators in tumour suppression. Nat Rev Cancer 11:317–323. doi:10.1038/nrc3051

Eastham AM, Spencer H, Soncin F et al (2007) Epithelial-mesenchymal transition events during human embryonic stem cell differentiation. Cancer Res 67:11254–11262. doi:10.1158/0008-5472.CAN-07-2253

Engl W, Arasi B, Yap LL et al (2014) Actin dynamics modulate mechanosensitive immobilization of E-cadherin at adherens junctions. Nat Cell Biol 16:587–594. doi:10.1038/ncb2973

Fang W-K, Liao L-D, Li L-Y et al (2013) Down-regulated desmocollin-2 promotes cell aggressiveness through redistributing adherens junctions and activating beta-catenin signalling in oesophageal squamous cell carcinoma. J Pathol 231:257–270. doi:10.1002/path.4236

Farahani E, Patra HK, Jangamreddy JR et al (2014) Cell adhesion molecules and their relation to (cancer) cell stemness. Carcinogenesis 35:747–759. doi:10.1093/carcin/bgu045

Faurobert E, Bouin A-P, Albiges-Rizo C (2015) Microenvironment, tumor cell plasticity, and cancer. Curr Opin Oncol 27:64–70. doi:10.1097/CCO.0000000000000154

Ferreira AC, Suriano G, Mendes N et al (2012) E-cadherin impairment increases cell survival through Notch-dependent upregulation of Bcl-2. Hum Mol Genet 21:334–343. doi:10.1093/hmg/ddr469

Floor S, van Staveren WCG, Larsimont D et al (2011) Cancer cells in epithelial-to-mesenchymal transition and tumor-propagating-cancer stem cells: distinct, overlapping or same populations. Oncogene 30:4609–4621. doi:10.1038/onc.2011.184

Flores ER, Sengupta S, Miller JB et al (2005) Tumor predisposition in mice mutant for p63 and p73: evidence for broader tumor suppressor functions for the p53 family. Cancer Cell 7:363–373. doi:10.1016/j.ccr.2005.02.019

Francavilla C, Maddaluno L, Cavallaro U (2009) The functional role of cell adhesion molecules in tumor angiogenesis. Semin Cancer Biol 19:298–309. doi:10.1016/j.semcancer.2009.05.004

Fuchs E, Raghavan S (2002) Getting under the skin of epidermal morphogenesis. Nat Rev Genet 3:199–209. doi:10.1038/nrg758

Fujita Y, Krause G, Scheffner M et al (2002) Hakai, a c-Cbl-like protein, ubiquitinates and induces endocytosis of the E-cadherin complex. Nat Cell Biol 4:222–231. doi:10.1038/ncb758

Gaggioli C, Hooper S, Hidalgo-Carcedo C et al (2007) Fibroblast-led collective invasion of carcinoma cells with differing roles for RhoGTPases in leading and following cells. Nat Cell Biol 9:1392–1400. doi:10.1038/ncb1658

Gasparini G, Longo R (2012) The paradigm of personalized therapy in oncology. Expert Opin Ther Targets 16(Suppl 1):S7–S16. doi:10.1517/14728222.2011.637921

Gayther SA, Gorringe KL, Ramus SJ et al (1998) Identification of germ-line E-cadherin mutations in gastric cancer families of European origin. Cancer Res 58:4086–4089

Giampietro C, Taddei A, Corada M et al (2012) Overlapping and divergent signaling pathways of N-cadherin and VE-cadherin in endothelial cells. Blood 119:2159–2170. doi:10.1182/blood-2011-09-381012

Gladden AB, Hebert AM, Schneeberger EE, McClatchey AI (2010) The NF2 tumor suppressor, Merlin, regulates epidermal development through the establishment of a junctional polarity complex. Dev Cell 19:727–739. doi:10.1016/j.devcel.2010.10.008

Goetz JG, Minguet S, Navarro-Lérida I et al (2011) Biomechanical remodeling of the microenvironment by stromal caveolin-1 favors tumor invasion and metastasis. Cell 146:148–163. doi:10.1016/j.cell.2011.05.040

Graff JR, Herman JG, Lapidus RG et al (1995) E-cadherin expression is silenced by DNA hypermethylation in human breast and prostate carcinomas. Cancer Res 55:5195–5199

Graff JR, Greenberg VE, Herman JG et al (1998) Distinct patterns of E-cadherin CpG island methylation in papillary, follicular, Hurthle's cell, and poorly differentiated human thyroid carcinoma. Cancer Res 58:2063–2066

Graff JR, Gabrielson E, Fujii H et al (2000) Methylation patterns of the E-cadherin 5′ CpG island are unstable and reflect the dynamic, heterogeneous loss of E-cadherin expression during metastatic progression. J Biol Chem 275:2727–2732

Gregory PA, Bert AG, Paterson EL et al (2008) The miR-200 family and miR-205 regulate epithelial to mesenchymal transition by targeting ZEB1 and SIP1. Nat Cell Biol 10:593–601. doi:10.1038/ncb1722

Gregory DJ, Mikhaylova L, Fedulov AV (2012) Selective DNA demethylation by fusion of TDG with a sequence-specific DNA-binding domain. Epigenetics 7:344–349. doi:10.4161/epi.19509

Grotegut S, von Schweinitz D, Christofori G, Lehembre F (2006) Hepatocyte growth factor induces cell scattering through MAPK/Egr-1-mediated upregulation of Snail. EMBO J 25:3534–3545. doi:10.1038/sj.emboj.7601213

Gurevich-Panigrahi T, Panigrahi S, Wiechec E, Los M (2009) Obesity: pathophysiology and clinical management. Curr Med Chem 16:506–521

Ha CH, Bennett AM, Jin Z-G (2008) A novel role of vascular endothelial cadherin in modulating c-Src activation and downstream signaling of vascular endothelial growth factor. J Biol Chem 283(11):7261–7270. doi:10.1074/jbc.M702881200

Haidari M, Zhang W, Caivano A et al (2012) Integrin α2β1 mediates tyrosine phosphorylation of vascular endothelial cadherin induced by invasive breast cancer cells. J Biol Chem 287:32981–32992. doi:10.1074/jbc.M112.395905
Hamidov Z, Altendorf-Hofmann A, Chen Y et al (2011) Reduced expression of desmocollin 2 is an independent prognostic biomarker for shorter patients survival in pancreatic ductal adenocarcinoma. J Clin Pathol 64:990–994. doi:10.1136/jclinpath-2011-200099
Hammers CM, Stanley JR (2013) Desmoglein-1, differentiation, and disease. J Clin Invest 123 (4):1419–1422. doi:10.1172/JCI69071
Hanahan D, Folkman J (1996) Patterns and emerging mechanisms of the angiogenic switch during tumorigenesis. Cell 86:353–364
Harmon RM, Simpson CL, Johnson JL, Koetsier JL, Dubash AD, Najor NA et al (2013) Desmoglein-1/Erbin interaction suppresses ERK activation to support epidermal differentiation. J Clin Invest 123(4):1556–1570. doi:10.1172/JCI65220
Hazan RB, Qiao R, Keren R et al (2004) Cadherin switch in tumor progression. Ann N Y Acad Sci 1014:155–163
Hendrix MJ, Seftor EA, Meltzer PS et al (2001) Expression and functional significance of VE-cadherin in aggressive human melanoma cells: role in vasculogenic mimicry. Proc Natl Acad Sci U S A 98:8018–8023. doi:10.1073/pnas.131209798
Herranz N, Pasini D, Díaz VM et al (2008) Polycomb complex 2 is required for E-cadherin repression by the Snail1 transcription factor. Mol Cell Biol 28:4772–4781. doi:10.1128/MCB.00323-08
Hirate Y, Hirahara S, Inoue K-I et al (2013) Polarity-dependent distribution of angiomotin localizes Hippo signaling in preimplantation embryos. Curr Biol 23:1181–1194. doi:10.1016/j.cub.2013.05.014
Hoffmann I, Balling R (1995) Cloning and expression analysis of a novel mesodermally expressed cadherin. Dev Biol 169:337–346. doi:10.1006/dbio.1995.1148
Hulit J, Suyama K, Chung S et al (2007) N-cadherin signaling potentiates mammary tumor metastasis via enhanced extracellular signal-regulated kinase activation. Cancer Res 67:3106–3116. doi:10.1158/0008-5472.CAN-06-3401
Hurteau GJ, Carlson JA, Spivack SD, Brock GJ (2007) Overexpression of the microRNA hsa-miR-200c leads to reduced expression of transcription factor 8 and increased expression of E-cadherin. Cancer Res 67:7972–7976. doi:10.1158/0008-5472.CAN-07-1058
Ihrie RA, Marques MR, Nguyen BT et al (2005) Perp is a p63-regulated gene essential for epithelial integrity. Cell 120:843–856. doi:10.1016/j.cell.2005.01.008
Jacobs K, Feys L, Vanhoecke B et al (2011) P-cadherin expression reduces melanoma growth, invasion, and responsiveness to growth factors in nude mice. Eur J Cancer Prev 20:207–216. doi:10.1097/CEJ.0b013e3283429e8b
Jamora C, Lee P, Kocieniewski P et al (2005) A signaling pathway involving TGF-beta2 and snail in hair follicle morphogenesis. PLoS Biol 3:e11. doi:10.1371/journal.pbio.0030011
Jemal A, Bray F, Center MM et al (2011) Global cancer statistics. CA Cancer J Clin 61:69–90. doi:10.3322/caac.20107
Jiang B-H, Liu L-Z (2008) PI3K/PTEN signaling in tumorigenesis and angiogenesis. Biochimica Et Biophysica Acta 1784(1):150–158. doi:10.1016/j.bbapap.2007.09.008
Jiang R, Shi Z, Johnson JJ et al (2011) Kallikrein-5 promotes cleavage of desmoglein-1 and loss of cell-cell cohesion in oral squamous cell carcinoma. J Biol Chem 286:9127–9135. doi:10.1074/jbc.M110.191361
Johnson SK, Ramani VC, Hennings L, Haun RS (2007) Kallikrein 7 enhances pancreatic cancer cell invasion by shedding E-cadherin. Cancer 109:1811–1820. doi:10.1002/cncr.22606
Kalluri R, Weinberg RA (2009) The basics of epithelial-mesenchymal transition. J Clin Invest 119:1420–1428. doi:10.1172/JCI39104
Kanai Y, Ushijima S, Tsuda H et al (2000) Aberrant DNA methylation precedes loss of heterozygosity on chromosome 16 in chronic hepatitis and liver cirrhosis. Cancer Lett 148:73–80

Katafiasz D, Smith LM, Wahl JK (2011) Slug (SNAI2) expression in oral SCC cells results in altered cell-cell adhesion and increased motility. Cell Adh Migr 5:315–322

Kaur H, Phillips-Mason PJ, Burden-Gulley SM et al (2012) Cadherin-11, a marker of the mesenchymal phenotype, regulates glioblastoma cell migration and survival in vivo. Mol Cancer Res 10:293–304. doi:10.1158/1541-7786.MCR-11-0457

Kershaw MH, Devaud C, John LB et al (2013) Enhancing immunotherapy using chemotherapy and radiation to modify the tumor microenvironment. Oncoimmunology 2:e25962. doi:10.4161/onci.25962

Kim SH, Li Z, Sacks DB (2000) E-cadherin-mediated cell-cell attachment activates Cdc42. J Biol Chem 275:36999–37005. doi:10.1074/jbc.M003430200

Kim N-G, Koh E, Chen X, Gumbiner BM (2011) E-cadherin mediates contact inhibition of proliferation through Hippo signaling-pathway components. Proc Natl Acad Sci U S A 108:11930–11935. doi:10.1073/pnas.1103345108

Kimura TE, Merritt AJ, Garrod DR (2007) Calcium-independent desmosomes of keratinocytes are hyper-adhesive. J Invest Dermatol 127:775–781. doi:10.1038/sj.jid.5700643

Klonisch T, Wiechec E, Hombach-Klonisch S et al (2008) Cancer stem cell markers in common cancers – therapeutic implications. Trends Mol Med 14:450–460. doi:10.1016/j.molmed.2008.08.003

Klucky B, Mueller R, Vogt I et al (2007) Kallikrein 6 induces E-cadherin shedding and promotes cell proliferation, migration, and invasion. Cancer Res 67:8198–8206. doi:10.1158/0008-5472.CAN-07-0607

Koehler A, Bataille F, Schmid C et al (2004) Gene expression profiling of colorectal cancer and metastases divides tumours according to their clinicopathological stage. J Pathol 204:65–74. doi:10.1002/path.1606

Kolegraff K, Nava P, Helms MN et al (2011) Loss of desmocollin-2 confers a tumorigenic phenotype to colonic epithelial cells through activation of Akt/β-catenin signaling. Mol Biol Cell 22:1121–1134. doi:10.1091/mbc.E10-10-0845

Korpal M, Lee ES, Hu G, Kang Y (2008) The miR-200 family inhibits epithelial-mesenchymal transition and cancer cell migration by direct targeting of E-cadherin transcriptional repressors ZEB1 and ZEB2. J Biol Chem 283:14910–14914. doi:10.1074/jbc.C800074200

Kotb AM, Hierholzer A, Kemler R (2011) Replacement of E-cadherin by N-cadherin in the mammary gland leads to fibrocystic changes and tumor formation. Breast Cancer Res 13:R104. doi:10.1186/bcr3046

Kowalczyk AP, Green KJ (2013) Structure, function, and regulation of desmosomes. Prog Mol Biol Transl Sci 116:95–118. doi:10.1016/B978-0-12-394311-8.00005-4

Kraemer A, Goodwin M, Verma S et al (2007) Rac is a dominant regulator of cadherin-directed actin assembly that is activated by adhesive ligation independently of Tiam1. Am J Physiol Cell Physiol 292:C1061–C1069. doi:10.1152/ajpcell.00073.2006

Kucharczak J, Charrasse S, Comunale F et al (2008) R-Cadherin expression inhibits myogenesis and induces myoblast transformation via Rac1 GTPase. Cancer Res 68:6559–6568. doi:10.1158/0008-5472.CAN-08-0196

Kurzen H, Münzing I, Hartschuh W (2003) Expression of desmosomal proteins in squamous cell carcinomas of the skin. J Cutan Pathol 30:621–630

Kyriakakis E, Maslova K, Philippova M et al (2012) T-Cadherin is an auxiliary negative regulator of EGFR pathway activity in cutaneous squamous cell carcinoma: impact on cell motility. J Invest Dermatol 132:2275–2285. doi:10.1038/jid.2012.131

Labelle M, Schnittler HJ, Aust DE et al (2008) Vascular endothelial cadherin promotes breast cancer progression via transforming growth factor beta signaling. Cancer Res 68:1388–1397. doi:10.1158/0008-5472.CAN-07-2706

Lammens T, Swerts K, Derycke L et al (2012) N-cadherin in neuroblastoma disease: expression and clinical significance. PLoS One 7:e31206. doi:10.1371/journal.pone.0031206

Lampugnani MG, Orsenigo F, Gagliani MC et al (2006) Vascular endothelial cadherin controls VEGFR-2 internalization and signaling from intracellular compartments. J Cell Biol 174:593–604. doi:10.1083/jcb.200602080

Lebeau S, Masouyé I, Berti M, Augsburger E, Saurat J-H, Borradori L, Fontao L (2005) Comparative analysis of the expression of ERBIN and Erb-B2 in normal human skin and cutaneous carcinomas. Br J Dermatol 152(6):1248–1255. doi:10.1111/j.1365-2133.2005.06687.x

Lee SW (1996) H-cadherin, a novel cadherin with growth inhibitory functions and diminished expression in human breast cancer. Nat Med 2:776–782

Lee M-Y, Chou C-Y, Tang M-J, Shen M-R (2008) Epithelial-mesenchymal transition in cervical cancer: correlation with tumor progression, epidermal growth factor receptor overexpression, and snail up-regulation. Clin Cancer Res 14:4743–4750. doi:10.1158/1078-0432.CCR-08-0234

Lei Q-Y, Zhang H, Zhao B et al (2008) TAZ promotes cell proliferation and epithelial-mesenchymal transition and is inhibited by the hippo pathway. Mol Cell Biol 28:2426–2436. doi:10.1128/MCB.01874-07

Leight JL, Wozniak MA, Chen S et al (2012) Matrix rigidity regulates a switch between TGF-β1-induced apoptosis and epithelial-mesenchymal transition. Mol Biol Cell 23:781–791. doi:10.1091/mbc.E11-06-0537

Leong KG, Niessen K, Kulic I et al (2007) Jagged1-mediated Notch activation induces epithelial-to-mesenchymal transition through Slug-induced repression of E-cadherin. J Exp Med 204:2935–2948. doi:10.1084/jem.20071082

Leroy P, Mostov KE (2007) Slug is required for cell survival during partial epithelial-mesenchymal transition of HGF-induced tubulogenesis. Mol Biol Cell 18:1943–1952. doi:10.1091/mbc.E06-09-0823

Li Y, Laterra J (2012) Cancer stem cells: distinct entities or dynamically regulated phenotypes? Cancer Res 72:576–580. doi:10.1158/0008-5472.CAN-11-3070

Li L, Ying J, Li H et al (2012) The human cadherin 11 is a pro-apoptotic tumor suppressor modulating cell stemness through Wnt/β-catenin signaling and silenced in common carcinomas. Oncogene 31:3901–3912. doi:10.1038/onc.2011.541

Liao F, Li Y, O'Connor W et al (2000) Monoclonal antibody to vascular endothelial-cadherin is a potent inhibitor of angiogenesis, tumor growth, and metastasis. Cancer Res 60:6805–6810

Libusova L, Stemmler MP, Hierholzer A et al (2010) N-cadherin can structurally substitute for E-cadherin during intestinal development but leads to polyp formation. Development 137:2297–2305. doi:10.1242/dev.048488

Liu-Chittenden Y, Huang B, Shim JS et al (2012) Genetic and pharmacological disruption of the TEAD-YAP complex suppresses the oncogenic activity of YAP. Genes Dev 26:1300–1305. doi:10.1101/gad.192856.112

Lochter A, Galosy S, Muschler J et al (1997) Matrix metalloproteinase stromelysin-1 triggers a cascade of molecular alterations that leads to stable epithelial-to-mesenchymal conversion and a premalignant phenotype in mammary epithelial cells. J Cell Biol 139:1861–1872

Ma L, Young J, Prabhala H et al (2010) miR-9, a MYC/MYCN-activated microRNA, regulates E-cadherin and cancer metastasis. Nat Cell Biol 12:247–256. doi:10.1038/ncb2024

Machado JC, Oliveira C, Carvalho R et al (2001) E-cadherin gene (CDH1) promoter methylation as the second hit in sporadic diffuse gastric carcinoma. Oncogene 20:1525–1528. doi:10.1038/sj.onc.1204234

Makrilia N, Kollias A, Manolopoulos L, Syrigos K (2009) Cell adhesion molecules: role and clinical significance in cancer. Cancer Invest 27:1023–1037. doi:10.3109/07357900902769749

Marchong MN, Yurkowski C, Ma C et al (2010) Cdh11 acts as a tumor suppressor in a murine retinoblastoma model by facilitating tumor cell death. PLoS Genet 6:e1000923. doi:10.1371/journal.pgen.1000923

Mateus AR, Seruca R, Machado JC et al (2007) EGFR regulates RhoA-GTP dependent cell motility in E-cadherin mutant cells. Hum Mol Genet 16:1639–1647. doi:10.1093/hmg/ddm113

Mateus AR, Simões-Correia J, Figueiredo J et al (2009) E-cadherin mutations and cell motility: a genotype-phenotype correlation. Exp Cell Res 315:1393–1402. doi:10.1016/j.yexcr.2009.02.020
May CD, Sphyris N, Evans KW et al (2011) Epithelial-mesenchymal transition and cancer stem cells: a dangerously dynamic duo in breast cancer progression. Breast Cancer Res 13:202. doi:10.1186/bcr2789
McCrea PD, Turck CW, Gumbiner B (1991) A homolog of the armadillo protein in Drosophila (plakoglobin) associated with E-cadherin. Science 254:1359–1361
McEwen AE, Escobar DE, Gottardi CJ (2012) Signaling from the adherens junction. Sub-Cell Biochem 60:171–196. doi:10.1007/978-94-007-4186-7_8
Minn AJ, Gupta GP, Siegel PM et al (2005) Genes that mediate breast cancer metastasis to lung. Nature 436:518–524. doi:10.1038/nature03799
Miotto E, Sabbioni S, Veronese A et al (2004) Frequent aberrant methylation of the CDH4 gene promoter in human colorectal and gastric cancer. Cancer Res 64:8156–8159. doi:10.1158/0008-5472.CAN-04-3000
Mo J-S, Meng Z, Kim YC et al (2015) Cellular energy stress induces AMPK-mediated regulation of YAP and the Hippo pathway. Nat Cell Biol 17:500–510. doi:10.1038/ncb3111
Mosesson Y, Mills GB, Yarden Y (2008) Derailed endocytosis: an emerging feature of cancer. Nat Rev Cancer 8:835–850. doi:10.1038/nrc2521
Nakagawa M, Fukata M, Yamaga M et al (2001) Recruitment and activation of Rac1 by the formation of E-cadherin-mediated cell-cell adhesion sites. J Cell Sci 114:1829–1838
Nakajima G, Patino-Garcia A, Bruheim S et al (2008) CDH11 expression is associated with survival in patients with osteosarcoma. Cancer Genomics Proteomics 5:37–42
Nass SJ, Herman JG, Gabrielson E et al (2000) Aberrant methylation of the estrogen receptor and E-cadherin 5′ CpG islands increases with malignant progression in human breast cancer. Cancer Res 60:4346–4348
Näthke IS, Hinck L, Swedlow JR et al (1994) Defining interactions and distributions of cadherin and catenin complexes in polarized epithelial cells. J Cell Biol 125:1341–1352
Nieman MT, Prudoff RS, Johnson KR, Wheelock MJ (1999) N-cadherin promotes motility in human breast cancer cells regardless of their E-cadherin expression. J Cell Biol 147:631–644
Niessen K, Fu Y, Chang L et al (2008) Slug is a direct Notch target required for initiation of cardiac cushion cellularization. J Cell Biol 182:315–325. doi:10.1083/jcb.200710067
Nishioka N, Inoue K I, Adachi K et al (2009) The Hippo signaling pathway components Lats and Yap pattern Tead4 activity to distinguish mouse trophectoderm from inner cell mass. Dev Cell 16:398–410. doi:10.1016/j.devcel.2009.02.003
Noë V, Fingleton B, Jacobs K et al (2001) Release of an invasion promoter E-cadherin fragment by matrilysin and stromelysin-1. J Cell Sci 114:111–118
Noren NK, Niessen CM, Gumbiner BM, Burridge K (2001) Cadherin engagement regulates Rho family GTPases. J Biol Chem 276:33305–33308. doi:10.1074/jbc.C100306200
Novak A, Hsu SC, Leung-Hagesteijn C et al (1998) Cell adhesion and the integrin-linked kinase regulate the LEF-1 and beta-catenin signaling pathways. Proc Natl Acad Sci U S A 95:4374–4379
Nyberg P, Xie L, Kalluri R (2005) Endogenous inhibitors of angiogenesis. Cancer Res 65:3967–3979. doi:10.1158/0008-5472.CAN-04-2427
O'Shea C, Fitzpatrick JE, Koch PJ (2014) Desmosomal defects in acantholytic squamous cell carcinomas. J Cutan Pathol 41:873–879. doi:10.1111/cup.12390
Oki Y, Issa J-PJ (2010) Epigenetic mechanisms in AML – a target for therapy. Cancer Treat Res 145:19–40. doi:10.1007/978-0-387-69259-3_2
Oliveira C, Pinheiro H, Figueiredo J et al (2013) E-cadherin alterations in hereditary disorders with emphasis on hereditary diffuse gastric cancer. Prog Mol Biol Transl Sci 116:337–359. doi:10.1016/B978-0-12-394311-8.00015-7

Oloumi A, McPhee T, Dedhar S (2004) Regulation of E-cadherin expression and beta-catenin/Tcf transcriptional activity by the integrin-linked kinase. Biochim Biophys Acta 1691:1–15. doi:10.1016/j.bbamcr.2003.12.002

Oshiro MM, Kim CJ, Wozniak RJ et al (2005) Epigenetic silencing of DSC3 is a common event in human breast cancer. Breast Cancer Res 7:R669–R680. doi:10.1186/bcr1273

Overholtzer M, Zhang J, Smolen GA et al (2006) Transforming properties of YAP, a candidate oncogene on the chromosome 11q22 amplicon. Proc Natl Acad Sci U S A 103:12405–12410. doi:10.1073/pnas.0605579103

Paredes J, Albergaria A, Oliveira JT et al (2005) P-cadherin overexpression is an indicator of clinical outcome in invasive breast carcinomas and is associated with CDH3 promoter hypomethylation. Clin Cancer Res 11:5869–5877. doi:10.1158/1078-0432.CCR-05-0059

Paredes J, Correia AL, Ribeiro AS et al (2008) Breast carcinomas that co-express E- and P-cadherin are associated with p120-catenin cytoplasmic localisation and poor patient survival. J Clin Pathol 61:856–862. doi:10.1136/jcp.2007.052704

Park S-M, Gaur AB, Lengyel E, Peter ME (2008) The miR-200 family determines the epithelial phenotype of cancer cells by targeting the E-cadherin repressors ZEB1 and ZEB2. Genes Dev 22:894–907. doi:10.1101/gad.1640608

Peglion F, Llense F, Etienne-Manneville S (2014) Adherens junction treadmilling during collective migration. Nat Cell Biol 16:639–651. doi:10.1038/ncb2985

Peinado H, Ballestar E, Esteller M, Cano A (2004) Snail mediates E-cadherin repression by the recruitment of the Sin3A/histone deacetylase 1 (HDAC1)/HDAC2 complex. Mol Cell Biol 24:306–319

Philippova M, Pfaff D, Kyriakakis E et al (2013) T-cadherin loss promotes experimental metastasis of squamous cell carcinoma. Eur J Cancer 49:2048–2058. doi:10.1016/j.ejca.2012.12.026

Pinho SS, Osório H, Nita-Lazar M et al (2009) Role of E-cadherin N-glycosylation profile in a mammary tumor model. Biochem Biophys Res Commun 379:1091–1096. doi:10.1016/j.bbrc.2009.01.024

Pinho SS, Seruca R, Gärtner F et al (2011) Modulation of E-cadherin function and dysfunction by N-glycosylation. Cell Mol Life Sci 68:1011–1020. doi:10.1007/s00018-010-0595-0

Qian X, Karpova T, Sheppard AM et al (2004) E-cadherin-mediated adhesion inhibits ligand-dependent activation of diverse receptor tyrosine kinases. EMBO J 23:1739–1748. doi:10.1038/sj.emboj.7600136

Ranscht B, Dours-Zimmermann MT (1991) T-cadherin, a novel cadherin cell adhesion molecule in the nervous system lacks the conserved cytoplasmic region. Neuron 7:391–402

Rezaei M, Friedrich K, Wielockx B et al (2012) Interplay between neural-cadherin and vascular endothelial-cadherin in breast cancer progression. Breast Cancer Res 14:R154. doi:10.1186/bcr3367

Ribeiro AS, Sousa B, Carreto L et al (2013) P-cadherin functional role is dependent on E-cadherin cellular context: a proof of concept using the breast cancer model. J Pathol 229:705–718. doi:10.1002/path.4143

Rodriguez FJ, Lewis-Tuffin LJ, Anastasiadis PZ (2012) E-cadherin's dark side: possible role in tumor progression. Biochim Biophys Acta 1826:23–31. doi:10.1016/j.bbcan.2012.03.002

Rudini N, Felici A, Giampietro C et al (2008) VE-cadherin is a critical endothelial regulator of TGF-beta signalling. EMBO J 27:993–1004. doi:10.1038/emboj.2008.46

Ruhrberg C, Watt FM (1997) The plakin family: versatile organizers of cytoskeletal architecture. Curr Opin Genet Dev 7:392–397

Sahlgren C, Gustafsson MV, Jin S et al (2008) Notch signaling mediates hypoxia-induced tumor cell migration and invasion. Proc Natl Acad Sci U S A 105:6392–6397. doi:10.1073/pnas.0802047105

Sanchez-Heras E, Howell FV, Williams G, Doherty P (2006) The fibroblast growth factor receptor acid box is essential for interactions with N-cadherin and all of the major isoforms of neural cell adhesion molecule. J Biol Chem 281:35208–35216. doi:10.1074/jbc.M608655200

Sarrio D, Rodriguez-Pinilla SM, Hardisson D et al (2008) Epithelial-mesenchymal transition in breast cancer relates to the basal-like phenotype. Cancer Res 68:989–997. doi:10.1158/0008-5472.CAN-07-2017

Schmitt CJ, Franke WW, Goerdt S et al (2007) Homo- and heterotypic cell contacts in malignant melanoma cells and desmoglein 2 as a novel solitary surface glycoprotein. J Invest Dermatol 127:2191–2206. doi:10.1038/sj.jid.5700849

Schober M, Fuchs E (2011) Tumor-initiating stem cells of squamous cell carcinomas and their control by TGF-β and integrin/focal adhesion kinase (FAK) signaling. Proc Natl Acad Sci U S A 108:10544–10549. doi:10.1073/pnas.1107807108

Schulenburg A, Ulrich-Pur H, Thurnher D et al (2006) Neoplastic stem cells: a novel therapeutic target in clinical oncology. Cancer 107:2512–2520. doi:10.1002/cncr.22277

Serrano I, McDonald PC, Lock FE, Dedhar S (2013) Role of the integrin-linked kinase (ILK)/Rictor complex in TGFβ-1-induced epithelial-mesenchymal transition (EMT). Oncogene 32:50–60. doi:10.1038/onc.2012.30

Shamir ER, Pappalardo E, Jorgens DM et al (2014) Twist1-induced dissemination preserves epithelial identity and requires E-cadherin. J Cell Biol 204:839–856. doi:10.1083/jcb.201306088

Shen Y, Hirsch DS, Sasiela CA, Wu WJ (2008) Cdc42 regulates E-cadherin ubiquitination and degradation through an epidermal growth factor receptor to Src-mediated pathway. J Biol Chem 283:5127–5137. doi:10.1074/jbc.M703300200

Shintani Y, Fukumoto Y, Chaika N et al (2008) ADH-1 suppresses N-cadherin-dependent pancreatic cancer progression. Int J Cancer 122:71–77. doi:10.1002/ijc.23027

Simões-Correia J, Figueiredo J, Oliveira C et al (2008) Endoplasmic reticulum quality control: a new mechanism of E-cadherin regulation and its implication in cancer. Hum Mol Genet 17:3566–3576. doi:10.1093/hmg/ddn249

Smith AG (2001) Embryo-derived stem cells: of mice and men. Annu Rev Cell Dev Biol 17:435–462. doi:10.1146/annurev.cellbio.17.1.435

Smutny M, Cox HL, Leerberg JM et al (2010) Myosin II isoforms identify distinct functional modules that support integrity of the epithelial zonula adherens. Nat Cell Biol 12:696–702. doi:10.1038/ncb2072

Soncin F, Mohamet L, Ritson S et al (2011) E-cadherin acts as a regulator of transcripts associated with a wide range of cellular processes in mouse embryonic stem cells. PLoS One 6:e21463. doi:10.1371/journal.pone.0021463

Sonnenberg A, Liem RKH (2007) Plakins in development and disease. Exp Cell Res 313:2189–2203. doi:10.1016/j.yexcr.2007.03.039

Soto E, Yanagisawa M, Marlow LA et al (2008) p120 catenin induces opposing effects on tumor cell growth depending on E-cadherin expression. J Cell Biol 183:737–749. doi:10.1083/jcb.200805113

Spaderna S, Schmalhofer O, Hlubek F et al (2006) A transient, EMT-linked loss of basement membranes indicates metastasis and poor survival in colorectal cancer. Gastroenterology 131:830–840. doi:10.1053/j.gastro.2006.06.016

Spencer HL, Eastham AM, Merry CLR et al (2007) E-cadherin inhibits cell surface localization of the pro-migratory 5T4 oncofetal antigen in mouse embryonic stem cells. Mol Biol Cell 18:2838–2851. doi:10.1091/mbc.E06-09-0875

Spoelstra NS, Manning NG, Higashi Y et al (2006) The transcription factor ZEB1 is aberrantly expressed in aggressive uterine cancers. Cancer Res 66:3893–3902. doi:10.1158/0008-5472.CAN-05-2881

Stahley SN, Kowalczyk AP (2015) Desmosomes in acquired disease. Cell Tissue Res 360:439–456. doi:10.1007/s00441-015-2155-2

Strathdee G (2002) Epigenetic versus genetic alterations in the inactivation of E-cadherin. Semin Cancer Biol 12:373–379

Sun L, Hu H, Peng L et al (2011) P-cadherin promotes liver metastasis and is associated with poor prognosis in colon cancer. Am J Pathol 179:380–390. doi:10.1016/j.ajpath.2011.03.046

Suyama K, Shapiro I, Guttman M, Hazan RB (2002) A signaling pathway leading to metastasis is controlled by N-cadherin and the FGF receptor. Cancer Cell 2:301–314

Symowicz J, Adley BP, Gleason KJ et al (2007) Engagement of collagen-binding integrins promotes matrix metalloproteinase-9-dependent E-cadherin ectodomain shedding in ovarian carcinoma cells. Cancer Res 67:2030–2039. doi:10.1158/0008-5472.CAN-06-2808

Tada H, Hatoko M, Tanaka A, Kuwahara M, Muramatsu T (2000) Expression of desmoglein I and plakoglobin in skin carcinomas. J Cutan Pathol 27(1):24–29

Takeuchi T, Ohtsuki Y (2001) Recent progress in T-cadherin (CDH13, H-cadherin) research. Histol Histopathol 16:1287–1293

Tambe DT, Hardin CC, Angelini TE et al (2011) Collective cell guidance by cooperative intercellular forces. Nat Mater 10:469–475. doi:10.1038/nmat3025

Tamura D, Hiraga T, Myoui A et al (2008) Cadherin-11-mediated interactions with bone marrow stromal/osteoblastic cells support selective colonization of breast cancer cells in bone. Int J Oncol 33:17–24

Tan H-X, Wang Q, Chen L-Z et al (2010) MicroRNA-9 reduces cell invasion and E-cadherin secretion in SK-Hep-1 cell. Med Oncol 27:654–660. doi:10.1007/s12032-009-9264-2

Tanaka H, Kono E, Tran CP et al (2010) Monoclonal antibody targeting of N-cadherin inhibits prostate cancer growth, metastasis and castration resistance. Nat Med 16:1414–1420. doi:10.1038/nm.2236

Taniuchi K, Nakagawa H, Hosokawa M et al (2005) Overexpressed P-cadherin/CDH3 promotes motility of pancreatic cancer cells by interacting with p120ctn and activating rho-family GTPases. Cancer Res 65:3092–3099. doi:10.1158/0008.5472.CAN-04-3646

Thiery JP (2002) Epithelial-mesenchymal transitions in tumour progression. Nat Rev Cancer 2:442–454. doi:10.1038/nrc822

Thiery JP, Acloque H, Huang RYJ, Nieto MA (2009) Epithelial-mesenchymal transitions in development and disease. Cell 139:871–890. doi:10.1016/j.cell.2009.11.007

Thuault S, Hayashi S, Lagirand-Cantaloube J et al (2013) P-cadherin is a direct PAX3-FOXO1A target involved in alveolar rhabdomyosarcoma aggressiveness. Oncogene 32:1876–1887. doi:10.1038/onc.2012.217

Tomita K, van Bokhoven A, van Leenders GJ et al (2000) Cadherin switching in human prostate cancer progression. Cancer Res 60:3650–3654

Trepat X, Fredberg JJ (2011) Plithotaxis and emergent dynamics in collective cellular migration. Trends Cell Biol 21:638–646. doi:10.1016/j.tcb.2011.06.006

Trojan L, Schaaf A, Steidler A et al (2005) Identification of metastasis-associated genes in prostate cancer by genetic profiling of human prostate cancer cell lines. Anticancer Res 25:183–191

Tsang SM, Liu L, Teh M-T, Wheeler A, Grose R, Hart IR et al (2010) Desmoglein 3, via an interaction with E-cadherin, is associated with activation of Src. PLoS One 5(12):e14211. doi:10.1371/journal.pone.0014211

Tsang SM, Brown L, Gadmor H, Gammon L, Fortune F, Wheeler A, Wan H (2012) Desmoglein 3 acting as an upstream regulator of Rho GTPases, Rac-1/Cdc42 in the regulation of actin organisation and dynamics. Exp Cell Res 318(18):2269–2283. doi:10.1016/j.yexcr.2012.07.002

Turashvili G, McKinney SE, Goktepe O et al (2011) P-cadherin expression as a prognostic biomarker in a 3992 case tissue microarray series of breast cancer. Mod Pathol 24:64–81. doi:10.1038/modpathol.2010.189

Van Marck V, Stove C, Van Den Bossche K et al (2005) P-cadherin promotes cell-cell adhesion and counteracts invasion in human melanoma. Cancer Res 65:8774–8783. doi:10.1158/0008-5472.CAN-04-4414

Van Marck V, Stove C, Jacobs K et al (2011) P-cadherin in adhesion and invasion: opposite roles in colon and bladder carcinoma. Int J Cancer 128:1031–1044. doi:10.1002/ijc.25427

van Roy F, Berx G (2008) The cell-cell adhesion molecule E-cadherin. Cell Mol Life Sci 65:3756–3788. doi:10.1007/s00018-008-8281-1 van Zijl F, Krupitza G, Mikulits W (2011) Initial steps of metastasis: cell invasion and endothelial transmigration. Mutat Res 728:23–34. doi:10.1016/j.mrrev.2011.05.002

Vandewalle C, Comijn J, De Craene B et al (2005) SIP1/ZEB2 induces EMT by repressing genes of different epithelial cell-cell junctions. Nucleic Acids Res 33:6566–6578. doi:10.1093/nar/gki965

Virani S, Virani S, Colacino JA et al (2012) Cancer epigenetics: a brief review. ILAR J 53:359–369. doi:10.1093/ilar.53.3-4.359

Vlahova L, Doerflinger Y, Houben R et al (2012) P-cadherin expression in Merkel cell carcinomas is associated with prolonged recurrence-free survival. Br J Dermatol 166:1043–1052. doi:10.1111/j.1365-2133.2012.10853.x

von Burstin J, Eser S, Paul MC et al (2009) E-cadherin regulates metastasis of pancreatic cancer in vivo and is suppressed by a SNAIL/HDAC1/HDAC2 repressor complex. Gastroenterology 137:361–71–371.e1–5. doi:10.1053/j.gastro.2009.04.004

Wang W, Xiao Z-D, Li X et al (2015) AMPK modulates Hippo pathway activity to regulate energy homeostasis. Nat Cell Biol 17:490–499. doi:10.1038/ncb3113

Werling AM, Doerflinger Y, Brandner JM et al (2011) Homo- and heterotypic cell-cell contacts in Merkel cells and Merkel cell carcinomas: heterogeneity and indications for cadherin switching. Histopathology 58:286–303. doi:10.1111/j.1365-2559.2011.03748.x

Wheelock MJ, Shintani Y, Maeda M et al (2008) Cadherin switching. J Cell Sci 121:727–735. doi:10.1242/jcs.000455

Wicha MS, Liu S, Dontu G (2006) Cancer stem cells: an old idea--a paradigm shift. Cancer Res 66:1883–1890–discussion 1895–6. doi:10.1158/0008-5472.CAN-05-3153

Williams EJ, Furness J, Walsh FS, Doherty P (1994) Activation of the FGF receptor underlies neurite outgrowth stimulated by L1, N-CAM, and N-cadherin. Neuron 13:583–594

Yashiro M, Nishioka N, Hirakawa K (2006) Decreased expression of the adhesion molecule desmoglein-2 is associated with diffuse-type gastric carcinoma. Eur J Cancer 42:2397–2403. doi:10.1016/j.ejca.2006.03.024

Zanetta L, Corada M, Grazia Lampugnani M et al (2005) Downregulation of vascular endothelial-cadherin expression is associated with an increase in vascular tumor growth and hemorrhagic complications. Thromb Haemost 93:1041–1046. doi:10.1267/THRO05061041

Zeisberg M, Hanai J-I, Sugimoto H et al (2003) BMP-7 counteracts TGF-beta1-induced epithelial-to-mesenchymal transition and reverses chronic renal injury. Nat Med 9:964–968. doi:10.1038/nm888

Zhang CC, Yan Z, Zhang Q et al (2010) PF-03732010: a fully human monoclonal antibody against P-cadherin with antitumor and antimetastatic activity. Clin Cancer Res 16:5177–5188. doi:10.1158/1078-0432.CCR-10-1343

Zhao B, Ye X, Yu J et al (2008) TEAD mediates YAP-dependent gene induction and growth control. Genes Dev 22:1962–1971. doi:10.1101/gad.1664408

Zhurinsky J, Shtutman M, Ben-Ze'ev A (2000) Plakoglobin and beta-catenin: protein interactions, regulation and biological roles. J Cell Sci 113(Pt 18):3127–3139

Zohn IE, Li Y, Skolnik EY et al (2006) p38 and a p38-interacting protein are critical for downregulation of E-cadherin during mouse gastrulation. Cell 125:957–969. doi:10.1016/j.cell.2006.03.048

Chapter 15
Cadherin-Related Diseases

Shinji Hirano and Keiko Imai-Okano

This is a list for quick reference of cadherins in human diseases.

Please note that many unproven candidates and possible involvements (some are marked by *) are also listed. In addition, this is not a complete list and there are many omissions of citations.

15.1 List of Cadherin-Related Diseases

G Mendelian genetic, *M* multifactorial, *E* epigenetic, *A* autoimmune, *I* infection, *P* Expression profile, *X* experimental

Cadherins/Gene type		Diseases	References
Classic cadherins			
E-cadherin (CDH1)	**G**	Gastric cancer (HDGC) (OMIM137215)	Guilford et al. (1998)
	G	Lobular breast cancer (OMIM114480)	Masciari et al. (2007)
	G	Endometrial carcinoma (OMIM608089)	Risinger et al. (1994)
	G	Ovarian cancer (OMIM167000)	Risinger et al. (1994)
	M	Prostate cancer (OMIM176807)	Jonsson et al. (2004)
	M	Ulcerative colitis (UC)	Barrett et al. (2009)
	M	Asthma	Ierodiakonou et al. (2011)
	M	*Crohn disease (OMIM266600)	Elding et al. (2011)

(continued)

S. Hirano (✉) • K. Imai-Okano
Department of Cell Biology, Kansai Medical University, 2-5-1 Shinmachi, Hirakata, Osaka 573-1010, Japan
e-mail: s-hirano@hirakata.kmu.ac.jp; imaikei@hirakata.kmu.ac.jp

S.T. Suzuki, S. Hirano (eds.), *The Cadherin Superfamily*,
DOI 10.1007/978-4-431-56033-3_15

Cadherins/Gene type		Diseases	References
	I	Candidiasis	Phan et al. (2007)
	I	Listeria infection	Mengaud et al. (1996)
	A	Pemphigus Foliaceus	Flores et al. (2012)
N-cadherin (CDH2)	**I**	Candidiasis	Phan et al. (2007)
P-cadherin (CDH3)	**G**	Hypotrichosis with juvenile macular dystrophy HJMD (OMIM601553)	Sprecher et al. (2001)
	G	EEM syndrome (OMIM225280)	Kjaer et al. (2005)
	M	*Crohn disease (OMIM266600)	Elding et al. (2011)
	M	Autism spectrum disorders (ASD)	O'Roak et al. (2012)
R-cadherin (CDH4)	**M**	Chronic kidney disease without diabetes mellius	Yoshida et al. (2010)
	M	Schizophrenia	Girard et al. (2011)
	E	*Colorectal and Gastric cancer	Miotto et al. (2004)
VE-cadherin (CDH5)	**M**	Central Serous Chorioretinopathy	Schubert et al. (2014)
	M	Autism spectrum disorders (ASD)	O'Roak et al. (2012)
	I	Infection by Leptospira	Evangelista et al. (2014)
	A	*Rheumatoid arthritis, Behçet's disease *Systemic lupus erythematosus	Bouillet et al. (2013)
K-cadherin (CDH6)	**M**	Autism spectrum disorders (ASD)	Butler et al. (2015)
	G	*Myopia 19	Ma et al. (2010)
CDH7	**M**	Bipolar disorders	Soronen et al. (2010)
	M	*Ectodermal dysplasia (OMIM 602401)	Tariq et al. (2008)
	M	Autism spectrum disorders (ASD)	O'Roak et al. (2012)
CDH8	**M**	Learning disabilities, autism	Pagnamenta et al. (2011)
CDH9	**M**	*Autism, ASD (OMIM209850)	Wang et al. (2009)
CDH10	**G**	*Myopia 19	Ma et al. (2010)
	M	*Autism, ASD (OMIM209850)	Wang et al. (2009)
CDH11	**E**	*Cancer metastasis	Carmona et al. (2012)
	M	Alcoholism	Johnson et al. (2006)
	M	Bipolar alcoholism	Lydall et al. (2011)
CDH12	**M**	Bipolar alcoholism	Lydall et al. (2011)
	M	Leptin level	Zhang et al. (2013)
	G	*Myopia 19	Ma et al. (2010)
	M	*Schizophrenia	Singh et al. (2010)
M-cadherin (CDH15)	**M**	Mental retardation (OMIM612580)	Bhalla et al. (2008)
	M	Autism	Willemsen et al. (2010)
CDH18	**M**	Metabolic syndrome	Zhang et al. (2013)
	M	Colorectal cancer	Venkatachalam et al. (2011)
	M	*Schizophrenia	Singh et al. (2010)
CDH19	**M**	*Ectodermal dysplasia (OMIM 602401)	Tariq et al. (2008)

(continued)

Cadherins/Gene type		Diseases	References
CDH20	**M**	Barrett's adenocarcinoma	Wiech et al. (2009)
PB-cadherin (CDH22)	**M**	Type2 diabetes	Bento et al. (2008)
	M	Kawasaki disease	Shendre et al. (2014)
Desmosomal cadherins			
desmoglein 1 (DSG1)	**G**	Palmoplantar Keratoderma (SPPK). (OMIM 148700)	Rickman et al. (1999)
	G	Congenital erythroderma with palmoplantar keratoderma, hypotrichosis, and hyper-IgE (EPKHE) (OMIM615508) = SAM syndrome	Samuelov et al. (2013)
	A	Pemphigus foliaceus	Eyre and Stanley (1987)
	A	Pemphigus vulgaris	Amagai et al. (1991)
	A	Paraneoplastic pemphigus	Amagai et al. (1998)
	I	Staphylococcal scalded-skin syndrome (SSSS)	Amagai et al. (2000)
desmoglein 2 (DSG2)	**G**	Arrhythmogenic right ventricular cardiomyopathy 10 (ARVC10) (OMIM 610193)	Pilichou et al. (2006)
	G	Cardiomyopathy, dilated, 1BB (CMD1BB) (OMIM 612877)	Posch et al. (2008)
	X	*Colon cancer progression	Kamekura et al. (2014)
desmoglein 3 (DSG3)	**A**	Pemphigus vulgaris	Amagai et al. (1999)
	A	Paraneoplastic pemphigus	Amagai et al. (1998)
	P	*Cancer progression	Brown and Wan (2015)
DSG4	**G**	Localized autosomal recessive hypotrichosis (LAH1) (Hypotrichosis 6 OMIM 607903)	Kljuic et al. (2003)
	G	Monilethrix-like congenital hypotrichosis	Shimomura et al. (2006)
	A	Pemphigus foliaceus	Nagasaka et al. (2004)
	A	Pemphigus vulgaris	Nagasaka et al. (2004)
desmocollin 2 (DSC2)	**G**	Arrhythmogenic right ventricular dysplasia-11 (ARVD11, OMIM 610476)	Syrris et al. (2006)
	G	ARVD with mild palmoplantar keratoderma and woolly hair (Naxos disease)	Simpson et al. (2009)
	X	*Cancer progression	Kolegraff et al. (2011)
desmocollin 3 (DSC3)	**G**	Hypotrichosis and recurrent skin vesicles (OMIM 613102)	Ayub et al. (2009)
	A	Pemphigus vulgaris	Mao et al. (2010)

(continued)

Cadherins/Gene type		Diseases	References
T-cadherin			
T-cadherin (CDH13)	**M**	Serum adiponectin level	Ling et al. (2009) and Jee et al. (2010)
	M	Metabolic syndrome	Fava et al. (2011)
	M	Hypertension	Org et al. (2009)
	M	Lung function in air pollution	Imboden et al. (2015)
	M	Height	Axenovich et al. (2009)
	M	*Hirschsprung disease	Carrasquillo et al. (2002)
	M E	Lung cancer	Brock et al. (2008)
	M	*Retinoblastomas	Gratias et al. (2007)
	M	Prostate cancer	Thomas et al. (2008)
	E	Non-small cell lung cancer (NSCLC)	Pesek et al. (2011)
	M	Alcoholism	Johnson et al. (2006) and Treutlein et al. (2009)
	M	d-Amphetamine response	Hart et al. (2012)
	M	Methamphetamine dependence	Uhl et al. (2008b)
	M	Addiction-related phenotypes	Uhl et al. (2008a)
	M	Autism, Autism spectrum disorders (ASD)	Chapman et al. (2011)
	M	Attention deficit hyperactivity disorder (ADHD)	Lasky-Su et al. (2008) and Lesch et al. (2008)
	M	Personality	Terracciano et al. (2010)
	M	Childhood asthma	Ding et al. (2013)
Protocadherin			
PCDH1	**M**	Bronchial hyperresponsiveness (BHR) (asthma-related traits)	Koppelman et al. (2009)
	M	Eczema	Koning et al. (2012)
PCDH cluster	**E**	*Wilms' tumor	Dallosso et al. (2009)
	E	*Colorectal cancer	Dallosso et al. (2012)
		*Cervical cancer	Wang et al. (2015)
PCDHA cluster	**M**	Schizophrenia and bipolar disorder	Lachman et al. (2008)
	M	Bipolar disorder	Pedrosa et al. (2008)
	M	Autism	Anitha et al. (2013)
PCDHA3	**M**	*Restless legs syndrome (RLS8) (OMIM 615197)	Weissbach et al. (2012)
PCDHB4	**M**	Autism spectrum disorders (ASD)	O'Roak et al. (2012)
	M	Microcephaly	Alazami et al. (2015)
PCDH7	**M**	Epilepsy	Genetic determinants of common epilepsies: a meta-analysis of genome-wide association studies (2014)
	M	Body shape and cholesterol level	Surakka et al. (2011)

(continued)

Cadherins/Gene type		Diseases	References
	M	Survival in non-small cell lung cancer	Huang et al. (2009)
	M	Sleep duration	Ollila et al. (2014)
PCDH8	**M**	Autism spectrum disorders (ASD)	Butler et al. (2015)
	M	Cognition in the presence of type2 diabetes	Cox et al. (2014)
	E	*Renal cell carcinoma	Morris et al. (2011)
	P	*B-cell chronic lymphocytic leukemia	Mittal et al. (2007)
	E	*Bladder cancer	Lin et al. (2013)
	G?	*Retinoblastoma and mental retardation microdeletion syndrome	Caselli et al. (2007)
PCDH9	**M**	ASD	Marshall et al. (2008)
PCDH10	**E**	Testicular cancer	Cheung et al. (2010)
	M	Autism	Morrow et al. (2008)
PCDH11X	**M**	Developmental dyslexia	Veerappa et al. (2013)
	M	*Schizophrenia, cerebral asymmetry	Levchenko et al. (2014)
	M	*Tourette syndrome (OMIM 137580)	Melchior et al. (2013)
	M	Late-onset Alzheimer's disease	Carrasquillo et al. (2009)
PCDH11X, 11Y	**M**	Languate delay	Speevak and Farrell (2011)
PCDH12	**M**	Schizophrenia	Gregorio et al. (2009)
PCDH17	**E**	Prostate cancer	Lin et al. (2014)
	E	Bladder cancer	Wang et al. (2014)
	G?	*Retinoblastoma and mental retardation microdeletion syndrome	Caselli et al. (2007)
PCDH18	**M**	Autism	Morrow et al. (2008)
PCDH19	**G**	Early infantile epileptic encephalopathy-9 (EIEE9) = epilepsy and mental retardation restricted to females (EFMR) (OMIM 300088)	Dibbens et al. (2008)
		with Dvavet syndrome-like features	Depienne et al. (2009) and Depienne and LeGuern (2012)
		with autism and cognitive impairment	Depienne and LeGuern (2012) and Camacho et al. (2012)
PCDH20	**E**	Nasopharyngeal carcinoma	Chen et al. (2015)
	M	Sphingolipid metabolism	Demirkan et al. (2012)

(continued)

Cadherins/Gene type		Diseases	References
FAT			
FAT1	**M**	*Facioscapulohumeral dystrophy-like	Puppo et al. (2015)
	M	Autism spectrum disorders (ASDs)	Cukier et al. (2014)
	M	4q-syndrome	Sadeqzadeh et al. (2014)
	M	*Cervical cancer	Chung et al. (2015)
	P	Intrahepatic cholangiocarcinoma	Settakorn et al. (2005)
	G P	Oral cancer	Nakaya et al. (2007)
	P	*Breast cancer	Kwaepila et al. (2006)
	M P	Various carcinomas and tumors	Sadeqzadeh et al. (2014)
	M	Bipolar affective disorder (BPAD)	Blair et al. (2006)
FAT2	**M**	Autism	Butler et al. (2015)
	M	*Restless legs syndrome (RLS8) (OMIM 615197)	Weissbach et al. (2012)
	M	Pancreatic cancer	Tang et al. (2014)
	M	*Colorectal cancer	Xie et al. (2014)
FAT3	**P**	*Lung cancer	Rohrbeck and Borlak (2009)
FAT4	**M**	*Lung cancer	Berndt et al. (2011)
	M	Esophageal cancer	Du et al. (2013)
	M	Gastric cancer	Wadhwa et al. (2013)
	M	Melanoma	Nikolaev et al. (2012)
	P	Breast cancer	Qi et al. (2009)
	G	Van Maldergem syndrome-2 (VMLDS2)	Cappello et al. (2013)
	G	*Hennekam lymphangiectasia-lymphedema syndrome-2 (HKLLS2; OMIM616006) (allele of VMLDS2)	Alders et al. (2014)
Dachsous			
DCHS1	**G**	Van Maldergem syndrome-2 (VMLDS2)	Cappello et al. (2013)
DCHS2	**M**	Alzheimer's disease	Kamboh et al. (2012)
	M	Osteoporosis	Han et al. (2012)
7D-cadherin			
LI-cadherin (CDH17)	**M**	Hepatocellular carcinoma	Wang et al. (2006)
	P	Gastric cancer	Oue et al. (2004) and Lee et al. (2010)
	M	Colorectal carcinoma	Chen et al. (2012)
	P M	Various cancers	Weissbach et al. (2012)
	M	Hypertension	Zhu et al. (2015)

(continued)

Cadherins/Gene type		Diseases	References
Calsyntenin			
CLSTN1 Calsyntenin-1	**M**	Pancreatic cancer	Tang et al. (2014)
	E	Prostate, colon, breast cancer	Chung et al. (2015)
	M	Lung cancer	Langer et al. (2010)
CLSTN2 Calsyntenin-2	**M**	Alzheimer's disease (AD15) (OMIM611155)	Liu et al. (2007)
Inner ear cadherins			
CDH23	**G**	Usher syndrome ID (OMIM 601067)	Bolz et al. (2001) and Bork et al. (2001)
	G	Usher syndrome type ID/F (OMIM 601067)	Zheng et al. (2005)
	G	Autosomal recessive deafness 12 (DFNB12) (OMIM601386)	Bork et al. (2001) and Schultz et al. (2005)
	M	Personality	Terracciano et al. (2010)
	M	Chronic kidney disease (CKD)	Gorski et al. (2015)
	E	Alzheimer's disease	De Jager et al. (2014)
PCDH15	**G**	Usher syndrome IF (OMIM 602083)	Ahmed et al. (2001)
	G	Deafness, autosomal recessive 23 (DFNB23) (OMIM 609533)	Ahmed et al. (2003)
	M	Extrapulmonary tuberculosis	Oki et al. (2011)
	M	Late-onset Alzheimer's disease (LOAD)	Fallin et al. (2010)
	M	Familial combined hyperlipidemia (FCHL)	Huertas-Vazquez et al. (2010)
	M	Retinal dystrophies	Coppieters et al. (2014)
	M	Antibody response to smallpox vaccine	Ovsyannikova et al. (2012)
PCDH21 (CDHR1)	**G**	Cone-rod dystrophy15 (OMIM613660)	Ostergaard et al. (2010)
		Retinitis pigmentosa 65	Henderson et al. (2010)
CDHR3 (CDH28)	**M**	Asthma	Bonnelykke et al. (2014)
CDH26	**M**	Asthma	Ferreira et al. (2009)
RET	**G**	Hirschsprung disease	Attie et al. (1995) and Angrist et al. (1995)
	G	Central hypoventilation syndrome (OMIM 209880)	Bolk et al. (1996)
	G	Multiple endocrine neoplasia, Type II (OMIM 171400, 162300)	Shirahama et al. (1998)
	G	Familial medullary thyroid carcinoma	Elisei et al. (2007)
	G	Pheochromocytoma (OMIM 171300)	Eng et al. (1995)
	G	Renal agenesis (OMIM 191830)	Skinner et al. (2008)

(continued)

Cadherins/Gene type		Diseases	References
	M	Vesicoureteral reflux (OMIM 193000)	Yang et al. (2008)
	G	Congential anomalies of the kidney or urinary tract (CAKUT)	Hwang et al. (2014)
	G	Various other cancers	Mulligan 2014)
CELSR1	**M**	Spina bifida	Lei et al. (2014)
	M	Ischemic stroke	Yamada et al. (2009)
	M	Hypertension	Ueyama et al. (2013)
CELSR2	**M**	*Serum lipid	Wallace et al. (2008), Kathiresan et al. (2008) and Samani et al. (2008)
	M	*Coronary heart disease	Ronald et al. (2009) and Kathiresan et al. (2009)
CELSR3	**E**	Oral squamous cell carcinoma	Khor et al. (2014)

Acknowledgment We often cited review articles as references for simplicity and due to space limitations, and we are sorry for any omissions of original works from the references. This work was supported by grants from and the Kansai Medical University Project for Young Researchers (to K.I.), the JSPS KAKENHI Grant Number 25430037, RIKEN cooperative research, Takeda Science Foundation, Sumitomo Foundation, and MEXT-Supported Program for the Strategic Research Foundation at Private Universities (S1201038) (to S. H.).

References

Ahmed ZM, Riazuddin S, Bernstein SL, Ahmed Z, Khan S, Griffith AJ, Morell RJ, Friedman TB, Wilcox ER (2001) Mutations of the protocadherin gene PCDH15 cause Usher syndrome type 1F. Am J Hum Genet 69(1):25–34. doi:S0002-9297(07)61442-6 [pii] 10.1086/321277

Ahmed ZM, Riazuddin S, Ahmad J, Bernstein SL, Guo Y, Sabar MF, Sieving P, Griffith AJ, Friedman TB, Belyantseva IA, Wilcox ER (2003) PCDH15 is expressed in the neurosensory epithelium of the eye and ear and mutant alleles are responsible for both USH1F and DFNB23. Hum Mol Genet 12(24):3215–3223. doi:10.1093/hmg/ddg358 ddg358 [pii]

Alazami AM, Patel N, Shamseldin HE, Anazi S, Al-Dosari MS, Alzahrani F, Hijazi H, Alshammari M, Aldahmesh MA, Salih MA, Faqeih E, Alhashem A, Bashiri FA, Al-Owain M, Kentab AY, Sogaty S, Al Tala S, Temsah MH, Tulbah M, Aljelaify RF, Alshahwan SA, Seidahmed MZ, Alhadid AA, Aldhalaan H, AlQallaf F, Kurdi W, Alfadhel M, Babay Z, Alsogheer M, Kaya N, Al-Hassnan ZN, Abdel-Salam GM, Al-Sannaa N, Al Mutairi F, El Khashab HY, Bohlega S, Jia X, Nguyen HC, Hammami R, Adly N, Mohamed JY, Abdulwahab F, Ibrahim N, Naim EA, Al-Younes B, Meyer BF, Hashem M, Shaheen R, Xiong Y, Abouelhoda M, Aldeeri AA, Monies DM, Alkuraya FS (2015) Accelerating novel candidate gene discovery in neurogenetic disorders via whole-exome sequencing of prescreened multiplex consanguineous families. Cell Rep 10 (2):148–161. doi:S2211-1247(14)01044-4 [pii] 10.1016/j.celrep.2014.12.015

Alders M, Al-Gazali L, Cordeiro I, Dallapiccola B, Garavelli L, Tuysuz B, Salehi F, Haagmans MA, Mook OR, Majoie CB, Mannens MM, Hennekam RC (2014) Hennekam syndrome can be caused by FAT4 mutations and be allelic to Van Maldergem syndrome. Hum Genet 133 (9):1161–1167. doi:10.1007/s00439-014-1456-y

Amagai M, Klaus-Kovtun V, Stanley JR (1991) Autoantibodies against a novel epithelial cadherin in pemphigus vulgaris, a disease of cell adhesion. Cell 67(5):869–877. doi:0092-8674(91)90360-B [pii]

Amagai M, Nishikawa T, Nousari HC, Anhalt GJ, Hashimoto T (1998) Antibodies against desmoglein 3 (pemphigus vulgaris antigen) are present in sera from patients with paraneoplastic pemphigus and cause acantholysis in vivo in neonatal mice. J Clin Invest 102 (4):775–782. doi:10.1172/JCI3647

Amagai M, Tsunoda K, Zillikens D, Nagai T, Nishikawa T (1999) The clinical phenotype of pemphigus is defined by the anti-desmoglein autoantibody profile. J Am Acad Dermatol 40 (2 Pt 1):167–170. doi:S0190-9622(99)70183-0 [pii]

Amagai M, Matsuyoshi N, Wang ZH, Andl C, Stanley JR (2000) Toxin in bullous impetigo and staphylococcal scalded-skin syndrome targets desmoglein 1. Nat Med 6(11):1275–1277. doi:10.1038/81385

Angrist M, Bolk S, Thiel B, Puffenberger EG, Hofstra RM, Buys CH, Cass DT, Chakravarti A (1995) Mutation analysis of the RET receptor tyrosine kinase in Hirschsprung disease. Hum Mol Genet 4(5):821–830

Anitha A, Thanseem I, Nakamura K, Yamada K, Iwayama Y, Toyota T, Iwata Y, Suzuki K, Sugiyama T, Tsujii M, Yoshikawa T, Mori N (2013) Protocadherin alpha (PCDHA) as a novel susceptibility gene for autism. J Psychiatry Neurosci 38(3):192–198. doi:10.1503/jpn.120058 [pii] 10.1503/jpn.120058

Attie T, Pelet A, Edery P, Eng C, Mulligan LM, Amiel J, Boutrand L, Beldjord C, Nihoul-Fekete C, Munnich A et al (1995) Diversity of RET proto-oncogene mutations in familial and sporadic Hirschsprung disease. Hum Mol Genet 4(8):1381–1386

Axenovich TI, Zorkoltseva IV, Belonogova NM, Struchalin MV, Kirichenko AV, Kayser M, Oostra BA, van Duijn CM, Aulchenko YS (2009) Linkage analysis of adult height in a large pedigree from a Dutch genetically isolated population. Hum Genet 126(3):457–471. doi:10.1007/s00439-009-0686-x

Ayub M, Basit S, Jelani M, Ur Rehman F, Iqbal M, Yasinzai M, Ahmad W (2009) A homozygous nonsense mutation in the human desmocollin-3 (DSC3) gene underlies hereditary hypotrichosis and recurrent skin vesicles. Am J Hum Genet 85(4):515–520. doi:S0002-9297(09)00389-9 [pii] 10.1016/j.ajhg.2009.08.015

Barrett JC, Lee JC, Lees CW, Prescott NJ, Anderson CA, Phillips A, Wesley E, Parnell K, Zhang H, Drummond H, Nimmo ER, Massey D, Blaszczyk K, Elliott T, Cotterill L, Dallal H, Lobo AJ, Mowat C, Sanderson JD, Jewell DP, Newman WG, Edwards C, Ahmad T, Mansfield JC, Satsangi J, Parkes M, Mathew CG, Donnelly P, Peltonen L, Blackwell JM, Bramon E, Brown MA, Casas JP, Corvin A, Craddock N, Deloukas P, Duncanson A, Jankowski J, Markus HS, McCarthy MI, Palmer CN, Plomin R, Rautanen A, Sawcer SJ, Samani N, Trembath RC, Viswanathan AC, Wood N, Spencer CC, Bellenguez C, Davison D, Freeman C, Strange A, Langford C, Hunt SE, Edkins S, Gwilliam R, Blackburn H, Bumpstead SJ, Dronov S, Gillman M, Gray E, Hammond N, Jayakumar A, McCann OT, Liddle J, Perez ML, Potter SC, Ravindrarajah R, Ricketts M, Waller M, Weston P, Widaa S, Whittaker P, Attwood AP, Stephens J, Sambrook J, Ouwehand WH, McArdle WL, Ring SM, Strachan DP (2009) Genome-wide association study of ulcerative colitis identifies three new susceptibility loci, including the HNF4A region. Nat Genet 41 (12):1330–1334. doi:ng.483 [pii] 10.1038/ng.483

Bento JL, Palmer ND, Zhong M, Roh B, Lewis JP, Wing MR, Pandya H, Freedman BI, Langefeld CD, Rich SS, Bowden DW, Mychaleckyj JC (2008) Heterogeneity in gene loci associated with type 2 diabetes on human chromosome 20q13.1. Genomics 92(4):226–234. doi:S0888-7543(08)00137-7 [pii] 10.1016/j.ygeno.2008.06.004

Berndt A, Cario CL, Silva KA, Kennedy VE, Harrison DE, Paigen B, Sundberg JP (2011) Identification of fat4 and tsc22d1 as novel candidate genes for spontaneous pulmonary adenomas. Cancer Res 71(17):5779–5791. doi:0008–5472.CAN-11-1418 [pii] 10.1158/0008-5472.CAN-11-1418

Bhalla K, Luo Y, Buchan T, Beachem MA, Guzauskas GF, Ladd S, Bratcher SJ, Schroer RJ, Balsamo J, DuPont BR, Lilien J, Srivastava AK (2008) Alterations in CDH15 and KIRREL3 in patients with mild to severe intellectual disability. Am J Hum Genet 83(6):703–713. doi: S0002-9297(08)00555-7 [pii] 10.1016/j.ajhg.2008.10.020

Blair IP, Chetcuti AF, Badenhop RF, Scimone A, Moses MJ, Adams LJ, Craddock N, Green E, Kirov G, Owen MJ, Kwok JB, Donald JA, Mitchell PB, Schofield PR (2006) Positional cloning, association analysis and expression studies provide convergent evidence that the cadherin gene FAT contains a bipolar disorder susceptibility allele. Mol Psychiatry 11 (4):372–383. doi:4001784 [pii] 10.1038/sj.mp.4001784

Bolk S, Angrist M, Schwartz S, Silvestri JM, Weese-Mayer DE, Chakravarti A (1996) Congenital central hypoventilation syndrome: mutation analysis of the receptor tyrosine kinase RET. Am J Med Genet 63(4):603–609. doi:10.1002/(SICI)1096-8628(19960628)63:4<603::AID-AJMG14>3.0.CO;2-M [pii] 10.1002/(SICI)1096-8628(19960628)63:4<603::AID-AJMG14>3.0.CO;2-M

Bolz H, von Brederlow B, Ramirez A, Bryda EC, Kutsche K, Nothwang HG, Seeliger M, del C-Salcedó CM, Vila MC, Molina OP, Gal A, Kubisch C (2001) Mutation of CDH23, encoding a new member of the cadherin gene family, causes Usher syndrome type 1D. Nat Genet 27 (1):108–112. doi:10.1038/83667

Bonnelykke K, Sleiman P, Nielsen K, Kreiner-Moller E, Mercader JM, Belgrave D, den Dekker HT, Husby A, Sevelsted A, Faura-Tellez G, Mortensen LJ, Paternoster L, Flaaten R, Molgaard A, Smart DE, Thomsen PF, Rasmussen MA, Bonas-Guarch S, Holst C, Nohr EA, Yadav R, March ME, Blicher T, Lackie PM, Jaddoe VW, Simpson A, Holloway JW, Duijts L, Custovic A, Davies DE, Torrents D, Gupta R, Hollegaard MV, Hougaard DM, Hakonarson H, Bisgaard H (2014) A genome-wide association study identifies CDHR3 as a susceptibility locus for early childhood asthma with severe exacerbations. Nat Genet 46(1):51–55. doi: ng.2830 [pii] 10.1038/ng.2830

Bork JM, Peters LM, Riazuddin S, Bernstein SL, Ahmed ZM, Ness SL, Polomeno R, Ramesh A, Schloss M, Srisailpathy CR, Wayne S, Bellman S, Desmukh D, Ahmed Z, Khan SN, Kaloustian VM, Li XC, Lalwani A, Bitner-Glindzicz M, Nance WE, Liu XZ, Wistow G, Smith RJ, Griffith AJ, Wilcox ER, Friedman TB, Morell RJ (2001) Usher syndrome 1D and nonsyndromic autosomal recessive deafness DFNB12 are caused by allelic mutations of the novel cadherin-like gene CDH23. Am J Hum Genet 68(1):26–37. doi:S0002-9297(07)62469-0 [pii] 10.1086/316954

Bouillet L, Baudet AE, Deroux A, Sidibe A, Dumestre-Perard C, Mannic T, Treillard B, Arboleas MA, Chiquet CA, Gulino-Debrac DG, Vilgrain IY (2013) Auto-antibodies to vascular endothelial cadherin in humans: association with autoimmune diseases. Lab Invest 93 (11):1194–1202. doi:labinvest2013106 [pii] 10.1038/labinvest.2013.106

Brock MV, Hooker CM, Ota-Machida E, Han Y, Guo M, Ames S, Glockner S, Piantadosi S, Gabrielson E, Pridham G, Pelosky K, Belinsky SA, Yang SC, Baylin SB, Herman JG (2008) DNA methylation markers and early recurrence in stage I lung cancer. N Engl J Med 358 (11):1118–1128. doi:358/11/1118 [pii] 10.1056/NEJMoa0706550

Brown L, Wan H (2015) Desmoglein 3: a help or a hindrance in cancer progression? Cancers (Basel) 7(1):266–286. doi:cancers7010266 [pii] 10.3390/cancers7010266

Butler MG, Rafi SK, Hossain W, Stephan DA, Manzardo AM (2015) Whole exome sequencing in females with autism implicates novel and candidate genes. Int J Mol Sci 16(1):1312–1335. doi: ijms16011312 [pii] 10.3390/ijms16011312

Camacho A, Simon R, Sanz R, Vinuela A, Martinez-Salio A, Mateos F (2012) Cognitive and behavioral profile in females with epilepsy with PDCH19 mutation: two novel mutations and review of the literature. Epilepsy Behav 24(1):134–137. doi:S1525-5050(12)00089-3 [pii] 10. 1016/j.yebeh.2012.02.023

Cappello S, Gray MJ, Badouel C, Lange S, Einsiedler M, Srour M, Chitayat D, Hamdan FF, Jenkins ZA, Morgan T, Preitner N, Uster T, Thomas J, Shannon P, Morrison V, Di Donato N, Van Maldergem L, Neuhann T, Newbury-Ecob R, Swinkells M, Terhal P, Wilson LC,

Zwijnenburg PJ, Sutherland-Smith AJ, Black MA, Markie D, Michaud JL, Simpson MA, Mansour S, McNeill H, Gotz M, Robertson SP (2013) Mutations in genes encoding the cadherin receptor-ligand pair DCHS1 and FAT4 disrupt cerebral cortical development. Nat Genet 45(11):1300–1308. doi:ng.2765 [pii] 10.1038/ng.2765

Carmona FJ, Villanueva A, Vidal A, Munoz C, Puertas S, Penin RM, Goma M, Lujambio A, Piulats JM, Mesia R, Sanchez-Cespedes M, Manos M, Condom E, Eccles SA, Esteller M (2012) Epigenetic disruption of cadherin-11 in human cancer metastasis. J Pathol 228 (2):230–240. doi:10.1002/path.4011

Carrasquillo MM, McCallion AS, Puffenberger EG, Kashuk CS, Nouri N, Chakravarti A (2002) Genome-wide association study and mouse model identify interaction between RET and EDNRB pathways in Hirschsprung disease. Nat Genet 32(2):237–244. doi:10.1038/ng998 ng998 [pii]

Carrasquillo MM, Zou F, Pankratz VS, Wilcox SL, Ma L, Walker LP, Younkin SG, Younkin CS, Younkin LH, Bisceglio GD, Ertekin-Taner N, Crook JE, Dickson DW, Petersen RC, Graff-Radford NR (2009) Genetic variation in PCDH11X is associated with susceptibility to late-onset Alzheimer's disease. Nat Genet 41(2):192–198. doi:ng.305 [pii] 10.1038/ng.305

Caselli R, Speciale C, Pescucci C, Uliana V, Sampieri K, Bruttini M, Longo I, De Francesco S, Pramparo T, Zuffardi O, Frezzotti R, Acquaviva A, Hadjistilianou T, Renieri A, Mari F (2007) Retinoblastoma and mental retardation microdeletion syndrome: clinical characterization and molecular dissection using array CGH. J Hum Genet 52(6):535–542. doi:10.1007/s10038-007-0151-4

Chapman NH, Estes A, Munson J, Bernier R, Webb SJ, Rothstein JH, Minshew NJ, Dawson G, Schellenberg GD, Wijsman EM (2011) Genome-scan for IQ discrepancy in autism: evidence for loci on chromosomes 10 and 16. Hum Genet 129(1):59–70. doi:10.1007/s00439-010-0899-z

Chen RY, Cao JJ, Chen J, Yang JP, Liu XB, Zhao GQ, Zhang YF (2012) Single nucleotide polymorphisms in the CDH17 gene of colorectal carcinoma. World J Gastroenterol 18 (48):7251–7261. doi:10.3748/wjg.v18.i48.7251

Chen T, Long B, Ren G, Xiang T, Li L, Wang Z, He Y, Zeng Q, Hong S, Hu G (2015) Protocadherin20 acts as a tumor suppressor gene: epigenetic inactivation in nasopharyngeal carcinoma. J Cell Biochem 116:1766–1775. doi:10.1002/jcb.25135

Cheung HH, Lee TL, Davis AJ, Taft DH, Rennert OM, Chan WY (2010) Genome-wide DNA methylation profiling reveals novel epigenetically regulated genes and non-coding RNAs in human testicular cancer. Br J Cancer 102(2):419–427. doi:6605505 [pii] 10.1038/sj.bjc.6605505

Chung TK, Van Hummelen P, Chan PK, Cheung TH, Yim SF, Yu MY, Ducar MD, Thorner AR, MacConaill LE, Doran G, Pedamallu CS, Ojesina AI, Wong RR, Wang VW, Freeman SS, Lau TS, Kwong J, Chan LK, Fromer M, May T, Worley MJ Jr, Esselen KM, Elias KM, Lawrence M, Getz G, Smith DI, Crum CP, Meyerson M, Berkowitz RS, Wong YF (2015) Genomic aberrations in cervical adenocarcinomas in Hong Kong Chinese women. Int J Cancer 137:776–783. doi:10.1002/ijc.29456

Coppieters F, Van Schil K, Bauwens M, Verdin H, De Jaegher A, Syx D, Sante T, Lefever S, Abdelmoula NB, Depasse F, Casteels I, de Ravel T, Meire F, Leroy BP, De Baere E (2014) Identity-by-descent-guided mutation analysis and exome sequencing in consanguineous families reveals unusual clinical and molecular findings in retinal dystrophy. Genet Med 16 (9):671–680. doi:gim201424 [pii] 10.1038/gim.2014.24

Cox AJ, Hugenschmidt CE, Raffield LM, Langefeld CD, Freedman BI, Williamson JD, Hsu FC, Bowden DW (2014) Heritability and genetic association analysis of cognition in the Diabetes Heart Study. Neurobiol Aging 35(8):1958 e1953-1958 e1912. doi:S0197-4580(14)00239-5 [pii] 10.1016/j.neurobiolaging.2014.03.005

Cukier HN, Dueker ND, Slifer SH, Lee JM, Whitehead PL, Lalanne E, Leyva N, Konidari I, Gentry RC, Hulme WF, Booven DV, Mayo V, Hofmann NK, Schmidt MA, Martin ER, Haines JL, Cuccaro ML, Gilbert JR, Pericak-Vance MA (2014) Exome sequencing of extended families with autism reveals genes shared across neurodevelopmental and neuropsychiatric disorders. Mol Autism 5(1):1. doi:2040-2392-5-1 [pii] 10.1186/2040-2392-5-1

Dallosso AR, Hancock AL, Szemes M, Moorwood K, Chilukamarri L, Tsai HH, Sarkar A, Barasch J, Vuononvirta R, Jones C, Pritchard-Jones K, Royer-Pokora B, Lee SB, Owen C, Malik S, Feng Y, Frank M, Ward A, Brown KW, Malik K (2009) Frequent long-range epigenetic silencing of protocadherin gene clusters on chromosome 5q31 in Wilms' tumor. PLoS Genet 5(11):e1000745. doi:10.1371/journal.pgen.1000745

Dallosso AR, Oster B, Greenhough A, Thorsen K, Curry TJ, Owen C, Hancock AL, Szemes M, Paraskeva C, Frank M, Andersen CL, Malik K (2012) Long-range epigenetic silencing of chromosome 5q31 protocadherins is involved in early and late stages of colorectal tumorigenesis through modulation of oncogenic pathways. Oncogene 31(40):4409–4419. doi: onc2011609 [pii] 10.1038/onc.2011.609

De Jager PL, Srivastava G, Lunnon K, Burgess J, Schalkwyk LC, Yu L, Eaton ML, Keenan BT, Ernst J, McCabe C, Tang A, Raj T, Replogle J, Brodeur W, Gabriel S, Chai HS, Younkin C, Younkin SG, Zou F, Szyf M, Epstein CB, Schneider JA, Bernstein BE, Meissner A, Ertekin-Taner N, Chibnik LB, Kellis M, Mill J, Bennett DA (2014) Alzheimer's disease: early alterations in brain DNA methylation at ANK1, BIN1, RHBDF2 and other loci. Nat Neurosci 17(9):1156–1163. doi:nn.3786 [pii] 10.1038/nn.3786

Demirkan A, van Duijn CM, Ugocsai P, Isaacs A, Pramstaller PP, Liebisch G, Wilson JF, Johansson A, Rudan I, Aulchenko YS, Kirichenko AV, Janssens AC, Jansen RC, Gnewuch C, Domingues FS, Pattaro C, Wild SH, Jonasson I, Polasek O, Zorkoltseva IV, Hofman A, Karssen LC, Struchalin M, Floyd J, Igl W, Biloglav Z, Broer L, Pfeufer A, Pichler I, Campbell S, Zaboli G, Kolcic I, Rivadeneira F, Huffman J, Hastie ND, Uitterlinden A, Franke L, Franklin CS, Vitart V, Nelson CP, Preuss M, Bis JC, O'Donnell CJ, Franceschini N, Witteman JC, Axenovich T, Oostra BA, Meitinger T, Hicks AA, Hayward C, Wright AF, Gyllensten U, Campbell H, Schmitz G (2012) Genome-wide association study identifies novel loci associated with circulating phospho- and sphingolipid concentrations. PLoS Genet 8(2):e1002490. doi:10.1371/journal.pgen.1002490 PGENETICS-D-11-01673 [pii]

Depienne C, LeGuern E (2012) PCDH19-related infantile epileptic encephalopathy: an unusual X-linked inheritance disorder. Hum Mutat 33(4):627–634. doi:10.1002/humu.22029

Depienne C, Bouteiller D, Keren B, Cheuret E, Poirier K, Trouillard O, Benyahia B, Quelin C, Carpentier W, Julia S, Afenjar A, Gautier A, Rivier F, Meyer S, Berquin P, Helias M, Py I, Rivera S, Bahi-Buisson N, Gourfinkel-An I, Cazeneuve C, Ruberg M, Brice A, Nabbout R, Leguern E (2009) Sporadic infantile epileptic encephalopathy caused by mutations in PCDH19 resembles Dravet syndrome but mainly affects females. PLoS Genet 5(2):e1000381. doi:10.1371/journal.pgen.1000381

Dibbens LM, Tarpey PS, Hynes K, Bayly MA, Scheffer IE, Smith R, Bomar J, Sutton E, Vandeleur L, Shoubridge C, Edkins S, Turner SJ, Stevens C, O'Meara S, Tofts C, Barthorpe S, Buck G, Cole J, Halliday K, Jones D, Lee R, Madison M, Mironenko T, Varian J, West S, Widaa S, Wray P, Teague J, Dicks E, Butler A, Menzies A, Jenkinson A, Shepherd R, Gusella JF, Afawi Z, Mazarib A, Neufeld MY, Kivity S, Lev D, Lerman-Sagie T, Korczyn AD, Derry CP, Sutherland GR, Friend K, Shaw M, Corbett M, Kim HG, Geschwind DH, Thomas P, Haan E, Ryan S, McKee S, Berkovic SF, Futreal PA, Stratton MR, Mulley JC, Gecz J (2008) X-linked protocadherin 19 mutations cause female-limited epilepsy and cognitive impairment. Nat Genet 40(6):776–781. doi:ng.149 [pii] 10.1038/ng.149

Ding L, Abebe T, Beyene J, Wilke RA, Goldberg A, Woo JG, Martin LJ, Rothenberg ME, Rao M, Hershey GK, Chakraborty R, Mersha TB (2013) Rank-based genome-wide analysis reveals the association of ryanodine receptor-2 gene variants with childhood asthma among human populations. Hum Genomics 7:16. doi:1479-7364-7-16 [pii] 10.1186/1479-7364-7-16

Du J, Ji J, Gao Y, Xu L, Xu J, Zhu C, Gu H, Jiang J, Li H, Ma H, Hu Z, Jin G, Guo W, Chen X, Shen H (2013) Nonsynonymous polymorphisms in FAT4 gene are associated with the risk of esophageal cancer in an Eastern Chinese population. Int J Cancer 133(2):357–361. doi:10.1002/ijc.28033

Elding H, Lau W, Swallow DM, Maniatis N (2011) Dissecting the genetics of complex inheritance: linkage disequilibrium mapping provides insight into Crohn disease. Am J Hum Genet 89(6):798–805. doi:S0002-9297(11)00484-8 [pii] 10.1016/j.ajhg.2011.11.006

Elisei R, Romei C, Cosci B, Agate L, Bottici V, Molinaro E, Sculli M, Miccoli P, Basolo F, Grasso L, Pacini F, Pinchera A (2007) RET genetic screening in patients with medullary thyroid cancer and their relatives: experience with 807 individuals at one center. J Clin Endocrinol Metab 92(12):4725–4729. doi:jc.2007-1005 [pii] 10.1210/jc.2007-1005

Eng C, Crossey PA, Mulligan LM, Healey CS, Houghton C, Prowse A, Chew SL, Dahia PL, O'Riordan JL, Toledo SP et al (1995) Mutations in the RET proto-oncogene and the von Hippel-Lindau disease tumour suppressor gene in sporadic and syndromic phaeochromocytomas. J Med Genet 32(12):934–937

Evangelista K, Franco R, Schwab A, Coburn J (2014) Leptospira interrogans binds to cadherins. PLoS Negl Trop Dis 8(1):e2672. doi:10.1371/journal.pntd.0002672 PNTD-D-13-01477 [pii]

Eyre RW, Stanley JR (1987) Human autoantibodies against a desmosomal protein complex with a calcium-sensitive epitope are characteristic of pemphigus foliaceus patients. J Exp Med 165 (6):1719–1724

Fallin MD, Szymanski M, Wang R, Gherman A, Bassett SS, Avramopoulos D (2010) Fine mapping of the chromosome 10q11-q21 linkage region in Alzheimer's disease cases and controls. Neurogenetics 11(3):335–348. doi:10.1007/s10048-010-0234-9

Fava C, Danese E, Montagnana M, Sjogren M, Almgren P, Guidi GC, Hedblad B, Engstrom G, Lechi A, Minuz P, Melander O (2011) A variant upstream of the CDH13 adiponectin receptor gene and metabolic syndrome in Swedes. Am J Cardiol 108(10):1432–1437. doi:S0002-9149 (11)02272-7 [pii] 10.1016/j.amjcard.2011.06.068

Ferreira MA, Zhao ZZ, Thomsen SF, James M, Evans DM, Postmus PE, Kyvik KO, Backer V, Boomsma DI, Martin NG, Montgomery GW, Duffy DL (2009) Association and interaction analyses of eight genes under asthma linkage peaks. Allergy 64(11):1623–1628. doi:ALL2091 [pii] 10.1111/j.1398-9995.2009.02091.x

Flores G, Culton DA, Prisayanh P, Qaqish BF, James K, Maldonado M, Aoki V, Hans-Filho G, Rivitti EA, Diaz LA (2012) IgG autoantibody response against keratinocyte cadherins in endemic pemphigus foliaceus (fogo selvagem). J Invest Dermatol 132(11):2573–2580. doi: jid2012232 [pii] 10.1038/jid.2012.232

Genetic determinants of common epilepsies: a meta-analysis of genome-wide association studies (2014). Lancet Neurol 13(9):893–903. doi:S1474-4422(14)70171-1 [pii] 10.1016/S1474-4422 (14)70171-1

Girard SL, Gauthier J, Noreau A, Xiong L, Zhou S, Jouan L, Dionne-Laporte A, Spiegelman D, Henrion E, Diallo O, Thibodeau P, Bachand I, Bao JY, Tong AH, Lin CH, Millet B, Jaafari N, Joober R, Dion PA, Lok S, Krebs MO, Rouleau GA (2011) Increased exonic de novo mutation rate in individuals with schizophrenia. Nat Genet 43(9):860–863. doi:ng.886 [pii] 10.1038/ng.886

Gorski M, Tin A, Garnaas M, McMahon GM, Chu AY, Tayo BO, Pattaro C, Teumer A, Chasman DI, Chalmers J, Hamet P, Tremblay J, Woodward M, Aspelund T, Eiriksdottir G, Gudnason V, Harris TB, Launer LJ, Smith AV, Mitchell BD, O'Connell JR, Shuldiner AR, Coresh J, Li M, Freudenberger P, Hofer E, Schmidt H, Schmidt R, Holliday EG, Mitchell P, Wang JJ, de Boer IH, Li G, Siscovick DS, Kutalik Z, Corre T, Vollenweider P, Waeber G, Gupta J, Kanetsky PA, Hwang SJ, Olden M, Yang Q, de Andrade M, Atkinson EJ, Kardia SL, Turner ST, Stafford JM, Ding J, Liu Y, Barlassina C, Cusi D, Salvi E, Staessen JA, Ridker PM, Grallert H, Meisinger C, Muller-Nurasyid M, Kramer BK, Kramer H, Rosas SE, Nolte IM, Penninx BW, Snieder H, Fabiola Del Greco M, Franke A, Nothlings U, Lieb W, Bakker SJ, Gansevoort RT, van der Harst P, Dehghan A, Franco OH, Hofman A, Rivadeneira F, Sedaghat S, Uitterlinden AG, Coassin S, Haun M, Kollerits B, Kronenberg F, Paulweber B, Aumann N, Endlich K, Pietzner M, Volker U, Rettig R, Chouraki V, Helmer C, Lambert JC, Metzger M, Stengel B, Lehtimaki T, Lyytikainen LP, Raitakari O, Johnson A, Parsa A, Bochud M, Heid IM,

Goessling W, Kottgen A, Kao WH, Fox CS, Boger CA (2015) Genome-wide association study of kidney function decline in individuals of European descent. Kidney Int. 87:1017-1029. doi: ki2014361 [pii] 10.1038/ki.2014.361

Gratias S, Rieder H, Ullmann R, Klein-Hitpass L, Schneider S, Boloni R, Kappler M, Lohmann DR (2007) Allelic loss in a minimal region on chromosome 16q24 is associated with vitreous seeding of retinoblastoma. Cancer Res 67(1):408–416. doi:67/1/408 [pii] 10.1158/0008-5472. CAN-06-1317

Gregorio SP, Sallet PC, Do KA, Lin E, Gattaz WF, Dias-Neto E (2009) Polymorphisms in genes involved in neurodevelopment may be associated with altered brain morphology in schizophrenia: preliminary evidence. Psychiatry Res 165(1–2):1–9. doi:S0165-1781(07)00298-3 [pii] 10.1016/j.psychres.2007.08.011

Guilford P, Hopkins J, Harraway J, McLeod M, McLeod N, Harawira P, Taite H, Scoular R, Miller A, Reeve AE (1998) E-cadherin germline mutations in familial gastric cancer. Nature 392(6674):402–405. doi:10.1038/32918

Han Y, Pei Y, Liu Y, Zhang L, Wu S, Tian Q, Chen X, Shen H, Zhu X, Papasian CJ, Deng H (2012) Bivariate genome-wide association study suggests fatty acid desaturase genes and cadherin DCHS2 for variation of both compressive strength index and appendicular lean mass in males. Bone 51(6):1000–1007. doi:S8756-3282(12)01163-5 [pii] 10.1016/j.bone. 2012.08.127

Hart AB, Engelhardt BE, Wardle MC, Sokoloff G, Stephens M, de Wit H, Palmer AA (2012) Genome-wide association study of d-amphetamine response in healthy volunteers identifies putative associations, including cadherin 13 (CDH13). PLoS One 7(8):e42646. doi:10.1371/ journal.pone.0042646 PONE-D-12-16689 [pii]

Henderson RH, Li Z, Abd El Aziz MM, Mackay DS, Eljinini MA, Zeidan M, Moore AT, Bhattacharya SS, Webster AR (2010) Biallelic mutation of protocadherin-21 (PCDH21) causes retinal degeneration in humans. Mol Vis 16:46–52, doi:6 [pii]

Huang YT, Heist RS, Chirieac LR, Lin X, Skaug V, Zienolddiny S, Haugen A, Wu MC, Wang Z, Su L, Asomaning K, Christiani DC (2009) Genome-wide analysis of survival in early-stage non-small-cell lung cancer. J Clin Oncol 27(16):2660–2667. doi:JCO.2008.18.7906 [pii] 10. 1200/JCO.2008.18.7906

Huertas-Vazquez A, Plaisier CL, Geng R, Haas BE, Lee J, Greevenbroek MM, van der Kallen C, de Bruin TW, Taskinen MR, Alagramam KN, Pajukanta P (2010) A nonsynonymous SNP within PCDH15 is associated with lipid traits in familial combined hyperlipidemia. Hum Genet 127(1):83–89. doi:10.1007/s00439-009-0749-z

Hwang DY, Dworschak GC, Kohl S, Saisawat P, Vivante A, Hilger AC, Reutter HM, Soliman NA, Bogdanovic R, Kehinde EO, Tasic V, Hildebrandt F (2014) Mutations in 12 known dominant disease-causing genes clarify many congenital anomalies of the kidney and urinary tract. Kidney Int 85(6):1429–1433. doi:ki2013508 [pii] 10.1038/ki.2013.508

Ierodiakonou D, Postma DS, Koppelman GH, Boezen HM, Gerritsen J, Ten Hacken N, Timens W, Vonk JM (2011) E-cadherin gene polymorphisms in asthma patients using inhaled corticosteroids. Eur Respir J 38(5):1044–1052. doi:09031936.00194710 [pii] 10.1183/09031936. 00194710

Imboden M, Kumar A, Curjuric I, Adam M, Thun GA, Haun M, Tsai MY, Pons M, Bettschart R, Turk A, Rochat T, Kunzli N, Schindler C, Kronenberg F, Probst-Hensch NM (2015) Modification of the association between PM10 and lung function decline by cadherin 13 polymorphisms in the SAPALDIA cohort: a genome-wide interaction analysis. Environ Health Perspect 123(1):72–79. doi:10.1289/ehp.1307398

Jee SH, Sull JW, Lee JE, Shin C, Park J, Kimm H, Cho EY, Shin ES, Yun JE, Park JW, Kim SY, Lee SJ, Jee EJ, Baik I, Kao L, Yoon SK, Jang Y, Beaty TH (2010) Adiponectin concentrations: a genome-wide association study. Am J Hum Genet 87(4):545–552. doi:S0002-9297(10) 00472-6 [pii] 10.1016/j.ajhg.2010.09.004

Johnson C, Drgon T, Liu QR, Walther D, Edenberg H, Rice J, Foroud T, Uhl GR (2006) Pooled association genome scanning for alcohol dependence using 104,268 SNPs: validation and use to identify alcoholism vulnerability loci in unrelated individuals from the collaborative study on the genetics of alcoholism. Am J Med Genet B Neuropsychiatr Genet 141B(8):844–853. doi:10.1002/ajmg.b.30346

Jonsson BA, Adami HO, Hagglund M, Bergh A, Goransson I, Stattin P, Wiklund F, Gronberg H (2004) 160C/A polymorphism in the E-cadherin gene promoter and risk of hereditary, familial and sporadic prostate cancer. Int J Cancer 109(3):348–352. doi:10.1002/ijc.11629

Kamboh MI, Barmada MM, Demirci FY, Minster RL, Carrasquillo MM, Pankratz VS, Younkin SG, Saykin AJ, Sweet RA, Feingold E, DeKosky ST, Lopez OL (2012) Genome-wide association analysis of age-at-onset in Alzheimer's disease. Mol Psychiatry 17 (12):1340–1346. doi:mp2011135 [pii] 10.1038/mp.2011.135

Kamekura R, Kolegraff KN, Nava P, Hilgarth RS, Feng M, Parkos CA, Nusrat A (2014) Loss of the desmosomal cadherin desmoglein-2 suppresses colon cancer cell proliferation through EGFR signaling. Oncogene 33(36):4531–4536. doi:onc2013442 [pii] 10.1038/onc.2013.442

Kathiresan S, Melander O, Guiducci C, Surti A, Burtt NP, Rieder MJ, Cooper GM, Roos C, Voight BF, Havulinna AS, Wahlstrand B, Hedner T, Corella D, Tai ES, Ordovas JM, Berglund G, Vartiainen E, Jousilahti P, Hedblad B, Taskinen MR, Newton-Cheh C, Salomaa V, Peltonen L, Groop L, Altshuler DM, Orho-Melander M (2008) Six new loci associated with blood low-density lipoprotein cholesterol, high-density lipoprotein cholesterol or triglycerides in humans. Nat Genet 40(2):189–197. doi:ng.75 [pii] 10.1038/ng.75

Kathiresan S, Voight BF, Purcell S, Musunuru K, Ardissino D, Mannucci PM, Anand S, Engert JC, Samani NJ, Schunkert H, Erdmann J, Reilly MP, Rader DJ, Morgan T, Spertus JA, Stoll M, Girelli D, McKeown PP, Patterson CC, Siscovick DS, O'Donnell CJ, Elosua R, Peltonen L, Salomaa V, Schwartz SM, Melander O, Altshuler D, Merlini PA, Berzuini C, Bernardinelli L, Peyvandi F, Tubaro M, Celli P, Ferrario M, Fetiveau R, Marziliano N, Casari G, Galli M, Ribichini F, Rossi M, Bernardi F, Zonzin P, Piazza A, Yee J, Friedlander Y, Marrugat J, Lucas G, Subirana I, Sala J, Ramos R, Meigs JB, Williams G, Nathan DM, MacRae CA, Havulinna AS, Berglund G, Hirschhorn JN, Asselta R, Duga S, Spreafico M, Daly MJ, Nemesh J, Korn JM, McCarroll SA, Surti A, Guiducci C, Gianniny L, Mirel D, Parkin M, Burtt N, Gabriel SB, Thompson JR, Braund PS, Wright BJ, Balmforth AJ, Ball SG, Hall A, Linsel-Nitschke P, Lieb W, Ziegler A, Konig I, Hengstenberg C, Fischer M, Stark K, Grosshennig A, Preuss M, Wichmann HE, Schreiber S, Ouwehand W, Deloukas P, Scholz M, Cambien F, Li M, Chen Z, Wilensky R, Matthai W, Qasim A, Hakonarson HH, Devaney J, Burnett MS, Pichard AD, Kent KM, Satler L, Lindsay JM, Waksman R, Knouff CW, Waterworth DM, Walker MC, Mooser V, Epstein SE, Scheffold T, Berger K, Huge A, Martinelli N, Olivieri O, Corrocher R, McKeown P, Erdmann E, Konig IR, Holm H, Thorleifsson G, Thorsteinsdottir U, Stefansson K, Do R, Xie C, Siscovick D (2009) Genome-wide association of early-onset myocardial infarction with single nucleotide polymorphisms and copy number variants. Nat Genet 41(3):334–341. doi:ng.327 [pii] 10.1038/ng.327

Khor GH, Froemming GR, Zain RB, Abraham MT, Thong KL (2014) Screening of differential promoter hypermethylated genes in primary oral squamous cell carcinoma. Asian Pac J Cancer Prev 15(20):8957–8961

Kjaer KW, Hansen L, Schwabe GC, Marques-de-Faria AP, Eiberg H, Mundlos S, Tommerup N, Rosenberg T (2005) Distinct CDH3 mutations cause ectodermal dysplasia, ectrodactyly, macular dystrophy (EEM syndrome). J Med Genet 42(4):292–298. doi:42/4/292 [pii] 10.1136/jmg.2004.027821

Kljuic A, Bazzi H, Sundberg JP, Martinez-Mir A, O'Shaughnessy R, Mahoney MG, Levy M, Montagutelli X, Ahmad W, Aita VM, Gordon D, Uitto J, Whiting D, Ott J, Fischer S, Gilliam TC, Jahoda CA, Morris RJ, Panteleyev AA, Nguyen VT, Christiano AM (2003) Desmoglein 4 in hair follicle differentiation and epidermal adhesion: evidence from inherited hypotrichosis and acquired pemphigus vulgaris. Cell 113(2):249–260. doi:S0092867403002733 [pii]

Kolegraff K, Nava P, Helms MN, Parkos CA, Nusrat A (2011) Loss of desmocollin-2 confers a tumorigenic phenotype to colonic epithelial cells through activation of Akt/beta-catenin signaling. Mol Biol Cell 22(8):1121–1134. doi:mbc.E10-10-0845 [pii] 10.1091/mbc.E10-10-0845

Koning H, Postma DS, Brunekreef B, Duiverman EJ, Smit HA, Thijs C, Penders J, Kerkhof M, Koppelman GH (2012) Protocadherin-1 polymorphisms are associated with eczema in two Dutch birth cohorts. Pediatr Allergy Immunol 23(3):270–277. doi:10.1111/j.1399-3038.2011.01201.x

Koppelman GH, Meyers DA, Howard TD, Zheng SL, Hawkins GA, Ampleford EJ, Xu J, Koning H, Bruinenberg M, Nolte IM, van Diemen CC, Boezen HM, Timens W, Whittaker PA, Stine OC, Barton SJ, Holloway JW, Holgate ST, Graves PE, Martinez FD, van Oosterhout AJ, Bleecker ER, Postma DS (2009) Identification of PCDH1 as a novel susceptibility gene for bronchial hyperresponsiveness. Am J Respir Crit Care Med 180(10):929–935. doi:200810-1621OC [pii] 10.1164/rccm.200810-1621OC

Kwaepila N, Burns G, Leong AS (2006) Immunohistological localisation of human FAT1 (hFAT) protein in 326 breast cancers. Does this adhesion molecule have a role in pathogenesis? Pathology 38(2):125–131. doi:Q450233Q75W047P3 [pii] 10.1080/00313020600559975

Lachman HM, Petruolo OA, Pedrosa E, Novak T, Nolan K, Stopkova P (2008) Analysis of protocadherin alpha gene deletion variant in bipolar disorder and schizophrenia. Psychiatr Genet 18(3):110–115. doi:10.1097/YPG.0b013e3282fa1838 00041444-200806000-00002 [pii]

Langer W, Sohler F, Leder G, Beckmann G, Seidel H, Grone J, Hummel M, Sommer A (2010) Exon array analysis using re-defined probe sets results in reliable identification of alternatively spliced genes in non-small cell lung cancer. BMC Genomics 11:676. doi:1471-2164-11-676 [pii] 10.1186/1471-2164-11-676

Lasky-Su J, Anney RJ, Neale BM, Franke B, Zhou K, Maller JB, Vasquez AA, Chen W, Asherson P, Buitelaar J, Banaschewski T, Ebstein R, Gill M, Miranda A, Mulas F, Oades RD, Roeyers H, Rothenberger A, Sergeant J, Sonuga-Barke E, Steinhausen HC, Taylor E, Daly M, Laird N, Lange C, Faraone SV (2008) Genome-wide association scan of the time to onset of attention deficit hyperactivity disorder. Am J Med Genet B Neuropsychiatr Genet 147B(8):1355–1358. doi:10.1002/ajmg.b.30869

Lee HJ, Nam KT, Park HS, Kim MA, Lafleur BJ, Aburatani H, Yang HK, Kim WH, Goldenring JR (2010) Gene expression profiling of metaplastic lineages identifies CDH17 as a prognostic marker in early stage gastric cancer. Gastroenterology 139(1):213–225 e213. doi:S0016-5085(10)00554-8 [pii] 10.1053/j.gastro.2010.04.008

Lei Y, Zhu H, Yang W, Ross ME, Shaw GM, Finnell RH (2014) Identification of novel CELSR1 mutations in spina bifida. PLoS One 9(3):e92207. doi:10.1371/journal.pone.0092207 PONE-D-13-40448 [pii]

Lesch KP, Timmesfeld N, Renner TJ, Halperin R, Roser C, Nguyen TT, Craig DW, Romanos J, Heine M, Meyer J, Freitag C, Warnke A, Romanos M, Schafer H, Walitza S, Reif A, Stephan DA, Jacob C (2008) Molecular genetics of adult ADHD: converging evidence from genome-wide association and extended pedigree linkage studies. J Neural Transm 115(11):1573–1585. doi:10.1007/s00702-008-0119-3

Levchenko A, Davtian S, Petrova N, Malashichev Y (2014) Sequencing of five left-right cerebral asymmetry genes in a cohort of schizophrenia and schizotypal disorder patients from Russia. Psychiatr Genet 24(2):75–80. doi:10.1097/YPG.0000000000000021

Lin YL, Ma JH, Luo XL, Guan TY, Li ZG (2013) Clinical significance of protocadherin-8 (PCDH8) promoter methylation in bladder cancer. J Int Med Res 41(1):48–54. doi:0300060513475571 [pii] 10.1177/0300060513475571

Lin YL, Xie PG, Wang L, Ma JG (2014) Aberrant methylation of protocadherin 17 and its clinical significance in patients with prostate cancer after radical prostatectomy. Med Sci Monit 20:1376–1382. doi:891247 [pii] 10.12659/MSM.891247

Ling H, Waterworth DM, Stirnadel HA, Pollin TI, Barter PJ, Kesaniemi YA, Mahley RW, McPherson R, Waeber G, Bersot TP, Cohen JC, Grundy SM, Mooser VE, Mitchell BD (2009) Genome-wide linkage and association analyses to identify genes influencing adiponectin levels: the GEMS Study. Obesity (Silver Spring) 17(4):737–744. doi:oby2008625 [pii] 10.1038/oby.2008.625

Liu F, Arias-Vasquez A, Sleegers K, Aulchenko YS, Kayser M, Sanchez-Juan P, Feng BJ, Bertoli-Avella AM, van Swieten J, Axenovich TI, Heutink P, van Broeckhoven C, Oostra BA, van Duijn CM (2007) A genomewide screen for late-onset Alzheimer disease in a genetically isolated Dutch population. Am J Hum Genet 81(1):17–31. doi:S0002-9297(07)62813-4 [pii] 10.1086/518720

Lydall GJ, Bass NJ, McQuillin A, Lawrence J, Anjorin A, Kandaswamy R, Pereira A, Guerrini I, Curtis D, Vine AE, Sklar P, Purcell SM, Gurling HM (2011) Confirmation of prior evidence of genetic susceptibility to alcoholism in a genome-wide association study of comorbid alcoholism and bipolar disorder. Psychiatr Genet 21(6):294–306. doi:10.1097/YPG.0b013e32834915c2

Ma JH, Shen SH, Zhang GW, Zhao DS, Xu C, Pan CM, Jiang H, Wang ZQ, Song HD (2010) Identification of a locus for autosomal dominant high myopia on chromosome 5p13.3-p15.1 in a Chinese family. Mol Vis 16:2043–2054, doi:220 [pii]

Mao X, Nagler AR, Farber SA, Choi EJ, Jackson LH, Leiferman KM, Ishii N, Hashimoto T, Amagai M, Zone JJ, Payne AS (2010) Autoimmunity to desmocollin 3 in pemphigus vulgaris. Am J Pathol 177(6):2724–2730. doi:S0002-9440(10)62901-X [pii] 10.2353/ajpath.2010.100483

Marshall CR, Noor A, Vincent JB, Lionel AC, Feuk L, Skaug J, Shago M, Moessner R, Pinto D, Ren Y, Thiruvahindrapduram B, Fiebig A, Schreiber S, Friedman J, Ketelaars CE, Vos YJ, Ficicioglu C, Kirkpatrick S, Nicolson R, Sloman L, Summers A, Gibbons CA, Teebi A, Chitayat D, Weksberg R, Thompson A, Vardy C, Crosbie V, Luscombe S, Baatjes R, Zwaigenbaum L, Roberts W, Fernandez B, Szatmari P, Scherer SW (2008) Structural variation of chromosomes in autism spectrum disorder. Am J Hum Genet 82(2):477–488. doi:S0002-9297(07)00035-3 [pii] 10.1016/j.ajhg.2007.12.009

Masciari S, Larsson N, Senz J, Boyd N, Kaurah P, Kandel MJ, Harris LN, Pinheiro HC, Troussard A, Miron P, Tung N, Oliveira C, Collins L, Schnitt S, Garber JE, Huntsman D (2007) Germline E-cadherin mutations in familial lobular breast cancer. J Med Genet 44 (11):726–731. doi:jmg.2007.051268 [pii] 10.1136/jmg.2007.051268

Melchior L, Bertelsen B, Debes NM, Groth C, Skov L, Mikkelsen JD, Brondum-Nielsen K, Tumer Z (2013) Microduplication of 15q13.3 and Xq21.31 in a family with Tourette syndrome and comorbidities. Am J Med Genet B Neuropsychiatr Genet 162B(8):825–831. doi:10.1002/ajmg.b.32186

Mengaud J, Ohayon H, Gounon P, Mege RM, Cossart P (1996) E-cadherin is the receptor for internalin, a surface protein required for entry of L. monocytogenes into epithelial cells. Cell 84 (6):923–932. doi:S0092-8674(00)81070-3 [pii]

Miotto E, Sabbioni S, Veronese A, Calin GA, Gullini S, Liboni A, Gramantieri L, Bolondi L, Ferrazzi E, Gafa R, Lanza G, Negrini M (2004) Frequent aberrant methylation of the CDH4 gene promoter in human colorectal and gastric cancer. Cancer Res 64(22):8156–8159. doi:64/22/8156 [pii] 10.1158/0008-5472.CAN-04-3000

Mittal AK, Hegde GV, Aoun P, Bociek RG, Dave BJ, Joshi AD, Sanger WG, Weisenburger DD, Joshi SS (2007) Molecular basis of aggressive disease in chronic lymphocytic leukemia patients with 11q deletion and trisomy 12 chromosomal abnormalities. Int J Mol Med 20 (4):461–469

Morris MR, Ricketts CJ, Gentle D, McRonald F, Carli N, Khalili H, Brown M, Kishida T, Yao M, Banks RE, Clarke N, Latif F, Maher ER (2011) Genome-wide methylation analysis identifies epigenetically inactivated candidate tumour suppressor genes in renal cell carcinoma. Oncogene 30(12):1390–1401. doi:onc2010525 [pii] 10.1038/onc.2010.525

Morrow EM, Yoo SY, Flavell SW, Kim TK, Lin Y, Hill RS, Mukaddes NM, Balkhy S, Gascon G, Hashmi A, Al-Saad S, Ware J, Joseph RM, Greenblatt R, Gleason D, Ertelt JA, Apse KA, Bodell A, Partlow JN, Barry B, Yao H, Markianos K, Ferland RJ, Greenberg ME, Walsh CA (2008) Identifying autism loci and genes by tracing recent shared ancestry. Science 321 (5886):218–223. doi:321/5886/218 [pii] 10.1126/science.1157657

Mulligan LM (2014) RET revisited: expanding the oncogenic portfolio. Nat Rev Cancer 14 (3):173–186. doi:nrc3680 [pii] 10.1038/nrc3680

Nagasaka T, Nishifuji K, Ota T, Whittock NV, Amagai M (2004) Defining the pathogenic involvement of desmoglein 4 in pemphigus and staphylococcal scalded skin syndrome. J Clin Invest 114(10):1484–1492. doi:10.1172/JCI20480

Nakaya K, Yamagata HD, Arita N, Nakashiro KI, Nose M, Miki T, Hamakawa H (2007) Identification of homozygous deletions of tumor suppressor gene FAT in oral cancer using CGH-array. Oncogene 26(36):5300–5308. doi:1210330 [pii] 10.1038/sj.onc.1210330

Nikolaev SI, Rimoldi D, Iseli C, Valsesia A, Robyr D, Gehrig C, Harshman K, Guipponi M, Bukach O, Zoete V, Michielin O, Muehlethaler K, Speiser D, Beckmann JS, Xenarios I, Halazonetis TD, Jongeneel CV, Stevenson BJ, Antonarakis SE (2012) Exome sequencing identifies recurrent somatic MAP2K1 and MAP2K2 mutations in melanoma. Nat Genet 44 (2):133–139. doi:ng.1026 [pii] 10.1038/ng.1026

O'Roak BJ, Vives L, Girirajan S, Karakoc E, Krumm N, Coe BP, Levy R, Ko A, Lee C, Smith JD, Turner EH, Stanaway IB, Vernot B, Malig M, Baker C, Reilly B, Akey JM, Borenstein E, Rieder MJ, Nickerson DA, Bernier R, Shendure J, Eichler EE (2012) Sporadic autism exomes reveal a highly interconnected protein network of de novo mutations. Nature 485 (7397):246–250. doi:nature10989 [pii] 10.1038/nature10989

Oki NO, Motsinger-Reif AA, Antas PR, Levy S, Holland SM, Sterling TR (2011) Novel human genetic variants associated with extrapulmonary tuberculosis: a pilot genome wide association study. BMC Res Notes 4:28. doi:1756-0500-4-28 [pii] 10.1186/1756-0500-4-28

Ollila HM, Kettunen J, Pietilainen O, Aho V, Silander K, Kronholm E, Perola M, Lahti J, Raikkonen K, Widen E, Palotie A, Eriksson JG, Partonen T, Kaprio J, Salomaa V, Raitakari O, Lehtimaki T, Sallinen M, Harma M, Porkka-Heiskanen T, Paunio T (2014) Genome-wide association study of sleep duration in the Finnish population. J Sleep Res 23 (6):609–618. doi:10.1111/jsr.12175

Org E, Eyheramendy S, Juhanson P, Gieger C, Lichtner P, Klopp N, Veldre G, Doring A, Viigimaa M, Sober S, Tomberg K, Eckstein G, Kelgo P, Rebane T, Shaw-Hawkins S, Howard P, Onipinla A, Dobson RJ, Newhouse SJ, Brown M, Dominiczak A, Connell J, Samani N, Farrall M, Caulfield MJ, Munroe PB, Illig T, Wichmann HE, Meitinger T, Laan M (2009) Genome-wide scan identifies CDH13 as a novel susceptibility locus contributing to blood pressure determination in two European populations. Hum Mol Genet 18 (12):2288–2296. doi:ddp135 [pii] 10.1093/hmg/ddp135

Ostergaard E, Batbayli M, Duno M, Vilhelmsen K, Rosenberg T (2010) Mutations in PCDH21 cause autosomal recessive cone-rod dystrophy. J Med Genet 47(10):665–669. doi: jmg.2009.069120 [pii] 10.1136/jmg.2009.069120

Oue N, Hamai Y, Mitani Y, Matsumura S, Oshimo Y, Aung PP, Kuraoka K, Nakayama H, Yasui W (2004) Gene expression profile of gastric carcinoma: identification of genes and tags potentially involved in invasion, metastasis, and carcinogenesis by serial analysis of gene expression. Cancer Res 64(7):2397–2405

Ovsyannikova IG, Kennedy RB, O'Byrne M, Jacobson RM, Pankratz VS, Poland GA (2012) Genome-wide association study of antibody response to smallpox vaccine. Vaccine 30 (28):4182–4189. doi:S0264-410X(12)00591-9 [pii] 10.1016/j.vaccine.2012.04.055

Pagnamenta AT, Khan H, Walker S, Gerrelli D, Wing K, Bonaglia MC, Giorda R, Berney T, Mani E, Molteni M, Pinto D, Le Couteur A, Hallmayer J, Sutcliffe JS, Szatmari P, Paterson AD, Scherer SW, Vieland VJ, Monaco AP (2011) Rare familial 16q21 microdeletions under a linkage peak implicate cadherin 8 (CDH8) in susceptibility to autism and learning disability. J Med Genet 48(1):48–54. doi:jmg.2010.079426 [pii] 10.1136/jmg.2010.079426

Pedrosa E, Stefanescu R, Margolis B, Petruolo O, Lo Y, Nolan K, Novak T, Stopkova P, Lachman HM (2008) Analysis of protocadherin alpha gene enhancer polymorphism in bipolar disorder and schizophrenia. Schizophr Res 102(1–3):210–219. doi:S0920-9964(08)00194-1 [pii] 10.1016/j.schres.2008.04.013

Pesek M, Kopeckova M, Benesova L, Meszarosova A, Mukensnabl P, Bruha F, Minarik M (2011) Clinical significance of hypermethylation status in NSCLC: evaluation of a 30-gene panel in patients with advanced disease. Anticancer Res 31(12):4647–4652. doi:31/12/4647 [pii]

Phan QT, Myers CL, Fu Y, Sheppard DC, Yeaman MR, Welch WH, Ibrahim AS, Edwards JE, Jr., Filler SG (2007) Als3 is a Candida albicans invasin that binds to cadherins and induces endocytosis by host cells. PLoS Biol 5(3):e64. doi:06-PLBI-RA-1883R1 [pii] 10.1371/journal.pbio.0050064

Pilichou K, Nava A, Basso C, Beffagna G, Bauce B, Lorenzon A, Frigo G, Vettori A, Valente M, Towbin J, Thiene G, Danieli GA, Rampazzo A (2006) Mutations in desmoglein-2 gene are associated with arrhythmogenic right ventricular cardiomyopathy. Circulation 113 (9):1171–1179. doi:CIRCULATIONAHA.105.583674 [pii] 10.1161/CIRCULATIONAHA.105.583674

Posch MG, Posch MJ, Geier C, Erdmann B, Mueller W, Richter A, Ruppert V, Pankuweit S, Maisch B, Perrot A, Buttgereit J, Dietz R, Haverkamp W, Ozcelik C (2008) A missense variant in desmoglein-2 predisposes to dilated cardiomyopathy. Mol Genet Metab 95(1–2):74–80. doi:S1096-7192(08)00169-8 [pii] 10.1016/j.ymgme.2008.06.005

Puppo F, Dionnet E, Gaillard MC, Gaildrat P, Castro C, Vovan C, Bertaux K, Bernard R, Attarian S, Goto K, Nishino I, Hayashi Y, Magdinier F, Krahn M, Helmbacher F, Bartoli M, Levy N (2015) Identification of Variants in the 4q35 Gene FAT1 in Patients with a Facioscapulohumeral Dystrophy-Like Phenotype. Hum Mutat 36(4):443–453. doi:10.1002/humu.22760

Qi C, Zhu YT, Hu L, Zhu YJ (2009) Identification of Fat4 as a candidate tumor suppressor gene in breast cancers. Int J Cancer 124(4):793–798. doi:10.1002/ijc.23775

Rickman L, Simrak D, Stevens HP, Hunt DM, King IA, Bryant SP, Eady RA, Leigh IM, Arnemann J, Magee AI, Kelsell DP, Buxton RS (1999) N-terminal deletion in a desmosomal cadherin causes the autosomal dominant skin disease striate palmoplantar keratoderma. Hum Mol Genet 8(6):971–976. doi:ddc116 [pii]

Risinger JI, Berchuck A, Kohler MF, Boyd J (1994) Mutations of the E-cadherin gene in human gynecologic cancers. Nat Genet 7(1):98–102. doi:10.1038/ng0594-98

Rohrbeck A, Borlak J (2009) Cancer genomics identifies regulatory gene networks associated with the transition from dysplasia to advanced lung adenocarcinomas induced by c-Raf-1. PLoS One 4(10):e7315. doi:10.1371/journal.pone.0007315

Ronald J, Rajagopalan R, Ranchalis JE, Marshall JK, Hatsukami TS, Heagerty PJ, Jarvik GP (2009) Analysis of recently identified dyslipidemia alleles reveals two loci that contribute to risk for carotid artery disease. Lipids Health Dis 8:52. doi:1476-511X-8-52 [pii] 10.1186/1476-511X-8-52

Sadeqzadeh E, de Bock CE, Thorne RF (2014) Sleeping giants: emerging roles for the fat cadherins in health and disease. Med Res Rev 34(1):190–221. doi:10.1002/med.21286

Samani NJ, Braund PS, Erdmann J, Gotz A, Tomaszewski M, Linsel-Nitschke P, Hajat C, Mangino M, Hengstenberg C, Stark K, Ziegler A, Caulfield M, Burton PR, Schunkert H, Tobin MD (2008) The novel genetic variant predisposing to coronary artery disease in the region of the PSRC1 and CELSR2 genes on chromosome 1 associates with serum cholesterol. J Mol Med (Berl) 86(11):1233–1241. doi:10.1007/s00109-008-0387-2

Samuelov L, Sarig O, Harmon RM, Rapaport D, Ishida-Yamamoto A, Isakov O, Koetsier JL, Gat A, Goldberg I, Bergman R, Spiegel R, Eytan O, Geller S, Peleg S, Shomron N, Goh CS, Wilson NJ, Smith FJ, Pohler E, Simpson MA, McLean WH, Irvine AD, Horowitz M, McGrath JA, Green KJ, Sprecher E (2013) Desmoglein 1 deficiency results in severe dermatitis, multiple allergies and metabolic wasting. Nat Genet 45(10):1244–1248. doi:ng.2739 [pii] 10.1038/ng.2739

Schubert C, Pryds A, Zeng S, Xie Y, Freund KB, Spaide RF, Merriam JC, Barbazetto I, Slakter JS, Chang S, Munch IC, Drack AV, Hernandez J, Yzer S, Merriam JE, Linneberg A, Larsen M, Yannuzzi LA, Mullins RF, Allikmets R (2014) Cadherin 5 is regulated by corticosteroids and associated with central serous chorioretinopathy. Hum Mutat 35(7):859–867. doi:10.1002/humu.22551

Schultz JM, Yang Y, Caride AJ, Filoteo AG, Penheiter AR, Lagziel A, Morell RJ, Mohiddin SA, Fananapazir L, Madeo AC, Penniston JT, Griffith AJ (2005) Modification of human hearing loss by plasma-membrane calcium pump PMCA2. N Engl J Med 352(15):1557–1564. doi:352/15/1557 [pii] 10.1056/NEJMoa043899

Settakorn J, Kaewpila N, Burns GF, Leong AS (2005) FAT, E-cadherin, beta catenin, HER 2/neu, Ki67 immuno-expression, and histological grade in intrahepatic cholangiocarcinoma. J Clin Pathol 58(12):1249–1254. doi:58/12/1249 [pii] 10.1136/jcp.2005.026575

Shendre A, Wiener HW, Zhi D, Vazquez AI, Portman MA, Shrestha S (2014) High-density genotyping of immune loci in Kawasaki disease and IVIG treatment response in European-American case-parent trio study. Genes Immun 15(8):534–542. doi:gene201447 [pii] 10.1038/gene.2014.47

Shimomura Y, Sakamoto F, Kariya N, Matsunaga K, Ito M (2006) Mutations in the desmoglein 4 gene are associated with monilethrix-like congenital hypotrichosis. J Invest Dermatol 126 (6):1281–1285. doi:5700113 [pii] 10.1038/sj.jid.5700113

Shirahama S, Ogura K, Takami H, Ito K, Tohsen T, Miyauchi A, Nakamura Y (1998) Mutational analysis of the RET proto-oncogene in 71 Japanese patients with medullary thyroid carcinoma. J Hum Genet 43(2):101–106. doi:10.1007/s100380050048

Simpson MA, Mansour S, Ahnood D, Kalidas K, Patton MA, McKenna WJ, Behr ER, Crosby AH (2009) Homozygous mutation of desmocollin-2 in arrhythmogenic right ventricular cardiomyopathy with mild palmoplantar keratoderma and woolly hair. Cardiology 113(1):28–34. doi:000165696 [pii] 10.1159/000165696

Singh SM, Castellani C, O'Reilly R (2010) Autism meets schizophrenia via cadherin pathway. Schizophr Res 116(2–3):293–294. doi:S0920-9964(09)00475-7 [pii] 10.1016/j.schres.2009.09.031

Skinner MA, Safford SD, Reeves JG, Jackson ME, Freemerman AJ (2008) Renal aplasia in humans is associated with RET mutations. Am J Hum Genet 82(2):344–351. doi:S0002-9297(08)00086-4 [pii] 10.1016/j.ajhg.2007.10.008

Soronen P, Ollila HM, Antila M, Silander K, Palo OM, Kieseppa T, Lonnqvist J, Peltonen L, Tuulio-Henriksson A, Partonen T, Paunio T (2010) Replication of GWAS of bipolar disorder: association of SNPs near CDH7 with bipolar disorder and visual processing. Mol Psychiatry 15 (1):4–6. doi:mp200986 [pii] 10.1038/mp.2009.86

Speevak MD, Farrell SA (2011) Non-syndromic language delay in a child with disruption in the Protocadherin11X/Y gene pair. Am J Med Genet B Neuropsychiatr Genet 156B(4):484–489. doi:10.1002/ajmg.b.31186

Sprecher E, Bergman R, Richard G, Lurie R, Shalev S, Petronius D, Shalata A, Anbinder Y, Leibu R, Perlman I, Cohen N, Szargel R (2001) Hypotrichosis with juvenile macular dystrophy is caused by a mutation in CDH3, encoding P-cadherin. Nat Genet 29(2):134–136. doi:10.1038/ng716 ng716 [pii]

Surakka I, Isaacs A, Karssen LC, Laurila PP, Middelberg RP, Tikkanen E, Ried JS, Lamina C, Mangino M, Igl W, Hottenga JJ, Lagou V, van der Harst P, Mateo Leach I, Esko T, Kutalik Z, Wainwright NW, Struchalin MV, Sarin AP, Kangas AJ, Viikari JS, Perola M, Rantanen T, Petersen AK, Soininen P, Johansson A, Soranzo N, Heath AC, Papamarkou T, Prokopenko I, Tonjes A, Kronenberg F, Doring A, Rivadeneira F, Montgomery GW, Whitfield JB, Kahonen M, Lehtimaki T, Freimer NB, Willemsen G, de Geus EJ, Palotie A, Sandhu MS, Waterworth DM, Metspalu A, Stumvoll M, Uitterlinden AG, Jula A, Navis G, Wijmenga C, Wolffenbuttel BH, Taskinen MR, Ala-Korpela M, Kaprio J, Kyvik KO, Boomsma DI, Pedersen NL, Gyllensten U, Wilson JF, Rudan I, Campbell H, Pramstaller PP, Spector TD, Witteman JC, Eriksson JG, Salomaa V, Oostra BA, Raitakari OT, Wichmann HE, Gieger C, Jarvelin MR, Martin NG, Hofman A, McCarthy MI, Peltonen L, van Duijn CM, Aulchenko YS, Ripatti S (2011) A genome-wide screen for interactions reveals a new locus on 4p15 modifying the effect of waist-to-hip ratio on total cholesterol. PLoS Genet 7(10):e1002333. doi:10.1371/journal.pgen.1002333 PGENETICS-D-11-01300 [pii]

Syrris P, Ward D, Evans A, Asimaki A, Gandjbakhch E, Sen-Chowdhry S, McKenna WJ (2006) Arrhythmogenic right ventricular dysplasia/cardiomyopathy associated with mutations in the desmosomal gene desmocollin-2. Am J Hum Genet 79(5):978–984. doi:S0002-9297(07)60842-8 [pii] 10.1086/509122

Tang H, Wei P, Duell EJ, Risch HA, Olson SH, Bueno-de-Mesquita HB, Gallinger S, Holly EA, Petersen G, Bracci PM, McWilliams RR, Jenab M, Riboli E, Tjonneland A, Boutron-Ruault MC, Kaaks R, Trichopoulos D, Panico S, Sund M, Peeters PH, Khaw KT, Amos CI, Li D (2014) Axonal guidance signaling pathway interacting with smoking in modifying the risk of pancreatic cancer: a gene- and pathway-based interaction analysis of GWAS data. Carcinogenesis 35(5):1039–1045. doi:bgu010 [pii] 10.1093/carcin/bgu010

Tariq M, Chishti MS, Ali G, Ahmad W (2008) A novel locus for ectodermal dysplasia of hairs, nails and teeth type maps to chromosome 18q22.1-22.3. Ann Hum Genet 72(Pt 1):19–25. doi:AHG391 [pii] 10.1111/j.1469-1809.2007.00391.x

Terracciano A, Sanna S, Uda M, Deiana B, Usala G, Busonero F, Maschio A, Scally M, Patriciu N, Chen WM, Distel MA, Slagboom EP, Boomsma DI, Villafuerte S, Sliwerska E, Burmeister M, Amin N, Janssens AC, van Duijn CM, Schlessinger D, Abecasis GR, Costa PT, Jr. (2010) Genome-wide association scan for five major dimensions of personality. Mol Psychiatry 15(6):647–656. doi:mp2008113 [pii] 10.1038/mp.2008.113

Thomas G, Jacobs KB, Yeager M, Kraft P, Wacholder S, Orr N, Yu K, Chatterjee N, Welch R, Hutchinson A, Crenshaw A, Cancel-Tassin G, Staats BJ, Wang Z, Gonzalez-Bosquet J, Fang J, Deng X, Berndt SI, Calle EE, Feigelson HS, Thun MJ, Rodriguez C, Albanes D, Virtamo J, Weinstein S, Schumacher FR, Giovannucci E, Willett WC, Cussenot O, Valeri A, Andriole GL, Crawford ED, Tucker M, Gerhard DS, Fraumeni JF, Jr., Hoover R, Hayes RB, Hunter DJ, Chanock SJ (2008) Multiple loci identified in a genome-wide association study of prostate cancer. Nat Genet 40(3):310–315. doi:ng.91 [pii] 10.1038/ng.91

Treutlein J, Cichon S, Ridinger M, Wodarz N, Soyka M, Zill P, Maier W, Moessner R, Gaebel W, Dahmen N, Fehr C, Scherbaum N, Steffens M, Ludwig KU, Frank J, Wichmann HE, Schreiber S, Dragano N, Sommer WH, Leonardi-Essmann F, Lourdusamy A, Gebicke-Haerter P, Wienker TF, Sullivan PF, Nothen MM, Kiefer F, Spanagel R, Mann K, Rietschel M (2009) Genome-wide association study of alcohol dependence. Arch Gen Psychiatry 66(7):773–784. doi:66/7/773 [pii] 10.1001/archgenpsychiatry.2009.83

Ueyama C, Horibe H, Fujimaki T, Oguri M, Kato K, Yamada Y (2013) Association of genetic variants of and 3q28 with hypertension in community-dwelling individuals. Biomed Rep 1(6):840–844. doi:10.3892/br.2013.168 br-01-06-0840 [pii]

Uhl GR, Drgon T, Johnson C, Li CY, Contoreggi C, Hess J, Naiman D, Liu QR (2008a) Molecular genetics of addiction and related heritable phenotypes: genome-wide association approaches identify "connectivity constellation" and drug target genes with pleiotropic effects. Ann N Y Acad Sci 1141:318–381. doi:NYAS1141018 [pii] 10.1196/annals.1441.018

Uhl GR, Drgon T, Liu QR, Johnson C, Walther D, Komiyama T, Harano M, Sekine Y, Inada T, Ozaki N, Iyo M, Iwata N, Yamada M, Sora I, Chen CK, Liu HC, Ujike H, Lin SK (2008b) Genome-wide association for methamphetamine dependence: convergent results from 2 samples. Arch Gen Psychiatry 65(3):345–355. doi:65/3/345 [pii] 10.1001/archpsyc.65.3.345

Veerappa AM, Saldanha M, Padakannaya P, Ramachandra NB (2013) Genome-wide copy number scan identifies disruption of PCDH11X in developmental dyslexia. Am J Med Genet B Neuropsychiatr Genet 162B(8):889–897. doi:10.1002/ajmg.b.32199

Venkatachalam R, Verwiel ET, Kamping EJ, Hoenselaar E, Gorgens H, Schackert HK, van Krieken JH, Ligtenberg MJ, Hoogerbrugge N, van Kessel AG, Kuiper RP (2011) Identification of candidate predisposing copy number variants in familial and early-onset colorectal cancer patients. Int J Cancer 129(7):1635–1642. doi:10.1002/ijc.25821

Wadhwa R, Song S, Lee JS, Yao Y, Wei Q, Ajani JA (2013) Gastric cancer-molecular and clinical dimensions. Nat Rev Clin Oncol 10(11):643–655. doi:nrclinonc.2013.170 [pii] 10.1038/nrclinonc.2013.170

Wallace C, Newhouse SJ, Braund P, Zhang F, Tobin M, Falchi M, Ahmadi K, Dobson RJ, Marcano AC, Hajat C, Burton P, Deloukas P, Brown M, Connell JM, Dominiczak A, Lathrop GM, Webster J, Farrall M, Spector T, Samani NJ, Caulfield MJ, Munroe PB (2008) Genome-wide association study identifies genes for biomarkers of cardiovascular disease: serum urate and dyslipidemia. Am J Hum Genet 82(1):139–149. doi:S0002-9297(07)00026-2 [pii] 10.1016/j.ajhg.2007.11.001

Wang XQ, Luk JM, Garcia-Barcelo M, Miao X, Leung PP, Ho DW, Cheung ST, Lam BY, Cheung CK, Wong AS, Lau SS, So MT, Yu WC, Cai Q, Liu KS, Hui CK, Lau GK, Poon RT, Wong J, Fan ST (2006) Liver intestine-cadherin (CDH17) haplotype is associated with increased risk of hepatocellular carcinoma. Clin Cancer Res 12(17):5248–5252. doi:12/17/5248 [pii] 10.1158/1078-0432.CCR-06-0558

Wang K, Zhang H, Ma D, Bucan M, Glessner JT, Abrahams BS, Salyakina D, Imielinski M, Bradfield JP, Sleiman PM, Kim CE, Hou C, Frackelton E, Chiavacci R, Takahashi N, Sakurai T, Rappaport E, Lajonchere CM, Munson J, Estes A, Korvatska O, Piven J, Sonnenblick LI, Alvarez Retuerto AI, Herman EI, Dong H, Hutman T, Sigman M, Ozonoff S, Klin A, Owley T, Sweeney JA, Brune CW, Cantor RM, Bernier R, Gilbert JR, Cuccaro ML, McMahon WM, Miller J, State MW, Wassink TH, Coon H, Levy SE, Schultz RT, Nurnberger JI, Haines JL, Sutcliffe JS, Cook EH, Minshew NJ, Buxbaum JD, Dawson G, Grant SF, Geschwind DH, Pericak-Vance MA, Schellenberg GD, Hakonarson H (2009) Common genetic variants on 5p14.1 associate with autism spectrum disorders. Nature 459(7246):528–533. doi:nature07999 [pii] 10.1038/nature07999

Wang XB, Lin YL, Li ZG, Ma JH, Li J, Ma JG (2014) Protocadherin 17 promoter methylation in tumour tissue from patients with bladder transitional cell carcinoma. J Int Med Res 42(2):292–299. doi:0300060513504364 [pii] 10.1177/0300060513504364

Wang KH, Lin CJ, Liu CJ, Liu DW, Huang RL, Ding DC, Weng CF, Chu TY (2015) Global methylation silencing of clustered proto-cadherin genes in cervical cancer: serving as diagnostic markers comparable to HPV. Cancer Med 4(1):43–55. doi:10.1002/cam4.335

Weissbach A, Siegesmund K, Bruggemann N, Schmidt A, Kasten M, Pichler I, Muhle H, Lohmann E, Lohnau T, Schwinger E, Hagenah J, Stephani U, Pramstaller PP, Klein C, Lohmann K (2012) Exome sequencing in a family with restless legs syndrome. Mov Disord 27(13):1686–1689. doi:10.1002/mds.25191

Wiech T, Nikolopoulos E, Weis R, Langer R, Bartholome K, Timmer J, Walch AK, Hofler H, Werner M (2009) Genome-wide analysis of genetic alterations in Barrett's adenocarcinoma using single nucleotide polymorphism arrays. Lab Invest 89(4):385–397. doi:labinvest200867 [pii] 10.1038/labinvest.2008.67

Willemsen MH, Fernandez BA, Bacino CA, Gerkes E, de Brouwer AP, Pfundt R, Sikkema-Raddatz B, Scherer SW, Marshall CR, Potocki L, van Bokhoven H, Kleefstra T (2010) Identification of ANKRD11 and ZNF778 as candidate genes for autism and variable cognitive impairment in the novel 16q24.3 microdeletion syndrome. Eur J Hum Genet 18(4):429–435. doi:ejhg2009192 [pii] 10.1038/ejhg.2009.192

Xie T, Cho YB, Wang K, Huang D, Hong HK, Choi YL, Ko YH, Nam DH, Jin J, Yang H, Fernandez J, Deng S, Rejto PA, Lee WY, Mao M (2014) Patterns of somatic alterations between matched primary and metastatic colorectal tumors characterized by whole-genome sequencing. Genomics 104(4):234–241. doi:S0888-7543(14)00127-X [pii] 10.1016/j.ygeno.2014.07.012

Yamada Y, Fuku N, Tanaka M, Aoyagi Y, Sawabe M, Metoki N, Yoshida H, Satoh K, Kato K, Watanabe S, Nozawa Y, Hasegawa A, Kojima T (2009) Identification of CELSR1 as a susceptibility gene for ischemic stroke in Japanese individuals by a genome-wide association study. Atherosclerosis 207(1):144–149. doi:S0021-9150(09)00249-4 [pii] 10.1016/j.atherosclerosis.2009.03.038

Yang Y, Houle AM, Letendre J, Richter A (2008) RET Gly691Ser mutation is associated with primary vesicoureteral reflux in the French-Canadian population from Quebec. Hum Mutat 29 (5):695–702. doi:10.1002/humu.20705

Yoshida T, Kato K, Yokoi K, Oguri M, Watanabe S, Metoki N, Yoshida H, Satoh K, Aoyagi Y, Nozawa Y, Yamada Y (2010) Association of genetic variants with chronic kidney disease in Japanese individuals with or without hypertension or diabetes mellitus. Exp Ther Med 1 (1):137–145. doi:10.3892/etm_00000023 etm-01-01-0137 [pii]

Zhang Y, Kent JW, Jr., Olivier M, Ali O, Cerjak D, Broeckel U, Abdou RM, Dyer TD, Comuzzie A, Curran JE, Carless MA, Rainwater DL, Goring HH, Blangero J, Kissebah AH (2013) A comprehensive analysis of adiponectin QTLs using SNP association, SNP cis-effects on peripheral blood gene expression and gene expression correlation identified novel metabolic syndrome (MetS) genes with potential role in carcinogenesis and systemic inflammation. BMC Med Genomics 6:14. doi:1755-8794-6-14 [pii] 10.1186/1755-8794-6-14

Zheng QY, Yan D, Ouyang XM, Du LL, Yu H, Chang B, Johnson KR, Liu XZ (2005) Digenic inheritance of deafness caused by mutations in genes encoding cadherin 23 and protocadherin 15 in mice and humans. Hum Mol Genet 14(1):103–111. doi:ddi010 [pii] 10.1093/hmg/ddi010

Zhu X, Feng T, Tayo BO, Liang J, Young JH, Franceschini N, Smith JA, Yanek LR, Sun YV, Edwards TL, Chen W, Nalls M, Fox E, Sale M, Bottinger E, Rotimi C, Liu Y, McKnight B, Liu K, Arnett DK, Chakravati A, Cooper RS, Redline S (2015) Meta-analysis of correlated traits via summary statistics from GWASs with an application in hypertension. Am J Hum Genet 96(1):21–36. doi:S0002-9297(14)00477-7 [pii] 10.1016/j.ajhg.2014.11.011

Index

A
Actin, 99, 102
α-Actinin, 117
Activity sensors, 332
Acto-myosin contraction, 261–262
Adherens junction (AJ), 46, 94
Adhesion, 226
Adhesion arm, 24
Adhesion-inhibitory antibody, 4
Adhesive strength, 342
Adhesome, 353
ADIP, 117
Adiponectin, 281
Afadin, 117
Akt, 131
Akt/β-catenin signaling, 381
Alcadein, 293
Allosteric changes, 55
Alopecia, 141
Alpha-catenin, 43, 46
Alzheimer's disease (AD), 142
AMP-activated kinase (AMPK), 383
Amygdale, 143
Amyloid precursor protein (APP), 293
Apex–apex junction, 135
Apical constriction, 127
Apical junctional complex, 135
Apico-basal polarity, 125, 324
apoe (apolipoprotein E), 142
Apoptosis, 170
ARM repeats, 28
Armadillo proteins, 28
A* strand, 77
Atypical cadherins, 17
Atypical protein kinase C (aPKC), 126
Auditory epithelium, 133
Autism spectrum disorders, 239, 332
Axon formation, 326–327
Axon patterning, 327–328
Axonogenesis, 328

B
Bacterial toxins, 180
Barrier function, 49–51
Bax, 209
Beta-catenin, 43, 59
Boundary formation, 52
Bovine herpes virus, 140

C
CA1, 142
CA3, 132
Ca^{2+} dependent, 165
Ca^{2+}-dependent cell adhesion, 5
Cadherin, 116, 117, 119, 120, 122–129, 131–135, 137–145
  diversity, 42
  extracellular domain, 94–97
  intracellular domain, 97–99
  introduction, 41–43
  nomenclature, 14
  refinement, 330
  regulation, 53–57
  subtype switching, 318
  switch, 373
Cadherin-13, 277–285
Cadherin-16, 300–301
Cadherin-17, 298–300

S.T. Suzuki, S. Hirano (eds.), *The Cadherin Superfamily*,
DOI 10.1007/978-4-431-56033-3

Cadherin-23, 85, 350
Cadherin major branch (CMB), 15
Cadherin-related (CDHR) proteins, 25
Cadherin-related major branch (CrMB), 15
Cajal–Retzius (CR) cell, 137
Calcium, 45
Ca^{2+} ligands, 73
Calsyntenins, 27, 293
Calyceal processes, 355
Cancer, 144, 284
 immunotherapy, 383
 therapeutics, 383
Canine distemper virus, 140
Ca^{2+}-switch, 168
Cardiomyopathies, 181
Cataract, 141
Catch bond, 104
Catenins, 6, 28, 42, 56, 196
α-Catenins, 32, 117
αE-Catenins, 102–105
β-Catenins, 30, 99–102, 117
β-Catenin subfamily, 30
δ-Catenins, 30
CCCTC-binding factor (CTCF), 206
CD155, 120
Cdc42, 126
CDH13, 19
CDHR5, 27
CDH(R)23, 27
Cell–cell adhesion molecules (CAMs), 116
Cell–cell contact, 125
Cell-cycle, 101
Cell communication, 51
Cell density, 106
Cell movement, 129
Cell planar polarity, 347
Cell polarity, 324
Cell proliferation, 131
Cell-rearrangements, 48
Cell sorting, 47, 52
CELSR/flamingo, 21, 22, 134, 252, 266
CELSR1, 256–263
Cerebral cortex, 201, 211
Checkerboard-like cell arrangement, 133
Cilia, 254, 256
Ciliary epithelium, 135
Cis-dimer, 120
Cis interactions, 81–85, 165
Classical cadherins, 17, 196
Claudin, 127
Clathrin-dependent endocytosis, 129
Cleft lip/palate, 141
Clustered Pcdh, 197
Clustered protocadherins, 24, 75, 86
Cochlea, 355
Collective movement, 316
Combinatorial code, 330
Compaction, 48, 53
Constant regions, 201
Contact inhibition, 60, 61, 129
Contextual fear memory, 143
Convergent-extension, 230, 347
Core functions, 43–46
Cortical tension, 52
Corticotropin-releasing hormone
  receptor 1, 143
Crb3, 127
Crystal structure, 45
c-Src, 131
Cutaneous syndactyly, 141
Cyclins D2 and E, 131
Cytoplasmic domains, 163
Cytoplasmic functions, 43
Cytoskeletal attachment, 48
Cytoskeleton, 43, 46, 50, 58, 354

D

Dachsous (Ds), 85, 252, 263–265, 268
7D cadherins, 17
DCHS1 (Dachsous 1, CDHR6), 25, 265–268
3D domain swapping, 77
Deafness, 353
Delta-catenin, 31
Delta-catenin subfamily, 30
Delta-protocadherins (δ-Pcdh), 22
Desmocollins, 8, 17
Desmoglea, 160
Desmogleins, 8, 17
Desmoglein 2 (Dsg2), 379
Desmoglein 3 (Dsg 3), 379
Desmosmal cadherin ectodomains, 74
Desmosomal cadherins (DCs), 17, 159
Desmosomes, 125, 159, 378
 assembly, 168
 junctions, 84
Development, 175
Differentiation, 173, 325
Diffusion trap, 84
Dishevelled, 134, 261
DNAX accessory molecule 1 (DNAM1 or
  CD226), 120
Dnmt3b, 206
Dorsal telencephalon, 142
Downregulation, 170
 E-cadherin, 369

E

E-cadherin (CDH1), 14, 19, 366
EC repeats, 14, 196
Ectodermal dysplasia–syndactyly syndrome-1 (EDSS1), 141
Ectodermal dysplasia syndrome, 141
Ectodomain, 264–265
Edelman, G.M., 4
Electromotility, 344
Endocytosis, 53, 97, 179
Endothelial nitric oxide synthase, 131
Enteric nervous system, 290
Ependymal cell, 142
Epidermal blistering, 178
Epidermal growth factor receptor (EGFR)-mediated signaling, 378
Epilepsy, 240
Epithelial cell, 125
Epithelial-to-mesenchymal transition (EMT), 238, 317, 369
Epithelial remodelling, 348
ERK, 131
Evolution, 21
Evolutionary history of β-catenin, 30
Extracellular cadherin (EC) domains, 72, 265
Extracellular domains, 45

F

Facial branchiomotor (FBM) neurons, 266
F-actin polymerisation, 353
Fascia adherens, 46
Fasciculation, 233
Fat, 9, 85, 252, 263–265, 268
FAT4 (CDHR11), 25, 265–268
FAT-like, 26
FGD-related Cdc42 GEF (FRG), 126
Fiber cell, 136
Fibrofatty replacement, 182
Flamingo/starry night, 252, 253, 257–259
Flexibility, 166
Focal adhesion kinase (FAK), 201
Four-jointed, 264
Frizzled (Fz), 134, 257, 266

G

Gastric cancer, 299
Gastrulation, 316–318
Gating spring, 351
GDNF family ligands, 288
Geiger, B., 6
Gene cluster, 171
Genetic mutations, 367
Giant cadherins, 85–86
Glia-dependent migration, 137
Glia-independent migration, 137
Growth factor receptors, 59

H

Hair bundles, 346
Hair cells, 342
Hair follicles, 259
Hedgehog signaling, 385
Herpes simplex viruses (HSV), 140
Herpes virus, 140
Heterophilic interactions, 330
Heterotypic cell–cell adhesion, 116
Hippocampus, 132
Hippo signaling pathway, 60
Hippo tumor suppressor signaling pathway, 371
Hirschsprung's disease, 291
Holtfreter, J., 3
Homophilic, 45, 167, 197, 227
Homotypic cell–cell adhesion, 116
Hydrocephalus, 142
Hyperadhesion, 164
Hypermethylation, 367

I

IgA pemphigus, 180
Inner ear, 344
Integrin, 129
Interdomain linkers, 77
Intrinsic disorder, 164
Intrinsically disordered proteins (IDPs), 228
Invasion, 371
Involution, 317

J

Junctional adhesion molecule (JAM), 127

K

Kallikreins (KLKs), 379
Kemler, R., 6
Kidney, 265
Kidney development, 292
Kif3a, 134
Kinocilium, 133
Ksp-cadherin, 300

L

LI-cadherin, 298
Lis1, 134
LMO7, 117
Long term potentiation (LTP), 332
Loss of cadherin adhesions, 324
Loss of heterozygosity (LOH), 367

M

MARCKS, 201
Matrix metalloproteinase, 199
Matrix metalloproteinase 9, 144
Measles virus, 140
Mechanical force, 106
Mechanoelectrical transduction (MET), 344
Mechanosensors, 372
MEK, 131
Membrane palmitoylated protein 3 (MPP3), 122
Metastasis, 364
Metazoan genomes, 15
Metazoans, 15
Methylation, 205
MF synapse, 132
Microenvironment, 372
MicroRNAs (miRNAs), 368
Microtubule dependent, 169
Midbrain, 142
Midline, 160
Migration, 371
Morphogenesis, 47–49
Motor neurons, 329
Motor neuron pool, 329
Motor pool, 75
Multiple PDZ domain protein (MUPP1/MPDZ), 122
MUPCDH, 27
Mutations, 180
Myosin, 261–262, 266

N

N-cadherin, 19
Nectins, 46, 51, 116, 117, 119, 120, 122–129, 131–135, 137–145
Nectin-like molecules (Necls/CADMs), 120
Nematostella vectensis, 15
Neural crest cell migration, 318–319
Neural disorders, 332–333
Neural progenitor cell maintenance and differentiation, 321–328, 330, 331
Neural progenitor cells, 265
Neural tissue morphogenesis, 316–320
Neural tube, 127
  closure, 266
  formation, 261–262
Neurocristopathies, 319
Neuroendocrine cancers, 290
Neuroepithelial cells, 127, 324
Neuroepithelial progenitors, 321
Neuronal-cadherin (N-cadherin), 373
Neuronal clustering, 329–331
Neuronal migration, 267
Neuron-restrictive silencer factor (NRSF), 206
Neurons, 24
Neurulation, 232, 318
NK cell, 144
Nonmetazoan, 33
Nonsensory supporting cells (SCs), 133
Notch, 376
Notch signaling, 385

O

OB-cadherin, 375
Occluding junctions, 50
Ocular lens, 136
Oncogenic signaling pathways, 374
Optic fissure, 320
Optic tectum, 236
Organ of Corti (OC), 133
Oriented cell division (OCD), 265
Otic placode, 262
Oviduct, 254, 256, 262–263
Ozawa, M., 6

P

p120-catenin, 44, 54, 56
p120^{ctn}, 117
p190RhoGAP, 126
p27Kip1, 131
p38MAPK signalling, 179
Pals1-associated tight junction protein (PATJ), 122
Par-6 (PARD6A in mammals), 126
Partitioning defective 3 homologue (Par-3/PARD3 in mammals, 122
P-cadherin, 259
Pcdh 15, 27, 85, 350
α-Pcdhs, 200
γ-Pcdhs, 200
Pemphigus, 177
Peste des petits ruminants virus, 140
Peyer's patch formation, 292

Pharmaceutical intervention, 382
Phosphorylation, 56, 98, 99, 170
Photoreceptor cells, 355
Phylogenetic analysis, 31
Pigment and nonpigment cell, 135
Placental-cadherin, 374
Plakoglobin, 30, 381
Plakophilins, 31
Planar cell polarity (PCP), 125, 252, 348
Plaques, 162
Platelet-derived growth factor (PDGF), 129
PMP-22 (PERP), 382
Poliovirus, 140
Poliovirus receptor (PVR), 120, 140
Ponsin, 117
Postsynaptic density (PSD), 207
Prickle, 134
PRKCA1 (PICK1), 122
Projection neuron, 137
Proliferation, 105, 321
Prometastatic processes, 364
Promoter choice, 204
Protein clustering, 169
Protein kinase C (PKC), 166, 201
Protein processing, 368
Protein tyrosine phosphatase, nonreceptor type 13 (PTPN13), 131
Protocadherin (Pcdh), 8, 22, 196, 223
Pseudorabies virus, 140
PTPμ phosphatase, 131
Puncta adherentia junctions (PAJs), 118
Purkinje cell, 204
Pyk2, 201

R
Rac1, 126
Radial glial cells (RGCs), 137
Radial glial fiber, 137
Raf, 131
Rap1, 127
Ras, 120
RE-1 silencing transcription factor (REST), 206
Reelin, 137
RET (CDHR16), 27, 285
RET signaling, 288–289
Retinal-cadherin (R-cadherin), 377
Retinal ganglion cells, 327
Retinal morphogenesis, 320
Rho GEF, 127
Rho kinase, 127
Rho1, 127
RhoA, 126
Rosettes, 323

S
SAM syndrome, 181
Schizophrenia, 333
Secretase, 200
Selectivity and sorting, 51–53
Self-avoidance, 213
Sensory hair cell, 133
Seven-pass transmembrane (7-TM) cadherins, 252, 263, 266–269
SH3P12, 126
Shroom, 127
Signaling, 49, 57–61
Signalling pathways, 171
sorbs1, 126
Specific adhesion hypothesis, 3
Sprouty, 131
Stereocilia, 256, 265, 349
Stereociliary bundle, 133
Stereociliogenesis, 355
Sternum, 266
Strain, 101
Strand-swap binding, 77–79
Stratum lacunosum moleculare, 142
Stratum lucidum, 132
Stress-induced cognitive dysfunction, 143–144
Subcortical band heterotopia, 143
Suzuki, S.T., 8
Synapse, 51, 331–333
Synapse formation and function, 331–332
Synaptic junctions (SJs), 132
Synaptic plasticity, 295
Synaptic vesicle, 133
Synaptosome, 207
SynCAM1, 120

T
T cell, 144
T cell immunoreceptor with Ig and ITIM domains (TIGIT), 120
Tactile (CD96), 120
Tage4, 120
Takeichi, M., 5
T-cadherin, 81, 277–285
Tension, 96
TGF-β signaling, 376
Tight junctions (TJs), 125
Tip-link, 75, 351
Townes, P.L., 3

Trans interactions, 94
Transcription factors, 172, 367
trans-dimer, 120
Transgenic mice, 173
Trichoplax adhaerens, 15, 31
Trp2, 77
Trp4, 78
Truncated-cadherin (T-cadherin), 377
Tumor microenvironment, 375
Tumorigenic stem cell populations, 384
Tumor-suppressing activities, 379
Turnover, 99
Type-I cadherins, 17, 74
Type-II cadherins, 17, 74
Type-III cadherins, 21
Tyrosine kinase receptor signaling pathways, 385

U
Ubiquitinated, 235
Urmetazoan, 21
Usher syndrome, 349
Uvomorulin, 5

V
Valve, 262, 267
Van Gogh/ Strabismus, 257
Vang-like (Vangl), 134
Vascular endothelial- (VE-) cadherin, 376
Vascular endothelial growth factor (VEGF), 131
Vascularization, 376
Vav2, 126
VE-cadherin, 78
VEGFR2, 376
Vertebrates, 15
Vinculin, 104, 117

W
Whole-genome duplications (WGD), 15, 33
Willin, 122
Wing Hairs, 253–255
Wnt, 214
  β-catenin, 174
  pathway, 59
  signaling, 385
Wound healing, 176

X
X-Dimers, 79–81

Z
Zebrafish, 175
Zipper model, 6
Zonula adherens, 46
Zyxin, 122

MIX
Papier aus verantwortungsvollen Quellen
Paper from responsible sources
FSC® C105338

If you have any concerns about our products,
you can contact us on
ProductSafety@springernature.com

In case Publisher is established outside the EU,
the EU authorized representative is:
Springer Nature Customer Service Center GmbH
Europaplatz 3, 69115 Heidelberg, Germany

Printed by Libri Plureos GmbH
in Hamburg, Germany